Prison and Jail Administration
Practice and Theory
Second Edition

Edited by:

Peter M. Carlson, DPA
Associate Professor
Department of Government
Christopher Newport University
Newport News, Virginia

Judith Simon Garrett, JD
Deputy Assistant Director
Information, Policy and Public Affairs Division
Federal Bureau of Prisons
Washington, DC

JONES AND BARTLETT PUBLISHERS
Sudbury, Massachusetts
BOSTON   TORONTO   LONDON   SINGAPORE

World Headquarters
Jones and Bartlett Publishers
40 Tall Pine Drive
Sudbury, MA 01776
978-443-5000
info@jbpub.com
www.jbpub.com

Jones and Bartlett Publishers
Canada
6339 Ormindale Way
Mississauga, Ontario L5V 1J2
Canada

Jones and Bartlett Publishers
International
Barb House, Barb Mews
London W6 7PA
United Kingdom

Jones and Bartlett's books and products are available through most bookstores and online booksellers. To contact Jones and Bartlett Publishers directly, call 800-832-0034, fax 978-443-8000, or visit our website www.jbpub.com.

Substantial discounts on bulk quantities of Jones and Bartlett's publications are available to corporations, professional associations, and other qualified organizations. For details and specific discount information, contact the special sales department at Jones and Bartlett via the above contact information or send an email to specialsales@jbpub.com.

Production Credits
Chief Executive Officer: Clayton Jones
Chief Operating Officer: Don W. Jones, Jr.
President, Higher Education and Professional Publishing: Robert W. Holland, Jr.
V.P., Sales and Marketing: William J. Kane
V.P., Design and Production: Anne Spencer
V.P., Manufacturing and Inventory Control: Therese Connell
Publisher—Public Safety Group: Kimberly Brophy
Acquisitions Editor: Jeremy Spiegel
Associate Managing Editor: Robyn Schafer
Production Director: Amy Rose
Production Editor: Renée Sekerak
Production Assistant: Julia Waugaman
Director of Marketing: Alisha Weisman
Marketing Manager: Wendy Thayer
Manufacturing and Inventory Control Supervisor: Amy Bacus
Composition: Auburn Associates, Inc.
Interior Design: Anne Spencer
Cover Design: Jonathan Ayotte
Cover Image: © Courtesy of S. Craig Crawford, U.S. Department of Justice. Special thanks to the United States Penitentiary in Marion, Illinois, and the Federal Bureau of Prisons.
Chapter Opener Image: © Masterfile
Photo Research Manager and Photographer: Kimberly Potvin
Text Printing and Binding: Malloy Incorporated
Cover Printing: Malloy Incorporated

Library of Congress Cataloging-in-Publication Data

Prison and jail administration : practice and theory / [edited by]
Peter M. Carlson and Judith Simon Garrett. — 2nd ed.
    p. cm.
  Includes bibliographical references and index.
  ISBN-13: 978-0-7637-2862-5
  ISBN-10: 0-7637-2862-4
  1. Prisons—United States—History. 2. Prison administration—United States—History.
3. Punishment—United States—History. 4. Prisoners—United States—Social conditions.
5. Correctional personnel—United States. I. Carlson, Peter M. II. Garrett, Judith Simon.
  HV9304.P725 2008
  365.068—dc22
                                                                        2007033655
6048

Printed in the United States of America
11 10 09 08   10 9 8 7 6 5 4 3 2

Contents

Part I   Corrections Past and Present   1

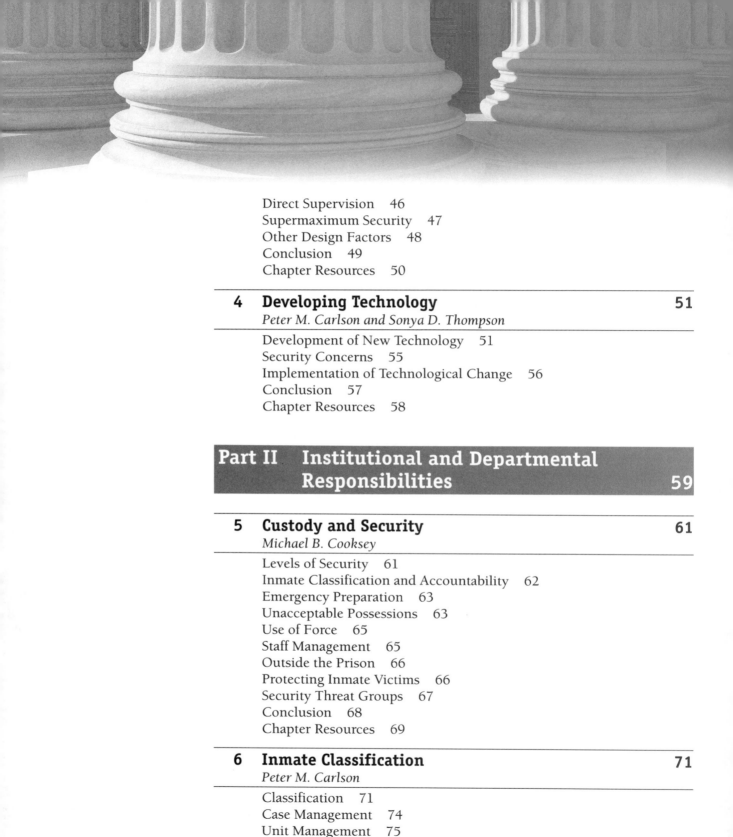

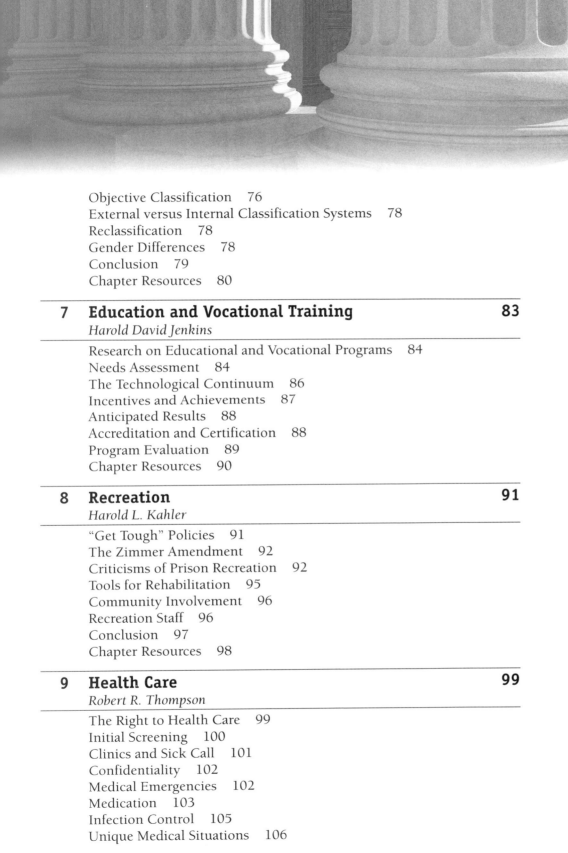

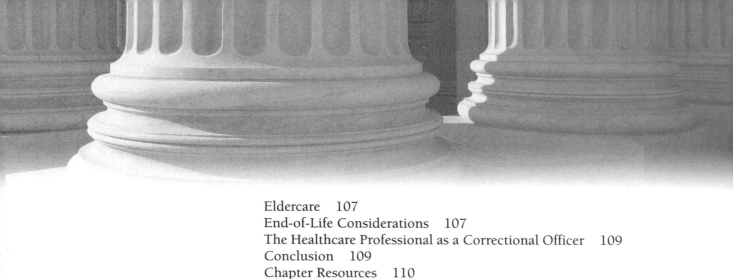

10   Mental Health   113
Sally C. Johnson

11   Religious Programming   127
Susan M. Van Baalen

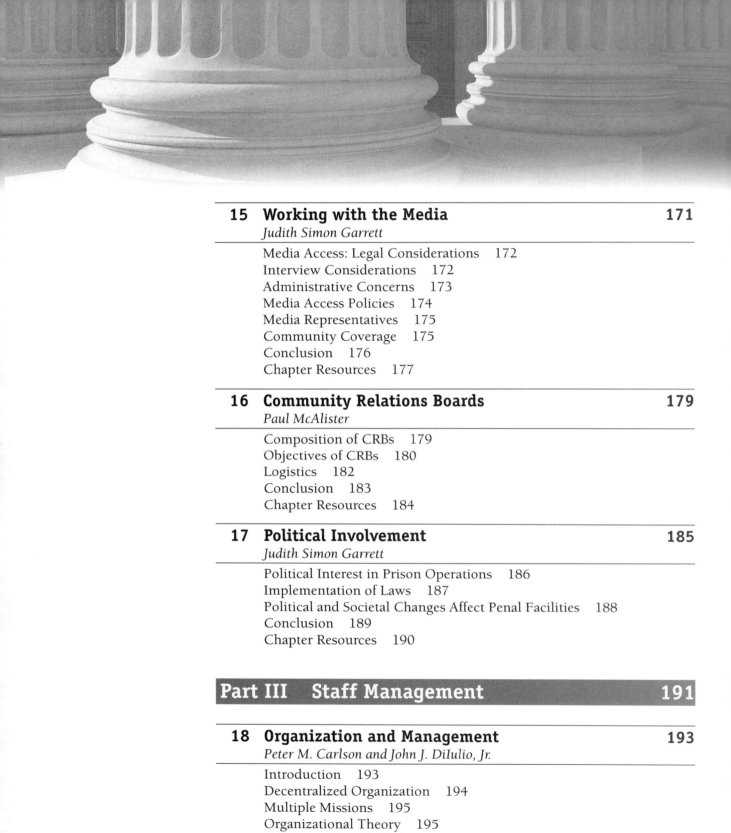

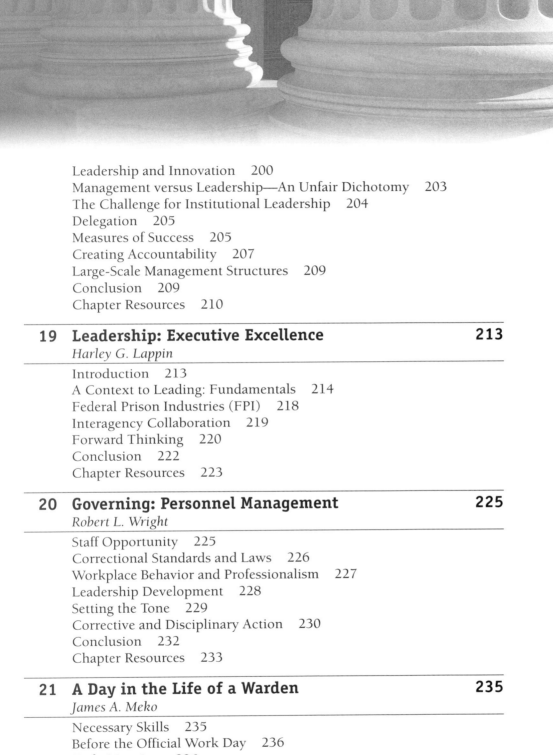

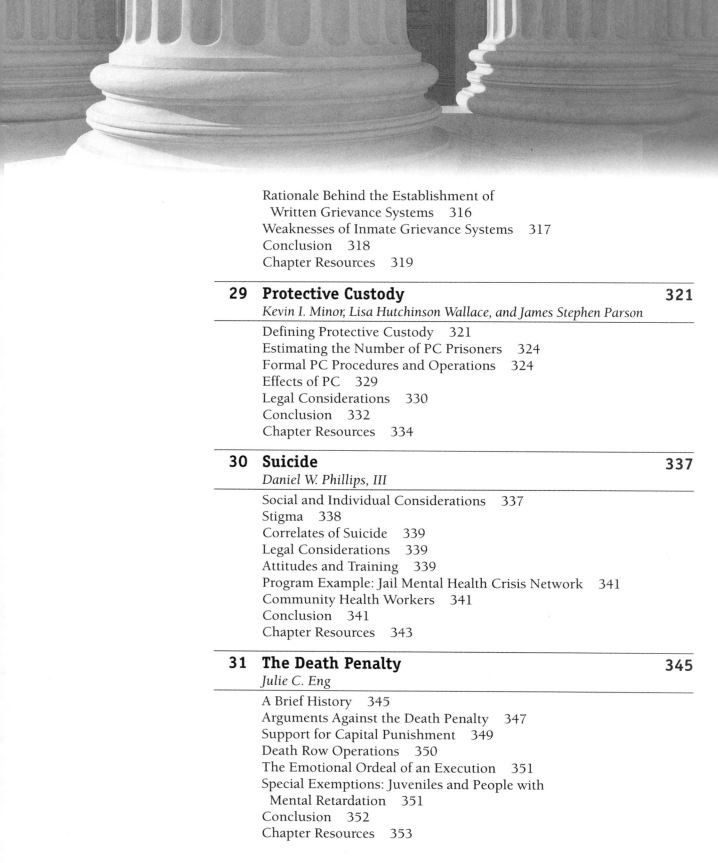

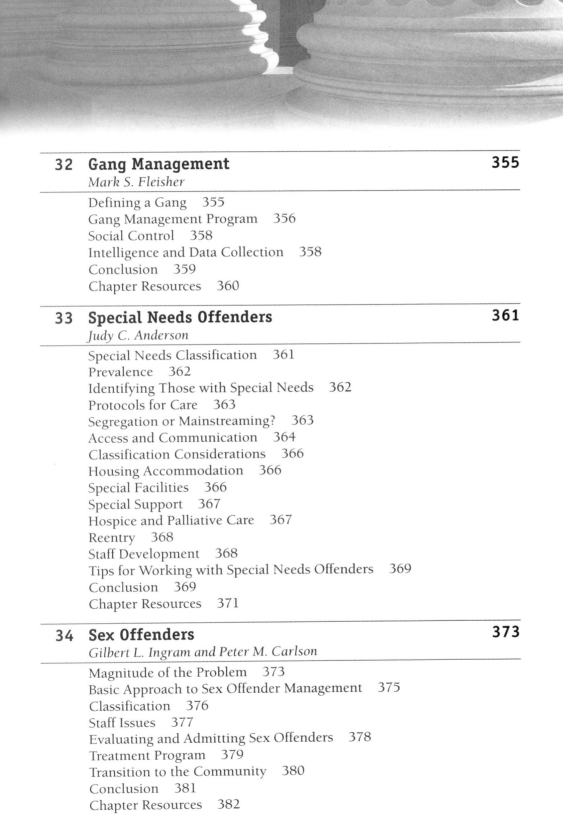

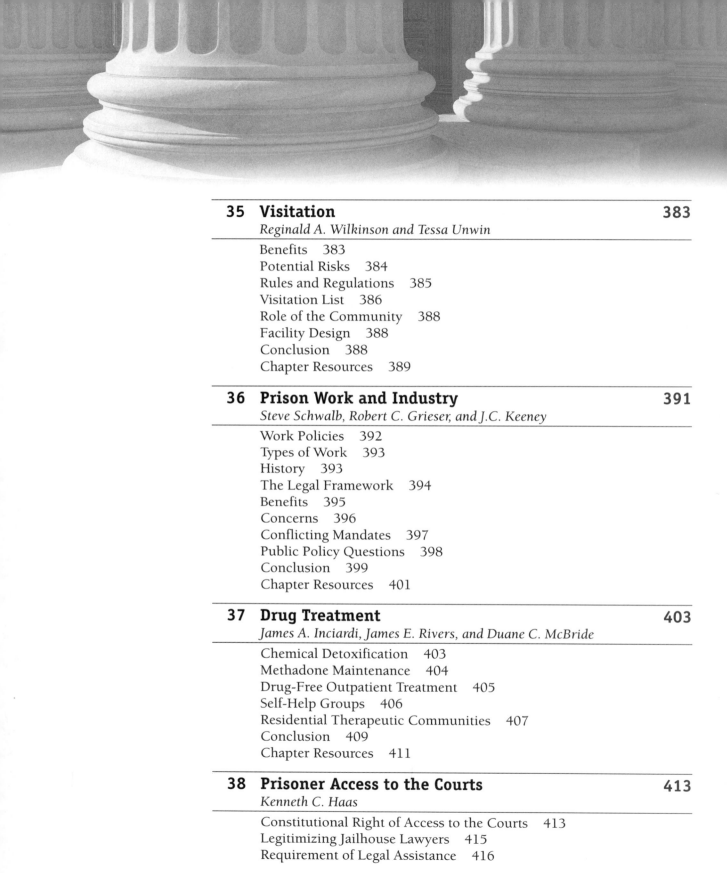

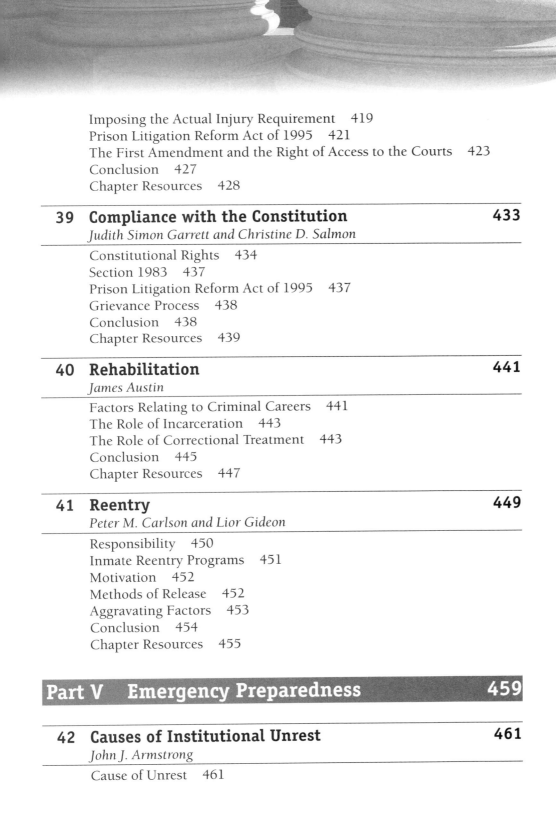

Dedication

For Rhonda, Julie, Fred, Lillemor, and Annika—For not complaining when I spend too much of my life in front of the computer!

Peter M. Carlson
Smithfield, Virginia

For Charlotte, Julian, Olivia, and Portia, who make it worthwhile, and for their father, Ed—my husband and best friend—without whom none of this would have been possible.

Judith Simon Garrett
Vienna, Virginia

Contributors

Judy C. Anderson is Warden for the South Carolina Department of Corrections at the Camille Griffin Graham Correctional Institution. Previously, Ms. Anderson served as the Chief of Institutional Operations for the South Carolina Department of Juvenile Justice, where she was responsible for secure institutions, classification, and disciplinary procedures. Prior to that assignment, she served in many capacities for the South Carolina Department of Corrections, including as warden of two institutions, deputy regional director, and chair of state classification. Ms. Anderson received her bachelor's degree from the University of Southern Mississippi and her master's degree from the University of South Carolina.

John J. Armstrong is retired from a 27-year career with the Connecticut Department of Correction where he began as a correctional officer and was ultimately appointed as the Commissioner of the agency, a position he held for eight years. He is currently a corrections consultant, has worked in Iraq, and continues to teach criminal justice at Naugatuck Valley Community College in Connecticut. Mr. Armstrong holds an MS degree from the University of New Haven.

James Austin is President of the JFA Institute in Washington, DC. Before taking this position, he was co-director of the Institute on Crime, Justice and Corrections at the George Washington University in Washington, DC, served as executive vice president of the National Council on Crime and Delinquency, and was employed by the Illinois Department of Corrections. Mr. Austin has authored numerous publications, was honored with the American Correctional Association's Peter P. Lejins Research Award, and received the Western Society of Criminology Paul Tappin Award for outstanding contributions to the field of criminology. Dr. Austin received his undergraduate and graduate degrees in sociology from Wheaton College, DePaul University in Chicago, and the University of California at Davis.

Armand R. Burruel has over 32 years of experience working with state civil service in California and over 20 years managing the California Department of Corrections and Rehabilitation's (CDCR) Human Resources Management division. Mr. Burruel worked in California's labor–management relations during the years 1978 through 1991. More recently, he held the position of Associate Warden at two prisons before accepting his current position at CDCR's Division of Reentry and Recidivism Reduction under the newly established Adult Programs area.

Peter M. Carlson is an Associate Professor in the Department of Government at Christopher Newport University, Newport News, Virginia. He is retired from a 30-year career with the U.S. Department of Justice, Federal Bureau of Prisons (BOP), where he served as the Assistant Director of the Agency, Regional Director of the Western Region, and Warden of three federal prisons. He received his BA degree from Willamette University in Salem, Oregon, and an MS in education from Western Oregon University. He also earned a master's degree and doctorate degree in public administration from the University of Southern California.

Michael B. Cooksey is a Senior Advisor for Corrections for the International Criminal Investigative Training Assistance Program (ICITAP), and works primarily with their Iraq program. Prior to his current position with ICITAP, Mr. Cooksey served as assistant director for the Correctional Programs Division at the Federal Bureau of Prisons. He provided bureau-wide oversight for inmate programs, custody and security, community corrections, and private prisons. During his career, he served as warden at four federal prisons. Mr. Cooksey also has served on advisory committees at several colleges and received his BS in business and MA in psychology from Middle Tennessee State University.

Clair A. Cripe is the retired General Counsel for the Federal Bureau of Prisons. After legal work as a JAG officer in the U.S. Navy, he entered government service in Washington, DC, first as a trial attorney in the Food and Drug Administration, and then as a lawyer in the Federal Bureau of Prisons. He worked for 28 years in the Prisons Bureau, the last 15 years as its General Counsel. He taught a course in Corrections Law for 15 years at the National Law Center at George Washington University and criminal justice classes at several other schools. He holds degrees from Oberlin College and Harvard Law School.

Julius Debro is the Associate Dean of the Graduate School at the University of Washington. Dr. Debro began his career in probation with Alameda County in Oakland, California. After seven years, he joined the California Department of Corrections at San Quentin, California, as a counselor, then joined the U.S. Probation Department in San Francisco on a special research project. He has conducted extensive research in the areas of corrections, juvenile delinquency, and policing. Dr. Debro received his bachelor's degree from the University of San Francisco, his master's in sociology from San Jose State University, and his doctorate in criminology from the University of California, Berkley.

John J. DiIulio, Jr., is the Frederic Fox Leadership Professor of Politics, Religion, and Urban Civil Society at the University of Pennsylvania. He is the author,

coauthor, or editor of 12 books and has written op-eds for many major news-papers and popular magazines. Dr. DiIulio received his bachelor's degree in po-litical science and economics, a master's degree in political science-public policy from the University of Pennsylvania, and a doctorate from Harvard University.

Julie C. Eng is a national board-certified teacher and reading specialist who main-tains a research interest in the field of social policy and criminal justice. She is an English teacher at Smithfield High School, Smithfield, Virginia. Ms. Eng has a bachelor's degree in sociology from Emory University and is currently com-pleting an MA in education at the College of William and Mary.

Mark S. Fleisher serves as a youth gang and corrections consultant to the De-velopment Services Group in Bethesda, Maryland, and is a member of the American Correctional Association's Education Committee. Previously, he was a Professor and Director of the Begun Center for Violence Prevention Research and Education at Case Western Reserve University and Special Assistant to the Regional Director for the U.S. Department of Justice, Federal Bureau of Prisons in the Western and North Central Regions. He was also a professor at Illinois State University, Washington State University, and Columbia University.

Jeffery W. Frazier has 18 years of experience in corrections, both at the state and local levels, and is currently employed as the superintendent of the Northern Neck Regional Jail. He served in the U.S. Army and the Virginia Air National Guard and is a veteran of Desert Storm. Mr. Frazier received his bachelor's de-gree in business administration from Strayer College and was one of the first in the nation to receive a certification as a jail manager through the Jail Manager Certification Commission. He was also the recipient of the 1998 American Jail Association Correctional Administrator of the Year award and is certified as a General and Firearms Instructor through the Virginia Department of Criminal Justice Services.

Judith Simon Garrett is the Deputy Assistant Director in the Information, Policy, and Public Affairs Division of the Federal Bureau of Prisons. She oversees the offices of Legislative Affairs, Public Affairs, Research and Evaluation, and Policy and Information Management. She has a law degree from Washington University in St. Louis and a bachelor's degree from the University of Wisconsin–Madison.

Robert S. George, FAIA, joined the Federal Bureau of Prisons as a staff architect and managed the design of several new correctional facilities in the concept of di-rect supervision. He moved to the Bureau's Western Regional Office in 1977 where

he served as a staff architect and facilities administrator. In 1984, he returned to private architectural practice, and, in 1998, he was elevated to the College of Fellows of the American Institute of Architects. He continues to practice in the San Francisco area. Mr. George received his BA in architecture from the University of California, Berkeley.

Lior Gideon is currently a full-time professor in the department of Law, Police Science, and Criminal Justice Administration at John Jay College of Criminal Justice in New York. He specializes in corrections-based program evaluation and focuses his research on rehabilitation, reintegration, and reentry issues, specifically by examining offenders' perception of needs. Dr. Gideon earned a PhD from the Faculty of Law, Institute of Criminology at Hebrew University in Jerusalem and completed a post-doctoral fellowship at the University of Maryland's Bureau of Governmental Research.

Robert C. Grieser is the Chief of Strategic Business Development and Marketing for Federal Prison Industries, Inc. He began his career in corrections in 1976 with the Virginia Department of Corrections and served as Director of Operations for the Institute for Economic and Policy Studies, a nonprofit consulting firm. He has served on the National Correctional Industries Association (NCIA) Board of Directors since 1989, serving as NCIA President in 1998. Mr. Grieser earned an MSW degree from Virginia Commonwealth University.

Marie L. Griffin is an Associate Professor in the School of Criminology and Criminal Justice at Arizona State University. Her research interests include issues of organizational climate in the correctional setting; use of force in corrections; prison and jail misconduct; and gender and crime. She received her PhD in Justice Studies from Arizona State University.

Kenneth C. Haas is a political scientist who holds positions of Professor in the Department of Sociology and Criminal Justice as well as in the Department of Political Science and International Relations at the University of Delaware. He has won the University's Excellence-in-Teaching Award three times. He specializes in constitutional law with an emphasis on corrections and capital punishment law. His articles have been published in law reviews, social science journals, and scholarly books. He was coeditor of *Challenging Capital Punishment* (1988) and *The Dilemmas of Corrections* (1995).

John R. Hepburn is a Professor in the School of Criminology and Criminal Justice and the Dean of the College of Human Services at Arizona State University. His

research has focused on issues of correctional administration and inmate control as well as offender reentry.

Martin F. Horn is Commissioner of Probation and Correction of the City of New York. Previously, Mr. Horn served as Secretary of Administration of the Commonwealth of Pennsylvania and as Pennsylvania's Secretary of Correction. He has spent over 35 years working in corrections and has served as Warden and Chief Operating Officer of New York's parole agency. He holds an MA in criminal justice from John Jay College of Criminal Justice.

James A. Inciardi is Director of the Center for Drug and Alcohol Studies and Professor of Sociology and Criminal Justice at the University of Delaware. He also holds positions as Adjunct Professor in the Department of Epidemiology and Public Health at the University of Miami School of Medicine and is a Distinguished Professor at the State University of Rio de Janeiro and guest professor in the Department of Psychiatry at the Federal University of Rio Grande do Sul in Porto Alegre, Brazil. Dr. Inciardi earned his PhD from New York University and has published numerous books, articles, and chapters in the areas of substance abuse, criminology, criminal justice, history, folklore, public policy, AIDS, medicine, and law.

Gilbert L. Ingram is a criminal justice consultant who retired after 35 years of service with the Federal Bureau of Prisons. His experience included serving as the warden of two federal prisons, director of two federal prison regions, and Assistant Director of Correctional Programs. He has also served as Adjunct Faculty Member at seven universities. Dr. Ingram received his PhD in psychology from the University of Maryland.

Michael H. Jaime is retired from the California Department of Corrections. He served in many responsible positions including the role of Chief of Labor Relations. He also has finance and labor relations experience in three other state departments in California. Mr. Jaime received his bachelor's degree in social work and master's degree in public administration from California State University, Sacramento.

Harold David Jenkins served for many years as the Maryland Division of Correction's Educational Liaison. He began his career in corrections in 1973 as a manager of a major department-wide research and planning project. He serves on Maryland's Use Industries Advisory Committee and the State Advisory Committee for Adult and Continuing Education. Mr. Jenkins holds bachelor's and master's degrees from Florida State University, a doctorate from the University of Maryland, and a certificate of advanced study in education from Johns Hopkins University.

Lavinia B. Johnson, CFP, is the Training Development Coordinator, Food Service Specialty for the Virginia Department of Corrections' Academy for Staff Development. Ms. Johnson is a former National President for the Academy of Correctional Food Service Administrators.

Sally C. Johnson is currently in private practice as a forensic psychiatrist and consultant. She is retired from career service as a United States Public Health Service physician assigned to the Federal Bureau of Prisons. She currently is Senior Lecturer in Law at Duke School of Law and a Consulting Associate Professor in the Department of Psychiatry at Duke University Medical Center.

Harold L. Kahler served in numerous educational positions with the Federal Bureau of Prisons, including an assignment as Director of Outreach at the National Academy of Corrections with the National Institute of Corrections. After he retired from the Bureau of Prisons, he taught criminal justice courses at the Community College of Aurora, Colorado, and now works with Social Services of Colorado. Mr. Kahler received his bachelor's degree from Howard Payne University, Brownwood, Texas and his MA in education from the University of Virginia.

J.C. Keeney has retired from a 36-year career in corrections. He began his career as a recreation officer in 1960 and was promoted through the ranks to become the superintendent of the Oregon State Penitentiary. He also served in the Arizona Department of Corrections as the Assistant Director of Adult Institutions and has worked in private corrections as the Warden of Arizona State Prison (Phoenix West).

Ken Kerle, a nationally recognized expert in the area of local corrections, is the Managing Editor of *American Jails Magazine*, the official publication of the American Jail Association. Mr. Kerle has worked in jails and has served as a consultant to correctional institutions over the past 30 years. He received his doctorate from American University in public administration and has authored numerous books and articles.

Gothriel "Fred" LaFleur is currently serving as the Director of Community Corrections for Hennepin County, Minneapolis. He served as Commissioner of the Minnesota Department of Corrections from 1996–1999, and as warden at

the Minnesota Correctional Facility–Lino Lakes. Mr. LaFleur is a graduate of Bethune–Cookman College in Florida, Mankato State University, and the University of Minnesota, Duluth.

Harley G. Lappin is the seventh Director of the Federal Bureau of Prisons. He is responsible for the oversight and management of the BOP's 114 institutions and for the safety and security of the more than 198,000 inmates under the agency's jurisdiction. He received a BA in forensic studies from Indiana University in Bloomington, Indiana, and an MA in criminal justice and correctional administration from Kent State University in Kent, Ohio.

Jim Lyons has been employed by the Minnesota Department of Corrections for over 20 years. He is a graduate of St. Cloud University.

Paul McAlister is Treatment Director for Zumbro Valley Mental Health Center's Residential Treatment Program in Rochester, Minnesota. He has taught for 28 years at Crossroads College, Augsburg College, and seven years at the Mayo Medical School in the areas of medical ethics, philosophy, and theology. He has spent many years working with Community Corrections boards and committees as well as the Community Relations Board at the Federal Medical Center, Rochester, Minnesota. His doctorate is in the areas of philosophy, ethics, and theology.

Duane C. McBride is Chair of the Behavioral Sciences Department and Director, Institute for Prevention of Addictions, Research Professor of Sociology at Andrews University in Michigan. His areas of expertise are criminology and drug abuse. His current research focuses on the areas of juvenile delinquency and AIDS infection of IV drug users. Dr. McBride currently teaches courses on Criminology, Introduction to Sociology, Drug Use in American Society, Theories of Addictive Behavior, and Juvenile Delinquency.

Douglas C. McDonald is a Senior Associate at Abt Associates Inc., a policy research organization headquartered in Cambridge, Massachusetts. He has directed research projects on a variety of criminal justice topics and has been published widely. His professional interests include research on correctional privatization, public finance of corrections, correctional health care, criminal sentencing policies and practices, and substance abuse treatment—especially in prisons and jails. Dr. McDonald has conducted research on sentencing reforms at the Vera Institute of Justice in New York City. He received his PhD in sociology from Columbia University, and his bachelor's degree from Columbia College.

James A. Meko retired from the Federal Bureau of Prisons in 1996, after a 25-year career as a warden of two correctional institutions as well as holding the positions of the Senior Deputy Assistant Director and the Chief of Staff Training. Following his retirement, he served as a faculty member in criminal justice at Gannon University, Erie, Pennsylvania. Mr. Meko received his bachelor's degree from Gannon University and his master's degree from the University of Notre Dame.

Kevin I. Minor has taught in the Department of Correctional and Juvenile Justice Studies at Eastern Kentucky University since 1992. His areas of professional interest are institutional and community corrections, criminological theory, juvenile delinquency and justice, and evaluation research. He holds a BS in psychology and criminology from Indiana State University, an MS in correctional psychology from Emporia State University, and a PhD in sociology/criminology from Western Michigan University. While completing these degrees, he worked at both a juvenile institution and an adult penitentiary. He also spent four years teaching criminal justice courses at Southwest Missouri State University.

Anadora Moss is President of The Moss Group, Inc., a criminal justice consulting firm based in Washington, DC. Ms. Moss has a long history of working on issues related to staff sexual misconduct in correctional facilities. She was involved in early strategies to address staff sexual misconduct as an assistant deputy commissioner in the Georgia Department of Corrections during the *Cason v. Seckinger* lawsuit in the early 1990s and as a Program Manager with the National Institute of Corrections from September 1995–February 2002.

James Stephen Parson is a research associate in the Department of Correctional and Juvenile Justice Studies at Eastern Kentucky University's College of Justice and Safety.

Daniel W. Phillips, III, is Assistant Professor of Sociology and Criminal Justice at Lindsey Wilson College in Columbia, Kentucky. He is currently serving as editor for a special edition of the *Journal of Offender Rehabilitation* regarding mental health issues in the criminal justice system and current issues in probation and parole. Dr. Phillips also serves as an adjunct graduate instructor in the College of Justice and Safety at Eastern Kentucky University. He specializes in courses dealing with correctional mental health issues and correctional rehabilitation.

Beverly Pierce is retired from a 20-year career with the Federal Bureau of Prisons. She has served in many financial management positions and as the Associate Warden at the Federal Correctional Institution, Terminal Island, California.

James E. Rivers is Deputy Director of the Comprehensive Drug Research Center at the University of Miami School of Medicine and is a Research Associate Professor in both the Department of Sociology and the Department of Epidemiology and Public Health. Dr. Rivers has research, teaching, and administrative/policy experience in a broad range of substance abuse areas and, in recent years, has been Director of the Metropolitan Dade Country Office of Substance Abuse Control and Loaned Executive to the Law Enforcement, Courts and Corrections Task Force of the Miami Coalition for a Safe and Drug Free Community. Dr. Rivers earned his PhD in sociology from the University of Kentucky.

Tom Roth is a Senior Consultant for MGT of America, Inc., a management research and consultant firm in Austin, Texas. He is retired from a 27-year career with the Illinois Department of Corrections where he served as Assistant Director of Administration and as Warden of four separate prisons. He was an auditor with the American Correctional Association for nine years and holds a master's degree from Michigan State University.

Christine B. Salmon is a student at the University of Richmond School of Law. She began her law school career at Tulane University but was forced to transfer as a result of the devastation wrought by Hurricane Katrina. Ms. Salmon completed a summer internship at the Federal Bureau of Prisons in 2006.

Steve Schwalb is the Executive Director of Pioneer Human Services in Seattle, Washington, a correctional services industry. He started his lengthy correctional career in 1973 and has served as the Assistant Director of the Federal Bureau of Prisons and as the Chief Operating Officer of Federal Prison Industries, a government-owned corporation that employs thousands of federal inmates in manufacturing and service tasks. He has also served as the warden of a federal prison and as the Director of the King County Department of Adult Detention in Seattle. Mr. Schwalb received his bachelor's degree in personnel management and labor relations from the University of Washington.

Sam S. Souryal is a Professor of Criminal Justice and Ethics in the College of Criminal Justice at Sam Houston State University in Texas. He is most interested in the areas of policing, corrections, and ethics. He has written several books on criminal justice administration and management, ethics and justice, as well as books on comparative cultures and religions.

Richard L. Stalder is the Secretary of the Louisiana Department of Public Safety and Corrections. He began his career with the department in 1971 as a correctional

officer and has served as superintendent and warden of major juvenile and adult facilities among other management roles. In 2002, he received the American Correctional Association's E.R. Cass Correctional Achievement Award and the Association of State Correctional Administrators' Michael Francke Award. He possesses bachelor's and master's degrees from Louisiana State University.

Louis Stender is retired from the Minnesota Department of Corrections where he served in many developmental positions and as warden of the Minnesota Correctional Facility in Faribault.

E.A. Stepp has held correctional leadership positions for many years. He is retired from the Federal Bureau of Prisons where he served as the warden of the Federal Correctional Complex in Coleman, Florida, and at the United States Penitentiary, in Marion, Illinois. He has also held the position of Chief of Emergency Preparedness for the agency. He is currently serving as the warden of the South Bay Correctional Facility, a private prison in South Bay, Florida. Mr. Stepp received his bachelor's degree in criminal justice from Indiana State University.

Robert R. Thompson is a graduate of Thomas Jefferson University and received his medical residency training at the Mayo Clinic. He is the author of 23 articles and a book dealing with a variety of medical and social issues. Prior to retirement, he worked as a physician at the Federal Medical Center in Rochester, Minnesota where he was medical director of the hospice program.

Sonya D. Thompson is the Deputy Assistant Director for Information Resource Management (and Chief Information Officer) for the Bureau of Prisons. She oversees the information technology (IT) department, systems development, network management, data communications, and user support. She has a BS in electrical engineering and a JD from Washington University in St. Louis.

Anthony P. Travisono is a former director of the master's degree program in Correctional Administration at Salve Regina University in Rhode Island and Trustee Emeritus of Correctional Properties Trust and Wackenhut Corrections. Prior to his retirement as Executive Director Emeritus of the American Correctional Association, he was the Director of Corrections, Director of Mental Health, and the Director of Social Welfare for the State of Rhode Island. Mr. Travisono has a bachelor's degree from Brown University and a master's degree from Boston University.

Tessa Unwin is the Public Affairs Liaison for the Ohio State Department of Rehabilitation and Correction. She has served in the public information depart-

ment for many years. Previously, Ms. Unwin worked as a producer in public radio. Ms. Unwin received her bachelor's degree in journalism from The Ohio State University.

Susan M. Van Baalen serves as the Chaplaincy Administrator for the Federal Bureau of Prisons (BOP), a position she has held since 1996. Prior to joining the BOP, she was a teacher and an administrator in Chicago elementary and secondary schools. She has taught theology at St. John's Provincial Seminary in the archdiocese of Detroit and Corrections Studies at the University of Michigan. Sister Van Baalen received her BA from Siena Heights College, Adrian, MI; an MA in social work in 1972 from DePaul University, Chicago, IL; an MDiv in 1980 from the Jesuit School of Theology, Chicago; an Honorary Doctor of Humane Letters from Siena Heights University in 2000; and is currently a doctoral candidate at Georgetown University, Washington, DC.

Lisa Hutchinson Wallace is Associate Professor in the Department of Criminal Justice at the University of Arkansas, Little Rock. Dr. Wallace earned her PhD at the University of New Orleans, and her MS and BS at the University of Southern Mississippi. Prior to entering academia, Dr. Wallace served as a research coordinator of the Pretrial Research Study of Orleans and Jefferson Parishes in Louisiana, where she researched national trends in pretrial programs and served as coordinator of the Judicial Watch program, evaluating the post-indictment phase of the criminal district courts in Orleans and Jefferson parishes. She also worked as a Predispositional Delinquency Counselor for the Jackson County Youth Court.

Arthur Wallenstein is the Director of Montgomery County Maryland Department of Corrections and Rehabilitation. He previously served as Director of King County Department of Adult Detention in Seattle and Director of Bucks County Pennsylvania Department of Corrections. He is a member of the National Institute of Corrections Advisory Board. He received his BS from Georgetown University and an MA from the University of Pennsylvania.

Reginald A. Wilkinson served as Director of the Ohio Department of Rehabilitation and Correction from 1991 until his retirement in 2006. He also served as President of the American Correctional Association (ACA) from 1995 to 1998. He was the first superintendent of Ohio's Corrections Training Academy, the first warden at the Dayton Correctional Institution, and the first regional director of the prisons in the southern half of Ohio. Mr. Wilkinson has served on the faculties at Wilmington College and the University of Cincinnati. Mr. Wilkinson holds bachelor's and master's degrees from The Ohio State University.

Robert L. Wright served 35 years in corrections. He held many progressively responsible positions with the Oregon Department of Corrections and served as the superintendent of the Eastern Oregon Correctional Institution. He also served as the superintendent of the Clallam Bay Corrections Center in Clallam Bay, Washington. He recently retired as the Executive Director of the Eastern Oregon Alcoholism Foundation. Mr. Wright received his bachelor's and master's degrees from Oregon State University.

Corrections Past and Present

I

YOU ARE THE ADMINISTRATOR
Wet and Wild

There was no question that (fictional) Mid River Jail inmate Bosco Givens was difficult to have in custody. He was awaiting trial on a charge of rape and had been confined in the county jail for 85 days (though to staff and inmates, it seemed like 85 years). He had been difficult from the moment he arrived. During the admission processing, the admitting officer noted that Givens was acting spacey. He had pupils that were constricted to tiny specks, and he could not walk without support. Staff took him to the county hospital's emergency room where the ER physician determined that Givens was high on heroin.

Once he was back at the jail, Givens was constantly creating problems in the cell house. He had to be segregated after picking a fight with another inmate and often started arguments with staff members. He cursed at officers, threw food at them, and occasionally spit in their direction. The jail administrator had a mental health professional evaluate Givens, but the psychologist reported that Givens was not mad . . . just bad.

Recently, Givens used his food tray to reach up and break the sprinkler head off the emergency fire suppression sprinkler in his cell. By the time the staff on duty managed to shut down the sprinkler system, Givens was soaked. The lieutenant on duty ordered that Givens be placed outside in the secure recreation cage while his cell and surrounding area were mopped up. This might have been a reasonable action, except that the temperature outdoors was 28°F, and Givens was left outside in cold, wet clothing for over an hour. Afterwards, Givens filed a complaint with the county jail authority through his attorney. The incident was reported in the local newspaper, and it seemed that everyone believed the jail staff had acted inappropriately. The local newspaper editor wrote an editorial condemning the jail personnel for their "Gestapo-like" punishment tactics.

- *How should jail personnel have handled this incident?*
- *How would inmates like Givens have been dealt with in the past?*
- *What aspects of prison architecture could offer potential solutions to dealing with difficult inmates?*
- *What technological developments could assist in similar situations in the future?*

History of Corrections

Peter M. Carlson, Tom Roth, and Anthony P. Travisono

Chapter Objectives

- Grasp why the concept of punishment has become such a major force in the American administration of justice.
- Identify the differences between the Auburn and Pennsylvania systems.
- Outline the history and trends of prison reform, including the significant change in the prison environment after World War II and the decline of the medical model.

It may be hard to imagine a time when there were no prisons in the United States, but America has responded to criminal behavior in a variety of ways throughout its history. The way society treats its scofflaws and crooks varies with the times and is a reflection of society's ideas and values.

◼ Sentencing

There are four predominant sentencing goals in the American judicial system:
- Rehabilitation—preparation for a law-abiding return to the community
- Specific deterrence—sentence that deters individual offenders from committing future crimes
- General deterrence—the effect on other citizens that prevents future crimes
- Punishment—creation of pain or suffering (emphasizes the negative aspect of sanctions)

Judges often are working to achieve one or more of these goals in sentencing offenders. In doing so, they may institute one or more of the following sanctions.

Fines

Along with probation and incarceration, fines are one of the most basic punishments found in the current criminal justice system in America. Fines are monetary sanctions imposed by the courts for offenses ranging from misdemeanors (violations like shoplifting) to felonies (crimes like murder). Fines may be the only sanction imposed by the court, or they may be combined with other alternatives such as probation, restitution, or confinement. The laws and guidelines that authorize the use of fines vary widely across criminal court jurisdictions and tend to be applied inconsistently.

Research is not clear as to whether fines are an effective punishment in deterring criminal behavior, because the effect of monetary loss strongly depends on individual financial resources. Some jurisdictions have tried to compensate for this concern by utilizing day fines that are based on an individual's income and assets. Day fines also vary based on the severity of the offense and are considered a much more equitable method of assessing monetary sanctions. However, as criminologist Douglas McDonald notes, the specific economic circumstances of an offender may be difficult to determine accurately.[1]

Restitution

Another monetary punishment that can be imposed on convicted persons is a requirement to make restitution to the victim or the community. Restitution is often required as a partial sanction and can be used as a condition of another punishment such as probation. This sanction involves paying a specified amount of money to the person damaged by a criminal act or repaying the local community by the performance of community services.

Many judges impose other economic punishments, including requiring the offender to pay for court costs or to forfeit certain assets that he or she may have. The forfeiture of owned property is often tied to the personal property connected in some manner to the crime. For example, in federal courts it is commonplace for an offender to forfeit an automobile or airplane if the vehicle was associated with criminal activity.

Probation

According to the Bureau of Justice Statistics, over 4.9 million adults were under federal, state, or local probation supervision in 2005, with approximately 4,162,500 on probation and 784,000 on parole.[2] Probation supervision allows the offender to remain in the community with special conditions and accountability requirements. Probation is generally associated with incarceration in the sentencing process; if the individual does not meet all conditions of probation, probation is revoked, and the sentence is served in prison or jail.

Intensive probation is another form of community supervision sanction. It is occasionally utilized by the courts for those individuals who may be considered high risk. In general, intensive probation means the supervising probation officer has a smaller caseload and therefore is able to spend more time supervising and assisting the offender. This variation of probation also demands more intense reporting requirements and often has more structured accountability of the probationer's whereabouts, as well as living and working conditions.

Incarceration

The courts utilize incarceration—a criminal sanction that involves the sentencing of an offender to a term of confinement in a prison or jail—when the offense or the individual's personal characteristics are such that the judge believes that society must be protected from the possibility of further victimization by the criminal. Confinement is typically placement in a jail or prison. Jails are generally used for shorter-term confinement for those serving misdemeanor sentences of one year or less. Prisons confine those sanctioned for felony convictions and sentenced to more than one year. Some states and the federal government have policies that permit the use of private, for-profit confinement facilities.

Incarceration has gained tremendous popularity in the modern American view of corrections. It is the most common (and most commonly expected) form of punishment and is almost exclusively the sanction for serious and repetitive offenders. The placement of a criminal behind bars—the taking of one's liberty—believed by many to have the most significant effect on crime and is the punishment that victims of crime typically expect from American jurisprudence.

Prisons and jails play a major part in punishment today, and short of capital punishment, confinement is the most serious sanction utilized by courts in the United States. Imprisonment as punishment is an American concept that has been adopted throughout the world. The rate of incarceration in the United States (491 per 100,000, as of December 2005) is believed to be the highest in the world, reliant on the credibility of international statistics. The nation's federal and state prison inmate population grew to nearly 2.3 million in 2005, with one out of every 108 men confined to prison or jail.[3]

Further, American criminal courts are heavy-handed in that prisoners serve longer sentences than in other countries. The stringent "get tough" policies enacted throughout the last 25 years continue to have a huge impact on the growth of the populations of our nation's correctional facilities. Present-day sentencing includes mandatory sentencing requirements for many drug offenses, truth-in-sentencing provisions that preclude release on parole, and three-strikes laws for repeat offenders.

■ Role of Law

Law is a social construct in that members of society trade some restriction in exchange for some benefit from the government. Laws are established by every society, and violations of the law are defined by those in power. The definition of what is legal or illegal changes over time in all societies as collective views and attitudes shift. For instance, the United States outlawed the manufacture, transportation, and sale of alcohol during Prohibition and subsequently legalized it again. When people band together as friends, family, society, or as a nation, they develop social rules and apply the rules to all members. This requires submission to the accepted mores, and, in turn, demands a sanction if one does not comply with expectations.

Nonobedience requires a price—punishment. Vengeance, both from the aspect of private retaliation and from public justice, is a key aspect of our human nature. When we are injured by another, figuratively or literally, our nature is such that we want and expect the offender to be dealt with in a just manner. This concept of *lex talionis*—an eye for an eye—has been well accepted in many cultures around the world.

Punishment is the infliction of a penalty and often includes a component of retributive suffering. Punishment for misbehavior and violations of the law, or any perceived malfeasance on the part of others, is a social response that can be found throughout the history of all major civilizations. American society believes in punishment. The concept of just deserts (justification for punishment) has its roots in the early history of the original colonies and even earlier in British jurisprudence.

■ Role of Religion

Retribution is also important in American culture, and the evolution of retributive punishment has intertwined with that of religion over the years. The Roman emperor Justin organized Roman law through a written document known as the Justinian Code which has had a lasting influence over the centuries. However, as the Roman Empire disintegrated, chaos ruled Europe, and there was no central authority other than the church, whose punishments were often bloody and violent.

The concept of free will has evolved from the religious belief that one should make the decision to follow the path of righteousness. The parallel thought is that one may also choose to violate the law and should therefore be held responsible for his or her actions. This idea has formed the center focus of the American criminal justice system, and it is the logic behind the development of rehabilitative programs within the judicial system.

■ History of Punishment

The earliest prisons in America were modeled after English houses of confinement, or gaols, which were utilized for short-term detention of law violators awaiting trial. They were not pleasant places but were a vast improvement over the Anglo-Saxon legacy of revenge in which felons were often publicly humiliated, banished, tortured, or killed.

Previously, corporal punishment had been imposed with impunity, and the severity of the sanctions was extreme. Offenders were buried alive, beheaded, drowned, burned at the stake, boiled, stoned, and otherwise mutilated in every imaginable way. All punishment was public, and even minor sanctions such as placement in the pillory were conducted in front of amused crowds. These sanctions were greatly valued and have contributed to the seeming blood-thirsty nature of revenge and retribution. Justice certainly has evolved from this patchwork of societal response to crime, but the American sense of justice is deeply steeped in the physical and aggressive punishments of early England.

In the 16th century, the Church of England began to use the bishop's facility at St. Bridget's Well for confining and beating misdemeanants for crimes such as prostitution and begging. Such institutions, known as bridewells, became commonplace.

As these facilities spread, they rapidly deteriorated and became known as "houses of darkness" because of the conditions of confinement. These British gaols were dimly lit, filthy, and disease-ridden. The English prison-reformer John Howard noted that more prisoners died of sickness and disease than execution.[4] All inmates, regardless of their gender, age, or offense, were confined together in these houses of pestilence.

England often utilized deportation (transportation from one's homeland) as punishment. Hundreds of thousands of lawbreakers were banished to the American colonies, and later to Australia, where they were forced into servitude for a number of years as part of their punishment.

Later, John Howard introduced the concept of a penitentiary, a prison in which incarcerated people are given the opportunity to repent for an extended period of time, and helped pass the Penitentiary Act in 1779. This Act provided several major reforms, including requiring secure and sanitary facilities and inspections, abolishing fees for basic services, and introducing a reformatory model.

■ Punishment in the American Colonies

Colonists in the United States tended to view crime as a sinful act, not a social problem.[5] Criminals were viewed as sinners, not as individuals who were led astray by imperfections in society. Punishments imposed in colonial times resembled those used in England. Justice demanded harsh penalties, and an extraordinary number of offenses were subject to death, banishment, or various forms of corporal punishment. Jails in America did exist, but imprisonment was rare and was primarily intended to detain those awaiting trial or sentencing or those who were unable to pay their debts. When jails were used for the detention of pretrial offenders, the conditions were often as bad as those in English gaols.

After the American Revolution, society began to turn away from many concepts and practices imported from England. As new ideas emerged in the fledgling states, prison and jail reformers had a major influence on the nature of punishment.

In the early 1680s, a Quaker named William Penn arrived in the colony later named after him—Pennsylvania. In response to the harsh and humiliating punishment used throughout the colonies, Penn focused on reforming existing criminal sanctions and offered colonists a penal code known as the Great Law. The Great Law advocated imprisonment, hard labor, and fines, not humiliating and violent punishments.[6] In the Great Law, Penn abolished all capital offenses except murder. Penn's code ultimately combined reform and rehabilitation with the existing philosophy of deterrence. Following Penn's death in 1718, the Great Law was immediately overturned. England's Anglican Code was reinstated, and Pennsylvania returned to using harsh corporal punishment and increased capital offenses.

However, Benjamin Rush, a doctor, and later a political statesman and prison reformer, spoke adamantly against portions of the new penal legislation instituted in 1786. Rush was opposed to public punishment of offenders and believed that punishment should be a means to reform the offender and should not be used as an act of revenge. Rush and others, including Benjamin Franklin, met in 1787 with the first known prison reform group, the Philadelphia Society for Alleviating the Miseries of Public Prisons (later known as the Pennsylvania Prison Society) to discuss potential changes in the penal code, prison management, and prison labor. As a result, the reform groups' first project was directed toward the Walnut Street Jail in Philadelphia.

America's First Prisons

The first recognized prison in America was located in an abandoned copper mine in Simsbury, Connecticut. This poorly conceived underground facility known as Newgate opened in 1773 and focused primarily on punishment and labor. Numerous riots, management disorganization, and constant chaos in the prison led to its closure in the 1820s.[7] The Pennsylvania legislature opened the next penitentiary, the Walnut Street Jail, in 1790. As a result of early reformers and the Quakers' significant influence, the legislature declared a wing of the Walnut Street Jail a penitentiary where convicted felons would be confined as punishment.

The Walnut Street Jail enjoyed a popular following because of its humanitarian approach. The Quakers introduced educational opportunities, religious and health care services, and prison industries. In addition, the Quakers established separate living quarters for felons, women, and debtors. The most dangerous felons were housed in solitary cells separate from others. Previously all felons, whether male or female, murderers, or debtors were crowded into common living quarters.

Eventually the jail experienced problems common throughout history—the issue of overcrowding, time management, conflicting external influence over prisoners and physical plant concerns. As a result, frequent disturbances and violence led to excessive staff turnover and the eventual closing of the Walnut Street Jail in 1835.

In view of the success in the early years experienced by the Walnut Street Jail, other states used it as a model for their prisons. The concepts of penitence, work, single cells, and separation by type of offender became important linchpins in the developing field of corrections in the United States. The Philadelphia Society for Alleviating the Miseries of Public Prisons established the first attempt at penal reform. In Pennsylvania, the legislative acts of 1818 and 1821 provided for the building of two new prisons—the Western State Penitentiary and the Eastern State Penitentiary.

American Prison Philosophy

In the 1820s, two prison models came to the forefront in the United States: the Pennsylvania system and the Auburn system. The Pennsylvania system was founded on Quaker beliefs that encouraged solitary reflection and solitary cells. Punishment or confinement in this system was meant to instill penitence and repentance in the hearts of those sentenced. Forced solitude was designed to reform the evil nature of those who had violated the laws of society. Prayer and interpersonal reflection were believed to be the answer to criminal behavior. Known as the silent system or separate system, prisoners were not allowed to see or speak to each other, reflecting the concern that they would contaminate one another. Rehabilitation in the Pennsylvania system was a goal that could be achieved through isolation, contemplation, silence, and prayer. Pervasive idleness and isolation throughout these prisons led many inmates to have emotional breakdowns, and mental illness was a serious problem that resulted from isolation. Given these problems and the considerable cost of operation due to individual cells, the Pennsylvania system gradually lost public acceptance.

An alternative system in New York, known as the Auburn system or congregate system, was a harsh program in which prisoners were kept in solitary confinement during the evenings but permitted to work together during the day. Throughout all activities, inmates were expected to maintain total silence. This program was initially implemented in New York state prisons at Auburn and Ossining (better known as Sing Sing). The Auburn style of institution was very significant: the state purposefully designed much smaller cells because inmates were allowed out of their cells to work every day. The Auburn facility was designed as an industry facility with a large factory inside the prison. This type of prison reflected the political desire of elected leaders who wanted a cheaper form of confinement where offenders could work to help pay for the cost of the prison's operation.

These principles of economical operation, restricted interaction among convicts, congregate work, extreme discipline, and tight control became enduring precepts of penal operation and punishment. Strict obedience to prison employees and institutional regulations was strongly enforced, using corporal punishment if necessary. Chains, beatings, solitary confinement, and limited food became instruments of punishment and control within the American prison environment.

Captain Elam Lynds was the disciplinarian who led the development of the Auburn system. He believed that all prisoners should be treated equally, and he used a highly regimented schedule of activities that included lockstep marching and extreme discipline. Prisoners were dressed alike in black and white striped uniforms and were expected to work and pray each day. Prisoners were not permitted to receive visitors, could not send or receive mail, and were not allowed reading material other than the Bible. Advocates of the Auburn system believed that the strict routine would transform violators into law-abiding citizens. Through hard work, religious instruction, penitence, and obedience, the prisoners could change their ways and become productive members of society. This system was well received. It was initially a financial success because of the prisoners' work, and it presented firm order and control in the institutional setting.

Most American reformers adopted Auburn's architectural style as well as operational system. Architecturally, prisons were designed with a central office building, dining halls, and chaplaincy services connected to cell houses. The cells were small and designed back to back with natural light provided by windows in an outer wall. Prison grounds included factories, health care units, and a power plant. The Auburn system emerged as the prison model adopted by most states at the time, due to its cost-effectiveness. While the Auburn system prevailed, the Pennsylvania system's emphasis on reformation still affected the modern philosophy of prison management. The desire to reform prisoners, as well as the widespread conclusion that incarceration helped resolve social disorder, propelled the growth of American prisons.

Regional Differences

Although the Auburn and Pennsylvania systems were popular, they were not universally accepted. Due to differing economic and social conditions across different regions, the philosophy of prisoner management varied.

In the South, there was not a tradition of reliance on states to provide custody for criminals, and southern governments did not build prisons to rehabilitate offenders. The southern economy was based on agriculture, and the plantation infrastructure was supported by slave labor. Southern states developed a program where citizens could lease prisoners from the state as laborers; this labor program was intended to serve the people of the South and to control those who had broken the law.[8]

As crime escalated in the West, settlers responded by establishing small local jails. As the need for additional space became more urgent, Western territories contracted with other states and the federal government to house their prisoners. This concept appeared ideal at the time: Western states were paying approximately 50 cents per day, per prisoner, for others to maintain custody of these offenders, and the states avoided the costs of building and maintaining large prisons.[9] Over time, Western territories developed governments and, eventually, their own penal systems. Generally, the Western states copied the design of the New York prisons but embraced a different philosophy in which cost-effective operations— not rehabilitation—was the primary focus.

■ Development of Reformatories

As prisons increased and the number of prisoners grew in the late 1800s, major problems emerged. Overcrowding, poor management, and insufficient funding, combined with the shifting of resources to the Civil War, led to the demise of the original models of prison management. Prison officials found the rule of silence difficult to enforce, and corporal punishment became more widely used and more severe. Rehabilitation efforts were overwhelmed by the increased focus on discipline and control, and prison management could not balance these goals. Eventually, the rehabilitation of prisoners became secondary, and the custodial concerns of operating a prison became dominant throughout the country.

In 1870, leaders from the National Prison Association (later known as the American Correctional Association) met in Cincinnati, Ohio, to plan the ideal prison system. There experts addressed issues of corporal punishment, overcrowding, and the physical conditions of prisons and ultimately decided to replace many prisons with reformatories. Reformatories were designed to assist young adult offenders by emphasizing educational and vocational programming.

The ideas and practices of Scotsman Alexander Maconochie and Irishman Sir Walter Crofton affected the outcomes of these meetings. Their philosophies undergirded the development of reformative rather than punitive services for prisoners in the United States. Maconochie believed that fixed sentences should be eliminated and replaced with a "mark system" in which prisoners could earn freedom through the earning of marks for industrial work and conformance with the institution's regulations. He also favored a classification system in which prisoners could progress through various stages of increasing responsibility and gradually earn enough marks to qualify for release.[10] Crofton's prison operational philosophy involved a series of stages of imprisonment, and progression through the stages moved an individual closer to release.[11]

Indeterminate sentencing, in which judicial sentences were served for an unspecified time with a maximum limit, allowed reformatory officials to decide when an inmate was ready to be released. Prisoners were to be released when they had made sufficient progress in education, either academic or vocational. Work performance and conduct played a significant role in the release decision as well.

Despite positive intentions, the reformatory movement was not successful. Benefits of this system were canceled out by overcrowding, poorly trained staff, and continued emphasis on control. Even though it ultimately failed, the reformatory era presented American corrections with a new model that included individual treatment, indeterminate sentencing, classification, parole, and a focus on education—all of which have been lasting reforms in corrections.

The Industrial Period and the Progressive Era

As the populations of prisons and reformatories continued to grow in the early 1900s, work programs that kept prisoners busy and productive became major assets to institutional administrators. The Auburn and reformatory models both focused on congregate work, and the funds earned by inmate labor were welcomed. Industrial programs—a natural extension of prison maintenance work assignments—offered a structured means of establishing a strong work ethic and vocational training while offsetting some institutional costs. Several different prison industrial systems developed during this period:

- *Contract system:* Private businesses established industries within the prison by providing raw materials and contracting for prisoner labor.
- *Lease system:* Prisoners were leased to private business people who would transport the prisoners to the work site and provide supervision and discipline; this usually included work in agriculture and mining and was very common in the South.
- *State-use system:* Products such as license plates, food, furniture, and clothing were manufactured or produced by prisoners but limited in distribution to state agencies or nonprofit organizations.
- *Public works and ways system:* Under supervision, prisoners worked outside the prison repairing, constructing, and maintaining public roads and properties.

During this era, industrial prisons produced a wide range of finished products for military, government, and private sectors. Industrial operations were, and remain, widely accepted in part because they employ prisoners productively, lower operational costs, and create positive incentives for prisoner management.

Industrial operations were eventually trimmed back as labor organizations began to complain about the competition. As labor organizations became more powerful, and examples of inmate abuse by the contract and lease systems became known, state and federal laws were enacted to control the use of prison labor. At the federal level, the Hawes-Cooper Act, passed in 1929, determined that prison-made goods would be subject to the laws of any state where the product would be sold. In 1935, the Ashurst-Sumners Act prohibited interstate shipment where state laws did not approve of the interstate transportation of prison-made goods. This Act was amended in 1940 to prohibit all interstate transportation of prison-made goods.

Reformers of the time continued to view the ideal prison as an environment that closely replicated the free community. They proposed normalization of the institutional setting, developing programs similar to those available outside. Classification became very important as staff separated prisoners by age, aggressiveness, and programmatic needs. Education and vocational training were reemphasized to promote literacy and prepare prisoners to compete in the job market after their release.[12] Again, prison staff attempted to structure prisoners' time to further the goal of rehabilitation.

Congress created the Federal Bureau of Prisons (BOP) in 1930—a significant decision that affected corrections in many ways. This new agency pursued innovative programs and operations that were eventually adopted by state and local government. The BOP developed a diagnostic and classification system that required the use of professional personnel, including psychiatrists and psychologists. The BOP also sought more humane treatment of prisoners and promulgated improved living conditions in well-managed facilities. Many state and federal prison administrators established professionalism as a new goal.

■ Post–World War II

After an extensive and exhaustive war, Americans were eager to reinvent society. Prior to the war, the prisoner was depicted as a despicable character, and the warden and staff, operating independently from any central authority, were seen as incapable leaders barely able to control the scene. But in the 1940s and 1950s the stereotype began to change. The inmates were seen as "good guys" who were grossly misunderstood, and warden and staff became the "bad guys." An agitated and sympathetic public called for rapid change and the expansion of programs to help inmates.

During World War II, the fairly new professional bodies of knowledge in psychiatry, psychology, and social work were used extensively to help treat the men and women who were traumatized by the war. These treatment programs were so successful that newly appointed wardens, many of whom were military veterans, suggested that these treatment programs would also be helpful to inmates in federal, state, and local systems. In 1945, Garrett Heyns, then the Director of Corrections in Michigan and president of the American Prison Association (APA) stated:

> Now we are in a transition between war and normal peacetime activity. We are beating swords into plowshares but we are not certain that we want those plowshares shaped exactly as they were before. The war found us sidetracking programs, postponing changes, and

*accepting substitutes. Lack of proper personnel, the need of utiliz-
ing prison productive capacities, and similar considerations resulted
in a less effective attack upon our problems. The time for transition
is with us.*[13]

Heyns began the new rallying cry for more sensible and humanitarian programs. During this transition period, prisoners were seen as individuals with social, intellectual, or emotional deficiencies; it was believed that they should be diagnosed carefully and that their problems should be defined clinically. Programs within the prisons were intended to help inmates return home and assume a productive, law-abiding role in the community. To achieve these goals, authorities believed it was important to require inmate participation in the treatment programs, continue to humanize institutional living, improve the educational level of officers, and increase the number of professional staff members responsible for treatment and training.

■ The Medical Model

During the years immediately following the war, the medical model gained strength. Correctional professionals were not only encouraged but also compelled—by public opinion and some legislators of state and federal governments—to adopt this model. The three components of the model—diagnosis, evaluation, and treatment—began to affect the everyday lives of staff and inmates.[14] Newly designed diagnostic centers were built in most state systems and within the Federal Bureau of Prisons. To ensure the success of the medical model, proper classification of offenders became important. (Although classification principles were developed in the 1930s, they did not really take hold until the early 1950s.) The designers of classification principles sincerely felt that these principles would become a useful tool in the rehabilitation and restoration process.[15]

Although classification was one of the greatest concepts invented during this period, it became at best a management process rather than a reliable tool to aid in rehabilitation. Classification, continually redefined, has been used extensively by all correctional professionals for the orderly management of institutions and the protection of staff and inmates.

Many leaders embraced this new correctional philosophy, but others did not accept these novel ideas about "gentle" incarceration. Many wardens and superintendents had a difficult time accepting the change due to the loss of power and authority and the new control from state and federal agencies and control boards. In spite of the erosion of their power in the postwar period, wardens had to adapt, or they would face the wrath of the central authorities attempting to gain a foothold in the new correctional leadership hierarchy.

Associations and Federal Guidance

In 1946, the APA published the *Manual of Suggested Standards for State Correctional Systems,* which revolutionized prison administration. These guidelines helped administrators maintain order, combat prison crowding, and reduce disturbances and rioting—rising problems nationwide. The APA also resolved to create a more positive view of prison life in the media. In 1951, the *Standards* were updated to apply to the wider correctional system, and, soon after, the organization was re-named accordingly—the American Correctional Association (ACA).

Congress also created a new federal agency to assist correctional agencies in their efforts to professionalize the field: the Law Enforcement Assistance Administration (LEAA) provided research grants to many colleges and universities to encourage students to enter the field of corrections and to help develop new programs for correctional institutions. However, LEAA offered little significant help. Professionals began to argue that there was never sufficient treatment staff to prove that programs could work, and many state and local systems provided only token acceptance for these programs. In the late 1980s the federal government began a fiscal restructuring and eliminated the LEAA, substituting a new agency named the Office of Justice Assistance (OJA).

The medical model carried with it the idea that the public had some responsibility for what happened in criminal justice and corrections. The good intentions of the professional practitioner and the acceptance by the public were helpful for corrections, increases in crime and the resulting prison crowding thwarted the ideals of the forward-looking correctional leadership.

Architectural Progress

One of the most fascinating aspects of the introduction of the medical model was the concept that new prison architecture had to be adopted. In this model, prisons included educational and vocational programs, extensive health care, counseling, and prison work industries; thus, the old-line fortress institution was no longer suitable.

In 1965, the Illinois Department of Corrections opened the first newly designed campus for a minimum security institution in Vienna, Illinois. The design was indeed revolutionary. In the new facilities, several prison features had been eliminated: cell houses with 5 galleries (each holding 50 cells on either side of the interior cell design), huge dining halls, and mass movement of inmates in the yard.

■ A Return to Punishment

As time passed and the ideas of the 1960s began to have a positive effect on corrections, the medical model came under serious scrutiny in many systems and, in some, was found wanting. Many young people were sentenced to prison and the conservative public, less than sympathetic to "un-American" activities, wanted to toughen the system against irreverent youth. The media, politicians, and the general public began to use words like "mollycoddling" and "country club prisons" to describe some correctional approaches and facilities.[16]

It was becoming increasingly evident that the American youth and their adult sympathizers would question any military or civilian authority. The prison systems became caught up in the debates surrounding society's ability to punish as it had for so many years. Politicians were forced to take a new look at the communities for which they were responsible, especially the civil rights of minorities in the communities. People questioned discriminatory practices and lack of minority staff in prison systems. Constitutional issues of cruel and unusual punishment were also affecting prison systems. The politicians were often silent and some refused to carry the responsibility, so correctional administrators were left to fend for themselves. Inmates, attorneys, and others who sought to destroy the prison system found a new ally in the federal court system. From the mid-1960s through the 1970s, thousands of lawsuits were filed. Because the courts now believed that inmates had complete access to and protection of the courts, inmates began to be treated very differently.

A sleeping giant began to cast a huge shadow over all civil and constitutional issues pertaining to prison operation: the federal courts.[17] What was commonly called the "hands-off" policy of the court rapidly became a strong "hands-on" approach. Civil rights attorneys and inmates began using the courts to bring prison systems into compliance with the United States Constitution and the Bill of Rights. The medical model, never wholly accepted, slowly began to change to a legal model of care.

Chaos and confusion reigned during this period, particularly after the infamous riot of the New York State correctional facility in Attica in September 1971. Prisoners in Attica were able to take control of a major segment of the prison and held many officers hostage as they sought relief from sordid conditions. It was perhaps the longest standoff and lengthiest negotiations that have ever taken place in correctional history. The state police finally attacked the prison, and a bloodbath ensued. Before the prison was under control, 20 prisoners and 10 correctional officers were killed.

Private Family Visiting

The furlough system and private family (conjugal) visits caught on in the late 1960s and lasted through the 1970s and early 1980s. Thousands of inmates were awarded furloughs for educational programs in community colleges and weekends with relatives and spouses. Furloughs were also granted to allow prisoners to pursue jobs. These furloughs were unescorted, and the rate of prisoner return without committing further crimes was 95 percent, if not higher. Occasionally, in a given state, an inmate would not return from or commit a crime during a furlough. When this happened, that state would be forced to reexamine and revise its eligibility criteria for furloughs.

The furlough program was seriously compromised in 1987 when inmate Willie Horton attacked a young couple while on furlough. He brutalized the man and repeatedly raped the woman. This incident was publicized by then-presidential hopeful George H. Bush. The public was outraged, and legislatures began to cut back or eliminate furlough programs. Within two years, almost all furlough programs were dismantled.

Family visiting, based upon the idea that spouses and children were vital to the effort to keep families together, began in the late 1970s. However, in the late 1970s, due to conservative correctional administrators who believed that sex and private family visits should not be a part of a punishment program, the idea never gained widespread popularity.

End of the Medical Model

The medical model lasted for a relatively short time in correctional history. It began to influence corrections in the early 1950s, and by the late 1970s its hold on corrections had practically vanished. Many of the early proponents lost faith in the ability of the model to function properly in a prison setting, and many inmates and staff could not accept the permissive aspects of the model.

A major blow to the medical model was a rather infamous negative report produced in the early 1970s by a researcher studying rehabilitation programs across the country. In 1974, Robert Martinson produced a report about what was commonly called the "nothing works" syndrome.[18] Correctional professionals believed the report was based on erroneous statements regarding the new programs. Considering recidivism a major factor in assessing the success of rehabilitation programs, this report attempted to show that there was no significant difference in recidivism rates for those who were in rehabilitation programs and those who were not. This report intrigued politicians and the public and ultimately helped bring down evolving rehabilitation programs. All these programs (i.e., individual therapy, group therapy, social work, and psychology intervention) were considered "soft" on crime and criminals, and it became difficult to continue to sell these new ideas to a society that was increasingly focused on punishment.

The American public grew very weary of violent crime in their home communities, and concerns with gangs and drugs on the streets began to affect the expectations of society in terms of how convicted offenders should be treated while confined. As the medical model faded, a public desire for punishment took its place. Politicians eager to please their constituencies began a "get tough on crime" platform. Elected officials became more interested in harsher sentencing legislation, and America's state and federal legislative bodies began to create tougher sentencing laws. Across America, correctional institutions became overcrowded with inmates with longer sentences and those not eligible for parole. Additionally, there was less public support for positive programming inside the institutions. However, treatment programs continued to be part of corrections (including educational and counseling programs) in limited scope.

■ Conclusion

The evolution of the prison and jail in the United States has followed the shifting social forces at work in the nation. Those who have advocated reform of these institutions believe that an individual's social deviance is a problem that can be addressed and corrected. They have advocated that correctional institutions provide a healthy environment and work toward the goal of reformed criminals.[19] Two opposing forces—punishment and rehabilitation—have driven the many changes that have beset the operation of American penitentiaries, yet the primary focus of confinement has remained a custodial function.

Prisons, jails, and other correctional environments have gone through many changes, as has American society in general. As the American public remodels the world outside, correctional facilities adopt new attitudes and philosophies. Although corrections has a major concern regarding security and serenity, fiscal resources have continued to be insufficient to meet the need. The future for significant changes may not be as probable as everyone may perceive. It would be extremely encouraging for professional corrections personnel to feel that sentencing and program enhancement could keep up with the exploding population; however, it has not kept up in the past and does not look like it will in the future.

DISCUSSION QUESTIONS

1. What are the primary forms of judicial punishment in the United States today?

2. Why does the American public not consider alternative forms of sanctions (sentencing options other than imprisonment) to be appropriate punishment?

3. What role does society have in developing attitudes which affect prison programs?

4. Will there ever be sufficient resources to manage safe, humane, and secure prison facilities?

5. What does the history of corrections suggest about its future?

ADDITIONAL RESOURCES

Z. Brockway, *Fifty Years of Prison Service: An Autobiography* (Montclair, NJ: Paterson Smith, 1969).

D. Clemons, *The Prison Community* (Boston: Christopher Press, 1940).

J. DiIulio, Jr., *Governing Prisons: A Comparative Study of Correctional Management* (New York: The Free Press, 1990).

G. Erickson, *Warden Ragen of Joliet* (New York: E.P. Dutton and Company, Inc., 1957).

P. Keve, *Prisons and the American Conscience: A History of US Federal Corrections* (Carbondale, IL: Southern Illinois University Press, 1995).

N. Teeters, *The Cradle of the Penitentiary: The Walnut Street Jail at Philadelphia* (Philadelphia: Temple University, 1955).

The National Archives, "Crime and Punishment," http://www.learningcurve.gov.uk/candp/

H. Toch, *Corrections: A Humanistic Approach* (Guiderland, NY: Harrow and Heston, 1997).

NOTES

1. D. McDonald, "Introduction: The Day Fine as a Means of Expanding Judge's Sentencing Options," in *Day Fines in American Courts: The Staten Island and Milwaukee Experiments*, (Washington, DC: U.S. Department of Justice, 1992).

2. Bureau of Justice Statistics, *Prisoners in 2005* (Washington, DC: U.S. Department of Justice, November 2006).

3. Bureau of Justice Statistics, *Prison and Jail Inmates at Midyear 2005* (Washington, DC: U.S. Department of Justice, 2006).

4. J. Howard, *The State of the Prisons* (E.P. Dutton & Co., 1929), p. 10.

5. D. Rothman, *The Discovery of the Asylum: Social Order and Disorder in the New Republic* (Boston: Little, Brown and Company, 1990).

Chapter Resources

6. H. Barnes and N. Teeters, *New Horizons in Criminology* (Englewood Cliffs, NJ: Prentice Hall, 1959), p. 326.

7. A. Durham, "Newgate in Connecticut: Origins and Early Days of an American Prison," *Justice Quarterly* 6, no. 1 (1989), pp. 89–92.

8. B. McKelvey, *American Prisons: A History of Good Intentions* (Montclair, NJ: Patterson Smith Publishing, 1977), p. 228.

9. J. Conley, "Economics and the Social Reality of Prisons," *Journal of Criminal Justice* 10 (1982), p. 27.

10. S. Walker, *Popular Justice: A History of American Criminal Justice* (New York: Oxford University Press, 1980), p. 95.

11. H. Abadinsky, *Probation and Parole: Theory and Practice*, 9th ed. (Englewood Cliffs, NJ: Prentice Hall, 2005), pp. 144–146.

12. D. Rothman, Conscience *and Convenience: The Asylum and Its Alternatives in Progressive America* (Boston: Little, Brown, and Co., 1980), pp. 123–128.

13. American Correctional Association, *The American Prison: From the Beginning . . . A Pictorial History* (College Park, MD: American Correctional Association, 1983), p. 193.

14. T.R. Clear and G.F. Cole, *History of Corrections in America* (Belmont, CA: Wadsworth Publishing Company, 1997), p. 69.

15. Clear and Cole, *History of Corrections in America*, p. 243.

16. H.E. Allen and C.E. Simonsen, *Corrections in America*, 7th ed. (Englewood Cliffs, NJ: Prentice Hall, 1995), pp. 80–81.

17. N. Morris and D. Rothman, *The Oxford History of the Prison: The Failure of Reform* (New York: Oxford University Press, 1995), p. 193.

18. R. Martinson, "What Works? Questions and Answers about Prison Reform," *The Public Interest* 35 (1974), pp. 22–54.

19. R. Johnson, *Hard Time: Understanding and Reforming the Prison* (Belmont, CA: Wadsworth, Inc., 1987), p. 19.

American Jails

2

Arthur Wallenstein and Ken Kerle

Chapter Objectives

- Identify the size and scope of the local jail population in the United States.
- Explain the system of health care in American jails and issues of mental illness, substance abuse, and infectious and communicable diseases.
- Comprehend the difficulties associated with housing long-term prisoners in local jails.

In the past, Americans have focused mainly on institutions housing sentenced offenders—namely, large prisons—when discussing corrections. Even though there are over 3,300 U.S. jails, they only rarely have been mentioned in any serious discussion of criminal justice practice, and the role of jail at the local level typically has not been well defined.[1] However, jails are central to any discussion of public safety and public policy. Interest in jails has grown as Americans have become more aware of domestic violence, drunk driving, mental illness and its relationship to public safety, victims' rights, increased levels of expenditures for public safety, and, above all, the staggering growth of jail population levels. This chapter focuses on the U.S. jail system and the many challenges that jail operations present.

Jail Populations

According to recent data generated by the Bureau of Justice Statistics, federal, state, and local correctional facilities house over 2.3 million persons on a daily basis. Because only 750,000 of these reside in local jails, some reach the conclusion that jails are a relatively minor part of the criminal justice structure, though nothing could be further from the truth.[2] Average daily population data, while instructive in some operational and policy discussions, do not adequately reflect the role of the jail and its vastly expanded importance in American criminal justice and public safety discussions. Between 10 and 15 million persons pass through jail systems during a calendar year.[3] This suggests that jails handle at least 10 times the population of prisons and means that they are in a position to influence public safety to a far greater degree than long-term sentencing facilities.[4] The jail should not be an afterthought; any institution, public service, or support program that touches between 10 and 15 million persons per year should be a central component of any analysis of public safety.

Most persons who are booked into jail remain for short periods of time, ranging from hours to months. While law enforcement efforts often focus on more serious crimes like murder or robbery, these are not the crimes that affect the largest number of persons in this country. Misdemeanor offenses involve public civility, domestic violence, substance abuse, and public safety and have an enormous effect on the public's perception of safety. Those who engage in domestic violence or endanger citizens while driving while intoxicated should influence public safety considerations considerably.

In their recent work on street-level civility issues and misdemeanors, George Kelling and Catherine Coles argue that local jails should be seen as a focal point of correctional strategy and public safety policy, given the vast numbers of people who pass through jails.[5] Jails receive, house, treat, affect, and release hundreds of thousands of citizens who do not meet community standards. Substance abuse offenders, members of the sex industry, juveniles who are declined for youth prosecution, and persons who fail to appear for court hearings or trials or to complete required programs or conditions all go to local jails. A small number of clearly serious and predatory offenders move through jails and eventually go to prison.

■ Health Care

One of the largest U.S. health care delivery systems is in jails. This system is staggering in size and scope because virtually all persons entering American jails receive some form of health care—initial screening, evaluation, treatment, or community-based referral. This approach meets constitutional, risk avoidance, and professional growth standards and has at times conflicted with public concerns that prisoners receive better health care than others. The costs are enormous and account for large portions of correctional expenses in many jurisdictions.

People in jails have a magnitude of health care problems, some of which did not exist in the past.[6] Health care issues include HIV/AIDS, hepatitis, sexually transmitted diseases, tuberculosis, problems related to alcoholism and substance abuse, heart disease, women's issues, and disorders relating to aging. Many people in jail have not received adequate care or preventive dental care, or information about family planning and pregnancy issues. People often arrive in jail with preexisting conditions and little or no relationship to community health care programs. The jail must triage extant issues, secure emergency interventions, and provide referrals to community-based health care delivery programs when people are released.

Jails must also address homelessness, substance abuse combined with mental illness, and medication concerns. Health care staff simply cannot permit incoming prisoners to continue to take whatever medications they bring in. To avoid problems of drug abuse, improper medication, or related issues, physicians must review medications before a new inmate may be authorized to continue a pharmaceutical protocol.

Standards of Healthcare Delivery

While public sentiment often reflects an ultraconservative view that inmates do not deserve quality medical treatment, it is clear that if a county, city, or regional government intends to operate a jail, it must provide constitutional levels of health care. In the past 15 years, standards for health care in correctional operations have risen. Since the mid-1970s, health care delivery has been mandated through federal case law, and state and local guidelines establishing acceptable levels of medical treatment are a basic and almost irrevocable aspect of institutional patient care.

National health care accreditation has been the driving force in providing standards, guidance, discussion, and an evolving sense of what is right in jail practice.[7] Beginning in the 1970s with the dynamically creative accreditation program under the auspices of the American Medical Association (AMA) and now firmly established through the National Commission on Correctional Health Care (NCCHC) and the American Correctional Association (ACA) accreditation program, every jail in America—whether it seeks accreditation or not—has had a guide to a full range of health care standards that apply to jails. In many larger jurisdictions, NCCHC standards form the basis of health care delivery programs in jails, and regular national inspections help jails maintain proper levels of health care. This practice has worked well in programs of risk avoidance emphasizing proactive efforts to meet standards in the following areas:

- Medication and pharmaceutical practices
- Intake screening
- Interviews
- Health assessments of longer-term jail prisoners
- Sick call procedures
- Emergency services
- Recordkeeping covering all aspects of health care delivery and referral
- Mental health evaluations, services, and treatment
- Other protocols covering a full range of health care practices and treatment regimes

Other critical issues include continuity of care, diet and exercise, special needs prisoners, suicide prevention, substance abuse protocols, restraint protocols, sexual assault issues, pregnancy, staffing standards and training requirements, and a broad range of essential data, confidentiality, and quality assurance requirements.

The standards of correctional health care have improved medical decision making significantly and have blended the security responsibilities of access to health care with the screening, evaluation, and treatment responsibilities of health care professionals. This blending of security and health care responsibilities and the demand for teamwork have been the greatest accomplishments of the standards movement, and enormous credit must be given initially to the AMA and now to NCCHC for helping maintain the quality of health care delivery in jails. These standards have evolved during a period of negative political rhetoric regarding prisoners and a lack of public support for quality health care in correctional facilities. Without a commitment to health care, local jurisdictions would have been subject to much legal intervention and potentially huge costs for negligence and failure to provide effective and constitutional health care. No area of detention practice is more important than health care, whether in the megajails of the largest jurisdictions or the smallest rural jails in the United States.

Mental Illness and Substance Abuse

It can be argued from the data that as outside community treatment and social and human service delivery systems are stretched thin, the jail systems increasingly become a human service provider of last resort. As medical assistance in other sectors of the community network ceases to exist or is limited severely, the burden often shifts to correctional organizations. These cost shifts have been created by several important changes in social policy: closure of state mental hospitals and placement of large numbers of former patients in the community, new approaches to health care delivery such as managed care and its relationship to the nature and extent of treatment, public civility ordinances and use of the jail for those who have numerous misdemeanor violations, homelessness and its relationship to mental illness and substance abuse, and a current political climate in which many in need of therapeutic intervention are not a high priority of our public policy process.[8] The growing use of jails for public health matters may not have been a conscious policy or the result of specifically targeted legislative efforts, but it has been a result of other policy decisions. Jails have faced the increased level of criminality that comes with housing more people with mental illness and substance abuse problems.

All across the country, local jurisdictions are reporting that more and more mentally ill persons are finding their way into the criminal justice system. Many prisoners have poor community relationships, a history of instability, and sporadic work records. Stays in jail occur more frequently and are of increasing duration. Studies have indicated that mentally ill prisoners may be held in jail longer than other prisoners who are booked for the same offense. Linkage to community-based mental health programs is difficult and often not a requirement of local and state funding contracts with community providers. In addition, there is a well-developed fear of what might happen in the community if a mentally ill person is released without a place to go for treatment. Jails have never been intended as mental health treatment centers, and many people rightly fear that if jails provide longer-term treatment strategies, they will attract even more mentally ill offenders who will remain for longer periods of time. This reflects the logic that cost shifts as social policy is driven by changes in funding and efforts to reduce the cost of traditional mental health services in the community.[9]

Larger metropolitan jails have developed and expanded their mental health evaluation and housing capacities, in some instances providing all the facets of hospitalization. But few argue that jail is an appropriate mental health treatment environment. Community linkages must be mandated decisively so that a jail is seen as a clear extension of the community continuum of service; support should follow offenders into jail and remove them as quickly as possible as a service delivery requirement for community-based programs. Many mentally ill offenders are in jail for minor offenses, yet they require an enormous expenditure of resources for evaluation, monitoring, medication issues, short-term therapeutic intervention, special court provisions, and ongoing efforts at suicide prevention for those focused on self-destruction. Broward County, Florida and King County, Washington have developed mental health courts to seek alternative means of responding to the vast numbers of mentally ill people in jail who could best be served someplace else.[10] The walls of the jails must come down so that jails can collaborate with community-based treatment providers. This is one of the most important directions for corrections and detention at the local level. Across the nation, states have overlooked the problems that state hospital closure has brought to local government.

Local governments are expressing more concern that state legislatures and executive branches often do not recognize how mental health service reductions in state hospitals and community programs affect jail populations. The local governments are focused on the political reality of cost shifts that are often disguised as service or program improvements and as fiscal prudence. Movement of persons from the state hospital to the community to the jail is not an improvement—it simply ignores the issue of mental illness.

Jails increasingly are housing people with a wide range of substance abuse charges, including offenses associated with personal use of alcohol and drugs. Simply put, jails are filled with people who abuse substances. Their criminal behavior may not be a direct and irrevocable result of the consumption of alcohol and other drugs, but the relationship is clear. Jail populations would be dramatically reduced if levels of substance abuse diminished in the community. The increased criminalization of alcohol-related issues and much tougher sanctions for those involved in the drug trade result in billions of dollars in jail and prison construction and far greater operating expenses over the life cycle of such facilities. The paltry amount spent on intervention, treatment, and system linkages is an embarrassment given the industry that has developed around incarceration.

Jails offer little or no deterrence to those involved in the drug world; alcohol-related offenses often reflect a deep pathology in which relapse will occur until some treatment success can be realized. Public safety issues and political reality have driven legislative and executive action to vastly expand jail populations, but jails are poorly equipped to offer significant substance abuse treatment. Drug courts and Bureau of Justice Assistance funding of some institutional treatment reflect the data generated from the programs evaluating the extent of drug abuse among serious offenders entering a broad range of large metropolitan jail systems. Public policy must determine whether jails will provide significant substance abuse intervention, treatment, and referral—or simply function as custodial holding facilities.

The National Center on Addiction and Substance Abuse reported that 80 percent of prisoners in Illinois in the mid-1990s were involved in drugs or alcohol—the largest percentage reported to date; in 2004, 68 percent of jail inmates throughout the country abused or were dependent on drugs and/or alcohol.[11] Jails might as well be substance abuse intake centers. It is unclear why massive treatment intervention through proven modalities has not been suggested as an alternative to vastly expanded jail construction. There has been discussion, but little concerted action has taken place to challenge the traditional use of incarceration at the local jail level. If the jail is to be the response of choice to substance abuse, then direction should be provided to facilitate treatment programs and demand a fair split between enforcement and treatment to diminish the likelihood of criminal relapse. Jails can provide effective short-term substance abuse intervention and treatment if programs are funded properly and there are requirements to refer people from jails to the local treatment system.[12] Drug overdose and alcohol-related medical crisis situations follow offenders into local jails throughout this country. Jail intake and booking units increasingly resemble hospital emergency rooms, with their focus on triage and emergency intervention.[13]

Co-Occurring Disorders

The number of people with co-occurring disorders (a combination of mental illness, substance abuse, and antisocial personality disorders) in jails is growing. Triage performed in jail intake and booking units has identified a growing population afflicted with co-occurring disorders, creating even greater need for more specialized services within local jails. As a facet of jail admissions and intake, this population represents in part an unintended aspect of deinstitutionalization. Given the disproportionate numbers of people with mental illness and substance abuse problems who find their way to local jails, a strong case can be made for the eradication of traditional boundaries separating elements of the treatment system. Staff must recognize that each element of these disorders demand attention if assistance is to be provided and jail populations reduced when other environments and modalities are available.

The presence of persons with co-occurring disorders in jails does in part reflect the general focus of single-dimension treatment programs, which finds it difficult to treat persons with multiple disorders. It has been suggested that "such persons may be arrested because they are too *mad* for substance abuse programs and too *bad* to be treated in mental health facilities."[14] This does not surprise jail practitioners, for the traditional "bad" versus "mad" debate has placed many persons in jail who appear otherwise to be appropriate candidates for community-based treatment or inpatient treatment through involuntary commitment or self-imposed hospitalization.

Far too often, police are unable to find treatment options or referral programs that offer alternatives to jail. Given the propensity of persons with co-occurring disorders to come in conflict with some aspect of the criminal justice system, the local government will either deal appropriately with these critical cases or experience severe crisis situations that develop during the jail intake process.[15]

Occupational Exposure and Communicable Disease Safety Measures

Staff are the single most valuable resource in jails, and they are finding that new issues are arising that did not exist in traditional jail correctional practice in the past. Fear gripped jail staff members as the HIV infection became more prevalent and a belief spread that occupational exposure would create personal risk for every person working in a local correctional facility. Information was scarce in the beginning, but gradually, through the work of groups such as the Centers for Disease Control, the principles of universal precautions were understood and accepted. Jails that received thousands of persons potentially infected with HIV took control of their own environment through staff training. Staff education and the implementation of guidelines under both federal and state mandates have been essential. While HIV infection still concerns staff somewhat, even more threatening communicable diseases have now come into local jail facilities.

Dealing with bloodborne and airborne pathogens are now a routine part of jail operations. Communicable diseases found in jails include hepatitis A, hepatitis B, hepatitis C, HIV, tuberculosis, measles, rubella, and varicella. Federal and state regulations and guidelines generally demand an exposure control plan in all jails for bloodborne pathogens and tuberculosis to minimize and manage staff exposure. To ensure staff and inmate protection, jails must have a written exposure control plan (updated annually) with the following elements:

- Engineering controls to isolate the pathogen
- Work practice controls such as universal precautions
- Personal protective equipment
- Staff training; hazard communication through proper labeling and identification of hazard items
- Availability of hepatitis B vaccine
- A postexposure evaluation and follow-up
- Detailed recordkeeping procedures for appropriate documentation of training, vaccinations, and postexposure care

Guidelines generally require written protocols for responding to spills, occupational exposures, and other possible means of transmission as well as confidential counseling and testing if an occupational exposure exists. Jurisdictions now have (or are developing) procedures to gain court orders to mandate testing after an exposure incident when the inmate is the source of the exposure and refuses testing.

For many years, jail staff understood the importance of following appropriate procedures in working with high-risk prisoners—those under the influence of alcohol or who were seriously mentally ill and potentially assaultive. Staff usually could identify those cases easily. This new aspect of jail operations concerns dangers that cannot be seen or heard and that require a commitment and belief in training for universal precautions. Jail staff who have direct contact with thousands of newly admitted prisoners do face risks. But more than a decade of practice has demonstrated clearly that when safety measures are implemented, staff are trained to know the procedures, proper equipment is provided, and the potential for transmission is minimal. The materials of jails and admission practices (e.g., negative pressure cells, blood spill containment kits, ultraviolet lamps, and latex gloves) are changing as a function of community and public health issues and developments. In the past, "sharps" might have referred only to inmates' prison-made knives, but now it also applies to the searching of incoming prisoners who might have a syringe in their clothing. Concerns about jail security and the prevention of communicable diseases are integral to jail operations.

These staff safety issues also affect prisons, but less severely. Offenders sent to state prisons almost always have completed a period of pretrial incarceration or review and have been evaluated for obvious health care problems. The millions who pass through U.S. jails have not been evaluated, and many are gone long before any laboratory results could possibly be received.

■ Victim Notification

Public policy has often been driven by the determined involvement of advocacy groups with special areas of focus such as domestic violence, violence toward women, and sexual violence. All across the nation, well-developed and knowledgeable victims' organizations have affected not only legislative policy but jail practices. Historically, victims have never been a focus of jail operations. Jails held persons for very short periods of time and did not seek out the names and locations of victims in pretrial situations. The fact that a conviction might not have occurred and that many people enter jails and stay for an uncertain duration, made such an effort to notify victims of an offender's status either very difficult or simply not part of local jail operations. Jails' approach to victims has changed dramatically with the high visibility of cases where persons accused of domestic violence have left jail prior to disposition of their cases and returned home or to the location of the initial violence and assaulted the victims again.

A nationwide movement has developed as victims' groups have demanded, with complete justification, the implementation of release policies that ensure a court hearing for most accused domestic violence perpetrators prior to pretrial release and a program of victim notification at both the pretrial and completion of sentence stages of the jail corrections process. Information technology and traditional recordkeeping techniques now have made linkages with victims possible even in situations where thousands of prisoners are released annually from a jail. Jail practice is moving toward both policy and legal requirements that, given the relationship between crimes of domestic violence and the potential for repetition upon release, victim notification be part of the operations of any jail release process. In the near future, there will be no justification for even the smallest short-term jail to fail to provide victim notification for sentenced offenders and pretrial detainees upon their release on bail or through personal recognizance after a court hearing. Such operations conducted by local jails enhance the development of linkages with other groups, including local victims' organizations, victim-witness agencies, and the media. The names of those in custody or the designation of their crimes or alleged crimes are not kept confidential when the release of basic information would help victims stay safe.

■ Information Technology and Integration

Historically, prisoners booked into jail often were released with additional charges pending and warrants outstanding because information systems were manual and subject to human error. Additionally, false names and incomplete basic data occasionally permitted persons with other serious charges pending to walk out of jail when the immediate charge was dismissed. Information technology and identification systems have vastly improved this situation through electronic imagery, swift fingerprint identification, sorting of names and other personal data and characteristics, and multiple systems to seek identity and related information.

Unfortunately, despite the vast technological growth that has had an impact on law enforcement, there has been very little systems integration among agencies. Most jurisdictions have proceeded on an individual basis with little or no real collaboration with other parts of the criminal justice system to facilitate real information sharing and joint information development. It is still not uncommon for data to be collected at numerous different points of access and to be entered separately into multiple systems—each of which has material about a particular person. Points of data entry might include the arresting police agency, jail booking unit, jail health care unit, pretrial release agency or unit, community supervision unit, numerous court elements, and community referral programs.[16] This in part explains the extent of incorrect information that systems collect and the vast amount of duplication. Duplication is costly, inefficient, and often the product of traditional bureaucratic processes.

Reengineering must occur within the jail field, or jails will drown under a sea of data that could legally and appropriately be shared on an ongoing basis with an array of system components and community-based stakeholders. Automation was an end in itself since 1990; planners of work systems must now see automation as a method that will be no more valuable than hard-copy record-keeping unless true integration of systems occurs. Efficient labor utilization as well as data accuracy, timeliness, and completeness are compelling reasons to move toward true system integration. With the few exceptions of health care and substance abuse information, most correctional data can be shared; those judgmental notations and evaluative decisions of concern can be shielded through existing firewalls. Jail booking units are a critical data point in the criminal justice system, and jails must participate as advocates for true collaboration in information system integration.

■ The Hidden Jail Population

Felons' criminal behavior drives the development of anticrime policies that focus on crimes of violence in the United States. State prisons have expanded enormously since 1985 in response to sentencing enhancements in nearly every state. The nature of misdemeanor offenses is not well understood. Many consider misdemeanor offenses to be insignificant criminal activity that is more of a nuisance than a threat to public safety. As mentioned before, this is one reason that local jails have not been seen as critical elements of the U.S. criminal justice system.

This logic has been challenged in recent years as a result of efforts to diminish criminal behaviors in large urban environments such as New York City. It has been learned through street experience that most public perceptions regarding street-level safety focus on misdemeanant behaviors. Experiences in New York City and now in other urban environments such as Seattle have demonstrated that when minor offenses are overlooked, felony behavior expands and felons are not apprehended. Public perceptions of personal safety are also diminished, and the urban environment is perceived as dangerous and not part of an open society.

George Kelling, James Q. Wilson, and Catherine Coles have publicized these concepts over a period of years. Their argument about "fixing broken windows" emphasizes the importance of having policies that take misdemeanant offenses seriously and attack street-level behaviors.[17]

In many U.S. jails, thousands of misdemeanant bookings do not occur (either because of formal local practice or day-to-day decision making) as a result of insufficient jail space. Many communities and their elected officials are careful to avoid informing local citizens that police agencies may not bring many classes of misdemeanant offenders to jail and that hundreds of thousands of existing warrants are not served because of lack of jail space. These decisions, made without voter approval, challenge the efficacy of criminal justice and law enforcement at the local level. Jail populations may be artificially low because warrants are not served and citations are issued in cases where jail is the appropriate response. Site visits to cities where misdemeanant offenses and a vast array of warrants are not served or booked have shown offenders on street corners waiving citations in the air. This demolishes the morale of police officers in the field and speaks poorly about a city's commitment to public safety and criminal justice.

This problem, which is far broader than generally understood, challenges many sheriffs. They are sworn to enforce the law yet must turn away prisoners because of lack of space and funds to provide adequate staffing for their jails. This is a partially hidden element of jail intake, but it reflects an evolving and growing practice in local jail operations when resources do not keep pace with need. Communities that do not restrict bookings or that release prisoners early often have encountered sizeable financial costs—at the expense of popular recreation, social service, and health care programs. However, other communities have led the way in developing nonjail options and alternatives to traditional incarceration.[18] At the local level, these efforts include efficiency in court operations and all aspects of the pretrial process to maintain the integrity of the criminal justice system.

■ Diversity in Jail Intake

Inmates processed in jail intake units reflect the cultural and linguistic diversity of this nation. Linguistic skills and cultural sensitivity are extremely important in the jail intake process, where language and cultural understanding may ensure the personal safety of staff or inmates or even facilitate critical life-sustaining decisions. These concerns are changing staff development policies and broadening traditional bases of recruitment. With large numbers of prisoners who are members of minority groups, jails must increase staff diversity. Jail administrators must develop a workforce that reflects our national community.

■ Systems Partnership and the Role of the Community

Because most jailed persons will be returning to the community in a very short time, jail populations offer opportunities for enhancements to public safety through networks of people, service providers, and systems of assistance. If our jails become part of expanding community-based service networks, then the jails will not be seen as simply a place where persons charged with serious offenses await disposition or a place where minor offenders spend a few hours or a few days. Local jails should be seen as an integral part of the community. Jails deal with their clientele in their home areas—where they grew up, reside, and will return, and where their families and associates continue to live.

Linkages are not defined just as calls to agencies about the release of a prisoner. Rather, they should be thought of as a collaboration with a service agency and information sharing between agencies from the time offenders arrive until they depart. Service delivery should be increasingly seamless. For years, co-occurring disorders have proved troublesome to treat for agencies that are focused on one dimension and that are unable to provide both mental health and substance abuse services. In the same way, the jail was unable to access key services in the community (e.g., education, human service, social and family services, vocational training and referral, employment assistance, and health care in its broadest construction) because jails were seen as formidable physical plants without an interest in collaboration. Jails must serve as a broker of services or find broker-type information capabilities to assist persons leaving the jail, whether to secure a room for the night or to find people treatment for chronic mental health problems. With the expansion of information technology and the understanding of the relationships between many public safety and human and family service issues, the only option other than collaboration is isolation and a lost potential of great proportions.

Growth of Jail Populations

Traditionally, most jails in the United States primarily kept pretrial offenders, prisoners in transit, parole or probation violators, people held in protective custody, or people serving sentences of one year or less. However, in the 1980s and 1990s, an increasing number of larger jails found themselves housing more people serving longer periods of time. Before the close of the 20th century, crowding pressures brought on by increased longer sentences (due to mandatory three-strikes and truth-in-sentencing provisions) caused some jails to become long-term institutions, and sentenced felons often became a part of the regular county jail populations. In 1983, jails held 223,000 prisoners. By the end of 2005, that figure had jumped to 747,529.[19]

Jail facilities have had trouble dealing with long-term populations, because most jails were not designed for this purpose. They often lack program space, do not have staff devoted to education and training, and generally have extremely limited recreational facilities. What has caused this surge in jail populations, and what can be done to deal with the problems presented by prisoners serving longer terms in facilities designed for short-term detention?

There are three broad reasons for the overall increase in the numbers of jail inmates, especially long-term cases.

1. More crime, more criminal arrests, less public tolerance for crime, and longer jail sentences have created larger numbers of typical jail inmates.

2. Many local and regional operations have been required to house prisoners who have been sentenced to state prisons, but the state facilities have been so overcrowded that they have been unable to accept the newly sentenced offenders.

3. Some local jurisdictions have attempted to "rent out" jail space to other jurisdictions. Such contractual agreements have created income for local authorities and an interest in seeking boarders from other jurisdictions—occasionally remote states.

In the mid 1990s, 25 states had jails that held prisoners for other jurisdictions. In some cases the counties were coerced by the state government to hold sentenced felons. In other cases the counties were trying to earn money to pay off jail construction bonds. Still others were trying to increase revenue in the country treasury. As the crowding in state prisons became a longer-term problem, many jails began to hold sentenced inmates for two, three, four, or five years, and sometimes longer. In one case, a county jail was home for an inmate serving a life sentence. In 2005, local jails held about five percent of state and federal prisoners (about 73,000 inmates). The South held the largest number of sentenced prisoners in local jails (about 60,000 inmates) (see **Table 2–1**).[20]

Table 2–1	State and Federal Prisoners Held in Local Jails

Total population of state and federal prisoners held in local jails in 35 states

2003	73,440
2004	74,445
2005	73,097

For the years 2004 and 2005, local jails held about 5% of the state and federal prisoners.

Total population of state and federal prisoners held in local jails in the Southern region (the region with the greatest number)

2004	62,966
2005	60,621

Percentage of state inmate population housed in local jails (in the three highest percentage states)

2005	Louisiana	45% of state inmate population
2005	Tennessee	27% of state inmate population
2005	Kentucky	29% of state inmate population

Source: P. Harrison and A. Beck, Prisoners in 2005, U.S. Department of Justice Office of Justice Programs, November, 2006.

Housing Long-Term Inmates

Most jails are not designed to house long-term inmates, and thus housing prisoners serving long sentences poses substantial challenges. Jails often lack program and recreational space, both essential for long-term inmates, and county jail budgets are not sufficient to cover the cost of housing state and federal inmates. A 1994 federal law added to the problem by making construction funds available to states and counties to build more institutions to ease prison crowding. This funding was contingent upon the recipients of the money agreeing to make sentencing reforms that would effectively lengthen prison terms ("truth-in-sentencing" grants).

Historically, the federal government housed prisoners in local jails, and federal agencies paid a per diem to the court. In most instances, federal prisoners were a welcome source of revenue for the county. At the state level, however, it proved to be more complicated. Some states negotiated agreements with the counties, and local officials were satisfied with the arrangements. Other states antagonized the counties by not paying them adequate per diem for holding state inmates. In an effort to cover jail expenses and make a profit, some counties enlarged their jails or built new facilities with the intent of housing other jurisdictions' prisoners.

The unprecedented growth in the jail population was fueled by changes in the way the country responded to crime. Crime trends, increased arrests, and patterns of offending proved to be only tenuously connected to the phenomenal inmate increase. State laws that increased sentence lengths for certain types of crimes (particularly violent crimes) proved to be the main factors driving increases in prison crowding, compelling some states to look to jails for assistance. Many local jails bore the brunt of both increased sentence lengths and increased use of pretrial detention.

Housing State Inmates

Many of the excess jail prisoners in recent years have been legally the responsibility of various state penal institutions; more often than they like to admit, state prison authorities allowed these state inmates to be backed up in jails to relieve the crowding at the state prisons. Some jail administrators greatly resented being dumped on by state governments; invariably, local officials felt cheated by the sometimes-paltry amounts paid to house state inmates.

Some states had more problems with overcrowding in their jail operations than others. The following are examples of states that experienced significant issues.

Texas

Probably the worst example of state departments of correction creating issues for local jails was found in the Texas prison system. In Texas, thousands of state inmates were crammed into numerous county jails. Texas found itself in this contentious situation because of federal litigation, *Ruiz v. Estelle*, which had led the court to place the entire state prison system into receivership. Part of the hotly contested case was directly related to serious overpopulation of the state penal facilities operated by the Texas Department of Correction. The lawsuit resulted in population capacity limitations being imposed on the corrections department; over 21,000 state-sentenced individuals were left detained in Texas local jails, as the TDC was unable to accommodate new prisoners at the rate the state courts were sentencing them.

In a later lawsuit, *The County Nueces, Texas v. Texas Board of Corrections*, the state was charged to meet its statutory obligation to receive state-convicted felons into its prison system. In the end, local authorities prevailed, and the state was required to pay the counties $20 per day for each convicted felon housed in local jails who should have been placed in the state prison system, amounting to over $100 million.[21]

New Jersey

New Jersey, a small state geographically, with 21 counties, had over 16,000 inmates in local jails in 1994. The New Jersey Department of Corrections worked closely with local governments to arrange for suitable prisoners to be incarcerated in county jails. The state agreed not to send "hardened" prisoners and gave the counties the option of sending back the troublemakers to state prisons. One of the problems that county jails discovered was that some inmates remained a very long time and became an accepted feature in the jail environment. This can be dangerous, because a familiar face can lull the staff into thinking that the individual can be trusted and cause correctional staff to let down their security scrutiny on the person in question. Returning people to the state prison system after a set time period has elapsed is one way of avoiding this possibility.[22] By the end of 2002, New Jersey jails held 3225 state inmates, which was 10.8 percent of the New Jersey jail population.[23] But by the end of 2005, that figure had dropped to 1754 or 6.4 percent of inmates; 2600 were held in private facilities.[24] When prisoners serve six years instead of six months, the orientation of the detention operation must shift accordingly. Program opportunities must be in place to help prisoners survive lengthy jail stays in a productive manner. Additionally, security measures must be enhanced. Too often, county officials do not think through the implications of renting out bed space to out-of-county prisoners. To most it appears to be a simple dollars-and-cents proposition with the bottom line of increasing the revenue for the county treasury.

New York

Similar to Texas, the state of New York kept a large number of state prisoners in local jails. In the late 1980s, the New York State Sheriffs' Association sued the New York State Department of Correctional Services and won. The New York Court of Appeals ruled that all state-ready inmates must be accepted by the state agency within ten days of state readiness. By the end of 2002, only 320 state prisoners were housed in local jails in New York, or 0.05 percent of the correctional population.[25] By 2005 that figure had dropped to only 11 inmates.[26]

Tennessee

Tennessee also had a similar problem. The Tennessee Department of Corrections (TDC) lacked bed space and allowed its inmates to back up in local jails throughout the state. In a short period of time, local county jails began to feel the impact of this backup. In this situation, the TDC worked with the local jails and attempted to limit inmates in jails to those considered less of a security risk; the state's diagnostic center took pains to see that those inmates left at Shelby County were classified as below maximum security. Restrictions were applied in the county jail system for these individuals. They were not permitted to work outside the correctional center, and they could not be housed at the adult offender center, which was designed for misdemeanant and low-security felons.

Virginia

In Arlington, Virginia, the sheriff and jail management staff noted that the jail was still not the place for state-ready inmates because it was a high-rise building with limited movement and minimal opportunities for family contact. Delays in the state moving out its state-ready inmates caused unacceptable and dangerous crowding problems. Eventually, the Arlington County sheriff joined with two other sheriffs in Virginia and threatened court action to compel the state to remove its state inmates whose confinement in county facilities placed the state in violation of its own laws.[27] The state voluntarily removed its inmates and avoided possible protracted litigation.

Resolution

By July 1994, state prison construction started to relieve the pressure on county facilities, many of which had been expanded. As the new state facilities accepted their prisoners, the inmate count at many county jails began to recede to normal and below-normal levels.

For example, Texas eventually built several new institutions, which created some excess of jail beds. By 1996, Texas jails were operating at 65 percent of their rated capacity, the lowest percentage since the Texas Commission on Jail Standards began compiling jail population figures in the 1980s. To utilize this bed space and create a source of revenue, the counties then chose to contract with federal agencies such as the United States Marshals Service, the Immigration and Naturalization Service, and the Federal Bureau of Prisons. At the end of 2000, figures indicated that 10 Texas counties housed 8402 federal inmates, and that figure has remained fairly constant to the present.[28]

A similar, but less successful, situation developed in 1996 when Missouri contracted to house 400 prisoners in the county facility of Brazoria, Texas; the government facility was contractually operated by a private corrections company. This business relationship was not of significant duration. A tremendous public outcry arose nationally when a staff member of the private corrections company videotaped, apparently for a training class, a scene of officers using police attack dogs on the inmates and inflicting kicks and punches on unresisting prisoners. Missouri quickly reclaimed its inmates.[29]

Considerations with Commitments to Other Jurisdictions

For many years, the Arlington County Detention Center, in Virginia—like many other jails—solely held pretrial detainees and sentenced misdemeanants. When longer-term inmates began to show up at the booking desk, it became clear that the old ways would no longer suffice. Arlington County went to direct supervision management and a new facility in 1994. They successfully developed new approaches to classification, academic and vocational education, substance abuse treatment, discipline, access to legal materials and counsel, maintenance of family ties, and reintegration into the community.

When prisoners are serving longer terms, the orientation of the detention operation must shift accordingly. Practitioners have noted that program opportunities must be in place to help these prisoners survive lengthy jail stays in a productive manner.[30] Additionally, security measures must be enhanced. For example, staff rotation should be given significant emphasis; over an extended period of time, an officer might become too comfortable and too familiar with specific inmates.

In New York, the Albany County jail administrator did assess the county's operations before seriously considering holding other jurisdictions' inmates. Since 1990, the Albany County Correctional Facility has generated more than $20 million in boarder revenues through the rental of vacant cells. This county worked hard to develop alternative programs to incarceration for local inmates to free up more jail cells for rent.

To succeed in the leasing of bed space, authorities must insist that for counties to participate in contracting for housing state inmates, they must provide certain services and examine other issues. For example, in Utah counties must affirmatively meet the following criteria before contracting with others:

- Are rehabilitation programs available to state inmates housed in local jails?
- Are adequate services such as recreation and visitation made available?
- Will legal services be provided to state inmates housed in local jails?
- Will state inmates be chosen for the program and who decides?
- Will steps be taken to discourage local jurisdictions from building new jails far too large for future needs?[31]

Despite these hurdles, the Utah placement program has been in effect for more than 20 years. It would appear that both the state and local county government found financial benefits worth the trouble of making the program succeed. Its success would indicate that it will be continued into the future.

■ Matrix Classification System

Large urban counties have found themselves under the gun in federal lawsuits over the past quarter of a century as expanding prisoner populations have increased the pressure to build more jail space. Local officials have responded by planning for the construction of new jail cells, which usually results in the issue being presented to voters in jail bond issues although voters sometimes refuse to approve such expenses.

Such was the case of Multnomah County, Portland, Oregon. In 1986, the jail system had 732 beds and nearly 1000 inmates. State voters had turned down four prison bond issues since 1980, and county voters had twice rejected local jail and inmate rehabilitation levies. The federal court had placed the county facility under federal court order in 1983. The county correctional officials developed a release matrix system to manage the crisis in overpopulation. In 1986, the federal court in Oregon empowered the sheriff to release inmates in order to maintain the established population limits set by the court.[32]

The matrix system was designed to release inmates early from the county institutions when the population exceeded capacity. The idea was to release the least dangerous people first, as identified by the objective, computer-based scoring system. A person booked into the jail was scored on the basis of the nature of the crime committed, with additional points awarded for felony charges and failure to appear after having been served with a warrant. The matrix system was used to control the custody population without increasing the physical danger to the community. The system was designed to meet the following goals:

- Be objective
- Consider all inmates equally
- Allow for the input of additional information related to danger that could not be measured objectively
- Be capable of being computerized
- Have the capacity of generating a list of prioritized inmates at any time whose releases might be necessary
- Seek to identify the physically dangerous inmates and limit their potential for release[33]

Since the beginning of the matrix system in 1986, Multnomah County has released over 10,000 inmates to reduce crowding. Combined with the difficulties of housing long-term prisoners in jails, overcrowding in jails is a more complex problem than overcrowding in prisons in the United States. Most jails are not designed for detention and lack the programs and staff that are critical to the successful confinement of long-term offenders.

■ Conclusion

Public administrators and elected officials should seize the moment to understand the unique role of the jail and its potential in broader public safety considerations for the future, with regard to issues of health care, victim notification, and information technology. Additionally, the upsurge of incarcerated inmates caused by new sentencing laws and other factors has made the county jail an option for holding prisoners remanded to state or federal institutions. This overcrowding is not well understood by many elected county officials and must be examined to ensure public safety and the continued functioning of America's jail system.

DISCUSSION QUESTIONS

1. What is the role of professional health care and treatment for mental health issues, and co-occurring disorders in county jails?
2. What emerging staff issues affect jail operations?
3. How are victims' issues important in jail operations?
4. What are the primary factors that have contributed to increases in the jail inmate population over the past decade?
5. What types of arrangements between state correctional departments and local jails help manage the "state-ready" inmate population?

ADDITIONAL RESOURCES

Criminal Justice/Mental Health Consensus Project,
   http://www.consensusproject.org
The National Institute of Corrections, http://www.nicic.org
The American Jail Association, http://www.corrections.com/aja

NOTES

1. Sourcebook of Criminal Justice Statistics Online, Bureau of Justice Statistics (2003), available at http://www.albany.edu/sourcebook, accessed August 6, 2007.
2. Bureau of Justice Statistics, *Prisoners in 2005* (Washington, DC: Department of Justice, November 2006).
3. P. Harrison and A. Beck, *Prison and Jail Inmates at Midyear 2004* (Washington, DC: Bureau of Justice Statistics, April 2005).
4. M. O'Toole and A. Wallenstein, "Jail Crowding: Bringing the Issue to the Corrections' Center Stage," *Corrections Today*, December 1996, pp. 76–81.
5. G. Kelling and C. Coles, *Fixing Broken Windows: Order and Reducing Crime in Our Communities* (New York: The Free Press, 1996).
6. B. Anno, *Prison Health Care: Guidelines for the Management of an Adequate Delivery System* (Washington, DC: National Institute of Corrections/ National Commission on Correctional Health Care, 1991), p. 5.
7. National Commission on Correctional Health Care, *Standards for Health Services in Jails* (Chicago: 1996), p. xi.
8. F. Butterfield, "Prisons Replace Hospitals for the Nation's Mentally Ill," *New York Times*, 5 March, 1998, A1; "Jails Nationwide Trying to Cope with Increasing Numbers of Mentally III Offenders," *Corrections Alert* 17, November, 1997, p. 1.

9. I. Miller, "Managed Care Is Harmful to Outpatient Mental Health Services: A Case for Accountability," *Professional Psychology: Research and Practice* 27, no. 4 (1996), pp. 349–363.

10. Mental Health Court Task Force, "Recommendations for the King County Mental Health Court" (Seattle, WA: King County Government, 1998), pp. 1–35.

11. J. Karberg and D. James, *Bureau of Justice Statistics Special Report: Substance Dependence, Abuse, and Treatment of Jail Inmates* (Washington, DC: U.S. Department of Justice, Office of Justice Programs, 2005).

12. T. Slyter, Jr., "Addicts in Our Jails—Do We Warehouse, Punish, or Treat Them?" *American Jails*, July/August (1998), pp. 41–43.

13. National Institute of Corrections, *Survey of Mental Health Services in Large Jails and Jail Systems* (Longmont, CO: National Institute of Corrections Information Center, 1995), pp. 1–15.

14. K. Abram and L. Teplin, "Co-Occurring Disorders among Mentally Ill Jail Detainees," *American Psychologist*, October 1991, p. 1042.

15. H. Steadman and B. Veysey, "Providing Services for Jail Inmates with Mental Disorders," *American Jails*, May/June 1997, pp. 11–23.

16. SEARCH (The National Consortium for Justice Information and Statistics), *System Integration: Issues Surrounding Integration of County-Level Justice Information Systems* (Washington, DC: Bureau of Justice Assistance, 1996).

17. J. Wilson and G. Kelling, "The Police and Neighborhood Safety," *The Atlantic Monthly*, March 1982, pp. 29–38; G. Kelling and C. Coles, *Fixing Broken Windows: Restoring Order and Reducing Crime in Our Communities* (New York: Free Press, 1996).

18. American Jail Association, *Jail Population Reduction Strategies: An Examination of Five Jurisdictions' Response to Jail Crowding* (Longmont, CO: National Institute of Corrections Information Center, 1997), p. 4; P. McGarry and M. Carter, eds., *The Intermediate Sanctions Handbook: Experiences and Tools for Policymakers* (Washington, DC: Center for Effective Public Policy, 1993).

19. P. Harrison and A. Beck, *Prisoners in 2005* (Washington, DC: Bureau of Justice Statistics, 2006).

20. K. Kerle, "Jails at the Crossroads," *American Jails*, November/December 1994 (Hagerstown, MD: American Jail Association, 1994); P. Harrison and A. Beck, *Prisoners in 2005* (Washington, DC: Bureau of Justice Statistics, 2006).

21. D. Gutierrez, "Texas Jails," *American Jails,* September/October 1993 (Hagerstown, MD: American Jail Association, 1993).

22. R. van den Heuvel, "When Jails Become Prisons," *American Jails* 8, no. 5, November/December 1994 (Hagerstown, MD: American Jail Association, 1994).

23. Bureau of Justice Statistics, *Sourcebook of Criminal Justice Statistics 1996,* (Albany, NY: The Hindelang Criminal Justice Research Center, U.S.

Department of Justice, 1997), p. 5.

24. Harrison and Beck, *Prisoners in 2005.*

25. *Sourcebook of Criminal Justice Statistics, 1996.*

26. Harrison and Beck, *Prisoners in 2005.*

27. D. Bogard, "State-Ready Inmates in Local Jails: Are You in Jeopardy?" *American Jails*, January/February 1995 (Hagerstown, MD: American Jail Association, 1995).

28. Bureau of Prisons, internal agency documents.

29. K. Bell, "Texas County Jail Criticized for Allegedly Abusing Prisoners," *Milwaukee Sentinel,* August 19, 1997.

30. van den Heuvel, "When Jails Become Prisons."

31. M. Norman and H. Locke, "Housing State Prisoners in County Jails— The Utah Experience," *American Jails*, January/February 2002 (Hagerstown, MD: American Jail Association, 2002).

32. W. Wood, "Multnomah County Sheriff's Office Population Release Matrix System," *American Jails*, March/April 1991 (Hagerstown, MD: American Jail Association, 1991).

33. *Ibid.*

Prison Architecture

3

Robert S. George

Chapter Objectives

- Understand the limitations of building correctional facilities with specific materials.
- Distinguish different housing unit models on the basis of their architectural characteristics.
- Differentiate among security levels that are suitable for various housing units.

To design a prison or jail facility, architects must consider many factors, including:
- Characteristics and numbers of inmates
- Management and punishment philosophy
- Availability of funding
- Site and utility characteristics
- Staffing requirements
- Type of housing unit

Correctional institutions are communities unto themselves and require many different services, including:
- Food service
- Medical support
- Maintenance
- Work and industrial areas
- Education and recreation facilities
- Isolation cells for rule violators

A correctional institution needs at least one basic housing configuration. To accommodate various correctional programs, however, institutions are often a collection of different types of housing units. The design of the elements that support the institution, collectively referred to as the institution core by some systems, tends to evolve from the architectural design of the inmate housing unit or units.

■ Housing Configurations

The evolution of prison housing concepts corresponds closely to the evolution of correctional management practices over the centuries. Prison architecture is influenced significantly by the operating agencies' policies and management styles. Societal attitudes toward incarcerated people affect decisions about the architectural details of a housing unit as well.

There have been few changes in housing unit design over the centuries, but most have been dramatic departures from their predecessors. Changes have tended to follow the philosophy and attitudes of the citizens of the country and the wishes of the elected representatives that authorize the design and construction. Sometimes architectural design changes follow a swing toward more punitive attitudes. At other times, they respond to a belief that the behavior of sentenced criminals can be improved through their living environment, and these facilities are designed as places of rehabilitation.

The desire to separate criminals from society and punish them has been the most consistent influence on correctional architecture through the years. Architecture is a language of symbols. Some prison architecture conveys a message of extreme punishment. In recent years, correctional architecture has come to reflect classification systems that assess inmates' behavior and attempt to forecast their needs while they are in custody.

■ Historical Models

The history of prison architecture has been greatly influenced by specific facilities and models.

The Bastille

This famous French fortress was built around 1370 as part of the fortifications for the wall around Paris. Its physical characteristics are linked clearly to the harsh approach to punishment practiced there. It was four levels high, with all levels contained within a continuous, stone masonry wall. Its massive form derived from eight cylindrical towers linked together with a series of straight wall sections. The walls had several windows on each level arranged directly over the ones beneath. Much like a medieval castle, it had two interior courtyards and it had a projected, crenellated parapet (a series of stone shields running along the top of the wall) to protect defenders stationed on the roof against arrows and other flying projectiles. Also like a castle, it was accessible only by drawbridge. In all likelihood, it had limited means for personal hygiene, and its walls allowed the weather elements to enter and circulate cold air and contaminants. The Bastille's architectural features symbolize its primary, and probably only, design goal of containing masses of people and resisting attackers bent on forcing prisoners' release.

The Second Western Penitentiary of Pennsylvania, an example of Bastille-like design on the other side of the Atlantic, was built near Pittsburgh in the 1830s. Also four levels high with thick walls and a crenellated parapet, its architectural proportions are different from its model in Paris, but its origins are undeniable. Secure, punitive, and gloomy, it operated throughout most of the 19th century to warehouse people in a hopeless, degenerative environment.

Convict Hulks

Numerous wooden ships docked in harbors such as Portsmouth were used widely in England in the last half of the 18th century to confine convicted persons. These crowded, dirty surplus boats served to separate England's convicted from their freedom regardless of their offense. They were infested with insects and diseases that were absorbed, incubated, and spread by means of their basic construction. These hulks were probably unsupervised, and their wooden construction made them a perpetual fire hazard. The convict hulk had a long-term influence on prison design. Eventually, after a few ghastly episodes stemming from the nature of their construction, prison reform advocates felt compelled to come up with a more humane design.

Panopticon

This housing unit concept was created by English architect Jeremy Bentham around 1790. Despite Bentham's English origins, no facilities with this housing design were ever built in England. The panopticon unit consists of two-person cells arranged side by side in a circular plan that generates a building in the form of a drum, now known as a roundhouse. At four tiers high and with a supervision tower in the center, it must have seemed at one time to be a highly efficient means for housing a large number of people under constant supervision (see **Figure 3–1**).

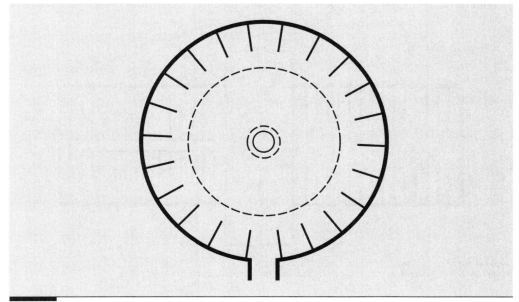

Figure 3–1 Panopticon Housing Unit

The cells in the panopticon cells face each other across a wide circular space and overlook an enclosed officer's observation station at the center. With the cells arranged along the thick masonry perimeter walls with narrow windows, if any, this configuration resembles the Pennsylvania model. Moving around on the ground level is simple and direct. On any of the upper tiers, however, officers must follow the curving balcony along the cell fronts for some distance to reach a stair. If officers in the observation station need to move quickly from the station to a problem they have seen from within the station, they also have a substantial distance to travel.

The panopticon concept includes another unattractive feature that undermines its use. Built with concrete or masonry, furnished with steel bunks and secured with steel cell fronts, the panopticon has extremely high normal, or ambient, noise levels because of reverberation and echoes within its hard walls. The circular plan shape, which generates the drum-shaped building, is a natural sound amplifier. Given the normal activity in a prison housing unit (e.g., talking, showering, closing doors, doing janitorial work), the ambient noise level in a panopticon at midday is so amplified by its shape that normal conversation sounds like shouting, which can add a great deal of stress to the environment.

Pennsylvania and Auburn Models

The Pennsylvania system (see Chapter 1) focused on imprisonment with hard labor, so prison design emerged as an important issue. To replace the open bay or congregate style of housing dozens of people that had prevailed in previous centuries, the Walnut Street Jail was erected in 1790 with small cells to house individual prisoners. In 1829, the Eastern Penitentiary at Cherry Hill in Philadelphia, PA, was developed based on cellular housing (see **Figure 3–2**).

At opposite ends of the 18th century, cellular imprisonment had been used in the papal prison of St. Michael in Rome and at Ghent in Belgium. These European models featured cells arranged along the exterior walls of the building, an arrangement now known as an outside-cell plan. Eastern Penitentiary's design borrowed

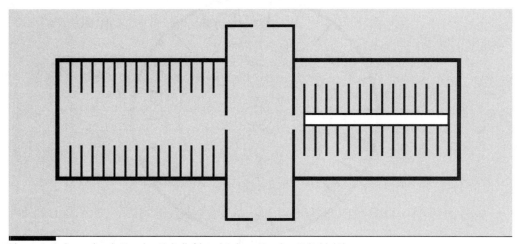

Figure 3–2 Pennsylvania Housing Unit (left) and Auburn Housing Unit (right)

this concept and took it a step further by organizing the cell buildings in a spoke pattern. In this plan, buildings enclosing a number of cells were arranged side by side in a linear pattern and in one or more levels radiating from a central hub space or rotunda. The application of this architectural configuration facilitated systematic identification and management of the institution's population in groups of predetermined size. This configuration has been used quite extensively in England, France, and other European countries.

Architecturally, the outside cell configuration is based on flanking cells arranged in a linear plan and facing a common central corridor and another row of cells on the other side. Depending on the number of inmates and floor area constraints, cells are stacked in one or more tiers accessible by stairs at either end of the range.

Cell-front design can be open with bars and a barred door, or they can be solid with a panel door. Because each prisoner can touch an exterior building wall in this configuration, the construction details of the wall and any windows it may include become essential to the institution's perimeter security. With the development of modern plumbing systems, outside-cell configurations now include toilet, lavatory fixtures, and showers. Typically, those fixtures are arranged along the fronts of the cells to permit a full view into the cell and to facilitate maintenance of the mechanical system from the walkway outside the cell.

The Pennsylvania concept led to the radiating wing organization for large housing units. In this organization, linear cellblocks are arranged like spokes in a wheel around a central hub.

The Auburn system used a slightly different model, featuring two back-to-back rows of multi-tiered cells arranged in a straight, linear plan. A typical cell for an individual prisoner could measure 3.5 by 7 by 7 feet. (By contrast, today's standards call for a room 7 or 8 feet wide, 10 feet long, and 8 feet high.) This housing concept dominated United States prison and jail design in the 19th and early 20th centuries. Like the Pennsylvania model, side-by-side cells extend far enough to accommodate the desired number of beds. The number of cells in a row can range from five or six to several dozen. Two of these blocks of cells can be joined at a central space that permits access to both.

Over the decades, electrical, plumbing, and ventilation systems were introduced into housing unit design. These systems were usually accommodated in the architectural design by separating the back-to-back rows of cells a few feet to form a continuous space called a chase. The chase can be entered from either end of the cellblock for maintenance of the systems and acts as a spine to serve the toilet and light fixtures at the back of each cell. The piping system in the chase tends to limit the overall length of the cellblock because of the relationship between a pipe's diameter and the volume of water it can handle. Other than this, there are no architectural or construction constraints to the length of a cellblock.

This concept allows the rows of cells to be stacked in tiers accessible by stairs, and these tiers can range as high as six levels. Multi-tiered applications of rows of cells have become commonly known as cellblocks. In the Auburn model or inside-cell plan configuration, cells are organized in the middle of the overall space with their fronts facing the building's exterior walls. Unlike in the Pennsylvania model, the cells and their occupants do not face each other. The distance between the cell front and the outside wall at the ground floor level usually equals or exceeds the depth of the cells to allow for a continuous balcony along the front of the cells on the levels above the ground floor. Because the occupants of the cells cannot reach the walls as long as the cell front is closed, the exterior walls can have windows for light and ventilation without compromising the building's security.

Auburn-style cellblocks were designed to provide a certain number of cells in one housing unit. However, in many of the larger institutions with long cellblocks, a crossover corridor has been incorporated at midpoint to allow movement to the cells on the other side of the unit without having to walk or run all the way to one end of the building. Group showers are often located at this same crossover corridor. The building is accessible from one end where it joins a corridor that, in turn, leads to more cellblocks or other components of the institution. Somewhere close to the other end, the building may have another door to the outside to permit entrance into the building by staff in the event of a disturbance. Beginning in the late 1970s, modern fire and life safety concerns have influenced correctional architecture significantly, and these second doors are more common and are considered emergency exits to allow evacuation. Some old cellblocks have been divided with fire-rated cross walls and doors so that one end can act as an area of refuge for the other in an emergency.

Dozens of Auburn-style housing units have been built throughout the United States. This housing concept dominated much of prison and jail design in the 19th century. Its features have become so familiar that when most people think of correctional facilities, they think of the inside-cell architectural model. There are still several functioning examples of the Auburn housing unit, probably because overcrowded conditions keep the demand for housing so high that replacement is not economically feasible. But it is interesting to note that many of them have been substantially remodeled and internally subdivided to upgrade their life-safety characteristics and convert them into more manageable modules.

When supervising in both the Auburn and Pennsylvania model housing units, officers patrol on the ground floor or the continuous balcony in front of the rows of cells. Officers need to look directly into each cell during their patrols. The application of gang-locking hardware systems in the early 20th century permitted officers to selectively open one or several doors in a range of cells at once to let certain inmates out for meals, work, or recreation. The linear housing unit configuration and the supervision practices it fostered has led in some instances to inadequate attention from staff and contributed to significant neglect and harsh treatment.

Later, institutions made up of combinations of the Pennsylvania and Auburn models arranged in the radial plan developed to take advantage of the merits of both (see **Figure 3–3**). Another site plan arrangement known as the telephone pole plan does much the same thing (see **Figure 3–4**) by attaching housing units of different configurations to opposite sides of a central corridor or spine.

Together, the Pennsylvania and Auburn configurations are now known together as the linear-indirect configuration because both of them feature the long, narrow organization and can only be effectively supervised by walking back and forth along its length. This supervision style means that staff can temporar-

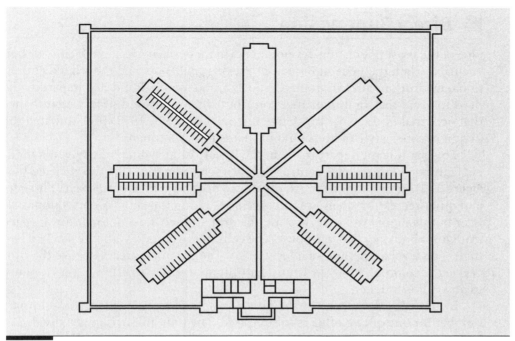

Figure 3–3 Radiating Wing Organization

ily lose awareness of some parts of the unit. This architectural style has been used extensively in American prisons and jails over the 20th century. However, other housing styles have emerged over the century that give the institution a wider variety of housing conditions within the same security perimeter.

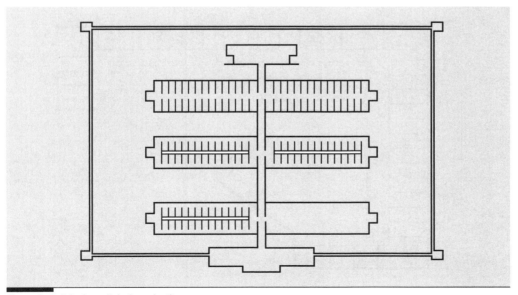

Figure 3–4 Telephone Pole Organization

■ Direct Supervision

One of the most interesting developments in modern correctional facility design occurred when the Federal Bureau of Prisons (BOP) opened the federal correctional institutions at Pleasanton, California, and Miami. Prompted in part by the need to abate the conditions that contributed to a long and deadly disturbance that occurred in the New York State Prison at Attica in 1971, BOP initiated the design of a new style of housing unit that is nearly a square.

This significant departure from the linear Auburn and Pennsylvania models features a large, open central indoor recreational or day room space (see **Figure 3–5**). Individual cells are organized around this square space. Showers and quiet recreation areas are interspersed among the cells. A correctional officer in this model can roam around the unit and see most of the interior space from just about any vantage point. The cells in the housing unit are stacked two high, one level above and one level below the common area. Because the officer on the common floor is within a half-flight of either level of rooms, response to any cell is quicker.

By design, the capacity of the unit was limited to 125 cells, an appropriate number for one or two officers to supervise. The unit, in turn, can be divided in half or quarters by means of sliding doors or temporary partitions. Like its predecessor, the Pennsylvania model, the building envelope (i.e., the exterior walls and the roof) has been detailed to provide the building's perimeter security. The secure envelope means that the interior partitions, doors, hardware, stairs, and other features could be built of lighter materials. This concept, which has become known as the direct supervision model (or new generation model), encourages a humane atmosphere by facilitating inmate-staff communication as well as security.

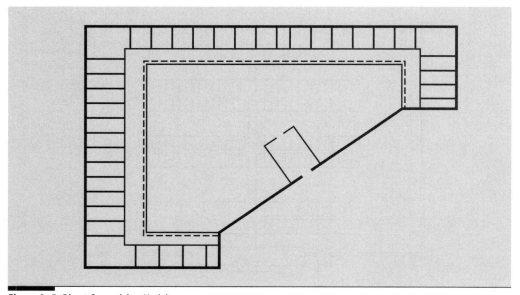

Figure 3–5 Direct Supervision Model

The exterior shape of this housing unit includes a sloping roof covered with conventional shingles so that persons can be seen on the roof. Inmate cell windows are quite large and the walls are trimmed with large wood beams. Because more systems from commercial construction can be used in this concept, it is more economical to build.

■ Supermaximum Security

Prison systems have found it necessary to develop high-security institutions to handle groups of inmates who are especially violent.[1] BOP operated the United States Penitentiary on Alcatraz Island in the San Francisco Bay for 30 years for very dangerous inmates. Architecturally, Alcatraz was a combination of the Auburn system (with stacked inside cells) and the Pennsylvania system (with rows of cells that face each other across an open corridor or range). It had manually operated gang-locking doors and the central plumbing chase characteristic of the Auburn model. The buildings and support structures around the island were constructed of reinforced concrete. Its capacity ranged from 200 to 250 inmates, each in single cells. The rows of cells were stacked two high, and the main roof over the housing unit was high enough to permit skylights that were well out of the reach of inmates. It also featured a central dining room. Outdoor recreation took place on the south side of the island in a large, open yard enclosed by a tall concrete wall. The institution included industries and some staff housing.

Marion included a control unit of about 70 cells for inmates within the federal system with dangerous and aggressive behavior, long sentence duration, or other administrative conditions that required that they be housed under constant segregation conditions. The unit's design is based on inside cells with a dedicated shower at one end. This shower and a small recreation yard adjacent to the unit are available to only one inmate at a time. Meals are delivered on trays to each cell, and all movement within the unit is under multiple escorts.

High-security institutions were taken to a new level in California in the late 1980s and in Colorado in the 1990s. The California Department of Corrections' Pelican Bay Prison near the Oregon border includes two security housing units (SHUs) totaling 1056 beds. The SHU is a new model of the administrative-maximum security facility to house management cases, habitual criminals, prison gang members, and the like. In these units, the inmate lives alone in a single cell. Each unit has its own grille-covered recreation yard. The inmate is permitted to use this yard for a short period each day. Doors to each cell are sliding, perforated steel plates with overhead, motor-operated sliding devices operated from a control center. With the exception of escorted, scheduled movements for recreation or other appointments, the inmate never leaves the cell.

The administrative-maximum security institution at the United States Penitentiary in Florence, Colorado, is the current federal edition of a supermaximum facility. As in Marion, the capacity of the Florence, Colorado, basic module is small to facilitate supervision. Each unit has 64 cells, each cell is accessible through its own sally port, and each cell includes its own shower as well as toilet and lavatory. Cells are arranged in the outside-cell configuration on two levels split at the unit entrance. But a wall down the middle of the unit screens the view of the cells across the corridor. Inmates can use a large outdoor recreation area between units on established schedules.

■ Other Design Factors

The housing unit of a prison or jail is the most important element of correctional design. Depending on the capacity of the institution, the collection of housing units typically account for at least half of the land covered by a correctional facility. The architecture of the inmate housing area has a way of steering the design of the rest of the institution. The other elements consist of the spaces needed to support the housing units. Space is needed to prepare and serve food, run programs, provide medical services, put out fires, ensure the institution's security, and provide for its sanitation and maintenance. In many modern institutions, large industrial buildings are included so inmates can manufacture goods or provide services for outside agencies.

A correctional facility is a large, expensive, and complex place to build and to maintain. It costs more to construct prisons than most other types of buildings. There are other significant operational costs that must be considered in the design of a prison or jail. Correctional facilities are in constant need of maintenance. Their various systems are used heavily each day, and they need to be repaired or replaced frequently. Institution facilities are never truly complete because their changing populations and space needs demand expansion or alteration, which leads to ongoing renovation. And, like all buildings, correctional facilities must face the devastating effects of earthquakes, hurricanes, floods, fire, and other natural disasters.

Prison design tends to react to new or changed conditions, operations, or programs. In the process, correctional design often employs technology originally devised for other building types. Rarely is a new technology invented for use in a correctional environment. Usually, a new technology is adapted slowly for prison use after first proving itself in some other arena. For example, most forms of the electronic life safety and surveillance and control systems now common in modern jails and prisons were common in schools, hospitals, dormitories and the like well before they made their way into corrections.

A correctional institution of any size is a place for people to live and for others to work. Correctional architecture needs to contribute to a sense of safety and health in both of these groups as they go about their lives. Obviously, correctional institutions can be dangerous places, and they can readily become a setting for the worst in mass human behavior. Almost every day in a correctional institution, life is a repetition of the day before, but on occasion, disturbances occur, exposing people—staff and inmates alike—to serious threats. Balancing the dichotomous nature of the culture it serves is a massive challenge in the process of designing a correctional facility.

Over the centuries, prison architecture has had the same central purpose—to separate convicted offenders from the rest of society—but different forces have influenced how correctional facilities are designed:

- Social reformers, dismayed at conditions they found in the justice system of their particular day
- Correctional science and classification systems that identify and separate offender types
- Management sciences used to train staff and manage resources
- Technological advances in construction systems, detention hardware, and electronic surveillance and control systems

■ Conclusion

Prison design and architecture are driven by societal attitudes and directly relate to the purpose for which the institution is designed. If the prevailing attitude is supportive of harsh punishment, institutions are designed and built to emphasize harsh control features. If citizens wish to emphasize rehabilitation, the design will reflect more normal-appearing, less-controlling architectural features. The history of prison design is a fascinating one and it parallels changes in expectations and attitudes that have shifted in American society.

DISCUSSION QUESTIONS

1. What kind of correctional programs would work best in each of the architectural configurations described in this chapter?

2. What are some of the possible health, safety, and emotional effects that might accrue from long-term confinement to any of these housing configurations?

3. In comparing possible goals for a correctional institution, what factors would you consider as you establish security and program requirements to be factored into the architectural design of the facility?

4. Which housing configurations lend themselves to good sanitation and maintenance?

5. Which housing configurations would be the most difficult to maintain and keep clean?

ADDITIONAL RESOURCES

American Correctional Association, *Design Guide for Secure Adult Correctional Facilities* (College Park, MD: American Correctional Association, 1983).

American Correctional Association, *The American Prison: From the Beginning . . . A Pictorial History* (College Park, MD: American Correctional Association, 1983).

L. Fairweather and S. McConville, *Prison Architecture Policy, Design and Experience* (New York: Architectural Press/Elsevier, 2000).

NOTES

1. R. Johnson, *Hard Time: Understanding and Reforming the Prison* (Belmont, CA: Wadsworth/Thomas Learning, 2002).

Developing Technology

4

Peter M. Carlson and Sonya D. Thompson

Chapter Objectives

- Examine technological developments in the field of corrections.
- Outline concerns about new technology and institutional security.
- Grasp implementation techniques to increase the acceptance of new technology.

Just as technology can be used to enhance people's daily lives, it can be used in correctional settings to aid staff in performing their jobs, rehabilitate inmates, and protect society. Yet, in the past, prisons and jails have not demonstrated an affinity for change due to increasing concerns about security. Prison and jail administrators must learn to appreciate the benefits of new technology in the correctional setting and find ways of developing and implementing this technology that further the overall goals of their institutions.

■ Development of New Technology

Most criminal justice agencies are struggling to do their work in the face of burgeoning inmate populations and increasingly stark budgets in support of the daily institution operations. Correctional practitioners are striving to do more with less, and new technology and automated processes can allow staff to cover more ground in an age of dwindling resources. Like it or not, today's leaders in prisons and jails must leverage technology to enhance productivity.

Despite the reticence about new concepts, corrections has undergone as much technological change as any other business or industry in the United States and the world. Daily prison and jail routines are very different today than they were just a few years ago.

Many new technologies have been adapted from applications developed by the National Aeronautics and Space Administration (NASA) and the Department of Defense. The National Institute of Corrections has been working at this adaptation since 1989. The Office of Law Enforcement Technology Commercialization has been a major force in bringing the research community together to offer affordable, market-driven technologies to work in state and federal prison systems.[1] Many military innovations have proven to be very adaptable to the prison and jail setting, including barrier wire, infrared night vision technology, and identification verification equipment.

Information Management Systems

Information management systems and computers have transformed many aspects of correctional facilities. Computers monitor many aspects of institutional life, ranging from tracking actual inmate counts to status reports on open security doors. They can alert staff to problems, help track and order supplies, and manage documentation. With computers, correctional institution staff can monitor inmates throughout their confinement and easily access information on an individual prisoner, a specific institution, or the entire correctional system. The advent of new technologies has increased availability of systemic information and efficiency in many aspects of prison and jail administration.

One of the best uses of technology within a correctional environment is the electronic management of inmate data. Depending on the size and complexity of the system, almost all aspects of an inmate's confinement can be managed, modified, and controlled via an information system. For example, an inmate's housing assignments can be tracked and monitored to ensure that he or she is not confined with persons who are potential victims or are potential threats. Similarly, inmates' disciplinary records, program and work assignments, educational needs, and medical needs can be managed via an information management system. Using such a system yields many benefits, because staff can access updated information and make real-time operational decisions. Additionally, in this centralized system, information travels with inmates as they move throughout the correctional system, and multiple members of inmate management teams can share and coordinate activities more easily. This information also can be shared with other law enforcement agencies to conduct criminal investigations or monitor intelligence.

After the terrorist attacks of September 11, information-sharing initiatives have become a necessity. Many individual states have coordinated data sharing among their own state agencies with the support of the Justice Technology Information Network (JNET); in other instances, states are developing and using regional information-sharing initiatives to share data among regional, state, and federal law enforcement entities. While these systems may be more difficult to develop and manage (because each participating agency wants a specific configuration and list of access and distribution), benefits are numerous:

- More available data, shared across agencies
- Encouraging other types of shared communications
- Shared costs, which helps smaller agencies gain access

Information systems may also interface with outside civilian agencies such as medical contractors. To ease information exchange, information systems should be built using commonly accepted protocols and standards so that health professionals in various medical facilities can understand and access the necessary data.

The availability of information to the public sector has also been enhanced by computers and the internet. For example, the public can view the names of violent inmates soon to be eligible for parole and contact the state division of parole to comment on an inmate's bid for parole. Victims may use computerized services such as Victim Information and Notification Everyday (VINE) to get up-to-date information about custody status, bond status, and court dates of convicted offenders. Geographic information systems (GIS) chart locations of assaults, review inmate demographics, and track cell and work assignments.[2]

Prisoner Identification

Prisoner identification also has been enhanced greatly by the development of new technology, increasing institutional security and cost-effectiveness. From digital mug shots to retina-imaging and iris-scanning equipment to barcoded wrist bands and electronic bracelets, technology is helping update prisoner identification, processing, and tracking throughout correctional facilities. Additionally, mobile barcode scanning equipment allows staff to inventory, record, and track all inmate personal property.

PRISM, the Prison Inmate and Safety Management System, requires inmates and staff to wear an electronic wristband that emits a radio signal every two seconds to receptors located throughout the prison. Using this system, inmates and staff members can be recognized and tracked individually. In the event of an emergency, this system allows personnel to retrieve and identify all offenders that were in the area at the time of the incident. If an inmate blocks the signal or takes the band off, the system notifies staff.[3]

Perimeter Security

Technology has improved perimeter security systems to lessen the likelihood of escapes with motion detectors, electric fences, closed-circuit television, and entrance and exit systems and procedures that enhance staff ability to monitor fence lines and other secure areas.

Motion Detectors

Motion detectors can monitor an institution's perimeter for unauthorized movement. Historically, the probability of detection (increased sensitivity of the sensor) led to false alarms, but through advanced digital signal processing, today's adaptive sensors can detect moving targets based on size or mass and movement to minimize false alerts. Additionally, these systems are now connected with electronically measured perimeter lighting.

Electric Fences

Electric fences are exceptionally effective means of deterring escape attempts. These offer the opportunity to cut back on staffing in towers and external mobile patrols. However, these fences may inadvertently injure or kill birds and other wild animals; special netting has been developed that allows the lethal fences to protect the perimeters as well as keep wildlife off the electrical grids.

Closed-Circuit Television

Closed-circuit television (CCTV) has been a great technological addition to staff supervision in all correctional environments. Most secure facilities have used camera supervision to monitor and record inmate visiting rooms to help supervise the areas and help prevent drug or other contraband from entering the institution. Monitoring can be accomplished from remote stations, giving management the ability to add to the duties of some posts. High-security penitentiaries have found that cameras in cell houses, dining rooms, recreation areas, and work production zones have cut back significantly on the level of violence. Video recordings also have been very effective in subsequent prosecutions of those inmates who chose to assault others under the eye of the camera. Staff members also appreciate the additional sense of security provided by the presence of the cameras.

Entrance and Exit Procedures

New technologies also focus on perimeter points of ingress and egress, including:

- Identification cards with magnetic barcodes
- Voice printing and hand geometry readers
- Biometric recognition (measures and compares a physical characteristic such as retina, iris, fingerprints, voice, or face to identification database)
- Heartbeat detectors in vehicle sally ports (detects the heartbeat of an intruder who may be hiding in vehicles)
- X-ray screening (detects metal contraband)
- Body orifice security scanners (detects concealed metal objects)
- Drug detection screening systems (screens for the scent or trace of up to 30 different illegal narcotics on visitors' clothing or bodies)

Inmate Programming

Because many inmates often come to prison with little education, minimal exposure to computers, and few job skills, inmate programming that teaches about technology and computer skills can greatly ease reintegration. Proficiency in basic computer skills such as word processing and data entry can make inmates more marketable for entry-level jobs. Some corrections systems even employ inmates in jobs that allow them use of more sophisticated computer programs that are the same one used by private companies in the community, including responding to customer assistance calls, data quality assurance, and directory assistance. Apprenticeships and vocational training programs may provide an opportunity for inmates to use sophisticated information systems, including computer-assisted drafting and desktop publishing. Fortunately, a plethora of opportunities can be made available to inmates at minimal cost due to computer labs outfitted with donated or recycled computers without internet capabilities, to minimize involvement with illegal behavior.

Medical Services

Management of inmate medical records is a substantial challenge for many corrections departments, particularly large systems where inmates transfer to many different institutions during their terms of incarceration. The use of electronic medical records in prisons allows healthcare providers to concentrate their limited resources on treating inmates rather than locating files, taking down histories, and requesting redundant lab tests. Many available electronic medical record systems allow providers to input information about the patient during the visit, thereby increasing the likelihood that the information is accurate and complete. This information can then be made available to all subsequent treating practitioners, even if the inmate transfers institutions, thus minimizing mistakes and opportunities for inmates to manipulate the system. Adding a pharmacy component to such a system that tracks prescribed and dispensed medications can also be a very useful enhancement.

Advancements in digital imaging and data transmission have enhanced the provision of medical care in prisons by allowing for remote diagnosis and treatment of a variety of conditions in specialties ranging from radiology to psychiatry to orthopedics. Specifically, this technology allows medical staff, or even the patient, to view records that may be located thousands of miles away, thereby avoiding the costs and security concerns associated with transporting inmates for medical care.

Crisis Management

Emergency response equipment has improved significantly in recent years. Non-lethal—or more appropriately, less-lethal—weapons are now the first level of response in crisis situations due to advances in technology. These include stun guns, flash bang distraction devices, and gas, instead of firearms. Additionally, battery-powered stun belts (which can issue electric shocks) provide a hidden passive restraint for inmates under escort out of the institution. Laser lights, which can cause momentary blindness, physically disorienting and stunning the subject, have also aided crisis management.

New Construction and Reconstruction

Pre-finished concrete modules have become widely accepted in prison and jail design and construction. Modular building has exceptional quality control, and budgets stretch further with much quicker construction schedules. It is estimated that correctional jurisdictions can save hard dollar costs by using pre-cast modular cells. Normal, onsite construction of prison cells can typically cost up to $90,000 per cell; modular construction costs approximately $20,000 per cell. Finally, these pre-cast technological improvements are expected to last longer and require less maintenance.[4]

■ Security Concerns

In providing staff with access to new technology, it is critical that they be trained in protecting access to systems and technology, as well as protecting information from being disclosed to unauthorized persons. Security breaches may result in inmate fraud, abuse, escapes, and potentially, the loss of innocent lives. Careful measures need to be taken to ensure that institution and physical security is not compromised.

The easiest control to put in place is developing procedures to secure technology when not in use by authorized persons. Computers, laptops, and mobile computing devices such as personal data assistants (PDAs) should be issued only to approved persons. Additionally, only specific persons should be authorized to approve the purchase of IT equipment to ensure that unapproved technology is not in use within the facility. Finally, staff should be trained thoroughly in security best practices to ensure that they understand when and how information can be disclosed.

Access

The National Security Administration advises that staff members should only be given access to information that is required to perform their jobs. Human resources staff should not have access to financial systems, and correctional services line staff should not be able to view sensitive medical data and systems. Strictly controlling who has access to each system ensures that the risks of unauthorized disclosures, particularly to inmates, are minimized.

Software systems or applications that store private or sensitive staff or inmate data should include user access controls and secure passwords.[5] Other simple measures that can be deployed include:

- Password-protected screensavers
- Privacy screens on monitors
- Well-enforced IT security policies
- Staff-training about how to protect data and report security breaches

Additional measures can be taken to ensure that the above-described practices actually are being carried out. One valuable tool is the use of audit trails; by incorporating audit capabilities into applications and systems, investigations can be supplemented to review when or if users viewed, edited, or deleted relevant data from a system. Property audits should be performed on a routine basis to ensure that equipment has not been stolen or lost.

■ Implementation of Technological Change

Despite all the ways in which new technology can enhance the correctional environment, correctional workers often express ambivalence towards new technology and change. On the one hand, they may be interested in high-tech equipment that makes their jobs easier or more efficient, but new technology is expensive and often represents a departure from the traditional direct supervision of offenders. These competing interests tend to suppress the consideration of new technology in the correctional world.

Because some ill-conceived technologies have resulted in security concerns that threatened the well-being of staff and inmates, this hesitation may not be unfounded. Additionally, new technology may be expensive and can be seen as a replacement for direct inmate care and surveillance. Lastly, new equipment and facility enhancements are often hyped solely as a means of improving efficiencies rather than as ways to help staff work more effectively.

For new technology to be accepted fully, it must be seen as a helpful tool that facilitates staff responsibilities, not as an additional burden or a threat to institutional personnel. Rather than replace personal interaction, high-tech products must make tasks easier and more efficient.

Staff reluctance to adapt to new procedures creates a significant management issue as institution administrators try to modernize their operations. Careful planning is necessary to effect development and integration of new technology to ensure that it is designed properly and helpful to staff, tailored to local operations, and user-friendly. Additionally, teamwork—including involvement of correctional staff in the process of developing and implementing technology—is a key to success.

Conclusion

As correctional systems have expanded, so have the interest and scrutiny of external constituencies. Corrections-related expenditures have grown exponentially and are often one of the largest expenses in local, state, and federal budgets. Elected representatives in state legislatures, government budget personnel, and representatives of the media all have become extremely interested in prison and jail operations. Specifically, they want to know the logic behind institutional management decisions. The implication for the world of prisons and jails is significant. New technology is not just important—it is critical for survival. Today's correctional leaders must overcome the inertia that slows the acceptance of change.

Changes in the administration of confinement facilities have come slowly over the years, but the last decade has brought extraordinary and astonishing new concepts to a very old business. To meet the demands of the future, correctional leaders will need to seek out new ways of doing this business without losing sight of the most basic goal—to operate safe, secure, and humane correctional programs for those who live and work inside.

DISCUSSION QUESTIONS

1. Are lethal fences an ethical means of providing perimeter security?
2. How can information systems be used to enhance management of correctional institutions and the criminal justice system?
3. What do you believe are the most promising areas of future technology that will evolve in prison and jail management?
4. Should citizens who are visiting confined family members or friends be subject to intrusive electronic security checks prior to receiving approval for visitation?
5. What are some of the challenges in protecting and securing information systems in the correctional environment?

ADDITIONAL RESOURCES

American Civil Liberties Union (2001), "Q & A on Facial Recognition," http://www .aclu.org/privacy/spying/14875res20030902.html
Corrections Telecommunications and Technology, http://www.lib.jjay.cuny.edu/ctt/
Electronic Frontier Foundation, http://www.eff.org/Privacy/Surveillance/RFID
Justice Technology Information Network, http://www.nlectc.org
National Institute of Corrections, http://www.nicic.org
Pennsylvania's Justice Network (JNET System, http://www.pajnet.state.pa.us/ pajnet/site/default.asp

NOTES

1. JustNet: Justice Technology Information Network, National Institute of Justice, available at http://www.nlectc.org/justnet.html, accessed June 17, 2007.
2. National Institute of Justice, *Tech Beat*, Summer 2001 (Washington, DC: National Law Enforcement and Corrections Technology Center, 2001).
3. *PRISM: A Prison Inmate and Safety Management System Brochure*, Technology Systems International, available at http://www.tsilink.com/images/ TSIPRISM.pdf, accessed June 6, 2005.
4. F. Becker and W. Sims, *Managing Uncertainty: Integrated Portfolio Strategies for Dynamic Organizations* (Ithaca, NY: Cornell University International Workplace Studies Program, 2000).
5. B. Thomas, *Simple Formula for Strong Passwords* (Bethesda, MD: SANS Institute, 2005).

Institutional and Departmental Responsibilities

II

YOU ARE THE ADMINISTRATOR
A New Approach

Recently, a women's prison in Vermont found that its educational and vocational programs had a dramatic effect on reducing recidivism (by as much as 25 percent). The state prison in Windsor used a grant from the U.S. Department of Education to fund its Workforce Development Program. This program offered inmates more than just an opportunity to work in shops and gain technical skills. In addition to producing signs and license plates, inmates participated in a program called Habits of Mind, which teaches inmates to use more than a dozen "patterns of thought" to manage impulsivity, gain a more positive outlook, and lead a more productive life.

This program aimed to help inmates communicate clearly with others, understand and empathize with others, take responsible risks, and get pleasure and enjoyment out of their relationships. The inmates' supervisors model the techniques they encourage the inmates to develop, including managing impulsivity and offering praise and support.

By focusing on positives and building on strengths, the staff created an inmate educational and vocational program that has been shown to reduce recidivism substantially, dramatically reducing prison costs. Additionally, the inmates who are released back to communities are better prepared to be productive workers and community members.

- *Would this type of program work in most correctional facilities?*
- *How would the demands of this program be balanced with other aspects of correctional management?*
- *Given the limited resources available in many correctional settings, how should prison and jail administrators decide which projects to fund and how to allocate their staff?*

Source: Associated Press, "Vermont Gets National Recognition for New Approach to Prison Work," March 25, 2007, available at http://www.boston.com/news/local/vermont/articles/2007/03/25/vt_gets_national_recognition_for_new_approach_to_prison_work/, accessed July 26, 2007.

Custody and Security

5

Michael B. Cooksey

Chapter Objectives

- Understand the role of classification, accountability, clutter control, and inmate personal property in maintaining security.
- Grasp the various components of security, including staff management, community awareness, and victim protection.
- Comprehend high-risk inmates and security threat groups.

Most inmates prefer a quiet, clean, and orderly prison where they can serve their time in a safe environment. A well-run institution has a certain feel about it: The quiet rumble of daily activities with no loud noises, clean and shining hallways, and lack of clutter in inmate cells signify that the staff is in charge and running the prison. Few inmates benefit from the disruption of daily activities. Proper security can ensure inmate safety and provide staff with good working conditions.

■ Levels of Security

Nearly all correctional agencies within the United States have prisons of varying levels of security. There are four general security levels of penal institutions:
- *Minimum Security:* These facilities often have a greatly reduced level of staff supervision and generally have no perimeter security such as a fence. These institutions are utilized to house offenders with no history of violence or sex offenses, and those confined in this type of facility generally are serving short sentences.

- *Medium Security:* Facilities in this category have secure perimeters (generally double fences with armed vehicles that patrol the outside of the fence line) and more security personnel. Inmates serving various sentences may be housed at this level. Those who have demonstrated good adjustment at higher levels (or misbehaved at minimum security facilities) may be transferred to medium security facilities.

- *High/Close Security:* These penitentiaries are for offenders with histories of violence and for those who represent a threat to others. Inmates serving lengthy terms of confinement, including life sentences, and those identified as gang members or escape risks are classified at this security level. Inmate housing is typically in cells, and perimeter security is significant (walls or double fences, often with armed towers).

- *Maximum or Supermaximum Security:* Many states have developed a locked-down facility for the small percentage of inmates who exhibit extremely disruptive behavior. This concept dates back to the 1930s when the Federal Bureau of Prisons created the United States Penitentiary at Alcatraz Island in the San Francisco Bay and later Marion penitentiary in Illinois. These last-resort facilities are extremely secure; inmate movement and activity is controlled tightly. Inmates in this type of institution typically are secured in their solitary cell 23 hours each day, with one hour for recreation and a shower.

■ Inmate Classification and Accountability

It is difficult to begin a discussion on institution security without first discussing proper classification of facilities and inmates. Institutions must be designed to house a certain type of offender. Violent, aggressive, and escape-prone inmates require more physical security features and staff resources.

Classification can best be defined as the systematic grouping of inmates into categories based on shared characteristics and behavioral patterns. Using the inmate's history, staff can make fairly accurate predictions about the inmate's future behavior and adjustment to incarceration. Inmates with similar characteristics living together in an appropriately designed facility are much easier to manage. Likewise, a strong inmate among a weaker population can wreak havoc. Escapes, assaults, and drug dealing very seldom occur in areas where the inmates are deliberately stratified, though inmates may find ways to manipulate the system so that they can be in areas of the institution where there is little staff supervision.

Therefore, accountability (knowing where inmates are at all times) is necessary in secure facilities. A system of callouts, passes, and controlled movement at prescribed times greatly assists staff with inmate accountability. Housing unit officers should know which inmates are in the unit and the destination of inmates leaving the housing unit. When inmates are given assignments outside the unit, such as work or educational programs, the work supervisor, education staff member, or some other staff person should be responsible for the inmate. A formal call-out system will greatly improve inmate accountability when an inmate is needed at a certain place for a short period of time such as for medical appointments, counseling sessions, etc.

In addition to formal counts at prescribed times, prison staff should conduct random census counts. During such counts, all institution activity stops and inmates are counted in place to determine quickly whether inmates are where they should be. If census counts are not practical, supervisory correctional staff can periodically check various work details, classrooms, or housing units to ensure that inmates are in their assigned areas.

Inmates should be informed of their responsibility to be in their authorized area. Disciplinary procedures should be established to deter inmates from being in unauthorized areas. Of equal if not greater importance are procedures that account for all staff and their approximate locations in the institution. Accounting for staff is difficult, as staff usually have more mobility than inmates within the institution. During emergencies, accounting for staff should be a top priority. If staff have been taken hostage, this has a tremendous impact on how the warden plans to resolve demonstrations, riots, or other emergencies.

Emergency Preparation

Even in the best-run prison, emergencies occur. At the very least, plans dealing with escapes, riots, work or food strikes, hostage situations, outside demonstrations, natural disasters, bomb threats, and evacuations are necessary to ensure that staff are prepared properly to deal with emergencies. Prison administrators should identify those areas that most concern them and prepare detailed plans to address these issues. If the prison is close to major roadways, shipping lanes, or railways, plans should be developed in case of toxic or chemical spills. In areas susceptible to natural disasters such as wildfires, hurricanes, or earthquakes, evacuation may be necessary to save lives.

Emergency plans should be easy to read and informative. Although brief, they should set out specific responsibilities. The plans should be updated periodically as situations change. Emergency plans are only as good as the preparation to implement them. All staff should be fully familiar with emergency plans. At least yearly, staff should read and discuss the plans with peers and supervisors. Periodic mock exercises improve staff knowledge and make them more comfortable with their role in emergencies. Developing memoranda of understanding and involving sister agencies and law enforcement in mock exercises will improve the outside agencies' knowledge of the correctional facility and foster good relationships.

Unacceptable Possessions

Controlling contraband should be a top priority in all correctional institutions regardless of security level. Contraband is any item or article that an inmate is forbidden to possess. All correctional facilities provide inmates with medical care, room and board, clothing, and basic hygiene items. Most facilities allow inmates to purchase items in the commissary or receive items through other authorized channels. Anything else that the inmate possesses is contraband.

Weapons and escape materials are all equally dangerous in the right circumstances. Most staff are acutely aware of the havoc that these items, as well as drugs and alcohol, can cause and the resultant danger for staff and inmates. Other items such as materials to make dummies, homemade rope or buffer cords, maps, and unauthorized clothing pose a danger by facilitating inmate escapes. Gambling paraphernalia may lead to inmate assaults by debt collectors.

Institutions should have regulations that restrict the amount of personal property that an inmate may possess. Cluttered cells and excess personal property are excellent hiding places for more serious contraband. In addition, these areas are much more difficult to search, tying up valuable staff time. Excess property can fuel fires and pose health-related hazards as breeding grounds for bacteria. Institution regulations should specify the amount of newspapers, magazines, pants, shirts, and even underwear an inmate may possess. Medications should be controlled tightly. Legal property provides great hiding places for contraband, as staff are reluctant to search legal items properly. The amount of legal property that an inmate can possess should be specified and controlled tightly. It is important to properly document seizure, confiscation, and disposition of contraband in case of civil lawsuits.

Staff must know what items enter and exit the prison. Incoming boxes and packages should be X-rayed before entering the correctional institution and searched prior to being given to inmates. Visitors should pass through a metal detector. Because most serious contraband is introduced by inmate visitors, visitors who behave suspiciously should be subject to a more thorough search prior to visiting. Thoroughly searching inmates following visits also will deter the introduction of contraband. All vehicles should be thoroughly checked, and trash receptacles should sit in the sally port and go through at least one count before being removed from the institution.

Random, frequent searches of inmate living areas can greatly reduce contraband. Inmates who have a history of hiding unauthorized items on their persons or in their living areas should be identified and searched more frequently. Common areas in the housing unit should be searched daily in a systematic manner to ensure all areas are covered. Likewise, inmate work areas should be searched daily not only to check for contraband but to make sure all equipment and fixtures are complete with no missing parts. Bars, windows, frames, and doors should be checked frequently to detect cuts and determine if the locking devices have been tampered with. It is imperative that staff account for all tools in the institution. Only authorized tools should be utilized by staff and inmates. Staff should never bring personal tools into the institution. Should a tool be lost, all activity in the area should cease until a thorough search is conducted and the tool is found. Limiting access to computers will protect the sensitive information they may contain. For example, two inmates at a federal penitentiary were able to obtain architectural drawings through a computer and plan an escape through a utility tunnel.

Drugs and alcohol are highly disruptive to the daily activities in a prison. Regular urinalysis and breathalyzer tests of suspected users and random tests of the entire population will determine the scope of use and deter abuse. During the holidays, inmates are more lonely and susceptible to temptation, and accordingly searches should be made even more frequently to control items that may lead to a disruption.

■ Use of Force

Occasionally it may be necessary to use physical force to gain an inmate's compliance. Naturally, the preferred scenario is for the inmate to comply with a verbal command, but in emotional and tense situations this does not always occur. Having a written use-of-force policy greatly increases the probability of gaining the inmate's compliance without injury to staff or inmate.

Immediate use of force occurs when an inmate acts out with little or no warning and staff are required to physically restrain the inmate. These are highly charged, emotional incidents for both staff and inmate. Proper training allows staff to gain control of the situation while controlling their own emotions and preventing inmate abuse. These incidents should be well-documented (e.g., in witness statements) by those involved or, preferably, videotaped to protect staff in the case of a civil lawsuit.

A calculated use of force occurs when inmates are confined in an area and do not present an immediate threat to themselves or others, yet are refusing to comply with staff orders. Staff should talk with these inmates to gain their voluntary compliance and allow time for staff to fully assess the situation. Staff should determine if the inmate has weapons and whether it is necessary to use gas, other less-than-lethal munitions, or a well-trained extraction team to move the inmate to the desired location. If more than one inmate is involved, the use of disturbance control or other tactical teams may be required.

Proper use of force has a great influence on staff and inmate morale. A highly professional attitude concerning use of force by administrative and supervisory staff will be modeled by line staff, prevent inmate abuse, and enhance inmate compliance with rules and regulations. Unfortunately, the history of corrections is marred by instances of staff physically abusing inmates. In many of these incidents, higher-echelon staff projected a cavalier or macho image that was imitated by line staff.

■ Staff Management

Accurate and reliable information about staff, inmates, the political landscape, and the local community is essential to running a well-organized and secure prison. The administrator who sits in the office waiting for information to arrive through the hierarchical organizational structure is doomed to be woefully uninformed. Administrative staff should tour the prison often to assess firsthand the atmosphere of the institution. Some inmates are chronic complainers, but others go about their daily activities in an orderly fashion while being respectful to staff and other inmates. When this latter group of inmates is unhappy, administrators should address the problems. Staff at all levels need to talk to and, more importantly, listen to inmates. If staff listen, inmates will tell them what is happening in the prison.

Staff who supervise inmate work details, teachers, counselors, and correctional officers working in the housing units are often trusted by the inmates and are excellent sources of intelligence. A mechanism that allows these staff to submit confidential reports of conversations and observations of inmates is critical to gathering accurate information. Once collected, this information can be analyzed and evaluated by specially trained intelligence staff. These informed judgments allow administrators to manage institutional security and forecast future security needs. Long-range strategic planning based upon accurate information allows the proper allocation of security assets.

Outside the Prison

Prisons do not operate in a vacuum but are integral parts of communities and larger correctional systems. Reading daily newspapers and professional magazines and maintaining good relationships with elected officials will keep prison administrators abreast of public sentiment and possible changes directed by politicians. Not too long ago, prisons were forgotten places to the public and political arena. Today, correctional institutions are major employers and very visible to local communities.

Prison and jail walls and fences are very permeable in the sense that the external world has a strong influence on these institutions. Televisions, radios, newspapers, telephone calls, visits, interaction with staff, and newly arriving inmates all carry information from the outside community into the correctional environment. It is critical that penal administrators stay tuned to events outside that may influence the attitudes and beliefs of those that are confined. Some issues move inside the facility rapidly, and others take longer to affect the population. Staff must constantly be alert to changes within and outside their institution.

Protecting Inmate Victims

Certain inmates present unique challenges to prison administrators. Weaker inmates and those who have committed especially heinous crimes may be victimized by other inmates. For example, child sex offenders find it especially difficult to serve their sentence in the general prison population once inmates learn of their offense. Weaker inmates usually spend great portions of their sentence in special housing units for protective custody and frequently are transferred between prisons. These inmates may act out against staff and are often litigious, filing institutional appeals and court documents, complaining about their conditions of confinement. Staff at all levels need to be trained properly in working with weaker inmates, as most cases of proven staff abuse occur in this area.

In many ways, the aggressive inmate is easier to manage than the weaker inmate. The highly assaultive, combative inmate lives best in a prison with other aggressive inmates. Aggressors seldom prey on aggressors. Many states and the federal system have developed supermaximum penitentiaries to house aggressive inmates, but often staff and inmates are victimized before aggressive inmates are placed in these facilities. In systems without supermaximum facilities, aggressive inmates often spend much of their sentences in special housing units. Policies and guidelines for handling aggressive inmates should be specific and followed by all staff. Ensuring staff safety is paramount when dealing with aggressive or combative inmates.

Sophisticated or manipulative inmates may target staff, other inmates, and the political system to gain items or favors that are otherwise prohibited. These inmates often have tremendous resources in the community—including finances and support groups. The media may follow their incarceration and show continued interest in their plight. These inmates may be leaders or quietly give advice and counsel to inmate leaders. They are experts at detecting and exploiting staff insecurities and procedural weaknesses.

There are also inmates who pose a risk of escape. These inmates tend to be smarter and more adept at recognizing weaknesses in physical structures and procedures. They may observe staff closely for any habits or consistent failures to follow policy that these inmates can exploit. For example, many successful escapes have involved inmates simply walking out the institution at visiting times or shift changes. Other escapes entail elaborate breaches of physical structures. Almost every investigative report following an attempted or successful escape reveals poor security procedures or staff failures to follow proper procedures.

■ Security Threat Groups

Security threat groups, or prison gangs, are responsible for the majority of homicides and assaults in prison. Well-organized, highly structured prison gangs have been around for decades. These gangs have strong leaders and exert pressure on other inmates through violence or the threat of violence. They are interested only in providing illicit drugs, alcohol, and contraband to other prisoners, and prison programs mean little to them.

Studies of prison gangs offer a great deal of qualitative and anecdotal evidence of numbers of gang members, but definitive numbers of gang members in prisons in the United States are not determined easily as each state and federal jurisdiction utilizes differing methods of classifying and validating gang membership. Additionally, different geographical regions have unique security threat groups.

In recent years, more street gang members have been incarcerated. Additionally, more inmates have sought membership in groups from a certain city or geographical area. These gangs are unpredictable, less structured, and, in many ways, more difficult to manage than the traditional prison gangs. Several state prison systems have developed strategies to deal with gangs. Some correctional systems just deny the existence of gangs. In still other systems, the problem is so complex that it defies solution. Controlling gangs and their disruptive activities will haunt many prison administrators until solutions are found.

■ Conclusion

The key to a prison or jail security system's success is well-trained staff who are alert and accountable to the inmate population. Maintaining control of a correctional environment is a daunting task, given the noncooperative nature of many of the inmates and the challenges they present to the staff. An institution's security staff perform a heroic and often dangerous duty. Prisons and jails must develop a culture that treats prisoners with respect, reinforces positive communication between staff and inmates, and offers inmates humane, safe, and sanitary housing. Good security is a product of good leadership, and results from a high-quality staff who believe their work makes a difference. Indeed, it does; those who work inside and contribute to the daily supervision of inmates are public servants in the finest sense of the term.

DISCUSSION QUESTIONS

1. Do you believe it important for administrators to see for themselves the day-to-day activities that occur in their prisons?
2. How can prisoner rights be balanced with the need for safety?
3. Do you believe that gangs are easier to control in a correctional environment?
4. How can staff members contribute to an inmate's ability to escape?
5. Which inmates are the most threatening to overall security?

ADDITIONAL RESOURCES

L. Bowker, *Prison Victimization* (New York: Elsevier, 1980).

P. Carlson, "Something to Lose: A Balanced and Reality-Based Rationale for Institutional Programming," in *Crime and Employment. Critical Issues in Crime Reduction for Corrections* (Walnut Creek, CA: Altamira Press, 2004).

M. Fleisher, *Warehousing Violence* (Newbury Park, CA: Sage Publications, 1989).

J. Fox, *Organizational and Racial Conflict in Maximum Security Prisons* (Lexington, MA: Lexington Books, 1982).

D. Garland, *The Culture of Control* (Chicago: University of Chicago Press, 2001).

K. Haas and G. Alpert, *The Dilemmas of Corrections* (Prospect Heights, IL: Waveland Press, Inc., 1995).

J. Irwin, *The Warehouse Prison: Disposal of the New Dangerous Class* (Los Angeles: Roxbury Publishing Company, 2005).

M. Jankowski, *Islands in the Street: Gangs and American Urban Society* (Berkeley, CA: University of California Press, 1991).

Inmate Classification

Peter M. Carlson

Chapter Objectives

- Explain classification, its role in the penal system, and the benefits of an objective classifications system.
- Outline the difference between case and unit management.
- Define a reliable and valid classification system.

Classification

Most penal facilities offer social service staff to provide classification and program/work advisory services for their inmate populations. As one of the preeminent responsibilities in a correctional facility, classification involves categorizing offenders by assessing an individual's social and criminal background and current programming needs and assigning him or her to an appropriately secure institution, housing area, work assignment, and program (see **Figure 6–1**). How classification is organized and conducted varies a great deal by jurisdiction, type of facility, and institutional staffing levels.

In earlier years, all decisions about an inmate's security and prison assignments for work and housing generally were made by one senior management official designated by the warden, often the deputy warden.[1] This individual controlled all aspects of life inside the institution and made unilateral decisions based strictly

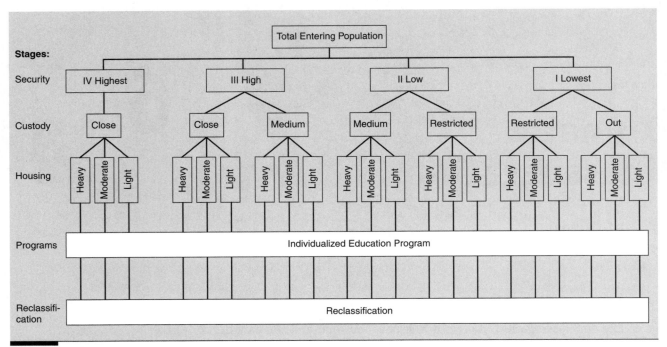

Figure 6–1 A Classification System

on his or her often-limited knowledge of inmates, their deportment, and their attitude. Although this was an effective method of establishing consistent governance in a very punitive environment, little attention was directed to interaction with the prisoners, and virtually no emphasis was given to the goal of positively influencing an inmate's life.

As this process evolved, this responsibility shifted from the deputy warden to a classification committee. In most correctional facilities, classification committees are large groups of subject-matter experts who gather regularly to evaluate new inmates or to reclassify inmates for custody, housing, work, and program assignments. The committee is often chaired by a senior management official such as an associate warden and comprises the heads of institution departments such as the captain of security, the chief of classification, the supervisor of education, and the inmate's case manager. Many case management committees require the attendance of the inmate being reviewed.

As the primary link of communication between inmates and staff, as well as an important connection to the individual's future life in the community, the case management team is responsible for a significant part of an institution's operation. It may be a small department of overworked case workers, or a large, organized network of social work and case management staff who consistently work with inmates, including unit managers, case managers (also known as case workers or social workers), counselors, education representatives, psychologists, and secretaries.[2] Responsibilities of the team include inmate classification, social service support, institution program planning, and release preparation.

Most correctional professionals believe that these tasks are critical to today's prison and jail operations.

The personnel who manage and work in prison and jail facilities recognize the need to separate the many different types of felons that are held in confinement facilities. Separating inmates (male from female, sick from healthy, youth from adult, and aggressive from passive) is a function that has important ramifications for all aspects of institution operations. Inmate classification is simply sorting inmates into appropriate categories. Once the correct category is determined, many other decisions can be made.

Accurate inmate classification is one of the primary factors that contribute to a safe and orderly penal environment. The classification of inmates is a process that ensures that a correctional system places inmates in an appropriate institution that can provide the necessary amount of security and supervision. All state prison systems, as well as the Federal Bureau of Prisons, utilize similar systems that separate inmates based on the level of security required to control and contain them. It would not be safe to place a hardened, violent offender in a correctional facility designed for minimum-security individuals. Conversely, it would be a waste of taxpayer funds to confine a non-violent offender in a maximum-security institution.

Many state correctional agencies operate one or more central reception centers; all newly committed inmates are placed initially at one of these institutions, where they are reviewed and classified. Once the classification process has been completed, staff know what the inmate's security requirements are as well as what programs may be important during the offender's confinement. The reception center staff are then able to select an appropriate prison that will meet the inmate's security and program needs.

History of Classification

Early penal facilities housed violators in the same detention facilities without any consideration of their gender, age, health, criminal history, or current offense. Over time, correctional practitioners recognized the value of separating offenders. Early classification simply involved separating male and female offenders, juveniles and adults, and first-time or non-violent offenders from more frequent or serious offenders.

Historically, institutional managers used their subjective judgment to assign inmates to the various security levels. Staff members would simply consider the offender's age, prior record, current offense and sentence, and institutional adjustment. Simply stated, these decisions were made based on intuition and experience. However, decisions could be affected by unacceptable factors such as an inmate's race, gender, or poor social skills. Accordingly, discrimination could cloud each individual decision.

Early classification committees were made up of only senior staff members who did not know the individual inmate well. Eventually, team classification developed; case management and security staff who worked with the inmate on a regular basis were charged with making initial and ongoing decisions related to classification and daily operational inmate requests. This system continued to evolve and became the case management system.

■ Case Management

Case management focuses on the provision of social support programs to an inmate population, and case management staff maintain the official classification documents for each inmate. In many jurisdictions, these staff members are not only responsible for determining the prisoner's custody and security needs but are also charged with helping the inmates plan their institution-based work and program assignments, representing the inmate to the parole board, offering counseling services, providing connections to the community, and handling release planning. Case managers perform myriad tasks that pertain to inmates' daily lives and guide inmates' activities with the ultimate goal of helping them make successful transitions back to their home communities after release.

Case managers or counselors are responsible for inmates from the time they first arrive in a correctional setting. Initial screening of new arrivals generally is accomplished by social service staff members who ask new prisoners about their needs. Interviewers ask if the inmate feels he or she needs protection from others that he or she may have testified against and try to identify other potential enemies within the institution's general population. Screening questions also seek information about the offender's physical and mental health and other pressing management issues.

Case management staff gather background information about new inmates in pre- or post- sentence reports (prepared for the sentencing courts by probation and parole officers) and seek other basic information about individuals. This attempt to gather information about the inmate is a direct result of a philosophical change in contemporary corrections. As prisons and jails began to do more than simply house offenders, most correctional agencies during the early 1960s began to implement rehabilitation programs—known as the medical model. As this name implies, supporters of the medical model believed that criminality was an illness and that inmates could be cured of their social deviance by program involvement during confmement.[3] The resulting emphasis on treatment required staff to focus on the criminal rather than the crime.

Once the case manager completed gathering information about the individual's prior arrest record, adjustment to earlier periods of incarceration, and social data about family and friends, he or she prepares a classification study report (in most jurisdictions). This document identifies the prisoner, mentions social factors that may have led to his or her offenses, and recommends institutional programs that may help prepare the individual for release. Details about the inmate's program participation or lack of progress are added to the record throughout his or her confinement.

In many correctional systems, the initial assessment is completed at a reception and diagnostic center over a period of four to eight weeks. In the jurisdictions that utilize these centers, the newly arrived prisoner is put through an extensive evaluation that often includes a complete personality assessment, intelligence and psychometric testing, review of past work habits and lifestyle, observation of how he or she interacts with staff and inmates, and identification of those factors that may have led the individual to crime. In other jurisdictions that do not have reception and diagnostic centers, inmates are committed directly to an institution and go through a similar classification process.

Once the background classification report is prepared, the inmate is then formally evaluated at a classification meeting. This meeting entails the development of an integrated work assignment, permanent housing, and educational, vocational, and social improvement programs for the offender.

Classification is the backbone of the security program of any prison or jail. It is imperative that staff know the background of each inmate and the threat each presents to the effective custodial management of the facility. Necessary basic information includes:

- Age
- Sex
- Social history
- Criminal sophistication
- History of violent or aggressive behavior
- Special needs (e.g., mental or medical issues)
- Potential challenges to security (e.g., escape history, gang membership)
- Special management factors (e.g., judicial recommendation, racial balance, program availability)
- Institutional capacity, availability, and security

If a realistic assessment is accomplished by case management staff, the inmate can be placed in housing with appropriate security and all other aspects of institution management will follow accordingly. It is important that inmates be placed in the least restrictive facility that is able to meet their security needs.

Once they classify and assign inmates, case managers track inmates throughout their confinement, offering assistance with the supervision of inmates, participating in discipline hearings, tracking progress, and continually assessing needs for program reassignment. Classification is not a one-time event but an ongoing procedure. The case manager serves as the offender's liaison to the classification committee for any changes in his or her program that are desired. Program modifications could include changes in work assignment, approvals for program participation, requests for transfer, consideration for custody reduction, or housing assignments. Case managers may also provide counseling support and approve and supervise outside visitors.

The final key component of case management staff is release planning, which actually begins at the time of initial classification. The appropriate goals of all institutional classification and programming should be to ensure the safety and security of all inmates and staff and prepare the offender for successful transition back to free society.[4] In all interactions with offenders, staff should encourage inmates to improve or increase their educational opportunities, job skills, self-sufficiency, and responsibility for their lives.

Unit Management

Many correctional systems recently have adopted a unit management approach to classification. Unit management involves dividing a large prison or jail population into smaller groups, often separated by housing unit. This decentralized form of management delegates much more decision-making authority to the staff that know the offenders the best—those who supervise inmates in the housing units. Having staff offices in the unit also serves the important goals of augmenting the day-to-day supervision of inmates and makes staff more accessible to the inmate population.

Functional unit management is, simply put, the decentralization of case management services to a diverse eclectic group of staff from different departments of the institution. The concept behind unit management was to subdivide the larger prison or jail into smaller groups of inmates, generally with their own housing unit, with staff offices within the unit. Similar to case management, inmate classification decisions are made by a unit management team. General policy establishes operational guidelines for these separate teams, and these staff members are empowered to make inmate classification, program, and housing decisions.

■ Comparison of Management Models

Unlike case management, the decentralized unit management model permits decisions about inmates to be made by the staff who know the inmates best. Clearly, when staff offices are next to inmate housing, staff can better supervise and get to know inmates. Positive, professional relationships are more likely to develop between inmates and staff. Daily interaction is helpful. Relationships among staff members are often greatly improved by unit management. Interdisciplinary staff of various departments who are assigned to a specific unit develop close working relationships that facilitate a productive working environment. In general, research has demonstrated that staff and inmate morale is improved with unit management. Inmates are much more pleased with responsive staff who know them, and staff are glad to have the authority to make program decisions.[5]

However, there are some negative aspects to the decentralization of prison management. It is much more difficult to maintain consistency in classification decision making when multiple teams are involved in inmate management determinations. It is critical that senior management establish overarching policy to guide the unit teams in their decision making. It is also important that penal institutions with unit management have open and effective lines of communication for staff and inmates. If unit staff are aware of inmate unrest or brewing tensions, they must share this knowledge with senior staff members.

There are three main functions of unit management: correction, care, and control.[6] Correction refers to the rehabilitative function of prisons and jails; care describes the assistance, resources, and support given to inmates; and control means the level of required custodial supervision. All of these functions are crucial to the administration of justice and successful prison or jail management. Unit management offers an efficient means of achieving these goals.

■ Objective Classification

A good, functional classification system is easy to use and sensitive enough to reflect the need for change as an offender progresses through the service of his or her sentence. A typical inmate classification system is designed to consider the inmate's criminal and social history, the current crime, the length of confinement, and, over time, reflect how the offender responds to confinement. Classification is designed to predict an individual's risk for violence or escape, and is based empirically on his or her past behavior. An individual's classification is made by reviewing that person's propensity for violent behavior.

Objective, fact-based classification facilitates agency-wide consistency that can be defended rationally and is perceived as equitable by all those involved in the process, including the inmates. Once the classification system has been validated, personal characteristics can be quantified, and each inmate can be scored accurately. Further, the system can facilitate rescoring based on an offender's progress while incarcerated.

An objective prison classification system must be both reliable and valid. Reliability means that the classification instrument consistently does what it purports to do; in other words, no matter which staff member utilizes the classification instrument for a specific inmate, the same result will be reached. Validity refers to the fact that the classification instrument is accurate in assessing a prisoner's future behavior and propensity for violence.

Contemporary correctional classification is based on the philosophy that an inmate is to be classified at the least restrictive security and custody level that meets the individual's needs. Correctional administrators must also ensure that an individual's classification considers the all-important issue of public safety. While practitioners do not want to overclassify an inmate, it is equally important not to underclassify the offender. Sometimes, this will mean that classification personnel will have to override the classification instrument. For example, a sex offender with a clean record throughout confinement may be eligible for consideration for a minimum security or trustee work assignment. But the nature of his or her offense may well preclude such consideration based on the potential serious threat to the community by the offender if he or she were to escape.

Research has demonstrated with significant validity and reliability that an individual's past behavior predicts future behavior.[7] Therefore, objective classification instruments should:

- Be validated on prison populations
- Utilize the same standards for all inmates
- Use a rational, uncomplicated process that is based upon factors that are related directly to the classification decision
- Recommend classification decisions that are based on the offender's background
- Promote consistent decisions for similarly situated offenders
- Use a process that is understood easily by staff and inmates
- Allow staff to monitor the inmate's progress efficiently and effectively[8]

Once the individual's security needs have been determined, staff members may develop plans to designate an appropriate institution for long-term incarceration. Inmates are generally assigned to an institution that meets their security needs, is as close to their home community as possible, and offers the appropriate programs to address their specific program needs. Each state correctional system utilizes its own institutional classification system, and there is not necessarily consistency between states as to the name of each security level.

While formal classification systems are based on firm data points, most systems also allow the final classification to be modified by the professional judgment of staff. It is important to allow staff judgment to enter the final decision and, if necessary, override the formula of the classification instrument.

External versus Internal Classification Systems

External classification is a process that determines how much security a specific inmate requires to assure his or her safety and the safety of others. This facilitates making a decision as to which level of security the inmate should be assigned. This classification decision is generally made at a central reception center that processes newly sentenced offenders into the correctional system.

An internal classification system is utilized to determine an inmate's housing, work, and program assignments. Many special programs such as drug abuse treatment or protective custody housing have specific criteria for inmate placement, and an internal classification system matches the right inmate population to the program resource.

Reclassification

Inmate classification and reclassification represent an outstanding means of keeping up to date with an inmate as he or she progresses. Recognizing that an inmate can and will change over time, for better or worse, staff must track these changes. If an offender's behavior deteriorates, staff should consider a transfer to a higher-security institution. If an inmate exhibits good behavior and a positive attitude, staff may want to consider a transfer to a less-secure facility at some point.

Initial classification should also be used to identify an inmate's program needs. Most prison facilities offer a large array of self-improvement programs that can have a positive effect on an inmate. Treatment programs include academic education, vocational education, mental health care, substance abuse programs, individual and group counseling, anger management, and many other opportunities. Many of these programs are not only effective in terms of changing behavior, they also play an important part in keeping inmates productively and positively engaged. The outcomes of positive and effective institutional programs serve all parties well—staff, inmates, and the general public.

Gender Differences

Female offenders are often subject to a separate classification standard than male inmates. Research and experience have established that men and women react differently in similar situations within a prison environment. In general, females are much less violent than males.[9] While women are not often involved in large amounts of violent behavior in prison, they do act violently on occasion. This violence is predictable with a valid and reliable classification system. Research also concludes that female offenders should be classified using a separate classification system from males to ensure that gender-specific predictors are identified.[10]

■ Conclusion

Inmate classification, if appropriately accomplished, serves all aspects of institutional management. Inmates and staff are safer, institutional organization by security level is cost effective, and program resources may be focused on the specific group of inmates that will most benefit from an activity. Classification of offenders will minimize the risk of escape and violence, provide a good rationale for the assignment of correctional staff, and provide the ability to minimize risk within the facility.

The ability to distinguish among groups of confined felons gives institutional personnel an outstanding tool in terms of staff safety and inmate accountability. Staff have an affirmative responsibility to operate correctional facilities in a safe manner and in such a way that protects the public's safety. Classification is at the center of these key obligations and provides staff members with the ability to execute their extremely important public service roles effectively.

DISCUSSION QUESTIONS

1. Why is a classification system important to inmate safety?
2. Do you believe that staff can combine effectively the roles of supervisor (and disciplinarian) with the supportive responsibility of being a counselor and support person?
3. What are the benefits of unit management?
4. Why is it important to have a case management team?
5. What are the goals of an effective classification system?

ADDITIONAL RESOURCES

J. Austin, P. Hardyman, and S. Brown, *Critical Issues and Development in Prison Classification* (Washington, DC: National Institute of Corrections, 2001).

T. Brennan, J. Alexander, and D. Wells, *Enhancing Prison Classification Systems: The Emerging Role of Management Information Systems* (Washington, DC: National Institute of Corrections, 2004).

T. Kane, "The Validity of Prison Classification: An Introduction to Practical Considerations and Research Issues," *Crime and Delinquency* 32 (Thousand Oaks, CA: Sage Publications, 1986).

R. Levinson, *Unit Management: The Concept that Changed Corrections* (College Park, MD: American Correctional Association, 1998).

NOTES

1. R. Levinson, "The Development of Classification and Programming," in *Escaping Prison Myths: Selected Topics in the History of Federal Corrections* (Lanham, MD: University Publishing Associates, Inc., 1994).

2. P. Keve, *Prisons and the American Conscience: A History of US Federal Corrections* (Carbondale, IL: Southern Illinois University Press, 1991), p. 234.

3. R. Balch, "The Medical Model of Delinquency: Theoretical, Practical, and Ethical Implications," *Crime and Delinquency* 21 (Thousand Oaks, CA: Sage Publications, 1975), pp. 116–129.

4. D. Glaser, *The Effectiveness of a Prison and Parole System* (Indianapolis, IN: Dobbs-Merrill Company, Inc., 1964).

5. H. Toch, *Corrections: A Humanistic Approach* (Guilderland, NY: Harrow and Heston, Publishers, 1997), pp. 30–34.

6. R. Levinson and R. Gerard, "Functional Units: A Different Correctional Approach," *Federal Probation* no. 37 (Thousand Oaks, CA: Sage Publications, 1973), p. 15.

7. R. Buchanan, et al., "National Evaluation of Objective Prison Classification Systems: The Current State of the Art," *Crime and Delinquency* 32, no. 3 (Thousand Oaks, CA: Sage Publications, 1986), pp. 272–290.

8. *Ibid.*

9. D. Steffensmeier and E. Allen, "The Nature of Female Offending: Patterns and Explanations," in *Female Offenders: Critical Perspectives and Effective Interventions,* ed. R. Zaplin (Sudbury, MA: Jones and Bartlett Publishers, 1998), pp. 5–29.

10. M. Harer and N. Langan, "Gender Differences in Predictors of Prison Violence: Assessing the Predictive Validity of a Risk Classification System," *Crime and Delinquency* 47, no. 4 (Thousand Oaks, CA: Sage Publications, 2001), p. 526.

Education and Vocational Training

Harold David Jenkins

Chapter Objectives

- Understand the need for education programs for inmates and the value of educational programming.
- Report how technological innovations can improve the delivery of educational services in prison.
- Outline the difficulties involved in making use of emerging technology.

Incarcerated persons are among the most educationally disadvantaged groups in the nation, and most inmates lack high school diplomas.[1] Approximately one in three inmates performed at the lowest level of literacy and would have difficulty reading a short newspaper article, completing an employment application, or locating a piece of information in a document.[2]

Despite the sobering statistics on the educational level of most inmates, recent research suggests that effective programs can improve inmates' educational performance significantly, and inmates who complete education and training programs exhibit improved institutional behavior and have greater success after release. Of course, facilities and resources vary greatly from facility to facility and from jurisdiction to jurisdiction. Even with limited resources, however, significant results can be obtained. Because state correctional systems generally have well-established correctional education programs, this chapter is directed primarily at offering suggestions on needs assessment, program design, program operation, evaluation, and community resources to local detention centers, jails, and smaller state facilities such as prerelease units.

◼ Research on Educational and Vocational Programs

Program availability differs widely in the various correctional systems. All federal prisons, 90 percent of state institutions, and 60 percent of local facilities offer educational programs. The low level of program availability in local facilities is a function of several factors, including the short stay of many offenders in these facilities, the size of the institutions, and the focus on other programs. In local institutions that do offer educational programming, General Equivalency Diploma (GED) lessons are the most widely offered service.

The impact of education and job training programs is far from uniform and exhibits various data collection and analysis problems. Data on the postrelease experiences of inmates who participated in correctional education programs have been difficult to collect and frequently available for only a small percentage of releasees. Despite these problems, an increasing body of research from federal, state, and local authorities indicates that a positive relationship exists between institutional education programs and a number of indicators of success for offenders.

A review of studies on the relationship between correctional education and recidivism determined that 85 percent of the studies reported a positive relationship between correctional education participation and lower recidivism.[3] Similarly, results from a study of Federal Bureau of Prisons releasees determined that inmates with training and/or work experience while imprisoned had better institutional adjustments, were less likely to relapse into crime, and were more likely to obtain employment upon release.[4] A U.S. Department of Education study determined that participation in education and training programs was consistently correlated with a lower three-year recidivism rate and that education program participation was related to lower rates of arrest, conviction, and reincarceration.[5] A UCLA study concluded that educating prisoners was nearly twice as cost-effective as incarceration in preventing crime.[6]

Although the results of this body of accumulating studies are not uniform, there is good reason to believe that the positive relationships between education and training and several positive outcomes are not simply coincidental. Educational attainment is a powerful indicator of income and employment for non-incarcerated persons, and it would be counterintuitive to assume that these variables are not positively related for incarcerated persons. Given this evidence, it seems beneficial to establish inmate education and training programs where they do not currently exist and to expand existing programs where available programs do not meet established need.

◼ Needs Assessment

The development of a successful education program depends on an assessment of inmate needs, community resources for education, and the labor market. Some of the data regarding inmate needs may already be available from intake interviews and other sources. Despite the evidence that most inmates exhibit serious educational deficiencies, inmate populations differ significantly among and within institutions. Intake interviews and presentence investigations frequently yield important data, including:

- Educational attainment prior to incarceration
- Functional performance
- Prior skill training
- Employment history
- Prior specialized treatment including special education services

This data can yield a profile of the population as well as the number of inmates with particular needs. In addition to education and training histories, the following questions should also be addressed:

- How long will the typical inmate be available for services?
- What other programs and services are inmates eligible to receive?
- Will the educational program compete with other programs for the same inmates?

In most instances, the development of an effective correctional education program will also require objective test data to determine the needs of the persons to be serviced. The Test of Adult Basic Education (TABE) is used frequently to assess the performance of adult learners. The TABE is used widely in adult education and correctional education programs. The California Adult Skills Assessment System (CASAS) also is used widely. Some systems use both assessment tools.

The potential students are frequently a good source of information. Unfortunately, because these adult students are incarcerated, their valuable input often is ignored. Adult education providers and community colleges actively solicit input from students and potential students to ensure that their institutions' educational offerings genuinely interest student customers. Correctional programs could benefit from this model; students themselves may have valuable input on which programs would benefit them the most.

Information on labor market conditions and trends in the community or communities to which inmates are to be released is vital to the development of any effective training program. Corrections is unfortunately the home to more than its share of programs that train inmates for jobs that no longer exist or are unlikely to be open to ex-offenders. Involvement of potential employers in the development phase is likely to pay off later. Employers can provide relevant information and advice to ensure training programs are current and valid.

Having established the need for educational programs, an institution must seek resources. Resources may be limited, but linking with community organizations can help. A thorough needs assessment frequently is the key element in accessing these community resources; it shows potential partners that an institution is serious and outlines a problem about which others may only be vaguely aware.

A results-oriented and goal-specific program helps institutions focus on this community-based information, which may be readily available from U.S. Department of Labor publications and state or local planning agencies, in addition to the following organizations:

- American Association for Adult and Continuing Education
- American Council on Education
- American Jail Association
- American Library Association
- Center for the Study of Correctional Education

- Correctional Education Association
- Correctional Learning Network
- Educational Research Information Center
- National Criminal Justice Reference Service
- National Institute of Correctional Education
- National Institute of Corrections
- Office of Juvenile Justice and Delinquency Prevention

Community resources such as literacy organizations, fraternal and religious organizations, local colleges and universities, labor and employee organizations, and senior and retiree organizations can provide additional assistance in materials. Lastly, local libraries have a wealth of information about community involvement and general references.

Community organizations can provide many forms of assistance for inmates, including:

- Financial support to establish and operate an education program
- Training of staff and inmates involved in educating inmates
- Sponsorship of various educational activities and facilities such as a learning lab, a family literacy center, or other special projects
- Volunteers to serve as instructors, tutors, aides, and office support
- Sponsorship of a graduation for those who complete a program or the loan of gowns for a graduation ceremony
- Volunteers to mentor students on postrelease opportunities, services, and various employment readiness activities and to provide advice on evaluation methods from local colleges and universities

■ The Technological Continuum

New technologies are available to deliver education and training programs to incarcerated persons. These include networked computer-assisted instruction and reference systems, interactive video classrooms, satellite broadcasting, and a variety of Internet services. Research suggests that measurable academic gains can be accomplished using these new technologies. In most cases, inmate users are positive about education programs incorporating these high-tech systems.

Unfortunately, these newer technologies can be quite costly to install and may be expensive to maintain. If an institution lacks the financial resources or infrastructure for these new technologies, positive results can be achieved using less sophisticated (and less costly) solutions. Older, but usable, computer equipment may be donated. Institution leaders should be careful about what equipment they accept, however, because appropriate software may not be available for some older systems.

Although the educational benefits of new technologies are clear, more traditional approaches can be effective as well. Video (VHS or DVD) format GED lessons such as those produced by several education curriculum providers, used in conjunction with workbooks, can provide quality instruction at a reasonable cost. A part-time instructor or qualified volunteer can supplement video lessons. A local public broadcasting station could broadcast the same series, although prerecorded video lessons offer more flexibility in that they can be shown whenever it is convenient. A recorded version also can be sequenced to fit the needs of inmates.

Institutions with very limited budgets can base instructional programs on inexpensive GED workbooks produced by several publishers. These workbooks frequently include diagnostic tests and a practice GED examination. Another low-tech delivery system involves the help of trained and motivated inmate tutors. Local literacy organizations such as Literacy Volunteers of America or Lauback Literacy may be able to train selected inmates to tutor other inmates. One-on-one tutoring can be a very effective educational tool, especially at the lower academic levels. The use of inmate tutors has the added benefit of providing useful, productive work for inmates who may otherwise be idle or underemployed. Inmate tutors need supervision and ongoing training by a staff member willing to devote a significant amount of time to this activity.

Although correctional systems remain strongly resistant to allowing inmate access to the Internet, there may be indications that some level of access may be allowed safely. The Maryland State Department of Education Correctional Education Office successfully operated an Internet-based post-secondary program at the Patuxent Institution in Jessup, Maryland for three years.[7] Their design, developed by a defense technology contractor, used secured servers in the prisons to receive course work and faculty communications that were downloaded to the student work stations. The Internet was disconnected during student use. Similarly, the onsite staff member reviewed student work and sent it to the college after the students left for the day. No security problems were reported in three years of operation in which more than nine inmates participated and nine students earned Associate of Arts degrees.[8]

Direct Internet use by adult prisoners have been implemented on a pilot basis in Norway where inmates are allowed to access pre-selected sites related to transition and community resources.[9] An instructor is in the computer lab to provide direct supervision at all times. If an inmate becomes aware of a site that he feels would be especially helpful to offenders, he can request that the instructor review the site and decide if it is appropriate to add to the approved sites for inmate access. Given the explosion of data on the Internet and the increasing use of the Internet as the only way to access certain data, a secure method must be identified to allow inmates access if they are not to become even further marginalized from mainstream society.

Effective education programs for inmates can be developed at various levels of technological sophistication (and cost). The more sophisticated technologies such as networked computer-assisted instruction excel at recordkeeping and reporting, and provide an attractive instructional format, but low-tech, less costly systems can produce satisfactory results as well. Program developers should review these various options and their costs.

■ Incentives and Achievements

Incentives for participation in education programs in state correctional systems are widespread. These incentives include good time (time off or early release), wages or stipends, parole considerations, priority transfers, extra visits, and telephone privileges. Although there may be resistance in some jurisdictions to rewarding inmates for joining programs, incentives are a reasonable way to encourage participation, especially when other prison assignments provide incentives as well. It is important to view school participation as a regular job with the benefits of other institutional assignments. Research strongly suggests that the completion of a major educational milestone such as GED or vocational training can improve an inmate's success after release significantly.

Inmates (and correctional institutions, for that matter) frequently have little to celebrate. Mention of prisons and detention centers in the media is usually associated with an escape, act of violence, disturbance, or some similarly sensational and thus "newsworthy" event. Educational achievement can provide justifiable reasons to celebrate both individual and institutional achievements. To the degree possible, major education achievements such as literacy program completions, high school diplomas, and vocational training certificates should be recognized with a graduation. Incarcerated students appreciate outside guest speakers, who show that the community is interested. Including other non-graduating students can motivate those students. Many inmate students doubt their ability to achieve major educational goals, but seeing their fellow inmates graduate helps them believe in themselves. Including parents and other family members in graduations can give them a more positive view of an institution. For many inmates, an institutional graduation will be the first public recognition of any educational achievement in their lives.

■ Anticipated Results

What results can a warden or facility administrator reasonably expect from the participation of inmates in an education program? Research at correctional systems at different levels (federal, state, and local) indicates that inmates who participate in education programs (academic and vocational) generally exhibit improved institutional adjustment and postrelease success, including a 10 to 25 percent reduction in recidivism. Education is not an inoculation against further criminal involvement; rather, it is an improvement in the offenders' chances to obtain legitimate employment and avoid further criminal involvement. Additionally, the consensus of the research is that positive results are correlated to the intensity and length of education involvement as well as the achievement of a major educational milestone. Some inmates have deep psychological problems and addictions and may not be helped by educational participation alone.

■ Accreditation and Certification

The certification of education staff and accreditation of schools are widespread practices in the public school system, although not necessarily at the adult education level or in some school systems with serious staff recruitment problems. In correctional education programs, which are offered by a variety of organizations (correctional agencies, local school systems, community colleges, correctional boards of education or state education agencies), issues of certification and accreditation are more varied. For example, community colleges have standards for instructional staff, but generally do not require public school teacher certification. On the other hand, teachers provided by state education agencies are more likely to be certified. The degree to which instructional staff is certified or meet other standards is a measure of program quality.

The American Correctional Association (ACA) and Correctional Education Association (CEA) have developed detailed standards for correctional education programs and offer trained staff to conduct onsite audits. They provide carefully developed points of reference to evaluate current programs as well as direction in the development of new programs.

■ Program Evaluation

Program evaluation may range from a relatively modest in-house design to an expensive professional evaluation by an outside consultant or university-based organization. Regardless of the level of the evaluation effort, an assessment of program impact is essential. An evaluation that includes both process and outcomes gives a balanced picture and provide feedback for program improvement.

The establishment of reasonable goals for institutional programs is critical. Programs must consider several important questions:

- What should the impact on the inmate clients be, and how will it be measured?
- If improved institutional behavior is a goal, how will that goal be achieved, and how will the data on that goal be collected?
- What rate of student progress has been identified?
- How many inmates are expected to achieve an established goal (such as a certificate of literacy, high school diploma, GED, or employment after release)?

Operational data are equally important in a correctional setting where instructional time can be affected by operational problems (e.g., the pass list did not get to the housing unit officer, the staff were held up entering the institution because a new officer was working the post, the meal was late because a popular item was on the menu and everyone showed up for dinner, etc.). Consistent and reliable data on operation and utilization of educational programs needs to be generated and provided to program managers for review and action.

One of the most overlooked areas of program operation in correctional facilities is the effect of the services on the postrelease success of inmates. Many administrators are especially averse to evaluation efforts that focus on postrelease success because of research difficulties and the fear of being held responsible for the inmates' performance in a community setting. Put directly, if correctional education programs cannot affect inmates' postrelease success, and if this success cannot be measured, why invest in these efforts at all? In an increasingly results-oriented environment, the impact of institutional programs on releasees' success is the bottom line for policymakers. Increasing pressures on public funding strongly suggest that programs without objective, verifiable measures of results are at risk, and research indicates that reasonable evaluations of postrelease success can be accomplished with limited resources.

DISCUSSION QUESTIONS

1. Should prisons require inmates to participate in education programs? Why or why not?
2. To what extent do prisoners' educational needs differ from those of individuals in the community?
3. What are some of the challenges that administrators face in trying to provide education to prisoners?
4. What are some of the tools educators have available to them and what are some of the difficulties in using these tools?
5. What role can emerging technologies play in educational and vocational training?

ADDITIONAL RESOURCES

National Institute of Correctional Education, http://www.iup.edu/nice
Center for Study of Correctional Education, http://www.csusb.edu/coe/csce
Correctional Education Association, http://www.ceanational.org
Education Resources Information Center, http://www.eric.ed.gov
American Council on Education, http://www.acenet.edu
American Association for Adult and Continuing Education, http://www.aaace.org
National Institute for Literacy, http://www.nifl.gov

NOTES

1. National Assessment of Adult Literacy, available at http://nces.ed.gov/NAAL/, accessed September 4, 2007.
2. *Ibid.*
3. T.A. Ryan and B. Mauldin, *Correctional Education and Recidivism: An Historical Analysis* (Columbia, SC: 1994).
4. M. Harer, "Recidivism among Federal Prisoners Released in 1987," (Washington, DC: Federal Bureau of Prisons, Office of Research and Evaluation, 2004), available at http://www.bop.gov/news/research_projects/published_reports/recidivism/oreprrecid87.pdf, accessed September 4, 2007.
5. S.L. Steurer and A. Tracey, *The Three State Recidivism Study* (Lanham, MD: Correctional Education Association, 2001).
6. A. Bazos and J. Hausman, *Correctional Education as a Crime Control Program*, prepared for U.S. Department of Education, Office of Correctional Education (March 2004).
7. Interview with Edward Duke, Learning Anywhere, Anytime Partnership Learning Lab, The Patuxent Institution (November 16, 2004).
8. *Ibid.*
9. B.D. Hansen, Presentation, European Prison Association Conference, Langesund, Norway (June 2003).

Recreation

Harold L. Kahler

Chapter Objectives

- Explain how the attitude of the public has affected prison recreation programs.
- Describe the positive and negative implications of recreational programming in penal institutions and explain how recreation programs can serve the goal of rehabilitation.
- Grasp how recreation programs affect the overall level of tension and safety within a correctional facility.

Having a well-run recreation program is a vital goal of every prison or jail administrator, yet prison recreation programs generally are not favored by the public. Society's tolerance of crime is now low, and there has been a return to the idea that incarcerated offenders should be punished for their crimes as an attempt to curb soaring crime rates. Within this context, recreation programs for inmates are perceived as frills that mitigate the pain of imprisonment and coddle inmates.

■ "Get Tough" Policies

"Get tough" policies developed in recent years attempt to crack down on crime by increasing sentence lengths and punishments. Another attempt to get tough on criminals focuses on prison recreation programs, because much of the public believes that only "bare bones" prisons—not the supposed "country club" prisons of today—will help lower crime. Prisons, many argue, should be a place of punishment, yet inmates make ceramic pottery, attend movies, watch television, or participate in sporting events or musical jam sessions. Voters have told their elected representatives that "enough is enough," and their representatives have heeded the message. In response to this outpouring of public concern, several states and the federal government have begun curtailing prison recreation programs. Many states (including Arizona, California, Georgia, Mississippi, North Carolina, Ohio, South Carolina, and Wisconsin) have restricted recreational activities, including television viewing and weightlifting.[1]

■ The Zimmer Amendment

The United States Congress approved the Zimmer Amendment to the 1996 Department of Justice Appropriations Bill. Provisions of this "no-frills" bill directly affect prison recreation activities by prohibiting the purchase and replacement of weightlifting equipment and musical instruments and the showing of R-rated movies. The effect of the Zimmer Amendment will be felt by every inmate in the federal prison system. Prison inmates share a variety of interests and abilities. While less active inmates spend their leisure time watching movies and reading magazines, more active inmates lift weights or participate in team sports, and musically inclined inmates may be in the music room. Correctional personnel argue that this amendment adversely affects not only the inmates' use of leisure time but also inmate management. For example, since most top films are rated R, inmates may reject movie viewing altogether and instead fill their idle time planning or pursuing activities that are not conducive to the orderly running of the institution. The prohibition against providing electronic musical instruments may also reduce the positive use of idle time by those inmates who currently spend leisure time in the music room.

Additionally, this amendment negatively affects many active federal inmates by eventually eliminating all weightlifting activities. The weight room is a popular recreation area in the institution, and many of the inmates whose leisure time is spent working out with weights are not interested in participating in other fitness activities. In a prison weight room, inmates act as "spotters" for others and encourage each other. Cooperation and *esprit de corps* are evident within the weight room. This recreation activity is supported by proponents because it fills idle time, reduces boredom, reduces tension, builds self-esteem, and teaches the necessity of cooperation.

The American Correctional Association's (ACA) *Standards for Adult Correctional Institutions* describes the provision of recreation and leisure time activities as essential elements of a correctional operation, along with an outdoor recreation area, an auditorium, a music room, and weightlifting equipment.[2] The Zimmer Amendment rejects these ACA standards but accurately reflects the public's view of prison recreation.

■ Criticisms of Prison Recreation

Many Americans are of the opinion that amenities provided prison inmates through recreation programs are too costly, coddle inmates, and create dangerous consequences for the personal safety of law-abiding citizens. If these charges are true, prison recreation programs should be curtailed. However, these views are not held by most prison administrators and their staff, and therefore, they should be examined in more detail.

High Cost

The history of United States correctional institutions shows that the cost argument is shortsighted. The two major problems plaguing prisons since their inception have been overcrowding and idleness. Today's inmates are serving longer sentences in vastly overcrowded institutions. This combination constitutes a potentially volatile environment. If inmates are not provided opportunities for the constructive use of idle time, they will fill that time in their own way, which may not be positive for the institution.

The President's Commission on Law Enforcement and Administration of Justice noted that as early as the 1800s, "mere restraint could not accomplish the purpose of corrections, and that many of the features of prison life actually intensified the problems of offenders." These prison officials also recognized that "recreation would ease institutional tensions and contribute to an atmosphere less detrimental to rehabilitation."[3]

Custodial and treatment staff recognize that prison recreation programs greatly reduce the probability of violence within institutions. Therefore, the cost of prison recreation programs to the taxpayer is minuscule when compared with the cost of repairing damages from wanton acts of violence and destruction or providing additional staff to supervise inmates who have no constructive alternatives for the use of idle time. Given these facts, the argument should be not that we cannot afford prison recreation, but that we cannot afford *not* to have it.

Coddling Inmates

If prison recreation programs were solely for the benefit of inmates, it might be appropriate to question them. However, as noted above, most correctional managers support these programs and consider them management and rehabilitation tools benefiting not only the inmates, but the staff and, ultimately, the public. The President's Commission on Law Enforcement and Administration of Justice recognized the negative impact of idleness in America's prisons when it concluded that "sometimes institutions foster conspicuously deleterious conditions—idleness . . . and moral deterioration." After listing a number of abuses found in American prisons, the commission declared "a more pervasive evil is idleness; this is especially destructive where there are no industries, no educational programs, no recreational facilities—only aimless loitering in corridors or yards." The commission described a model institution as one having classrooms, recreational facilities, day rooms, and perhaps a shop and library.[4]

This judgment is echoed by line staff who work in these institutions. As one correctional officer at a state prison remarked, "I feel that giving the inmates recreation time—time for weightlifting, jogging, basketball, and other activities to release stress—is good. It makes it safer for the staff, reduces inmate idleness, reduces tension, and contributes to the orderly running of the institution." And a correctional supervisor in a privatized prison recently commented that "recreation activities are the things that keep the lid on here."

Consequences for Public Safety

Many people fear that prison recreation programs diminish the deterrent effect of incarceration. They would argue that by offering recreation to a confined individual, society is making prison a positive experience. Most criminologists would argue, however, that deterrence and punishment as primary goals of prisons are ineffective, mainly because criminals do not expect to be caught.

Others fear that recreation programs—especially weightlifting programs—encourage intimidation, robbery, or assault of innocent citizens by released offenders who have used prison recreation programs to develop themselves physically. Correctional recreation specialists assert that there is no reliable research confirming that the bodybuilding efforts of incarcerated inmates lead to victimization of citizens when the inmates are released. These specialists suggest that individuals committing street crimes usually rely on weapons, not brute strength, to coerce their victims into submission.

Therefore, it is difficult to conclude that prison recreation programs diminish the deterrent effect of incarceration or that weightlifting programs result in an increase of violence against law-abiding citizens. But it cannot be denied that current recidivism rates indicate that the prison experience of many released offenders has not deterred them from subsequent criminal activities—including assault upon innocent citizens.

Myths of Weightlifting and Violence

Antagonists of prison recreation programs argue that allowing inmates to become more fit increases the potential that they will act more aggressively in the future. Essentially, the belief is that fitness leads to the ability to harm others—inmates, staff members, or citizens in their home community—after release. Specifically, some argue that correctional institutions should not facilitate the growth of muscles through a weightlifting program.

These same antagonists would point to the possibility of inmates assaulting other inmates with weight bars and weight discs. One federal high security prison, the United States Penitentiary at Lompoc, California, did experience such an incident on two occasions in 1998. In the first incident, one inmate hit another with a weight bar, and the warden warned that all weight equipment would be removed if there was another assault involving weights. A second incident occurred two weeks later, and the warden made good on his promise: every piece of weightlifting equipment was removed from the penitentiary and was never replaced. Other inmates throughout the federal system are aware that these incidents caused the removal of weight equipment, and this has reinforced everyone's interest in maintaining good behavior with such exercise gear.

Most correctional workers are proponents of a structured weightlifting program. Their argument is that weightlifters are among the best-behaved prisoners because they do not wish to jeopardize their relative freedom within the prison to work out every day as part of their daily schedule. Prison workers report that these inmates release their energy, frustration, and other pent-up emotions by a strenuous exercise routine at the weight pile in the recreation yard or gymnasium. Correctional officers and other staff members typically believe that weights and other exercise apparatus are positive aspects of an inmate's prison life and that exercise routines promote good behavior.

■ Tools for Rehabilitation

Various strategies to rehabilitate offenders have been adopted in U.S. prisons. These strategies have sought to provide insight into the inmate's criminal behavior, facilitate change in that behavior, increase the inmate's self-esteem, and empower the inmate to become a law-abiding, productive citizen when released.

In most correctional institutions, counseling is the main treatment strategy. However, if counseling is to be successful, it must include voluntary participation and active involvement that is open, honest, and free of any attempt to manipulate staff or other inmates. These conditions are rarely met. Many inmates, forced to attend counseling sessions, are reluctant participants whose presence jeopardizes the effectiveness of the program for those inmates who hope to benefit from it. Other inmates consider participation in these programs an opportunity to manipulate staff to obtain increased institution privileges. For these reasons, no program yet developed has been overwhelmingly successful, although a few have yielded positive results for some individuals.

Many citizens believe that offenders should not be sent to prison to attend college classes. In fact, offenders are not sent to prison to improve their education. They are, however, sent to prison for the protection of society, and, if possible, to be changed for the better. Criminals are not occupying prison cells because they lack college credits. There are many competing theories about why people commit crimes, but the inability to deal with frustration, control anger, and overcome a poor self-image combined with misguided use of leisure time almost always contribute greatly to criminal activity.

If the recidivism rate of offenders and the victimization of the innocent are to be reduced, prison inmates who are determined to continue their criminal activities when released must become active participants in programs that enable them to experience situations in which their usual way of responding does not work. When behavior as usual does not produce the coveted payoff, offenders are encouraged to learn new behavior.

A properly administered institution recreation program—although not generally recognized as such—can be an effective tool for teaching and developing the social skills that are necessary for participation in free society. Inmates involved in recreation activities learn the value of teamwork, fair play, anger management, and amiable conflict resolution. These programs also provide opportunities for the constructive use of idle time, relieve the stress and tension of incarceration, and help increase self-esteem. Traditional institution recreation programs include:

- Intramural sports (flag football, basketball, baseball, soccer, volleyball, handball)
- Games (card games, bingo)
- Individual activities (weightlifting, jogging, music)
- Arts and crafts (ceramics, painting)
- Media (television, radio, movies)
- Talent and drama shows
- Volunteer work
- Clubs (Key Club, Jaycees, Toastmasters)

With the possible exception of prison industries, no other institution-based program is as enthusiastically received by the inmates, and no other institution-based program is a more effective management tool. Threat of exclusion from participation in recreation programs is usually taken seriously by inmates, so they are encouraged to consider the consequences of their behavior before they act. The threat of exclusion from typical treatment programs may be ignored; if exclusion does occur, the inmates may consider themselves rewarded.

Community Involvement

Prison recreation activities that involve the community provide the institution administrator the opportunity to achieve an important goal of community corrections: preparing the public and the inmate for the inmate's successful reintegration into the community. Community members may help teach inmates various skills such as chess, ceramics, or sewing, or they may interact with inmates in athletic competitions, art shows, or music festivals. These are all opportunities for positive interaction between the offender and the community. This interaction may result in an empathetic awareness that the similarities between inmate and community member outnumber the differences. Thus, the activities help facilitate the reintegration of the offender into the community. Failure to achieve this reintegration guarantees a continuing recidivism rate unacceptable to everyone.

Recreation Staff

Recreation staff are role models and often parental figures to inmates. Because recreation staff speak the inmates' language and are at the institution after hours and on holidays, many inmates turn to them when they experience problems. Recreation staff also serve an outlet for inmates to relieve the anger and tension they experience when they feel that staff or other inmates have wronged them. In these interactions, recreation staff develop rapport with inmates that is seldom attained by professional therapists. Unlike staff in other disciplines, who often see behavior designed to mask the inmates' real selves, recreation staff have a unique opportunity to observe inmates as they really are. On the playing field intent on making a play, or in the craft shop striving to perfect a ceramic item, the inmates' defense mechanisms may be dropped momentarily in a spontaneous reaction to failure and frustration. This reaction provides recreation staff an accurate portrait of the inmates' ability to cope with these unwanted but common experiences.

Recreation staff and counselors observe inmates from different perspectives. Input from recreation staff may help create more effective inmate rehabilitation programs if it is sought and considered by the inmate's institution classification committee prior to initial classification and subsequent program reviews. Therefore, it is important that institution staff not treat recreation as an unimportant add-on to daily institution operations. Staff must recognize it as an important resource for maintaining a safe and secure institution. Management staff should support recreation programs by occasionally attending activities sponsored by the recreation department.

Senior managers must be selective while staffing a recreation department. Recreation programs are important to inmate rehabilitation and should be led by carefully chosen staff. In addition to a background in the management of leisure activities (e.g., field sports, arts and crafts), staff should have some knowledge of basic counseling. Wardens and superintendents should highlight the recreation program and stress its value to the community, the institution, and the inmate when addressing public gatherings, groups of judges, and legislative bodies. They should emphasize the importance of balancing work and free time in correctional institutions.

■ Conclusion

Institution recreation programs are currently under siege by the public and the public's elected representatives. The public decries the cost of these programs, but the advantages far exceed the costs. In fact, if there is any truth in the correctional supervisor's statement that "recreation activities are the things that keep the lid on," they are a bargain at almost any price. While some people believe that prison recreation programs coddle inmates and are frills that should be curtailed, professional corrections organizations and prison administrators support prison recreation programs because they reduce inmate idleness and the tensions produced by incarceration.

Recreation is an effective management tool. Team sports assist in the development of social skills as they teach the value of teamwork, anger management, and amiable conflict resolution. Developing skills in sports and arts and crafts activities increases the inmate's self-esteem. Inmates whose leisure time is consumed in the arts and crafts room, in the multipurpose room, and on the field are more easily accounted for and have less time and energy to engage in prohibited activities. Recreation programs play an extremely important role in the professional management of today's correctional institutions.

DISCUSSION QUESTIONS

1. Do you believe that inmates should be permitted to participate in organized recreational programs?

2. What are the positive and negative implications of prison weightlifting programs?

3. Why does the concept of inmates participating in a recreation program offend some members of the public?

4. What are other ways to achieve a "get tough" approach to crime without limiting recreation?

5. What role should recreation play in effective inmate management?

ADDITIONAL RESOURCES

"Correctional Recreation: An Overview," http://www.strengthtech.com/correct/overview.htm

M. Carter and K. Russell, "What is the Perceived Worth of Recreation? Results from a County Jail Study," *Corrections Today* (June 2005), pp. 80–83.

M. Dobie, "Sports in Prison," *Newsday*.com (July and August 2004), http://www.newsday.com/sports/ny-prisonsport,0,2605324.flash

The National Correctional Recreation Association, http://www.strengthtech.com/correct/ncra/ncra.htm

NOTES

1. S. Clayton, "Weight Lifting in Corrections: Luxury or Necessity?" *On the Line* 20, no. 5 (American Correctional Association, 1997), available at http://www.strengthtech.com/correct/issues/mediais/clayton.htm, accessed August 13, 2007.

2. American Correctional Association, *Standards for Adult Correctional Institutions* (Washington, DC: American Correctional Association, 1981).

3. President's Commission on Law Enforcement and Administration of Justice, "The Challenge of Crime in a Free Society," (Washington, DC: Government Printing Office, 1967), p. 163.

4. *Ibid.*

Health Care

9

Robert R. Thompson

Chapter Objectives

- Appreciate the legal basis for standards of medical care within prisons.
- Identify the similarities and differences between community and correctional medicine and be familiar with unique end of life issues present in correctional environments.
- Explain the process of intake screening and the importance and exceptions of confidentiality.

Medical practice in the correctional environment shares many similarities with the community at large, but inmates present some unique challenges to health care professionals. Thousands of health care professionals work inside prisons and jails and do so with care and diligence. It is their responsibility to provide prisoners with appropriate and competent care to the best of their ability in a fair and ethical manner.

▉ The Right to Health Care

Numerous United States Supreme Court decisions have established that an inmate's right to health care is essentially the same as the right of a non-incarcerated person. Specifically, in *Estelle v. Gamble* the Court held that any attempt to withhold or unduly delay medical care constituted cruel and unusual punishment. In the words of the Court, "the public [is] required to care for the prisoner, who cannot by reason of the deprivation of his liberty, care for himself . . . We therefore conclude that deliberate indifference to serious medical needs of prisoners constitutes the 'unnecessary and wanton infliction of pain,' proscribed by the Eighth Amendment."[1] Other Supreme Court and lower court decisions have likewise upheld the rights of both prisoners and pretrial detainees to dental care, chronic disease management, prenatal care, mental health counseling, communicable disease screening, Acquired Immune Deficiency Syndrome (AIDS) treatment, and medical care during self-imposed starvation (i.e., food strikes).[2] It is noteworthy that an inmate's right to health care in each of the above areas does not address the issue of malpractice or the lack of medical competence; the latter is adjudicated in the courts as a civil matter, and the standard utilized by the courts is that of deliberate indifference.

The American Correctional Society, the United States Public Health Service, and the American Medical Association collaborated to form the National Commission on Correctional Health Care (NCCHC), now the primary body that sets standard and accredits correctional facilities. The NCCHC offers technical assistance, educational programs for correctional staff, and clinical guidelines for the health practitioner. The results of these efforts have helped ensure that inmates receive medical, dental, mental health, and addiction treatment in a timely and safe manner by qualified health professionals in essentially the same manner as non-incarcerated persons.

It is important to note that correctional law as it relates to health care is an evolving subject, and standards of care in the community and in the correctional environment change constantly. The health care provider in the prison setting needs to be ever vigilant to this changing landscape to ensure that he or she is meeting the current standards of care while caring for patients.

■ Initial Screening

Good medical care begins with the inmate's arrival at the correctional facility. It is during this all-important interview that the following questions should be addressed:

1. What is the inmate's overall health status?

2. Does he or she have communicable disease risk factors (e.g., tuberculosis, AIDS) or diseases that might put other inmates or staff members at risk (e.g., hepatitis, sexually transmitted diseases)?

3. Are there obvious signs of mental illness or a history of any severe psychiatric disorders?

4. Are medications (if any) in order and does the inmate understand how to take them?

5. Has a physical examination been completed to assess any pre-existing injuries or deformities?

6. Are there physical or behavioral signs of recent or past alcohol or drug abuse that might suggest impending chemical withdrawal? (Note: Some inmates "self surrender" to an institution and to have one last "celebration" engage in prolonged drinking. Such inmates are apt to have severe

withdrawal and need to be monitored carefully or placed in a detox facility if one is present within the institution.)

7. Are there any other reasons that this inmate might need to be housed separately?

During this initial encounter the examiner would do well to remember that he or she is a correctional worker as well as a health care provider. This may require that the examiner is familiar with the inmate's correctional history and is aware of any behavioral issues that would put prison staff at risk. A review of the inmate's file and awareness of his or her custody status is part of the prison health care worker's job description. If restraints are required, they may need to remain in place during the interview and correctional officers may need to be present. As in all health care encounters with prisoners, staff, and other inmates, safety is of foremost importance.

During the arrival interview, the inmate may appear disheveled and unkempt due to a long travel interval from another facility. His or her mental status and level of cooperation may reflect stress due to a long court case, incarceration at another facility, frustration with the justice and legal system, and disappointment over the sentence. He or she may feel insecure in the new facility and be concerned for his or her safety around other inmates. All of these things set the tone for the intake encounter and must be taken into account in an evaluation.

Plans for follow-up care should be made if chronic conditions are present. For example if heart or lung disease is present, arrangements should be made for the inmate to be seen in one of the chronic care clinics as soon as possible after arrival. This is also a good time to make sure that all outside records, X-rays, and tests are available for review. If a contagious disease such as tuberculosis (TB) is suspected, the inmate must be placed in a negative air flow room (a sealed room in which fresh air is re-circulated and air moves out when the door is opened) and appropriate procedures utilized. This may involve placement in an outside medical facility that is equipped for dealing with infectious disease.

In many ways the initial screening interview sets the stage for the inmate's cooperation with medical personnel and future testing and compliance with medication during the period of incarceration. Trust is the key word. In the end, professional demeanor on the part of the interviewer will overcome hostility and help ensure compliance.

■ Clinics and Sick Call

It is important that the inmate understand the procedure for sick call. This is best accomplished in the inmate orientation session during which written material is handed out and the inmate has an opportunity to ask questions. Because sick call procedure varies from one institution to another, one cannot assume the inmate is familiar with the processes of a particular facility.

To discourage abuse of sick call, many institutions have instituted nominal inmate co-pay. This has been viewed as successful by many jurisdictions, but must be well thought out at each institution in order that it is not seen as punitive. While fee-for-service payments have been demonstrated to reduce inmate abuse of sick call procedures, these fees have also been perceived as preventing genuinely sick inmates from seeking care.

There should be as much continuity and teamwork as possible among the sick call staff so that the same staff regularly care for the same inmates. In that way, potential drug seekers and abusers can be identified more easily. To facilitate this continuity, some outpatient staff may meet as a group for a brief period to review the sick call list for the day and share any information about the inmates that might be helpful. This is also a good opportunity for consultation with each other about particularly vexing problems concerning an inmate.

As in all aspects of correctional medicine, teamwork among the medical disciplines is central to dealing with difficult inmates. Malingering, work avoidance behavior, pseudo-seizures, and drug seeking can best be dealt with by nursing, pharmacy, physical therapy, and mental health staff, and physicians all working together.

The use of the outpatient clinic or sick call is a good opportunity to further patient education. Low back pain, asthma management, and diabetic care are just a few examples of conditions that can be improved through patient education films, question and answer sessions, and individual counseling.

■ Confidentiality

All information in the medical record is to be treated with the same confidentiality that exists in community practice. Convicted offenders and pretrial detainees have a right to medical privacy, which means that medical personnel may not share details about inmate's medical records with other correctional staff. Nonmedical personnel may gain access to protected medical information in the process of doing their jobs, such as in conversations with inmates, checking inmates' property, or reviewing inmates' mail. There are, however, several exceptions to this general privacy rule, as outlined in the Health Insurance Portability and Accountability Act (HIPAA). Medical personnel may, on a limited basis, disclose protected health information about an individual to correctional personnel as necessary for:

- The provision of health care to such individuals
- The health and safety of an individual or inmate
- The health and safety of correctional officers or others at the institution
- The health and safety of staff members responsible for transporting inmates

Any disclosure of medical information must be based on verified need for personnel to have access to the data; this justification must include the need to ensure the safety, security, and good order of the correctional institution.[3] It is key that all personnel be trained and that every staff member know that it can be a violation of federal law to disclose protected health care information if the release is not specifically justified.

■ Medical Emergencies

Inevitably there will be emergencies within the walls or fences of every prison and jail. With the aging inmate population, heart disease, strokes, and peripheral vascular emergencies are increasingly common. In addition to medical emergencies, non-medical emergencies occur as a result of inmate fights. Staff also will be called to deal with suicide attempts and completions.

Unless the correctional facility has a very well-equipped hospital, and most do not, the main function of staff in such emergencies is to:

- Rapidly respond to the scene
- Stabilize the patient
- Transport the patient to a facility equipped to deal with the emergency

Rapid response almost always dictates some radio communication through central control to alert medical staff to the nature of the emergency and the location. Staff should respond as a group and must use good judgment in approaching a "down" inmate, particularly one who is surrounded by other inmates. Such inmates must be dispersed after noting which ones are present. The latter is particularly important if the emergency is due to an assault.

Institutional medical staff who are expected to respond to emergencies should be certified in advanced cardiac life support (ACLS) and in advanced trauma life support (ATLS). All staff should have yearly refreshers and stay current in cardiopulmonary resuscitation (CPR). Recently the use of automated external defibrillators (AEDs) has become widespread; if an institution is so equipped, these should be brought to the scene and appropriate staff trained in their use.

Transporting an injured or unconscious inmate should be done on a stretcher or backboard. If the transfer is to be made to an outside hospital, notification should have been made to the facility in advance of the transport. The receiving medical facility should be notified that an inmate is en route. As many escape attempts have occurred on medical trips to outside institutions, security remains a high priority. The inmate's history and security/custody classification will dictate the use of restraints and how many correctional personnel are needed for the trip.

As soon as possible after the medical emergency, an incident report should be written by correctional staff and the inmate's medical record updated by medical staff to include times and all medications and procedures administered. To keep track of these issues, a recorder is often appointed early in the emergency to note times, medications, and dosages, as well as any inmate patient response to these ministrations. This ensures a complete medical record and is of great assistance if the institutional medical response is questioned later.

■ Medication

Prescription medication is commonplace among prison populations. Most institutions will have a formulary that will dictate which medications within a class may be prescribed. For a new inmate, the use of generic medication equivalents may generate some angst and must be dispensed with a good deal of patient education and reassurance. This is best done by a pharmacist if one is available.

The cost of providing medication for inmates is a major expense for correctional facilities. Many specialized, mandatory drugs are required for care and treatment. State and federal governments spend huge amounts of monetary resources through Medicaid and other aid programs to support citizens who work for the government, retirees from government agencies, and prison inmates. The State of California alone spent $133 million on prison medications in 2002–2003.[4] Prison and jail systems try to save on their pharmaceutical purchases by joining with other agencies to create bulk purchasing power, using generic drugs, and developing procedures for electronic purchases from distant, less-expensive providers.

Good recordkeeping is essential to monitor a specific inmate's appropriate use of medication. In some cases, an inmate may be responsible enough to take the medication without supervision; in other cases he or she may need direct supervision even to the point of examining the mouth to make sure the medication was swallowed. Various institutions will handle this problem in different ways, and various monitoring mechanisms have been successful in monitoring inmate compliance.

Abuse and Dependency

Opiate abuse is a potential problem during incarceration just as it is in the community. Many inmates come to incarceration with chronic pain due to injuries and need an effective analgesic. Some inmates, however, have become quite adept at feigning pain; therefore, the best approach to pain management is an inclusive team who meet regularly to discuss to pain control and administration of opiates. All such meetings should be documented in the medical record and a judgment rendered as to whether such medication is habitually necessary and dictated by the medical condition.

In some cases, correctional personnel can be helpful during this assessment. For example, they may have seen an inmate who is allegedly suffering from chronic pain playing vigorously on the ballfield and not appearing to be in discomfort. Such an observation should be documented in the record and form the basis for an informed discussion within the committee. In no case should an inmate be denied pain medication without proper documentation in the record or as a punitive gesture.

Inasmuch as drug and alcohol use are widespread in the general population, they are often associated with criminal activity. That is to say that many offenders commit the crime or crimes for which they are incarcerated while either under the influence of alcohol or drugs, or the crime is committed in order to support the dependency. For this reason, most long-term incarceration facilities have a chemical dependency treatment program. In some cases, the sentencing judge will have predicated the inmate's sentence on the completion of such a program.

Whether or not a given institution has a complete chemical treatment facility within its walls will depend on its mission, level of custody and security, budget, and trained personnel. In a comprehensive and fully funded program, inmates will have access to a complete initial evaluation, psychological testing and evaluation, lectures, group therapy, and counseling. In many cases, outside self-help groups (such as Alcoholics Anonymous) will be allowed inside the prison and share their personal observations and experiences with the inmates. Without a comprehensive treatment approach for drug and alcohol abuse and followup, the inmate recidivism rate is considerably higher.[5]

Preventing drug and alcohol use depends on the vigilance and cooperativeness of the correctional staff. Inmates may ferment fruit and vegetables stolen from the kitchen to make alcohol. Drugs such as cocaine, heroin, and marijuana may come into the institution through mail, visitors, or staff. Only the most intense awareness and monitoring by correctional and medical staff will keep an institution drug free. Random urine monitoring programs, when carried out conscientiously by staff and supported by administration, are effective tools in managing substance abuse within the institution.

■ Infection Control

Communicable diseases are a concern in every confinement situation due to three factors:

1. Confinement lends itself to the spread of infectious organisms whether by airborne or direct contact.
2. Prior needle drug use, which is high in the prison population, predisposes inmates to certain infectious diseases (e.g., hepatitis).
3. The incidence of human immunodeficiency virus (HIV) is high in certain prison populations (although less than 5 percent overall).

Each institution needs an infection control manual that is clearly written and communicated to all staff on a regular basis. All people within the walls need to understand the principles of communicability, and most importantly, understand universal precautions, which are outlined by the NCCHC. Such precautions include disposable gloves and CPR masks; all personnel should have immediate access to this equipment.

Public health infection control nurses or equivalent medical staff are invaluable in monitoring tuberculosis and other communicable diseases within an institution. It is also important to keep other staff informed about procedures and techniques and to develop educational programs for both inmates and staff. Many inmates come to confinement with many misconceptions about the spread of HIV, and such misinformation can lead to institutional unrest. In addition to testing for TB, hepatitis B, and hepatitis C, each institution should develop a policy regarding the routine testing for HIV. This can be a screening test on blood with the Elisa test and appropriate followup for positive results. Not all institutions test routinely for HIV. In all cases, pre- and posttest counseling must be done if the testing for the HIV virus is performed. Strict confidentiality regarding test results must be maintained.

Today's treatment of HIV is expensive, time consuming, and complex, and it requires vigilance on the part of both the inmate and his or her caregivers. Frequent testing and monitoring of viral blood levels, CD4 counts, and secondary infections nearly always dictate that infected inmates be seen at regular intervals. Compliance with medications, which often have unpleasant side effects, is variable and often poor. This can usually be dealt with by medication adjustments and education, but even with these efforts sometimes the course of the illness is relentless. All AIDS patients should be monitored carefully for TB.

Medical and correctional staff who have direct contact with inmates should use discretion in obtaining a yearly blood test for HIV for themselves, but all should have a yearly test for TB. Multi-drug-resistant TB has recently become a problem in AIDS patients, especially among those in prison. This is a very serious issue and requires consultation from an infectious disease expert.

■ Unique Medical Situations

In every incarcerated population, medical situations occasionally arise from protests. In these instances, nearly all inmates will relent and respond to the professional, nonjudgmental demeanor of the staff when inmates realize that the particular form of protest they have chosen is not going to be effective. Punitive and abusive behavior on the part of staff can only serve to exacerbate the situation and make inmates more resolved to carry out their protest.

Use of Force

Occasionally, the use of force may be necessary to subdue or restrain an inmate who is a threat or danger to himself or herself, other inmates, or staff. The procedures for this are very familiar to correctional staff and usually prescribed in a manual. Medical staff typically are not involved in the application of restraining force. However, whenever possible, medical staff should be present when force is used and should perform a careful examination of the inmate after the force or restraint has been applied.

Particular attention should be paid to positional asphyxia (restriction of breathing due to positioning) during situations of restraint. Historically, most serious injuries in such emergencies have occurred as a result of airway obstruction. In an older inmate or one with known cardiac risk factors, particular attention should be paid to blood pressure, pulse, and questions of chest discomfort. If instability is suspected, the inmate should be transported to the infirmary and monitored and a physician notified. After the application of force and an exam, medical providers should document their findings carefully and note times and other staff who were present.

Fights or Riots

Disturbances within a prison or jail setting can vary from a fight involving two inmates to a full-blown riot with multiple injuries to staff and inmates. Sometimes fire, explosives, and wounds of various kinds can add to injuries. Medical staff's involvement will vary with the seriousness of the disturbance, and many of the principles that dictate their role will be the same as described above for emergencies. The important general principle is to have a plan to deal with such disturbances (e.g., disaster drills) and to practice the plan at regular intervals. Simple things like triage, proper use of radio frequencies, and transportation of injured persons will be difficult at best under actual disturbance conditions. However, successful medical intervention (and, for that matter, security) is much less likely if no forethought is given to developing and practicing relevant procedures.

Hostage Situations

Hostage taking is fortunately a relatively rare circumstance that may occur and in which medical staff may be asked to play a role. This role will likely be dictated by the manager of the hostage situation, usually the warden or a designed staff member. Again, forethought and planning are part of the educational process of everyone who works within the facility, and medical staff should prepare for their respective roles in emergency management, triage, and transport.

Hunger Strikes

Food strikes may occur among inmates as a group form of protest or by individuals as a protest against some alleged mistreatment or perceived unfairness. Food strikes are best managed by medical staff in a professional and nonjudgmental way. The inmate who declares a food strike is placed in a locked cell, and water to the room is shut off to preclude surreptitious hydration. Food and water should be offered on a continual basis. As the inmate becomes weaker and dehydrated, frequent exams should be done and results recorded with particular attention being paid to skin turgor (tenting) and dryness of mucus membranes. All urine output should be noted and all urine saved for specific gravity determination. After several days of no food or water, blood electrolytes must be drawn and tested on a daily basis to ensure against serious injury. If there has been no progress in resolving the inmate's issues in a reasonable period of time, at some point (medically determined) an intravenous (IV) line is to be infused and IV fluids given to the protesting inmate. All interventions and exams must be documented carefully in the medical record.

■ Eldercare

The aging inmate population across the United States is beginning to take its toll on institutional operations. Because inmates are locked up for longer periods of time and, in many jurisdictions, without the possibility of parole, more offenders age and die in penal institutions. It is projected that by 2030 there will be 33,000 geriatric prisoners in California alone, and that one third of the overall United States prison population will be geriatric inmates (those over the age of 55).[6]

The cost of specialized care for the elderly is high. Older inmates require more medical attention, sometimes need special diets, and eventually will be in need of other services that are necessary for geriatric individuals. Indeed, some states have established the equivalent of nursing home facilities that function both as prisons and medical-support programs.[7]

■ End-of-Life Considerations

There are several important end of life issues that arise in correctional environments, including in-custody deaths, living wills, hospice care, family visitations, compassionate release, handling the remains, and executions.

In-custody deaths (inmate deaths during imprisonment) are not rare, especially due to recent increases in lifelong sentences. These deaths may be due to natural or unnatural causes, which include suicide, homicide, mysterious surroundings, or death due to work injuries. All in-custody deaths are investigated carefully by a neutral body or commission. Death caused by unnatural causes should require that an autopsy be performed by an outside medical examiner.

At the time of death, the site of the death should be treated as a potential crime scene: The area should be sealed, and no one should move or touch the body after medical staff has declared that death has occurred. A scene investigation and photographs most likely will be taken by correctional staff or the medical examiner. The medical provider's role is to pronounce the death and record relevant physical findings in the medical record. If rigor mortis has occurred, it should be noted, but most likely the outside medical examiner will determine the time of death from other information. Medical staff should personally communicate any information they have to the medical examiner.

In some cases of natural death, the inmate will have a living will. If so, it should be properly drawn and notarized as part of the inmate's medical record. If, for example, the inmate has an incurable terminal disease and does not wish to have supportive care or cardiac resuscitation in the event of cardiac arrest, that information should be communicated to all staff caring for the inmate and should be documented in the inmate's medical record.

Hospice care for terminally ill inmates is not as common as in the community at large, but it does exist in some prisons that have a large percentage of medical cases. This innovative care has allowed other carefully screened and trained inmates to participate as hospice volunteers and sit by the bedside to bring comfort to the dying. In addition, multiple professional staff may be involved, including physicians, nurses, pharmacists, social workers, psychologists, chaplains, and interested correctional staff. This team approach to hospice has been shown to be effective in terminal care management.

Any care of terminally ill inmates usually will involve some communication with their families. Most families, even those estranged from the inmate, need to be informed on a periodic basis, perhaps weekly, of the inmate's status. They may have questions about the inmate's state and care, ranging from the amount of suffering to what will happen with his or her body after death. These are real issues and must be addressed by medical or case management staff members either in person or on the phone. Again, a professional, nonjudgmental manner is the key to good communication with the family. In some cases, a bedside visit within the facility can be arranged after appropriate security and custody clearance. All such visits are generally directly supervised by staff. Many times, a family's anger may be abated by such a visit when they see for themselves the helpful and supportive nature of the medical staff. A frank discussion also should be held with the family about disposition of the body at the time of death. It should be made clear that an autopsy most likely will be performed unless one's religious beliefs oppose it and the coroner accepts this direction. The individual's body will be shipped to the family, or the remains will be cremated and the ashes shipped.

Some state and federal facilities allow for a compassionate release in the event of terminal illness. Each jurisdiction allowing this has different criteria that must be met to attain a release. In most cases, the inmate will have only a few months to live and must be incapable of reoffending. Inmates must also have some place that is willing to receive them and the means to provide care. Often the sentencing judge and prosecuting attorney are involved in the decision to release the inmate and, depending on the crime, may be reluctant to do so.

The Healthcare Professional as a Correctional Officer

Health care providers and correctional staff have overlapping functions. Communication between and among all the professional elements within a custody environment is essential to ensure a safe and effective workplace. Occasionally tensions will develop between custody and medical staff over issues of inmate management. These issues invariably pertain to health questions and offender management. If such tensions or disputes arise, they should be resolved in a professional manner in the best interests of both inmate and staff. It is essential for all staff members to realize that everyone in corrections has the same end goal, namely the secure and humane custody of offenders in a safe environment. If the sight of that goal is lost, then the mission is compromised.

One of the best ways to ensure that this tension does not develop between custody and medical staff members is to have training sessions together. All medical, dental, mental, and health care staff should be familiar with the principles of security and custody. Correctional staff should be made familiar with basic medical procedures such as universal precautions in infection control and be proficient in CPR. When all staff train together in these procedures, a spirit of cooperation and camaraderie may develop to help medical and correctional staff work together to fulfill the mission.

Conclusion

The percent of those incarcerated relative to the general population is rising each year as society makes changes in sentencing laws and demands harsher sentences for offenders. The need for well-trained health care professionals is great within the prison system. Physicians, nurses, pharmacists, physical therapists, dentists, and mental health workers are in great demand. Medical practice within a correctional environment offers many exceptional opportunities for professional growth and a challenging, teamwork-driven environment.

As with any institution, there are those who criticize the prison system and the health options and surveillance available to inmates. While some of these criticisms are valid, others are the product of biased inmate advocacy groups. It is hard to refute such criticism in a general way because there are so many state and federal prisons, and each has its own health care delivery system. In addition, county and city jails often contract with the private sector, thus making it even more difficult to evaluate general health care of inmates. Just as in society as a whole, there is not one unifying health care system for incarcerated persons. National standards, or a common standard of care such as developed by the NCCHC and referred to earlier, will eventually serve to make health care more uniform. In the meantime, it is important to remember that access to health care within a correctional facility is a constitutional guarantee, and prisoners are the only segment of our society who have that right.

DISCUSSION QUESTIONS

1. Do you believe convicted offenders should have the same standard of medical care as law-abiding citizens in the community?
2. Should inmates have access to medical personnel daily at institutional sick call?
3. What are some problems in the health care area within prisons and jails?
4. Should inmates have to pay a small amount of money for seeing health care professionals in the correctional environment?
5. What emergency medical situations can arise in a correctional environment, and should these situations be handled by medical professionals?

ADDITIONAL RESOURCES

K.L. Faiver, *Health Care Management Issues in Corrections* (Lanham, MD: American Correctional Association, 1998).

E. Kempker, "The Graying of American Prisons: Addressing the Continued Increase in Geriatric Inmates," *Corrections Compendium* 28, no. 6 (2003).

J. Moore, ed. *Management and Administration of Correctional Health Care* (Kingston, NJ: Civic Research Institute, 2003).

National Commission on Correctional Health Care, http://www.ncchc.org

I. Robbins, "Managed Health Care in Prisons as Cruel and Unusual Punishment," *Journal of Criminal Law and Criminology* 90, no. 1 (September 22, 1999).

NOTES

1. *Estelle v. Gamble*, 429 U.S. 97 (1976).
2. *Bowring v. Goodwin*, 551 F.2d 44, 48 (4th Cir. 1977); *Lareau v. Manson*, 651 F.2d 96 (2nd Cir. 1981).
3. U.S. Department of Health and Human Services, "Title 45–Public Welfare," Code of Federal Regulations (Washington, DC: US Government Printing Office, 2004), available at http://a257.g.akamaitech.net/7/257/2422/12feb20041500/edocket.access.gpo.gov/cfr_2004/octqtr/45cfr164.512.htm, accessed August 13, 2007.
4. S. White, *Survey of California Department of Corrections Pharmaceutical Expenditures* (Sacramento, CA: Office of the Inspector General, 2003), available at http://www.oig.ca.gov/reports/pdf/corrPEsuvery0703.pdf, accessed August 13, 2007.
5. B. Pelissier et al., *TRIAD Drug Treatment Evaluation Project* (Washington, DC: Federal Bureau of Prisons, 2000), available at http://www.bop.gov/news/PDFs/TRIAD/TRIAD_pref.pdf, accessed August 13, 2007.

6. S. Enders, D. Paterniti, and F. Meyers, "An Approach to Develop Effective Health Care Decision Making for Women in Prison," *Journal of Palliative Medicine* 8, no. 2 (2005), pp. 432–439.

7. A. Stingley, "North Carolina Tackled Geriatric Question Early," *The Shreveport Times*, November 30, 1996.

Mental Health

Sally C. Johnson

Chapter Objectives

- Outline issues inherent in the provision of mental health care in the correctional setting.
- Determine if inmate participation in mental health care and treatment should be required.
- Explore the right to privacy with regard to mental health records.

■ Introduction

Provision of mental health services is a necessary but complex and often inadequately addressed part of any correctional operation. Adequate attention to the planning and implementation of quality services to address the mental health needs of the inmate population can contribute greatly to the smooth running of a correctional facility; inattention can lead to management problems, negative publicity, and even litigation. Many correctional administrators do not have training or experience in delivery of mental health care and thus may need to familiarize themselves with the literature about the needs and demands of mentally ill inmates and the standards and guidelines within the field to assure that their mentally ill offender populations are served adequately. Attitudes and skills developed to deal with the mentally ill population can be useful in management of the institution as a whole.

The process and programs of caring for society's mentally ill have changed dramatically in the United States. With the advent of new medications for mental illness (known as psychotropic drugs), much of the population suffering from illness no longer requires the strict supervision and structure of mental hospitals. Many of the institutionalized populations have been released to their home communities with the intent of having them supported by the intermittent supervision of community-based mental health clinics. This process of de-institutionalizing people with mental illness in the United States—now a half century in the making—has created a major decline in the populations of state and county mental hospitals; however, planning to treat offenders in the community does not necessarily result in successful care. Many of those with mental illness do not responsibly visit community clinics as required, and many are not compliant with taking medications to maintain their mental balance. When symptoms of illness return and the offender's behavior deteriorates, resulting antisocial behavior easily can result in criminal violations. Accordingly, with many mental institutions now closed, commitment to a state or federal correctional facility becomes a satisfactory alternative to the court. Indeed, current data reflect that there are more mentally ill people in penal institutions than in mental institutions.

At midyear 2005, more than half of all prison and jail inmates had mental health problems, including 56 percent of state prisoners, 45 percent of federal prisoners, and 64 percent of jail inmates.[1] Recent estimates indicate that between 5 and 16 percent of male inmates (and even a higher percentage of female inmates) meet criteria for diagnosis of a major mental disorder.[2]

The *Diagnostic and Statistical Manual of Mental Disorders, Fourth Edition-Text Revision* (DSM IV-TR) is the most commonly used classification system of mental diseases and defects. It defines a mental disorder as a "clinically significant behavioral or psychological syndrome or pattern that occurs in an individual and that is associated with present distress or disability," and lists over 300 different disorders, many of which may be evident within the inmate population.[3] Of principal concern are the more serious or major disorders that are identified as Axis I disorders. Considerable time and attention, however, may be delegated to meeting the needs of inmates with Axis II disorders, which include diagnoses of mental retardation and a variety of personality disorders.

■ Guidelines and Standards

Provision of mental health services is not an optional activity within the correctional environment. Unlike the nonincarcerated population, incarcerated individuals have rights of access to and provision of health care, including mental health care. Since 1980, through litigation in the courts at state and federal levels, the responsibility for provision of care to those who are denied the ability to access it on their own has been defined generally.[4] Little differentiation exists between the right to assessment and treatment of medical problems. The current standard is for a basic but clinically current level of care. The crucial issue is avoiding deliberate indifference toward the medical and mental health needs of the incarcerated population. Examples of deliberate indifference include lack of access to care, failure to follow through with care, insufficient staff resources, and poor outcome due to negligent care.[5] Ensuring the basic level of services does not fall below acceptable standards requires adequate health care facilities, a well-defined program structure, understandable written policy, and a quality assurance program.

A number of correctional and health care organizations have established minimum guidelines and standards to support the care and treatment of this population. These include:

- American Correctional Association (ACA)
- American Medical Association (AMA)
- American Public Health Association (APHA)
- American Psychiatric Association (APA)
- Joint Commission for the Accreditation of Healthcare Organizations (JCAHO)
- National Commission on Correctional Health Care (NCCHC)
- National Institute of Corrections (NIC)

Formal reviews and accreditation can now be conducted under the ACA, NCCHC, and JCAHO, and involve site visits, review of written policy and procedures, and observation and review of health records and other documentation of care. Institution services are reviewed against the standards established by each organization. Although there is considerable overlap of intent in the standards, the degree of fit between standards and a particular program and the degree of detail addressed by the standards varies greatly. The ACA standards are the least specific of the three, and thus many mental health programs within correctional facilities also choose to undergo accreditation review by the NCCHC or JCAHO in addition to the ACA.

The importance of external review and accreditation should not be underestimated. In addition to providing concrete guidelines for establishing and maintaining programs, preparation for the accreditation process forces internal auditing and review. Accreditation visits and reports frequently identify existing or potential problems in service care delivery and in the quality or quantity of services, and mandate establishment of a time frame for correction of those deficiencies. Successful accreditation provides support for the correctional program when questions about adequacy of care arise. The accreditation process also allows integration of the health care staff into the larger health care community and provides staff support and gives guidance to the correctional health care mission.

Access to Care

An adequate care delivery system for mental health services in a correctional environment must address a range of functions and needs within the environment. As early as 1980, in the case of *Ruiz v. Estelle*, the court focused on six issues required to meet minimally adequate standards for mental health care in a correctional environment:

1. A system to ensure mental health screening for inmates
2. Provision of treatment while inmates are in segregation or special housing units
3. Use of the training of mental health staff to ensure individualized treatment planning
4. Accurate and confidential medical record system
5. Suicide prevention program
6. Monitoring for appropriate use of psychotropic medication[6]

These issues remain essential for any correctional health care program. They form a continuum of services to meet the demands of the correctional population.

Correctional administrators must identify and implement an adequate health care delivery system that can ensure inmate access to health care and health care providers. This access must exist from the point of arrest to the point of discharge from correctional supervision or oversight. Staff and physical plant resource needs must be determined, and a plan must be developed and maintained. Administrators must ensure that adequate policies and procedures are established and utilized to provide a framework for care delivery. These policies and procedures then serve as standards against which individual episodes of care can be compared as part of continuous performance improvement efforts.

■ Screening

Mental health care services begin at the screening stage. Given the volume of patients entering the correctional environment and the fact that the presentation of these individuals is often unplanned and occurs outside of usual working hours, it is important that a good initial screening system is in place. Inmates should be screened immediately, before they are placed in a housing situation without direct staff observation. Initial screening should be done on an individual basis. Correctional staff should screen for current or recent medical or mental health care, hospitalizations, and medications. Initial health care review should be accomplished by health care providers who have been trained in detecting symptoms of potentially serious mental disorders and who have mastered good interviewing techniques. Individuals entering the criminal justice system are often angry, upset, frightened, anxious, or confused, and may have difficulty providing information, or present as uncooperative. Health care providers, however, must persist in obtaining necessary information to screen incoming inmates adequately.

The goal of screening is to identify quickly emergency situations and inmates who might require more extensive intervention prior to placement in the population of their assigned housing areas. During screening, staff members should identify themselves and the purpose of the interview. They should assess the inmate's general presentation and understanding of the situation. The staff member should provide correct orientation as to place, time, and situation, and reassure and educate the individual if necessary. The same questions should be addressed routinely with each inmate, with further exploration and data collection where indicated. The response to each question should be recorded on a screening intake form, and this form should then be placed in the inmate's correctional health record after any necessary referrals have been made. Proper mental health intake screening will determine the type and immediacy of need for other mental health services.

Inmates requiring further assessment or evaluation should be housed in an area with staff availability and observation appropriate to their needs. The assessment or evaluation should be assigned to a specific staff member and service, and normally includes interviews, record review, physical examination, laboratory studies, observation, and possibly psychological testing. A differential diagnosis should be established, recognizing that it may take time for the diagnostic picture to become clear.

Treatment and Followup Care

Treatment within the correctional environment can occur in a variety of settings:

- Outpatient—housed in general population with mental health services provided in a medical clinic or the housing unit
- Inpatient—housed in the hospital facility within the correctional environment or transferred to the community
- Transitional or intermediate—housed in one area of a general population housing unit or a separate housing unit

Inmates may move from one level of treatment to another. Outpatient services involve counseling, consultation, medication management, and ongoing screening to identify any change in treatment needs. Much of the counseling is supportive and provides a cost-effective mental health intervention that may prevent escalation to a higher level of care. The frequency of visits varies from every week to every 90 days. Psychiatric inpatient services often are established on site in correctional facilities to avoid the need for movement into community settings and to ensure adequate security measures remain in place. Some systems, however, continue to use a community-based hospital system, often the state mental health hospital system. Admission is voluntary or by civil commitment. Each potential patient must be reviewed for competency to consent to hospitalization.[7] Common admission criteria for inpatient hospitalization include:

- Presence of significant psychiatric symptomatology
- Inability to be handled in a less restrictive environment
- Need for complex or comprehensive assessment services not available in an outpatient setting
- Court-ordered inpatient evaluation
- Imminent danger to self or others

In recent years, in part due to limited resources, improved screening, and efforts to manage rising costs, the concept of transitional, intermediate, or habilitative care has developed.[8] Inmates who have known chronic mental disorders, are prone to relapse, have significant behavioral problems, or are unable to integrate well into the general prison population often benefit from this type of intervention. The usual goal is stabilization during a three-to-six-month period of less-intensive treatment in a somewhat sheltered environment. Treatment focuses on learning to cope with chronic symptoms and learning to adapt to the general prison environment. The goal of treatment is eventual integration into a regular prison population and preparation for eventual release to the outside community.

Treatment interventions are related directly to diagnoses and specific symptoms presentations and may be delivered or coordinated by various members of the mental health care treatment team. Documentation of a treatment plan in the inmate's health record helps available staff to better direct their interventions toward accomplishment of the inmate patient goals. **Table 10–1** provides a simplified outline of typical treatment interventions across the spectrum of psychopathology likely seen in a correctional setting.

Because of the relapsing nature of many mental disorders and the chronicity of others, discharge planning and followup care are crucial both when returning

Table 10–1	**Generalized Classification of Mental Disorders**			
Category	**Symptom Presentation**	**Treatment**	**Typical Settings**	**Provider**
Psychosis	Hallucinations Delusions Bizarre behavior	Medication Supportive treatments	Hospital Outpatient Transitional care	MD Psychologist
Mood Disorders	Increased or decreased mood Sleep disturbance Appetite disturbance	Medication Counseling	Hospital Outpatient Transitional care	MD Psychologist
Situational Problems	Anxiety Mild Depression	Counseling Environment support	Outpatient	MD Counselor Psychologist
Substance Abuse	Drug-seeking behavior Sleep disturbances Anxiety	Education Relapse prevention		Social Worker Counselor
Mental Retardation Organicity	Adjustment difficulties Viewed as vulnerable	Assisted living Supportive counseling	Transitional care	MD Social Worker Unit Team
Sexual Offenders	Sexual offense by history	Education Relapse prevention	Outpatient	MD Psychologist Counselor
Personality Disorders	Problem behaviors	Counseling Environment support	Outpatient	Unit Team Psychologist MD

inmates to a general population facility and when releasing them to the community. Successful discharge planning is facilitated by sound diagnostic assessment. Assuring adequate and cost-effective continued treatment is the goal. Education of inmates about their illness and treatment needs during incarceration increases the likelihood that treatment will continue outside the hospital/prison setting. Assessment of dangerousness to self and others must be reviewed to ensure that discharge from a hospital setting is the correct decision or that adequate safeguards in the community are implemented. To complete the process, there should be sufficient exploration of followup resources to ensure that resources are adequate to meet the inmate's needs, and proper communication, verbally and through records, occurs with followup care providers.

Crisis Intervention

The need for crisis intervention or suicide prevention may arise with inmates at any level of treatment, as well as those not identified as in need of treatment. Crisis intervention consists of short-term interventions to deal with acute mental distress. The level of distress varies from acute anxiety or anger, to that associated with psychotic decompensation. Often the crisis involves suicidal ideation or a suicide attempt. The frequency of suicidal thinking or behavior presenting as a crisis requires that all correctional staff be familiar with suicide prevention. The administration must ensure that an adequate suicide prevention program is in place within the correctional environment. A suicide prevention program includes adequate training of staff in identification of signs and symptoms of potentially suicidal inmates, ensuring the availability of a safe environment in which a suicidal inmate can be housed, and ensuring that the suicidal inmate is kept under constant observation. The latter requirement is costly and time intensive. This has led some systems, such as the Federal Bureau of Prisons, to try using inmate companions or peer watchers who can access staff rapidly if intervention is needed.

Issues Arising from Confinement

Mental health problems and the types of services needed to address these problems are similar throughout the phases of the criminal justice operation. Individuals may enter the system at any phase of an illness or may present new symptomatology once in custody. The stress of being involved in the criminal justice system can add to the symptom picture. Stressors include involvement in the legal system, separation from existing community support systems, peer-generated problems, and internal stressors associated with loss of control and individual decision making. Incarceration disrupts sleep and eating routines. Access to anxiety-reducing activities such as television, exercise, socialization, and smoking may be limited severely. Additional stress can result from a general lack of familiarity with or understanding of the legal or criminal justice process and the forced position of dependence consistent with being an inmate. A functioning mental health service is crucial to addressing the impact of these issues adequately.

■ Legal Requirements

Hospitalization in a psychiatric facility is voluntary unless inmates are so impaired that they present a danger to themselves or others or could clearly benefit from care in a hospital and are unable to function in a general population. Each state and the federal government by statute, directs how an individual can be civilly committed for involuntary hospitalization in a psychiatric facility. Civil commitment in a prison facility parallels the process for nonprisoners and involves due process, legal representation, and judicial decision making. Voluntary psychiatric hospital admission must be agreed upon by a competent inmate. The hospital record should contain documentation of the informed consent to hospitalization and assessment of competency to give consent.

Hospitalization alone, whether voluntary or involuntary, does not grant the care provider authority to treat.[9] Treatments, as well as the absence of treatment, carry varying degrees of risks and benefits. The inmate must give informed consent to any type of psychiatric treatment, except emergency treatment or treatment allowed involuntarily after adequate judicial or administrative review of a court-committed individual. Informed consent requires that the care provider give adequate, understandable information regarding the proposed treatment to a patient capable of understanding the information. It is required that the care provider discuss alternative treatments (as well as no treatment) as options. The risks and benefits of proposed treatment must be discussed thoroughly with the patient. Any common or severe side effects must be reviewed. Patients accepting voluntary treatment must be made aware that they may elect to discontinue the treatment any time simply by withdrawing their consent.

■ Privacy

Every day, correctional health care providers confront the issue of confidentiality. Patients in prison and outside of prison come to their health care providers with the expectation that information shared will be kept confidential. Medical and mental health information can be shared among health care providers only on a need-to-know basis. Records should be kept secure on the treatment units, with access limited to health care providers involved in the patient's care. Internal policy should define the members of the health care team. Patients should be advised as to the limits of confidentiality that may apply in special situations such as court-ordered forensic evaluations or injury assessment exams. Likewise, patients have the right to access copies of their records and to review their health care records unless their health care providers deem such reviews could be detrimental to the patient.

■ Dual Roles of Staff

At times, clinical staff may feel caught between their roles as caretakers and as correctional workers. Explaining these roles to the patient population at the onset of each significant encounter can be helpful in eliminating instances where the health care provider feels conflict between these roles. Patients need to be advised that any information that affects the security of the institution, requires intervention to prevent harm to the patient, other inmates, or staff, or concerns situations where serious damage to property will occur cannot be kept confidential.

Special Treatment Procedures

The use of special treatment procedures, such as seclusion or restraint, requires close attention in the correctional environment. Both interventions may be used in correctional situations that are not considered clinical in nature. They may also be used as part of the spectrum of treatment interventions for acutely disturbed psychiatric patients. A closely monitored review system must be put into place to ensure that special treatment procedures are used only as necessary and that the inmate's physical and psychological needs are addressed on an ongoing basis during the time he or she is secluded or restrained. Monitoring should ensure that patients are kept in these more restrictive situations for the minimum amount of time. Guidelines for the psychiatric use of seclusion and restraint have been available to guide practitioners since 1985, and current accreditation standards continue to focus attention on this high-risk area.[10] Limited reasons for the use of these special treatment procedures exist and generally are restricted to cases in which inmates pose imminent risk of harm to themselves or others or pose a threat of serious damage to property.

Inmates in seclusion require enhanced monitoring, to the point of continuous observation if risk of harm to self is an issue. Inmates in restraints must be assessed regularly to ensure that circulation has not been compromised and that toileting, meals, and repositioning are accomplished as necessary. Health care providers must guard against the inappropriate use of special treatment procedures.[11]

Medications

Medication prescription is another high risk area within a correctional environment. Appropriate use of psychotropic medication can be defined as using the correct medication at the right dosage for the right period of time. Pharmacotherapy is a mainstay of current mental health treatment. Unfortunately, medication use can also be problematic in a correctional environment. As in any population, there will be some medication-seeking clientele. Frequently these will include chronic pain patients, insomniacs, and substance abusers without access to their drug of choice. Several guidelines in this area may be useful:

- Medication prescription should be kept to a minimum. All prescriptions should be time limited, and the need for continuing a particular medication should be assessed thoroughly.
- Sleep medication should be avoided except in acute situations and limited to three days without further review. The cause of the sleep disturbance should be sought, taking care to uncover any underlying depression.

- Medication compliance should be followed closely. Non-compliance that persists should result in getting documentation of treatment refusal and discontinuing the prescription. Mouth checks and blood and urine screens should be used to determine compliance.
- Cost and ease of use as determined by route and dosing frequency may make a great deal of difference in whether the patient remains compliant after release.
- Guard against polypharmacy (use of multiple medications from the same class or from similar classes). Inmates are known to seek out multiple care providers, and correctional staffing patterns with frequent rotation of staff may worsen this problem.
- Pay close attention to side effects of the medications prescribed. Stay alert to potential drug interactions. Be aware of the potential to overdose or use drugs in non-conventional ways. Document any adverse reactions.

Mental Illness

In any correctional health care setting the topic of antisocial personality disorder arises frequently. Antisocial personality disorder is diagnosed almost exclusively on the basis of historical information. Criteria to support the diagnosis include a pattern of disregard for others demonstrated by breaking the law and lying, as well as impulsive, irresponsible, and aggressive behavior. Although not every incarcerated individual carries this diagnosis, a significant number of incarcerated individuals do. Inmates with other mental disorders may also demonstrate features of this personality disorder diagnosis. Antisocial personality disorder is difficult to treat. Despite its inclusion in psychiatric classification systems, this diagnosis is often not viewed as a mental disease or defect for legal purposes.

Malingering is a behavior that involves an individual falsely claiming and consciously faking symptoms of an illness. People malinger to avoid the consequences of being held responsible for their behavior or because it will enhance their situation. Malingering may result in an inmate getting referred for mental health evaluation or treatment. Staff in correctional treatment settings should not be too quick to label a patient as malingering. Doing so can prevent a psychiatrically disturbed inmate from receiving necessary care. Malingering is always a diagnosis of exclusion, and the diagnosis is made only after bona fide psychopathology is ruled out.

Integration of Mental Health and Medical Care

The success of any care delivery system for mental health services is integrally related to the adequacy of the general medical services available to the correctional population. It is not uncommon for medical illnesses to present with psychiatric symptoms. Anxiety, disorientation, confusion, and hallucinations can herald the onset or unmask the existence of physical illness or disease. A significant number of psychiatric patients have been found to have concurrent medical illnesses. Too often patients are dealt with only at the first level of symptom review, which may be the presentation of disturbed behavior. Each inmate entering the prison system should have a complete physical examination and baseline laboratory studies. Each inpatient admission to a psychiatric hospital facility within the correctional environment requires a current physical exam, review of medical history, and infectious disease screening. More extensive laboratory studies to rule out organic causes of a symptom picture should be conducted. These will also serve as a baseline against which potential side effects from medication treatment can be assessed. Screening for infectious diseases (including TB, HIV, hepatitis, and syphilis) should be completed.

■ Conclusion

Staffing the correctional health care setting can be a difficult task. Few clinicians and other providers are trained during their professional education to work in correctional environments. Most enter the field by chance and ultimately come to view it as either a challenge or a curse. In the correctional environment, provision of adequate care constantly competes with maintaining adequate security. Security demands often put limitations on how efficiently or effectively programs can be run. Salaries for clinicians may lag behind community levels. Professionals may view the setting as less than desirable, and this, in turn, can create low morale and prevent successful recruiting.

Ironically, despite these barriers, clinicians are realizing that the correctional environment may be one of the last public strongholds for adequate care of the seriously ill and treatment-resistant mentally ill patient. The structure of the environment, the absence of third-party payers, and the impact of externally imposed motivation to change, create a unique setting in which to provide mental health care. The correctional mental health care environment raises interesting questions about clinician-patient relationships, adequate data collection, personal responsibility for behavior, and the roles of genetics, economics, and education in the onset of illness. The experiences of correctional health care clinicians potentially have much to offer to the broader field of mental health care.

DISCUSSION QUESTIONS

1. Why must mental health practitioners have an offender's approval to proceed with mental health care?

2. What is the role of inmate screening and who does this process serve to protect?

3. Why is it important that staff members receive training in the recognition of mental illness?

4. How is the correctional environment unique in the field of mental health care?

5. What role should medication play in mental health care within prisons and jails?

ADDITIONAL RESOURCES

APHS Task Force on Correctional Health Care Standards, *Standards for Health Services in Correctional Institutions* (Washington, DC: APHA, 2003).

J. Fagan and R. Ax, *Correctional Mental Health Handbook* (Thousand Oaks, CA: Sage Publications, 2003).

H. Hills, C. Siegfried, and A. Ickowitz, *Effective Prison Mental Health Services: Guidelines to Expand and Improve Treatment* (Washington, DC: Department of Justice, National Institute of Corrections, 2004).

J. Moore, *Management and Administration of Correctional Health Care* (Kingston, NJ: Civic Research Institute, 2003).

National Commission on Correctional Health Care, *Correctional Mental Health Care: Standards and Guidelines for Delivering Service* (Chicago: National Commission on Correctional Health Care, 2003).

American Psychiatric Association, *Psychiatric Services in Jails and Prisons*, 2nd ed. (Washington, DC: American Psychiatric Publishing, Inc., 2000).

NOTES

1. Bureau of Justice Statistics, *Mental Health Problems of Prison and Jail Inmates* (Washington, DC: US Department of Justice, Office of Justice Programs, 2006).

2. American Psychiatric Association, *Diagnostic and Statistical Manual of Mental Disorders*, 4th ed. (Washington, DC: American Psychiatric Association, 2000).

3. *Ibid.*

4. *Estelle v. Gamble,* 429 U.S. 97, 98 (1976); *Bowring v. Godwin,* 551 F.2d 44 (4th Cir. 1977); *Farmer v. Brennan,* 51 U.S. 85, 832 (1994); *Ruark v. Drury,* 21 F.3d. 231, 216 (8th Cir. 1997).

5. R. Shanski, "Identifying and Correcting Constitutional Violations in Correctional Settings: The Role of Physician Experts" (Paper presented at the annual meeting of the American Public Health Association, Chicago, November 1989).

6. *Ruiz v. Estelle*, 53 F. Supp. 1265 (S.D. Texas 1980).

7. *Zinermon v. Burch*, no. 87-1965 (US, February 27, 1990).

8. W. Condelli et al., "Intermediate Care Programs for Inmates with Psychiatric Disorders," *Bulletin of the American Academy of Psychiatry Law*, vol. 22, no. 1 (1994), pp. 63–70.

9. *Washington v. Harper*, 110 S. Ct. 1028 (1990).

10. W. Condelli et al., "Intermediate Care Programs for Inmates with Psychiatric Disorders."

11. American Psychiatric Association, *Task Force Report 22: Seclusion and Restraint: The Psychiatric Uses* (Washington, DC: American Psychiatric Association, 1985).

Religious Programming

Susan M. Van Baalen

Chapter Objectives

- Explore the role of prison chaplains and understand the challenges of providing religious programs in the correctional environment.
- Explain the major programs coordinated by prison chaplains and the available resources for religious programming.
- Describe the general legal framework within which prison religious programs operate.

Religious services departments and programs play an important role in the correctional environment. Programs typically provide three essential services: religious accommodation, spiritual growth and development, and pastoral care.

Historical Background

Religion has played an important role in the life of American prisoners since the inception of penitentiaries in this country.[1] Many of the earliest prison facilities in the United States were influenced by the Quaker belief that time away from society—time to meditate on one's sinfulness and transgressions of the law— would change the hearts of offenders.[2] The correctional programs of these institutions sought both to reform the lives of the penitents and to protect society from offenders' aberrant behavior. Study of the Bible and spiritual counseling from local clergy were instituted to help prisoners achieve the goals of penitence and changed behavior. One of the problems with this early Quaker model was that these end goals were not necessarily shared by the offender. In addition, the practice of forced solitude caused many inmates to suffer emotional distress and, in some cases, led to mental illness.

The role of religion in jails and prisons has changed as correctional institutions continue to realign their mission and goals to societal pressures. Correctional institutions continue to turn to religious programming as one way to decrease the crime and antisocial behavior.

■ Religious Accommodation and Freedom Legislation

The First Amendment to the Constitution identifies freedom of religion as a fundamental right of every American citizen. This amendment guarantees the right to embrace the beliefs of one's chosen religion and to unfettered practice of that religion, even behind bars. Religious accommodation is a practice that ensures the prisoner's constitutional right to freedom of religion and is the justification for maintaining a religious services program in a correctional institution, regardless of jurisdiction.

Nevertheless, prison administrators can and do regulate religious practices within prisons in order to protect against threats to the security and good order of the institution or public safety. What administrators are permitted to regulate has evolved and continues to be affected by the interpretation of new and existing legislation around religious freedom. During the 1970s and 1980s, the U.S. Supreme Court found some government-imposed limits to religious practice to be appropriate as long as these limits furthered a reasonable penological interest.[3]

The Religious Freedom Restoration Act of 1993 (RFRA) and the Religious Land Use and Institutionalized Persons Act of 2000 (RLUIPA) complement the First Amendment, clarifying the application of the law to persons in the custody of the federal government or state or local institutions. The full intent of these laws is to ensure that the religious rights of all persons confined to long-term care facilities are not abridged or denied.

With the passage of the RFRA, governments were prevented from interfering with individuals' religious observances unless the interference was the least restrictive means of furthering a compelling government interest. Nearly all departments of corrections around the country opposed this legislation, fearing it would require them to succumb to all sorts of religious requests by inmates, stretching staff and budgetary resources, as well as threatening institution security. Of particular concern were requests from radical and/or militant organizations whose beliefs and practices are greatly offensive to other inmates.

In 1997, the Supreme Court ruled in *Boerne v. Texas* that the RFRA violated the Constitution. As a result, state departments of corrections were free to return to applying the previous standard in determining whether to permit various inmate religious practices. This was pursuant to the Supreme Court's decisions in *O'Lone v. Shabazz* and *Turner v. Safley*, which argued that prison regulations that interfered with inmates' religious practices were valid so long as they were reasonably related to legitimate penological objectives.

In response to the Supreme Court's ruling that the RFRA was unconstitutional, many states proposed and enacted laws intended to protect their citizens from encroachments by state and local governments upon religious observances. The new state laws required these governments to meet the RFRA standard in justifying laws and actions that interfere with individual religious practices.

The Religious Land Use and Institutionalized Persons Act of 2000 (RLUIPA) prohibits the government from restricting prisoners' religious worship opportunities unless the government can demonstrate that the restriction furthers a compelling government interest and is the least restrictive means of furthering that interest. This act was challenged in *Cutter v. Wilkinson*, a case involving five prisoners from Ohio including a Wiccan, a Satanist, and a member of a racist Christian sect. In this case, the U.S. Supreme Court unanimously ruled that RLUIPA was a permissible accomodation of religion justified by the fact that the government had severely burdened the prisoners' religious rights through the act of incarceration.

These laws (and in at least one case—Alabama—a proposed constitutional amendment) are widely supported by various religious organizations that believe that additional protections are necessary to ensure that religious observance is protected (for example, allowing Muslim students to wear head scarves to school despite a "no hats" policy and permitting Seventh Day Adventists and Jews to work shifts that do not conflict with their sabbath). Religious organizations also argue that the protection of religious freedom provided by RFRA or similar state legislation is necessary to prevent corrections officials from denying legitimate, important religious requests, particularly because of the reputed connections between religious beliefs or observance and inmates' successful reintegration into society.

In contrast, many corrections officials oppose these legislative initiatives for the same reasons they opposed the RFRA—they are concerned that inmates will file lawsuits requesting a variety of religious accommodations and prevail. They believe certain types of accommodations will threaten institution security as well as place substantial demands on staff and budgetary resources. While supporters of the proposed legislation argue that in many states the number of lawsuits filed by inmates did not increase following passage of RFRA, in some states the number of inmate grievances did increase dramatically. Moreover, many departments of corrections counter that many potential suits were not filed because institutions felt compelled to grant requests that would have been denied before the passage of RFRA, based on advice from their counsel.

The enactment of the RFRA and RLUIPA statutes raised the threshold required for the denial of religious practices in institutions from a threshold of the furtherance of a reasonable penological interest to that of a compelling government interest—a threshold more favorable toward religious freedom and less favorable toward correctional administrators. Landmark cases upholding the RFRA and RLUIPA standard are yet to be decided, as these laws are only beginning to be challenged in state and federal courts. Nevertheless, it is important for all who work in a correctional environment to know and understand the impact of these laws on religious freedom in prisons. Also, correctional staff should be aware that earlier attempts by state and federal prison administrators and professional correctional organizations to exempt prisons from these laws resulted in a clause emphasizing prisons (rather than the original draft's silence with respect to application in jails and prisons). A 2005 Supreme Court decision, *Cutter v. Wilkinson*, tested RLUIPA in the courts. In this case, inmates of small and generally unfamiliar religious groups prevailed in their Supreme Court appeal for opportunities to meet for religious worship and programs.[4]

■ Professional Standards

Federal statutes and laws of this country require quality religious programs in prisons and jails. Professional standards such as those established by the American Correctional Association (ACA) require prison administrators to strike the proper balance between providing sufficient and appropriate religious programs and restricting religious practices because of concerns for safety of staff and inmates or the security of the institution. Religious programs are developed within the framework of the pertinent laws and the religious needs of the inmates. Staff generally accommodate individual religious practices as long as the accommodation will not interfere with the security, safety, and orderly operation of the institution.

Inmates may be encouraged to participate in religious programs as a means of preparing for reintegration into the community following release from prison. On the other hand, administrators and other staff must respect the beliefs of inmates who decline to participate in religious programs. All religious programs must be voluntary; inmates should not be required or coerced into participating in religious activities of any kind, nor should those participating in religious programs be granted any preference with respect to housing, job placement, or consideration for parole or early release. To attach rewards or benefits to the practice of religion interferes with the freedom one has not to exercise or hold religious beliefs. Religion must not be used as a tool for manipulation.

Chaplains employed by the government are responsible for working with inmates of all faith groups, administering services for some and accommodating services for all, regardless of religious preference. This requires the management of programs in a fair and consistent manner and the supervision of all contract and volunteer religious service personnel. There may be an inclination on the part of the chaplain to provide more or better services for some groups (e.g., Christians and Jews) while making only a minimal accommodation for faith groups which may be smaller or less familiar (e.g., Islam, Hinduism, Rastafarianism, or Native American Spirituality).

Because the chaplains or religious coordinators are expected to accommodate inmates' religious beliefs and practices covering a wide spectrum of faith traditions, they necessarily rely on the expertise of community religious leaders to complement the care and services they can provide. Religious service providers (e.g., chaplains, contractors, and volunteers) administer the sacred rites, ordinances, or sacraments of the faith; provide religious education and counseling; assist with the development and provision of religious diets; respond to specific requests for religious objects, apparel, and literature; and coordinate the observance of religious holidays for each faith group.

Role of Chaplains

A chaplain may serve as spiritual guide, preacher, teacher, dietitian, counselor, and advocate, often all at the same time. Even volunteer and part-time chaplains must be committed to public safety and security of staff, inmates, and visitors.

The religious impact in the institution depends, not on merely what the chaplain does, but on the unique pastoral manner in which a chaplain helps inmates and staff deal with the experience of incarceration. In some religious circles, and in institutionalized settings like prisons, this careful and tender care of souls is often called pastoral care—the care a religious or spiritual guide provides to believers or those seeking a spiritual transformation. It seems important to note that the term "pastoral care" is a Christian term that has been somewhat generalized within institutions in the United States to describe the qualities of caregiving over a broader spectrum of religions.

Perhaps the gravest problem among religious service providers is that, in their passion for souls, they lose sight of their role as correctional workers. The ACA standards for adult and juvenile detention centers and prisons require that the religious services department be directed by a professional chaplain or religious coordinator.[5] In an ideal world, there would be a chaplaincy corps of full-time trained professional chaplains—but there is hardly any facet of jails and prisons that fall into the category of the "ideal world." In many cases, philosophical differences and budget constraints prevent the use of a professional chaplaincy corps; instead, jurisdictions rely on part-time chaplains or volunteers to ensure that religious accommodation meets the threshold of the law.

The complexity of the multi-faith correctional environment requires a high level of professionalism to ensure that inmates of all faiths have the opportunity for maximum benefit from religious programs. Successful chaplains are generally well-educated and integrated professionals, trained and experienced in both ministry and correctional management. The integration of ministry and management is essential because of the added dimension and challenges of ministry with an incarcerated congregation.

For many years, a pastor with strong preaching and teaching skills was used to "do church" in the correctional setting; therapeutic and administrative skills were secondary. But penal institutions have changed, making it essential for the religious services provider to have strong administrative and counseling skills as well. This is particularly true where chaplains are part of the prison treatment staff along with social workers, substance abuse counselors, case workers, psychologists, teachers, and medical personnel. Accordingly, clinical training and certification are beneficial qualifications for prison chaplains.

Religious Pluralism

Ideally, the chaplaincy corps should reflect accurately the beliefs of the inmate population. However, time in prison often pushes inmates to new horizons, even with respect to religion. Inmates who came to prison having been raised in one faith, or no faith, may find the loneliness and alienation of the prison setting an opportune time to seek some spiritual grounding. Religious issues or questions raised by inmates whose religious beliefs are outside the chaplain's expertise, training, or ecclesiastical endorsement are referred to qualified spiritual leaders in the community.

Using part-time contract chaplains and volunteers is a reasonable alternative in small jails and prisons where chaplains from various faith traditions are essential to accommodating a wide spectrum of religious beliefs and practices of the inmate population. Contract chaplains can be used as spiritual leaders for particular groups of inmates whose beliefs are different from the full-time professional chaplain. For example, in many institutions, a Catholic priest is contracted to provide the sacraments, religious education, and guidance to Catholic inmates, and an Islamic imam is contracted to serve as a teacher and spiritual leader for Muslim inmates. These contract chaplains do not function as religious coordinators or chaplains whose commitment is to accommodate all inmates' religious needs. As religious leaders within specific faith traditions, the contractors and volunteers are experts on specific religious subject matter. Accordingly, they need only be recognized by the denominations or faith groups they represent. The role and function of contract chaplains needs to be described in the prison policy, which should also specify the education, training, and other necessary qualifications required of contractors.

■ Congregate Services

Congregate services—the meeting of several people to worship, study, or pray—should be provided at all correctional institutions unless specific security or safety issues present a legitimate management concern. In these cases, arrangements must be made to provide inmates with the least restrictive alternative means of practicing their religion. The services must not only meet a reasonable professional standard but also demonstrate good correctional management. True spirituality is often discovered and developed through religious study and practice and may help prepare inmates for returning to the community. Inmates who develop or deepen their relationship with a deity and become involved in religious programs have an improved attitude and can draw upon a support group (e.g., a religious congregation) when they leave prison. After inmates are released, congregations and religious organizations frequently assist inmates and their families with jobs, clothes, housing, training, and other practical needs.

Most correctional institutions provide a variety of worship services due to the broad spectrum of religious beliefs the inmates hold. Whichever the case, the diversity can be addressed somewhat if congregate services are designed to appeal to the widest range of persons who share the basic tenets and beliefs of a particular religion. Congregate services are always led by a person with proper credentials. For example, a priest recognized by the Roman Catholic church should conduct mass for Roman Catholic inmates. When there are only a few inmates from a particular religious persuasion, a spiritual advisor can provide assistance to the individual practitioners. If the group is large, then the services of a local minister can be used on a weekly basis. At a minimum, all inmates should have access to spiritual leaders, regular opportunities for worship on weekly and special holy days, access to religious study materials, and, within reason, a religious diet accommodation commensurate with the tenets of their faith.

■ Religious Needs

Religious Programming

The religious coordinator or chaplain should develop a religious program that includes more than weekly worship services. Study groups and religious education classes serve a dual purpose in the correctional environment. From a public safety perspective, these programs provide opportunities for personal growth and change that deter the individual from criminal behavior. From a prison management perspective, these programs fill inmates' time that might otherwise be idle (which often leads to negative consequences). Therefore religious programs, generally offered during inmates' time free from scheduled activities, serve as a management tool protecting against the dangers that can ensue from extended periods of idleness. In 2005, 60,000 inmates (more than 35 percent of the total prison population) participated in weekly religious programs in Federal Bureau of Prison chapels.[6] This far exceeds the estimated 20 percent of American believers who attend church services weekly in the United States.[7]

Development of individual talents should be one objective of the religious program. For example, music education and choir provide an outlet for inmates to express themselves creatively. Special events that bring in choir groups, entertainers, and evangelists usually attract large groups and provide an emotional and spiritual outlet for inmates.

Religious Diets and Holy Days

Inmates of many faiths may request special diets mandated by their religion. It has been the general practice in jails and prisons to ask inmates to document that the requested diet is a basic requirement or tenet of their faith group; they also may be asked to prove their membership in the group. The institution should not be expected to accommodate dietary requests based solely on personal preferences so long as nutritionally adequate meals are provided that do not violate the inmates' religious dietary requirements. However, legitimate requests based on sound religious principles must be accommodated. For this reason, most correctional institutions offer a vegetarian or, at least, a pork-free menu. The broadest interpretation of the RFRA and RLUIPA in the courts suggests that requesting inmates to prove their religious affiliation or their religion's requirements for certain dietary accommodations may be too restrictive.

Some religious observances (e.g., Passover) require consumption of particular foods, and some observances require consumption of food at particular times (e.g., Ramadan, during which Muslim inmates must fast during daylight hours). Special accommodations should be made for these occasions. In some instances, particular religious observances include congregate ceremonial meals and days free from work. These requests generally can be accommodated as long as the prison staff is given sufficient notice to make arrangements. The chaplain or spiritual leader often is called upon to explain to staff the significance of the religious observance and the specific accommodations that are required (for example, completion of all work before sundown of a Jewish holiday).

Religious Literature, Apparel, and Objects

Inmates should have access to religious literature, including the sacred writings or scriptures of their religion. Security, safety, and sanitation concerns may limit the amount of literature inmates may possess within their living areas, but the inmate library or chapel should contain religious literature and make it available to inmates.

Religious clothing and headgear may be permitted as long as it is consistent with the security and good order of the prison. In some correctional facilities, inmates are permitted to wear the religious garb or accessories only while in the chapel and participating in religious worship. Religious items such as a medallion pose only minimal security concerns as long as policy controls are in place to ensure that they are inexpensive and small.

Religious Counseling

Inmates may be provided religious or spiritual counseling from the institution chaplain, community volunteer, or religious leader from the community. Through institution visits, phone calls, and correspondence, religious representatives from the community often minister to inmates who were a part of their church or other religious organization prior to incarceration.

Special Rites

Special rites are formal religious ceremonies of initiation such as baptism, confession, or individual communion. Special rites should be performed by the appropriate religious leader with proper credentials. The institution should approve special objects or supplies necessary to conduct such rites.

■ Religious Volunteers

A corps of specially selected, trained, and supervised volunteers can greatly enhance the effectiveness of a chaplain. On the other hand, it is important to remember that volunteers generally have their own agenda, and volunteers may not have the clinical skills to counsel inmates. While volunteers should not be the source of all institution religious programs, they often represent a vital and necessary part of the overall ministry.

The role of the volunteer should be made clear from the beginning. Faith groups that are represented in the inmate population will often provide pastors who are interested in working with inmates from their faith. This will be especially true of smaller groups and nontraditional faiths. For example, Muslim volunteers will work with Muslim inmates, Jehovah's Witness volunteers will work with Jehovah's

Witness inmates, and so forth. Volunteers need to understand that they are prohibited from recruiting converts to their faith. They can, however, work with those who seek them out. Volunteers must understand their role, and the chaplain must ensure that the volunteers operate within appropriate boundaries. Community religious resources are the best sources of volunteers. The chaplain should establish and maintain a good relationship with religious leaders in the community, such as through a community advisory board. Additionally, most states have prison ministry groups and representatives from national groups such as Prison International, Prison Fellowship, and Pious.

ACA standards and sound correctional practice mandate training and orientation of volunteers, including an institution tour. The volunteer should be given materials including a handbook about the institution, its policy on religious practice and confidentiality, information about inmate characteristics and needs, a list of what makes volunteers successful, and a list of "dos and don'ts."

■ Unique Requests for Recognition or Accommodation

Processing religious requests is an important aspect of the chaplain's job. An institution religious request review board and a community advisory board can be very helpful. The institution board might include the chaplain, a social worker, the food service director, the legal advisor, a security officer, and one of the assistant wardens. The board should consider several factors in handling requests:

- Whether the requested accommodation is a basic tenet and required of all of the religion's members
- Whether the inmate meets the religion's requirements for this practice
- Whether the inmate shows good faith in the discussion of a solution and accommodation

Additionally, it may be helpful to learn what other correctional institutions and the community are doing to accommodate certain practices.

Occasionally, prison administrators are requested to make special provisions for inmates who claim to belong to religions or faiths previously unknown, or to make unusual provisions for inmates who adhere to well-known religions (e.g., special dietary requirements for Protestant inmates). Generally it is best to rely on the classic definition of religion: an activity that concerns a person's relationship to a deity, to other people, and to him- or herself. That said, not all requests made in the name of religion will be accommodated. Behavior that threatens or harms others cannot be permitted. Of particular concern within a correctional institution is the creation of a religious organization that includes a hierarchy that would give some inmates authority over others (e.g., one inmate appointing himself the leader, spiritual or otherwise, of a new religion). Also of concern are religions that espouse intolerance or even hatred of persons of particular races or ethnicities, and religions that proclaim the superiority of a particular group. To maintain the security and good order of the correctional institution, such organizations cannot be permitted to practice.

■ Conclusion

The religious programs in correctional institutions should be tailored to the mission and resources of the institutions. Small prisons and those with limited staff and resources may be able to provide only the basic elements, but larger facilities may have the ability to provide well-rounded programs that can affect large numbers of prisoners. Regardless of the extent of religious programming, such programs should be administered in a fair and consistent manner and provide inmates with an adequate opportunity to prepare themselves for return to the community. The chaplain should look to the community for contract chaplains, volunteers, consultation, and support of the inmates' individual faith development.

The responsibilities of institution chaplains have grown over the years, and correctional clergy today are significant members of the management and program team of prisons and jails. The work is important, the opportunities are great, and the challenges are immense.

DISCUSSION QUESTIONS

1. What is the role of prison chaplain?
2. How does a prison chaplain differ from religious providers in the community at large?
3. What are some of the challenges that prison chaplains face?
4. What types of religious programs are often provided to inmates?
5. What have the courts demanded of corrections with respect to the provision of religious programs for inmates?

NOTES

1. B. McKelvey, *American Prisons: A History of Good Intentions* (Glen Ridge, NJ: Patterson Smith, 1977).
2. *Ibid.*
3. *O'Lone v. Shabazz*, 482 U.S. 342 (1987); *Turner v. Safley*, 482 U.S. 78 (1987).
4. *Cutter v. Wilkinson*, 544 U.S. 709 (2005).
5. American Correctional Association, *Standards for Adult Correctional Institutions*, 4th ed. (College Park, MD: ACA, 2000), pp. 155–157.
6. Federal Bureau of Prisons, *Religious Services Report, 2005*, internal agency document (Washington, DC: 2006).
7. A. Walsh, "Church, Lies, and Polling Data," *Religion in the News*, Fall 1998 1, no. 2, available at http://www.trincoll.edu/depts/csrpl/RIN%20Vol.1No.2/Church_lies_polling.htm, accessed August 22, 2007.

Intake, Discharge, Mail, and Documentation   **12**

Jeffery W. Frazier

Chapter Objectives

- Explain the critical nature of receiving and discharging offenders from jail or prison.
- Identify some of the complexities associated with receiving and discharge operations at prisons and jails.
- Differentiate between the various types of records maintained on arrestees and inmates.

The initial reception of prisoners into an institution, whether it is a jail or prison, is a critical process that individuals will remember throughout their confinement. The process begins as the citizen becomes a prisoner and is thrust into a new environment. Previous roles (e.g., father, wife, community leader) are all but eliminated. This identity change affects both the tangible and intangible. The citizen, now prisoner, is relieved of personal possessions as well as personal routines and activities. He or she is now told what to wear; when to get up in the morning; when to eat breakfast, lunch, and dinner; and when to use the telephone or watch television. While some jails and prisons have policies that allow prisoners to keep their personal belongings, which may reduce the humiliation of incarceration somewhat, such policies will do little to ameliorate the substantial impact incarceration has on an individual's life.

■ Intake

Staff who work in intake and booking should be mature, well-trained personnel who are skilled in interpersonal communication. Furthermore, staff should be thoroughly familiar with the institution's policies and procedures, committal documents, confinement orders, and other such documents to ensure there is a legal basis for confining the prisoner.

If not handled properly, the admission process can create undue humiliation and stress that can lead to disciplinary problems along with safety, security, legal, and health concerns. The basic intake process has several important goals:

- To prevent contraband from entering the institution
- To gather the necessary information about the offender
- To orient the offender to the policies and procedures of the institution
- To assess the offender's physical and mental health
- To perform an accurate inventory of the offender's personal property
- To promote personal cleanliness and minimize the risk of infestation or infection

Search

An initial search should be conducted immediately upon the arrival of a new prisoner, preferably in an area that prevents the introduction of contraband into the secure perimeter of the institution. Conducting a search requires tact and diplomacy on the part of the searcher. Furthermore, extreme caution should be exercised during any search due to the close proximity of the prisoner to the searching officer. Upset, aggressive, or agitated prisoners should be given an ample "cooling down" period before a search is conducted, unless doing so would create a greater safety or security concern.

The complete search should be conducted in a private area. It should be performed by a member of the same sex as the person being searched. The officer conducting the search should explain to the prisoner, in a calm and respectful manner, the purpose of the search and the procedure that will be followed. The search should be conducted slowly and methodically, with instructions given to the prisoner throughout the process. If a strip search is required or allowed by policy or law (strip searches may not be conducted on individuals charged with certain crimes), the same procedures should be followed. However, touching the prisoner during a strip search is not necessary and should be avoided. Under no circumstances should a strip search be conducted by a member of the opposite sex. Body cavity searches, if and when necessary, should be conducted only by trained medical personnel in an area that affords privacy. Under no circumstances should a body cavity search be conducted within the view of other inmates.

Gathering Information

It is critical to gather personal information about each inmate, but staff should be cautious not to ask questions related directly to inmates' criminal charges. All of the data gathered will assist classification staff in determining at what custody level the prisoner should be held. Furthermore, the data will help in determining housing assignments. The National Institute of Corrections has made available a number of forms, free of charge, that can assist in the intake and receiving process.

Orientation

Upon intake, all prisoners should also be oriented to the basic rules and regulations of the institution. The arrestee should be given a copy of those rules, and staff should carefully go over the rules, answering all questions that the prisoner may have. It is important that intake and booking staff verify that the prisoner can read and comprehend the rules and regulations before proceeding.

Inventory

In a jail setting, intake and booking usually begin with a complete inventory of the prisoner's personal property. Each item should be carefully noted; the inventorying officer should not omit any items (e.g., nails, staples, washers, and gum wrappers) regardless of how trivial they may seem. All jewelry should be described by color, not type of precious metal or stone. Rings, necklaces, and bracelets should be described as gold or silver in color. A ring, for example, might be described as "one gold in color wedding band containing a single clear stone." All identifying inscriptions should also be recorded. All clothing should be described as thoroughly as possible using sizes, brand names, and any other identifiable markings (e.g., stains, rips, tears). All money should be counted in the presence of the prisoner.

After the inventory, the prisoner should be required to sign a property/money slip indicating that he or she agrees with the inventory list. This slip should then be signed by the inventorying officer. A copy of this inventory list should be given to the inmate, a copy placed with the property, and a copy forwarded to the records department to be filed in the inmate's institutional file. A well-conducted inventory of all personal property will help to reduce or prevent false claims of damaged, lost, or destroyed property.

Health and Psychological Screening

All prisoners should be asked several basic questions about their current health condition, history, and medications. Such screening helps correctional and medical personnel address the individual's personal health needs and protects the health and well-being of others incarcerated within the jail. This screening should also include questions related to the individuals' psychological health, including any history of suicide attempts. The importance of the health and psychological screening process cannot be overstated. The first 48 hours of a prisoner's initial incarceration are the most critical, because this is the period when most suicides occur.

Most health screening forms are divided into two categories: observations and questions. Observations are details that the officer may notice, such as obvious bleeding, open sores or lesions, vermin infestation, intoxication from alcohol or drug use, signs of drug or alcohol withdrawal, convulsions, or seizures. Personal observations can be as detailed as the agency policy dictates. Questions should be asked in an area that affords maximum privacy so that other prisoners do not overhear. Questions should be worded so that the prisoner may respond initially with a simple "yes" or "no" answer. All "yes" answers should be followed up with additional inquiries to determine the exact nature and extent of the problem. For example:

(Q) "Are you allergic to any medications?"
(A) "Yes."
(Q) "What medications are you allergic to?"
(A) "Penicillin."

Responses should be documented. All screening forms should give the intake and booking officer specific directions on what to do with critical "yes" or "no" observations and answers. If an officer notes that an individual is bleeding, for instance, he or she would immediately notify the appropriate medical and supervisory personnel. If the prisoner describes a history of or has positive test results for a highly infectious disease such as tuberculosis, the officer may be required to notify medical personnel immediately and medically isolate the individual from the rest of the population.

Once completed, the health screening report should be placed in the individual prisoner's file. A thoroughly completed health screening form is a valuable tool in preventing frivolous litigation, especially if an individual arrives with multiple superficial cuts and bruises and later claims that he or she was assaulted by jail staff. Further, this report will help medical personnel during their initial medical evaluation of the prisoner.

Showering and Dress

Jails and prisons have different personal cleanliness standards and procedures. However, many jails and prisons require inmates to shower and change into jail or prison clothing. This reduces the likelihood of introducing body lice and other insects into the institution. The change from personal clothing to jail or prison clothing also reduces the possibility of theft, gambling, bartering, and the strongarming of inmates. Further, it reduces the problems that are associated with the laundering of personal clothing.

During the showering and exchange of clothing process, the intake and booking staff should search visually for rashes, cuts, abrasions, scars, tattoos, and so forth, all of which should be documented. At the conclusion of this process, inmates should be given clean linens, towels, and washcloths.

■ Release

The process of discharge from jail or prison is very similar to the intake and booking process. Discharge is a major responsibility—probably one of the most critical assignments that an officer can undertake. It is of the utmost importance that the releasing officer verifies the identity and sentence of the subject being released. Failure to do so could result in the wrong inmate being released or an inmate being released before the complete sentence is satisfied. Further, if a criminal background (i.e., wanted persons) check is not performed, a person who is wanted by a sister agency or state could be wrongfully released.

There are many reasons that an inmate may be released from custody. Special documentation and procedures apply to each case.

- *Personal Recognizance.* Release on personal recognizance is granted by a judge or other judicial officer based on the promise that the subject will appear in court on the scheduled date. Many factors are considered prior to releasing someone on personal recognizance, such as the nature of the crime, the subject's criminal history, and the subject's standing within the community.
- *Bail.* Bail is a specified amount of money, usually established by the court or other judicial officer, that must be presented before the inmate can be released. The money is held by the court to ensure that the subject appears in court when scheduled.
- *Bond.* A bond is something that is posted by a licensed bonding company or bondsman within a state. The subject is released into the custody of the bondsman, who will ensure that the subject appears in court on the scheduled date. If the subject fails to appear, the bonding company will be required to pay the entire bond amount to the court. Most bondsmen charge at least a 10 percent fee of the original bond amount for incurring the risk that the subject might not appear in court.
- *Court Order.* Inmates may be released by the court for reasons such as time served, sentence reduction, or dismissal of charges, or they may be granted a temporary release to attend the funeral of a loved one. When this occurs, the releasing officer should verify that a copy of such order has been obtained prior to release.
- *Time Served.* The inmate has satisfied the conditions of his or her sentence and can be released.
- *Release to Other Law Enforcement Personnel.* In some cases, a subject is released into the custody of other law enforcement personnel who may have pending charges against the subject, or to another agency because the subject has already been tried and convicted of committing a crime and now must satisfy that sentence. When a prisoner is released to another agency, the transporting officer's identity should be verified and a receipt should be obtained indicating that the transporting officer has accepted custody of the prisoner.
- *Release Documentation.* Regardless of the reason for release, the releasing officer should verify that the release documents are in order and properly signed. If the subject is being released because of time served, the computation of his or her sentence should be verified. A "wanted persons" check should be made through the agency's computer system that is tied into the National Criminal Information Center network. Finally, the subject's identity should be verified to prevent the wrongful release of an offender. Release documents should contain, at a minimum, date of release; time of release; reason for release; if released to another agency, the name of the agency and the person to whom the subject was released; the officer who released the subject; and a description of all personal property that was released with the subject.

▨ Mail

The First Amendment to the United States Constitution gives inmates the right to send and receive mail. However, prisons and jails can place reasonable restrictions on these rights. The Supreme Court established that a restriction on inmate mail is acceptable "if [the restriction] furthers an important or substantial governmental interest; and if the incidental restriction on alleged First Amendment freedoms is no greater than is essential to the furtherance of that interest."[1]

Thus, local regulations must show that a regulation that authorizes mail censorship furthers an important or substantial governmental interest such as security, order, or rehabilitation. In order to establish appropriate institution rules that comply with federal law, it is very helpful to delineate two types of inmate mail: legal and social.

Legal Correspondence

Legal or official correspondence is an inmate's correspondence with police, probation and parole officers, judges and attorneys, and so forth. Official correspondence should be inspected in front of the inmate to ensure that contraband is not enclosed. But it must not be read—inmates have a right to confidentiality with their attorneys and other public officials. All official correspondence, both incoming and outgoing, should be documented in a logbook. For incoming mail, the name and address of the sender and the date received should be logged. There should also be a place for the inmate to sign that he or she received the official mail. For outgoing mail, the logbook should list the date mailed and the name of the addressee.

Social Correspondence

Social correspondence is the personal letters of a prisoner to and from family and friends. Personal correspondence can be inspected outside the presence of the inmate; however, as with official mail, it should not be read unless the legal tests of the previously cited Supreme Court decision can be met and staff follow the basic requirements outlined in that case. Many institutions open all personal correspondence, inspect for contraband, and remove money orders so that the inmate's personal account can be credited. The mail is then forwarded to the inmate.

■ Documentation

Today, it is important that each institution maintain accurate, up-to-date records on all inmates, from their reception into the institution until their departure or ultimate release. It is only through sound records management that a foundation can be created to protect the staff and agency from inmate litigation for alleged violations of constitutional rights.

While in recent years the courts have been willing to entertain inmate complaints, they are not always willing to interfere with the operations of an institution unless there are clear constitutional violations, such as living conditions that are extremely barbaric and inhumane. There are a number of routine records that all institutions must maintain; others are determined by the institution's own needs or the statutory requirements of the locality and state.

Mandatory records are as follows:

- Admission and release records
- Medical records
- Disciplinary records
- Inmate grievance records
- Visitation records, both personal and professional
- Criminal justice system records (e.g., court orders and time computation records)
- Personal property records
- Inspection records that reflect the conditions of the institution, both from a security standpoint and from a life, health, and safety standpoint
- Logs that reflect all activities within the institution (e.g., meals served, recreation given, counts conducted, medication dispensed, mail delivered)

Documentation and records management provide administrators with data that can be used in making policy decisions, forecasting trends, performing staffing analyses, evaluating the climate of the institution, and projecting future budget needs. As the old saying goes, if it is not documented, it did not occur. Therefore, documentation is the first defense in preventing litigation. Once litigation has been filed, proper, complete, and up-to-date records will greatly improve an institution's chance of prevailing in the courts.

■ Conclusion

The reception and discharge of individuals is a critical process that requires the attention of responsible staff with excellent interpersonal communication skills and an eye for detail. It is the initial reception process that usually sets the tone of the prisoner's behavior during his or her stay at the institution and acts as the first physical barrier to contraband in the institution. The information-gathering phase of the receiving process provides the institution with a personal history of the offender, which ultimately will be used to assist the classification department. Finally, through the promotion of cleanliness, institutions can better identify infectious disease and reduce the infestation of body lice.

During the discharge of individuals from confinement, the releasing officer must be extremely cautious to prevent the unlawful or wrongful release of prisoners. All release documents must be examined for authenticity. The releasing officer must indicate the date and time of release, the reason for the release, and a description of all property released. If a prisoner is released to an individual from another agency, the officer must indicate the name of the agency and the individual. Wanted persons checks must be performed to verify that the individual being released is not wanted by another agency.

Inmates have the right to send and receive mail, but the institution can place reasonable restrictions upon that right. Only official mail must be opened in front of the inmate. An institution can censor an inmate's mail if it can show that the censorship furthers an important or substantial government interest and meets certain established legal minimum standards.

Finally, it is through the maintenance of accurate records documenting all aspects of an inmate's stay that institutions are able to reduce the likelihood of litigation and increase their chance of prevailing in court if litigation is filed. Accurate records also provide managers with the necessary data to forecast trends, prepare budgets, and make policy decisions that affect an institution's entire operation.

DISCUSSION QUESTIONS

1. In what ways are the receiving and discharge operations at jails and prisons critical to the operation of those facilities?
2. What are some of the complexities associated with institutional operations as covered in this chapter?
3. What are the principles that underpin the policies and practices for processing mail for inmates?
4. What types of information about arrestees or prisoners are maintained by jails and prisons?
5. What roles do mail processing and documentation management play in the overall operation of a penal institution?

ADDITIONAL RESOURCES

American Correctional Association, http://www.aca.org

American Jail Association, http://www.aja.org

C. Cripe, *Legal Aspects of Corrections Management* (Sudbury, MA: Jones and Bartlett Publishers, 1997).

P. Johnson, *Understanding Prisons and Jails: A. Corrections Manual*, 2nd ed. (Jackson, MS: Correctional Consultants, Inc., 1991).

National Institution of Corrections, *NIC Jail Resource Manual*, 4th ed. (Kents Hill, ME: Community Resource Services, Inc., 1989).

National Sheriffs' Association, *Jail Officers' Training Manual* (Alexandria, VA: National Sheriffs' Association, 1980).

J. Palmer, *Constitutional Rights of Prisoners*, 4th ed. (Cincinnati, OH: Anderson Publishing Company, 1996).

NOTES

1. *Procunier v. Martinez,* 416 U.S. 396, 94 S. Ct. 1800 (1974).

Food Service

13

Lavinia B. Johnson

Chapter Objectives

- Explain the training requirements for food service departments and outline the process of food preparation and storage.
- Describe the main security challenges posed by food service operations in prisons and jails.
- Explore the various methods used to serve meals to inmates.

In no other setting do food service personnel face the demands and conditions found in prisons and jails. The kitchen is one of the key areas in the prison; if personnel do not do their jobs correctly and on time, negative reactions from the inmate population are likely to occur.

Inmates look forward to their meals. For the inmate, eating is a major diversion from the stresses of prison life. Because of the major role food plays in the inmates' daily existence, the food service operation is subject to more scrutiny than most other aspects of the facility. Institutions' senior management and the public tend to hear about food service in correctional facilities only when there is a problem. Private food service contractors operate the food service area in many prisons and jails, but with or without a contract, the same standards apply to all food service providers. The food service staff has to ensure that meals taste good, that meals meet appropriate nutritional requirements, that portions are appropriate, and that sanitary conditions are maintained.

Approximately 2.3 million people are incarcerated in the United States today.[1] Because of the large number of inmates to be fed, the cost of providing quality food service to inmates nationwide is substantial. According to the Bureau of Justice Statistics, prison food service costs exceed $1 billion each year.[2] The average amount spent on an inmate's daily meals varies by state (see **Table 13–1**).

Table 13-1 **The Cost of Feeding Prisoners**
In 2004 the Department of Justice's Bureau of Justice Statistics reported on state spending for prisons in 2001, including how much states spend daily per inmate on food service. The lowest and highest state averages are listed below with the national average: 1: North Carolina $0.52   2: Alabama $0.72   3: Mississippi $0.81   48: Maine $5.03   49: Washington $5.68   50: Pennsylvania $5.69   National Average: $2.62
Source: J. Stephan, *State Prison Expenditures, 2001* (Washington, DC: Bureau of Justice Statistics, 2004), available at http://www.ojp.gov/bjs/pub/pdf/spe01.pdf, accessed August 21, 2007.

■ Extensive Training Requirements

The combination of external demands by legislators, law enforcement officials, and citizens' rights groups and internal demands of administrators and inmates have placed heavy pressure on corrections food service departments. Today, only the most highly trained professional can direct an operation that meets these demands.

The person responsible for this department must fully understand planning, preparing, and serving nutritious meals under sanitary and safe conditions. In addition, the food service manager and staff members must have highly developed interpersonal skills and the ability to act as role models for and trainers and supervisors of inmate workers. The effect of the food service staff on the correctional population is a dramatic one; effective management of this department positively influences the overall function of the institution. The alert food service professional is acutely aware of the potential for disruptive action from dissatisfied inmates and the calming effect a good food program can have on both the inmates and the custodial force.

■ The American Correctional Food Service Association

The American Correctional Food Service Association (ACFSA), formed in 1969, works to enhance, represent, and promote the correctional segment of the food service industry. The organization encourages standards of excellence and professionalism among its members to enhance food service operations in prisons. ACFSA provides education seminars, tours of correctional food service operations, written educational materials, and opportunities to discuss issues of common interest and establish a network among correctional food service professionals.

The ACFSA operates a certification program intended to raise the professional standards of food service personnel. Individuals who meet the rigorous standards for certification are designated Certified Food Service Professionals, and may use this designation on letterhead, business cards, etc. The certification pertains to individuals only and not to institutions wth which the person may be affiliated.[3]

Historical Background

Early in U.S.history, punishment did not rely upon confinement. Rather, there were many physical sanctions: flogging, death, slavery, and exile. Toward the end of the 18th century, the concept of confinement as punishment began to gain prominence; with incarceration came the need to feed prisoners. Early prisons in the United States were private, and in some, inmates had to pay for their food, bedding, and water. In other prisons, inmates had to earn their way. Inmates were provided with the minimum food necessary to sustain life. Porridge, bread and water or beans, stew, and bitter coffee were typical menus. This attitude persisted even as state and local jurisdictions began to develop jails and prisons.

Although reform actually began in 1790, it was not until the early 1970s with the major uprising and riot at the New York State Prison at Attica and subsequent disturbances across the country that the public became actively concerned about the state of correctional facilities. As the courts started to abandon their previous hands-off policy, judges found that some correctional facilities did not meet basic standards for human decency. Inmates and concerned citizens groups took some institutions to court and won.

Lacking established professional standards to guide them, the courts attempted on a case-by-case basis to define inmates' rights and to impose standards for correctional institutions. Unfortunately, these standards were often inconsistent and unrealistic. To bring order out of chaos and to guide correctional administrators, in 1977 the American Correctional Association (ACA) developed a set of standards that included food service.[4]

ACA standards provide a general picture of how a well-organized food service department operates. In almost all institutions, a full-time staff member who is experienced in food service management supervises the operation. This individual is given the resources, authority, and responsibility to manage the department effectively in terms of both labor and financial resources. The administrator of the food service department ordinarily supervises all the food service and related staff such as dietitians, bakers, and butchers. Depending on the system, a number of correctional officers also may be assigned directly to the department. More commonly, however, correctional staff are posted in the preparation and dining area but still work for the custody department.

Food Service Facilities and Equipment

Food service facilities and equipment vary from institution to institution. Some facilities have state-of-the-art equipment. Other institutions have kitchens where the equipment and space are outdated. In some remote local jails, one still could find a live-in sheriff or deputy whose spouse cooks for the inmates from the same kitchen in which she prepares meals for her family.

The most common serving method in today's prison systems is the cafeteria system, where inmates are systematically fed three meals each day. Some of the facilities serve on open lines, some on blind lines where the inmates cannot see who is serving.

In many jail facilities and some prisons there are no cafeterias; meals are preplated in the kitchen and sent to the housing areas on plastic insulated trays, a hot/cold cart, or a system that utilizes rethermalization equipment. Rethermalization involves microwaving food platters to boost the temperature. This term also refers to new technology that involves quick-chilling cooked meals and then reheating the food in a special thermalization cart in the satellite feeding area.

■ Menu Preparation

ACA standards for prisons and jails require that there be some form of menu preparation in advance. For jails, one week of advance preparation is required. The jail standard is shorter than the prison standard because some jails have very limited storage and it becomes difficult to maintain on-hand supplies for scheduled meals. For prisons, the 28-day menu rotation cycle is the most commonly used, although the Federal Bureau of Prisons uses a 35-day cycle. The menus are generally planned by the food service manager and/or the dietitian. If the dietitian does not actually plan the menu, then a registered dietitian should review it to ensure compliance with applicable nutritional standards.

All menus are planned according to the recommended dietary allowances of the National Research Council of the National Academy of Sciences, which is a national authority that recommends a balance of specific food groups. Caloric requirements will vary with sex, age, and general activity level of the inmates. Daily caloric levels usually range from 2600 to 3000 within an institutional setting, with an average of 2900 calories.

Menus should reflect the inmate population's cultural and ethnic preferences as well as provide for their medical and religious needs. Many jurisdictions actually ask their inmate population for their food preferences. These surveys are generally completed once each year.

■ Food Supplies and Storage

Food should be of the best quality possible within the institution's budget and of sufficient quantity to guarantee a wholesome diet. Available sources of food depend largely on the nature and location of the institution. Common sources are state purchasing warehouses, state contracts, local wholesale food distributors, and local vendors.

Some prison systems have food and farm operations that provide meat, vegetables, milk, grains, and other items for their institutions. Items such as cakes, pies, and breads can be purchased or prepared locally. Fresh vegetables and fruits are purchased seasonally, and availability depends on the region. All food should meet or exceed government standards. No longer are home canning or uninspected meat slaughtering acceptable.

The delivery and storage system should ensure that food supplies are fresh and delivered in a suitable condition. All incoming food not immediately used or processed in some way should be stored properly to prevent spoilage or waste. Proper storage should be available immediately for perishables such as meat, milk, eggs, and fresh vegetables and fruits. Semi-perishable foods such as canned goods may be kept in temperature-controlled storage rooms.

Shelf goods should be stored at 45°F to 80°F, refrigerated foods at 35°F to 40°F, and frozen foods at or below 0°F. Each refrigerator or walk-in storage unit should have a thermometer on the door of the exterior wall so that staff can check these temperatures easily. Temperatures should be checked and recorded as required.

■ Food Preparation

Food is usually prepared according to a recipe system that the jail or prison adopts. This recipe system ensures that the quantity and quality of meals are uniform. For this purpose, many institutions use the Armed Forces Recipe Cards as guides for food preparation. Sanitation measures in the kitchen are absolutely critical. Each institution must have a daily cleaning and inspection system that ensures that the food preparation, storage, serving, and dining areas are sanitary. This is far more than a cosmetic issue. The health of every inmate and staff member in the institution hinges on the cleanliness of the food service area.

Each institution should consider having a hazard analysis critical control point system in place to ensure that proper food-handling procedures are being followed for receipt, storage, preparation, holding, and serving of foods. This is an optional program but may help ensure that good personal habits and cleanliness are a focus for all food service staff and inmate workers. Adequate hand-washing facilities should be provided in the kitchen area. Clean uniforms and aprons should always be available, and food handlers should be required to wear head coverings, hairnets, and beard guards while cooking or serving food.

■ Special Diets

Medical diets should be made available to inmates based on medical authorization only. Each prison system should know the medical diets that it will prepare. Diet orders should be specific and complete, furnished in writing to the food service manager. Medical diets should be planned so that they are as close to the main menu as possible.

Prison and jail systems also have to deal with specific religious beliefs that require inmates to eat or not eat certain food. For example, Islamic and Jewish inmates are forbidden by their religions from consuming pork or pork products.

Because of this, many institutions now have a pork-free menu. Religious diets are ordinarily approved by a chaplain. They should be specific and furnished in writing to a food service manager. In many systems, they are reviewed periodically. If a staff member observes an inmate approved for a religious diet eating from the regular bill of fare, the staff member should report this to the chaplain so that the religious diet status of the inmate can be reviewed.

A separate area in the kitchen under staff control may be used to store all special diet trays, or they may be kept in a single hot cart behind the line for issue under staff direction. Many institutions use a diet card or pass system to ensure that only authorized inmates receive these meals.

■ Meal Service

When a new jail or prison is under construction, an important consideration is what type of serving system will be used. The degree of staff supervision required and the institution design will determine the system used for serving. Food can be served cafeteria style or preplated and carted to the inmates. Both systems have challenges.

Gathering a large number of inmates together in a cafeteria presents a security risk under any circumstances. The dining rooms create a potential site for serious disturbance and other incidents. As a result, in a cafeteria setting it is critical that correctional staff enforce an orderly system of food lines and seating as well as portion and utensil control.

To the extent possible, the dining rooms should be designed to enhance the attractiveness of the mealtime atmosphere. Meals assume a magnified importance in inmates' daily routine and are important to staff. Thus, the condition and cleanliness of the kitchen and dining area can influence an institution's entire atmosphere.

Food should be served as soon as possible after preparation and at appropriate temperatures. Temperatures are ordinarily maintained by keeping the food in warmers of some type, either cabinet or pan style. Direct service is usually from a steam table or some other type of cafeteria-style warming equipment. Food distribution should be supervised at all times. Frequently, inmates serving food will take advantage of an officer's temporary absence to be overly generous with their friends or to not give other inmates their entitled portions.

Selection of eating utensils should be dictated by the type of population confined in an institution. Many institutions now use highly durable, washable plastic utensils. Control of eating utensils can be maintained by requiring inmates to dispose of them in a carefully positioned and supervised receptacle when inmates drop off their trays.

■ Dining Room Routine

Inmates should be given enough time to wash before eating. Inmates working outside or in other active occupations should be allowed to change clothing before entering the dining room. Inmates must be fully clothed (in their issued uniforms) while in the dining room. Staff supervising the entrances should enforce the dress and smoking codes before inmates come into the area. Random inmate searches for weapons and contraband should be performed.

The dining rooms should provide normal group eating areas and permit conversation during dining hours. Whenever possible, there should be open dining hours to reduce the traditional waiting line. Many facilities have eliminated forced seating based on housing unit, shop assignment, and so on. Serving and dining schedules should offer a reasonable amount of time for inmates to eat. Tables and chairs should be arranged for good traffic flow and supervision.

Line cutting can become a problem in a crowded dining room. Close staff supervision can deter this activity and prevent major confrontations. The inmate dining room should not be used as a shortcut to other areas, and inmates should not be allowed to linger in the dining area.

■ Unit Dining

Unit dining is used primarily in jails or regional adult detention centers or during prison lockdowns. The meals are preplated in the kitchen and sent to the housing areas. This can be done by using an insulated tray, hot/cold carts, or a rethermalization system. Specific correctional management principles apply to all types of unit dining:

- All food carts should be thoroughly searched by staff for contraband being sent into the unit from cooperating inmates in the main kitchen area.
- Inmates should not be used to serve food to other inmates in segregation status. This is a prime opportunity for inmates to try to pressure and manipulate other inmates, for instance by tampering with the food of unpopular inmates. It can also provide opportunities to pass contraband into the unit.
- Inmates must be required to give back all utensils and other items on the food tray. This is done not only for the safety of the staff, but also because keeping food in cells can attract vermin and insects.
- Food service supervisory personnel should tour the locked units regularly during mealtime to ensure that inmates are being served properly, and meals are at proper temperatures: above 140°F for hot foods and below 41°F for cold foods.

■ Commissary Options

Home-cooked foods are not allowed in an institution. However, a commissary or inmate store is usually available in most locations for inmates to purchase a wide variety of discretionary food and other items.

Selecting the articles to be sold in a commissary requires careful study. Most commissaries limit the selection to snacks and light foods that are not in conflict with the regular food program.

■ Supervision Issues

Traffic control in and out of the food service area is important. The unrestricted movement of inmates not only presents an accountability problem but also permits theft of food items and pilferage of other contraband from the kitchen. The more traffic there is, the harder it will be to detect these problems. For that reason, the kitchen area should be out of bounds for all non-kitchen workers, and correctional staff should enforce that rule.

Also, controlling items coming into the kitchen through the loading dock is always a concern. The possibility of contraband coming in through regular food shipments from fixed sources of supply is quite high. Therefore, each institution should have a specific system for searching vehicles, loads, and drivers moving supplies into the compound.

Trash control is another security issue that pertains particularly to escape attempts. Any trash truck or dumpster load should be kept locked in a sally port area through one or more counts, to be sure that inmates are not hidden inside. Probing and other search techniques also may be used. The same concerns and search techniques should be used for boxes or containers of food prepared inside and sent to satellite camps or other locations; inmates can be hidden in these containers.

Trash compactors are another avenue for escape. Inmates have successfully (and at times unsuccessfully) fabricated skeletal frameworks of crushproof containers to hide in to escape via a dumpster. Using a sally port is the safest way to eliminate this avenue for escape attempts.

■ Controlling Kitchen Tools

Tool control in the kitchen consists primarily of knife control. However, in facilities with butcher shops, movement of saws and other tools will also need to be controlled. In most facilities, kitchen tools are stored in a locked cabinet in a secured area.

Inside the secure cabinet the tools are hung on hooks that are painted with shadow of each tool so that it is easy to see if any are missing. When an inmate or staff member checks out a tool, a durable metal tag with that person's name or other identifier is placed on the hook so that it is clear who has the item. A written inventory of all items in the cabinet should be kept in it, and at each shift change the responsible staff member should check the inventory and initial the list. This inventory also should be checked jointly with a correctional staff member on a regular basis—not less than monthly.

Equipment should be constantly checked to make sure no parts are missing. Inmates are very innovative when it comes to making weapons.

■ Yeast, Sugar, and Extract Control

Yeast, sugar, and extract control is another major concern in an institution. Yeast and sugar can be used to make "home brew," also referred to as "mash" or "pruno." Also, extracts have alcohol in them, so inmates may drink these instead of alcoholic beverages. These items should be kept under lock and key with a strict inventory maintained.

Inmates with unusual amounts of sugar or fruits, or even small amounts of unbaked bread in their possession should be viewed with suspicion as potential brew makers. Even if the institution does not have a possession limit on these food items, an officer encountering them in large amounts should refer the matter to a supervisor for advice.

■ Food as an Incentive

Staff should never be allowed to use food as payment for work or as a special privilege. However, when inmates work in outlying jobs or in odd shifts, it often is necessary to provide lunches or extra food to cover the shift portion of the day. Some institutions take into account the fact that some work assignments are more physically demanding than others and provide extra rations.

■ Conclusion

Food service operations are very important to an institutional routine, and positive operations are contingent upon quality staff, effective training, and vigilant supervision. Nutritious and flavorful meals served in a pleasant and safe environment are not possible in a correctional setting without adequate resources and senior-level management support. The food service staff must observe proper nutrition, cost controls, security, and supervision practices in the food preparation area. The distribution of food has to be fair, and teamwork is essential between food service staff and security staff in order to ensure a smooth operation. A well-run food operation will greatly enhance the morale, safety, and security of the prison or jail setting.

DISCUSSION QUESTIONS

1. What are some of the relevant factors to consider in developing menus for prison or jail food service operations?

2. What are the various options for how to serve meals to inmates in prison or jail?

3. What are the advantages and disadvantages of each meal service option?

4. What are some of the major security risks associated with food service operation in prisons or jails?

5. What role does food service play in the overall operation of a penal institution?

ADDITIONAL RESOURCES

American Correctional Food Service Association, http://www.acfsa.org

NOTES

1. P. Harrison and A. Beck, *Prisoners in 2005* (Washington, DC: Bureau of Justice Statistics, 2006).

2. J. Stephan, *State Prison Expenditures, 2001* (Washington, DC: Bureau of Justice Statistics, 2004), available at http://www.ojp.gov/bjs/pub/pdf/spe01.pdf, accessed August 21, 2007.

3. American Correctional Food Service Association, available at http://www.acfsa.org/, accessed August 14, 2007.

4. American Correctional Association, *Foundation/Core Standards for Adult Correctional Institutions* (Washington, DC: St. Mary's Press, 1989).

Financial Operations

Beverly Pierce

14

Chapter Objectives

- Describe key concepts applying to the fiscal management of penal institutions.
- Name at least three approaches to prudence in the utilization of public funds.
- Explore the issues that can cause balanced budget failure.

Few correctional administrators have a professional background in the financial management of institutional operations. Traditionally, wardens and jail administrators have learned their craft through the apprenticeship system and earned incremental promotions up the ranks of the institution. Upon reaching senior management, most individuals are fiscally unskilled and totally unprepared for the significant responsibility of jail or prison financial management.

The cost of running the country's prisons is becoming a major issue in today's world of expanding institutional populations. As more and larger prisons and jails become necessary, these institutions garner more of the public and media attention. As budgets take up an increasing amount of legislators' discretionary allotments, the issue of confinement can and will become the center of attention for influential third parties.

■ Understanding Financial Operations

The budget is best explained in three phases: budget development, budget execution, and budget oversight. Budget development is the beginning of the cycle—

the formulation of a funding request. Budget execution is the administration of those funds through expenditures and distribution. Budget oversight is the implementation of systems and internal controls that ensure that funds are used in a manner consistent with budgetary goals while protecting the integrity of the disbursements.

The results of making uneducated financial decisions can challenge the best-intentioned administrator. Poorly thought-out decisions can lead to disastrous reactions from staff, the inmate population, the governor's office, and the state legislature. In private corrections, the bottom line is critical to the senior administrator's survival.

The following problems can result from poorly made financial decisions:

- Allowing expenditures to be made that cannot sustain public scrutiny
- Failing to manage program funds within funds allocated
- Failing to understand the fundamentals of budgeting
- Establishing insufficient internal controls to prevent fraud, waste, and abuse
- Failing to provide adequate oversight for early detection of budgeting problems

Financial management of a multimillion-dollar institution budget requires specific knowledge and abilities such as understanding the concept of a budget, analyzing and comparing data, and differentiating between bona fide requirements and incidentals.

Mastering financial management means modifying many already-acquired management and administrative skills to apply to the financial arena, but nothing will serve the new administrator better than good old-fashioned common sense. Simply stated, do not spend more money than was allocated in the budget. Institutional managers always must consider the public perception of their financial decisions. A penal administrator should not spend money on programs that appear to provide inmates with a better quality of life than the general public enjoys. Public administrators must also be cautious not to spend money for buildings and landscaping that are so aesthetically pleasing that the correctional facility looks like a country club. Public stewardship also mandates that institution administrators only spend money on approved purchases, using the correctly appropriated funds. Every dollar spent should be able to sustain public scrutiny.

An administrator must have a planning staff. Staff should include both program and financial managers. These individuals are responsible for analyzing past expenditures and factoring in future budget requests to create a budget that will cover necessary operations and projects over the planning horizon (future fiscal years). Budget-tracking staff ensure that money is spent as planned and appropriately conserved throughout the budget year. Anyone assigned budgeting or purchasing responsibilities should be required to participate in a financial training course. Program managers also need to understand and apply sound financial principles to maintain budget accountability on a project basis. Clearly defined budgeting expectations and spending parameters should be established and documented in the institution's policies and procedures. Finally, the chief executive officer of the facility must insist that common sense and good public stewardship are exercised by any staff member given signature authority to expend funds.

■ Political Influence

Correctional staff should never forget that typically public funds are utilized to create, operate, and otherwise manage all correctional facilities. Even if an institution is built and/or managed by a private for-profit correctional company, the revenue stream for the facility originates with public funding—taxes. Because the institution's financial support originates with local, state, or federal government, these entities also exercise control over institution budgets.

Correctional staff also should understand the power of publicly elected officials. City councils, state legislatures, and the U.S. Congress are composed of elected representatives of the people. In a democratic republic, the public elects representatives to make decisions on their behalf. The decisions of those representatives then affect the governance of all public institutions, including prison and jail facilities. These elected representatives are expected to make logical and informed decisions as to how public funds should be used. They are also charged with making and changing policies that specify how public money is to be spent. In summary, elected representatives have been given the power and authority to provide broad policy guidelines within the criminal justice system and to make sure that an institution's financial decisions reflect those policies.

Virtually every aspect of correctional management is subject to law, policy, guidelines, rules, or other controls that conform to the broad policy philosophy of the government representatives elected to serve the people. No correctional facility can create a budget in a vacuum, free from government influence or control. All must be aware that correctional leaders are controlled by politicians who manage the policy priorities and purse strings of government operations. To restate the obvious, all policy and funding decisions are political.

■ Budget Development

In government budget cycles, performing strategic planning and properly preparing budget justifications are vital functions of the senior institution executive. Many state budget cycles require multiple-year lead times for budget submission, so anticipating and planning future requirements are critical. Per capita costs of inmate management (day-to-day costs associated with inmate housing, security, programs, and food and health expenses) must be factored into the daily cost projections along with adjustments for anticipated inflation. Capital outlay (equipment, furniture, machinery) must be included, as well as capital improvement expenses (maintenance and new construction).

Poor planning inevitably leads to crisis management—a mode of operation that can quickly sabotage a balanced budget. All too often, a lack of planning translates into less value for the dollar. Urgent or emergency purchasing of goods and services sacrifices price savings for shorter delivery times. Because the financial resources for prison operations are precious, every effort must be made to allow the maximum time practical to find and negotiate the best values.

Planning is critical in the budget development cycle and can make the difference between success and failure in maintaining a balanced budget.

Budgeting for Human Resources

The most important and expensive part of the institution budget is human resources—the institutional staff. Salaries make up the greatest percentage of institution costs, and agreeing on the number of staff members needed to efficiently, yet economically, operate each institutional department can be difficult. Line staff members and union officials always want to increase rosters. Senior administrators, constantly under pressure to reduce operating expenses, seek to do more with less. Roster management of operations in larger institution departments (such as the correctional security staff) can be a full-time job; supervisors must cover all posts while providing days off, sick leave, and training time, as well as loaning out staff for special projects.

Inevitably, managers must use overtime hours to cover all critical areas of the jail or prison with reasonable supervision. Obviously, the payment of unplanned overtime can destroy a carefully balanced financial plan. Caution should be exercised when attempting to implement cost containment measures by reducing the correctional security personnel. These efforts are often nothing more than smoke and mirrors. If staff rosters are reduced so drastically that overtime is the only alternative to handle special circumstances, no real dollars are saved. Overtime is a variable factor in budget planning and an expensive factor in budget administration.

In the event that major changes to human resources budget are anticipated, such as a new plan to avoid the use of overtime in the new budget year, it is important to provide adequate planning and lead time for those who must implement that program change. An overtime policy change can cause large scale ramifications to many personnel within the institutional setting. Morale can decrease if staff security is affected negatively. Ironically, morale can also take a negative turn if a decision is made to curtail the use of overtime, because some staff members depend on this additional income.

Although employee salaries constitute the largest portion of a budget, they do not have to be difficult to project, provided the staffing pattern remains consistent and there are no policy changes that significantly affect institutional operations. Usually, salary increases are negotiated, or at least predictable, prior to the budgeting cycle. Financial staff can provide reasonably accurate salary projections by taking the current work-year cost of each employee and adding a pay increase. A typical work-year for a full time employee consists of 2080 hours of paid employment—that is, 52 weeks times 40 hours per week. A specific amount of funding to support overtime, incentive awards, and premium pay should be included in the financial plan. Once staff have calculated salary projections, they have developed 60 to 70 percent of a budget.

Budget and Planning Committees

The establishment of a budget and planning committee is a key step in building a solid financial plan for the institution. Such a committee ideally would consist of a financial manager, subject matter experts at the department head level, the warden, and administrators. The subject matter experts should provide the rationale behind the funding requirements needed to operate their departmental programs and properly justify any need for increased funding levels. Conversely, they should be able to explain the reason for requesting less funding. In a zealous attempt to employ cost containment measures, staff may underestimate requirements, compromising the integrity of the entire institution's budget. Compensating for shortfalls in underfunded areas can hinder the institution's ability to meet budgetary goals.

The committee should serve in an advisory capacity to the chief executive officer and should meet periodically throughout the fiscal year. The committee should make recommendations to reallocate funds as needed to handle surpluses and deficits in different program areas.

Administrators must understand the political environment in which their agency operates, because changes in this environment may require the institution to ask for additional funding or alter the budget execution process. The budget should incorporate enough flexibility to allow staff to shift emphasis as missions change. Legislative changes in areas such as sentencing guidelines, environmental issues, accessibility for persons with disabilities, and life safety issues can influence funding requirements.

Often financial managers must consider more unknowns when formulating the operations portion of the budget. Prior year spending is the best starting point for developing the operations budget. Financial staff can provide an estimate of anticipated increases based on historical data and the consumer price index. After considering routine operations, staff must factor in any new requirements. A history of past obligations, anticipated new costs, and projections of a funding source for uncontrollable or unanticipated expenses (e.g., utility increases, institution emergencies, and catastrophic medical care events) should be well-substantiated.

Multi-year formal contracting for purchasing goods and services can be a very useful budgeting instrument. Contracts ensure competitive pricing—the best value for the dollar—and can guarantee prices. Using contracts can increase the accuracy of the budgeting forecast, as it removes some of the guesswork. Warranties and maintenance agreements for equipment can also be a good budgeting tool. They have recognized costs and can reduce unanticipated expenditures.

One of the greatest threats to the integrity of a budget request is the organization's own financial philosophy. Too often, staff believe that if they do not use all approved funding in one budget cycle, it will not be appropriated in future budgets. This belief perpetuates wasteful spending and discourages cost-containment initiatives.

Finally, to make a successful budget request, one must present a clearly defined budgeting goal. Whether the goal is cost containment, enhancing programs, or renovation, staff need to know what they want to achieve with the funds requested. Only when staff have identified the budgeting objective do they have a solid foundation for the request.

■ Managing the Institutional Budget

Once a budget has been approved and funded, the institution is accountable for managing that budget. Laws, statutes, and administrative rule generally govern most institutional financial matters. It is therefore critical that the senior administrator know the spending limits of the institution budget. These limits are documented in a legislature-approved budget that has been certified by the corrections department. The individual with signature authority to expend jail or prison financial resources must know the limits that apply to him or her along with the rules governing the transfer of discrete amounts of funds between fund categories (salaries, operations, capital outlay, and capital improvement) and ensure that all expenditures are reasonable and justified. **Table 14–1** shows the fiscal year 2001 budgets for adult facilities throughout the United States.

The management of planned expenditures is a dynamic process. The ebb and flow of jail and prison management requires some degree of flexibility in financial management, so it is critical that the senior administrator be prepared to shift funds between cost centers to the extent that the law or regulation will permit.

Additionally, senior administrators must ensure that there are internal controls that effectively prevent fraudulent or deficit spending. A checkpoint can be as simple as examining the percentage of the budget expended as a proportion of the full budget cycle. For example, if staff have used 70 percent of the yearly budget and are only halfway through the budget cycle, there should be a reasonable explanation. Sometimes contracts fees covering the entire budget cycle must be paid in advance, which can skew a given checkpoint. However, department heads should be aware of such situations and be able to explain any departure from what was planned.

Financial staff should be on the lookout for any invalid obligations, as they will distort the budget picture by overstating expenditures and underestimating available funds. Invalid obligations are funds that were overestimated for the purchase of goods or services and then not deleted in the accounting system after payment. This situation occurs when cost estimates exceed actual expenses. Often, accounting staff do not know when an order is complete and fully invoiced. Unused funds associated with the order or contract remain encumbered. For example, this often occurs with medical expenses. Frequently, the treatment or procedure is different than expected or even deemed unnecessary after the obligation has occurred. The opposite situation can arise when funds are encumbered for a procedure or service that was more expensive than originally planned; this can cause deficit spending or an overstated budget balance.

Reducing the chance of budget errors due to invalid obligations requires the input of subject matter experts. The program manager must review the open obligation records in the accounting system to verify their accuracy. When searching for explanations for budget shortfalls or surpluses, invalid obligations are a good place to start.

Early detection of disparities in the budget is critical to an administrator's ability to take corrective action. Identifying potential surpluses in one area is equally important if staff need to compensate for budgeting shortfalls or unforeseen fund expenditures in other areas.

While some correctional administrators may argue the point, most believe there are truly very few large-scale unforeseen expenses associated with prison and jail management. There are some obvious exceptions: natural disasters, inmate disturbances, and catastrophic medical care for inmates. Special contingency funding must always be set aside to prevent such events from breaking the bank and damaging an administrator's career.

Table 14-1 Actual Expenditures, 2001

Fiscal Year 2001 Correctional Agencies' Actual Expenditures

	Capital	Operating	Total
Alabama	$7,300,000	$238,500,000	$245,800,000
Alaska		$167,928,500	$167,928,500
Arizona	$5,383,500	$569,580,500	$574,964,000
Arkansas	$7,203,864	$201,493,630	$208,697,494
California	$154,000,000	$5,240,635,000	$5,240,635,000
Colorado	$31,536,879	$468,578,857	$500,115,736
Connecticut	$315,982	$493,635,298	$493,951,280
Delaware		$181,609,700	$181,609,700
District of Columbia	$1,731,337	$212,207,000	$213,938,337
Florida	$47,588,953	$1,344,858,986	$1,392,447,939
Georgia	$9,403,221	$914,101,287	$923,504,508
Hawaii	$938,000	$121,431,110	$122,369,110
Idaho	$2,952,500	$88,854,800	$91,807,300
Illinois	$188,420,000	$892,537,400	$1,080,957,400
Indiana	18,232,336	$385,632,662	$403,864,998
Iowa		$253,485,570	$253,485,570
Kansas	$12,319,520	$188,184,572	$200,504,092
Kentucky	$17,610,400	$268,619,100	$286,229,500
Louisiana	$19,693,878	$412,098,955	$431,792,833
Maine	$174,093	$94,633,655	$94,807,748
Maryland		$532,707,761	$532,707,761
Massachusetts	$4,715,658	$409,965,901	$414,681,559
Michigan	$4,475,000	$1,606,999,900	$1,611,474,900
Minnesota	$5,459,000	$191,849,570	$197,308,570
Mississippi	$2,971,926	$255,215,596	$258,187,522
Missouri	$131,606,785	$455,129,086	$586,735,871
Nebraska	$27,854,511	$107,578,155	$135,432,666
Nevada	$21,251,745	$168,707,160	$189,958,905
New Hampshire		$68,602,589	$68,602,589
New Jersey	$19,015,000	$903,523,000	$922,538,000
New Mexico	$6,000,000	$185,399,600	$191,399,600
New York	$256,834,838	$1,890,751,940	$2,147,586,778
North Carolina	$6,372,906	$899,584,323	$905,957,229
North Dakota	$2,754,294	$21,384,133	$24,138,427
Ohio	$59,420,255	$1,505,722,810	$1,565,143,065
Oklahoma		$419,697,150	$419,697,150
Oregon	$345,170,277	$366,850,306	$712,020,583
Pennsylvania	$43,526,000	$1,171,568,000	$1,215,094,000

continued

Table 14-1 Actual Expenditures, 2001 continued			
Fiscal Year 2001 Correctional Agencies' Actual Expenditures			
	Capital	Operating	Total
Rhode Island	$2,473,245	$130,719,870	$133,193,115
South Carolina	$22,257,232	$333,899,549	$356,156,781
South Dakota		$32,364,861	$32,364,861
Tennessee		$421,806,583	$421,806,583
Texas	$55,587,900	$2,027,932,394	$2,083,520,294
Utah	$879,859	$216,492,230	$217,372,089
Vermont	$6,150,000	$69,907,145	$76,057,145
Virginia	$21,052,000	$734,583,118	$755,635,118
Washington	$64,474,054	$525,205,882	$589,679,936
West Virginia	$647,795	$88,045,579	$88,693,374
Wisconsin	$72,584,400	$759,835,200	$832,419,600
Wyoming	$2,744,995	$98,633,123	$101,378,118
FBOP	$777,084,000	$3,472,401,000	$4,249,485,000

Capital Expenditures: *Money spent on new construction, physical plant improvements, and equipment.*

Operating Expenditures: *Money spent on routine expenses (e.g., staff, food, clothing, medical services, programs utilities, maintenance).*

Source: Courtesy of Camille and George Camp of the Criminal Justice Institute Inc., Middletown, CT.

Appropriate levels of reserve funding can be calculated by developing an equitable formula that sets aside a percentage of each discipline's budget for contingencies. This should be done at the beginning of the cycle. The reserves can be used to correct budget problems or reappropriated to other projects. However, senior administrators must realize that using funds appropriated for inmate care for another purpose may draw criticism. For example, funding for medical care, food, and inmate comfort items should remain in the salaries portion of the budget. The budget and planning committee should be able to identify personnel changes and salary variations in advance of a catastrophe.

Accountability in any financial process is important, but in the budgetary environment it is mandatory. While it is acceptable, and generally desirable, to decentralize cost center management control to the department head level within the institution, it is critical that the overall budget manager assign specific tracking responsibility along with the authority to spend. Individuals with signature authority to spend must be required to justify expenditures in writing, account for all fund outlays, and keep spending within preset limits.

Compliance monitoring or financial auditing is critical to the integrity of an agency's budgeting process. In many cases, the parent organization has an official financial auditing system in place. However, an in-house review should be conducted periodically using specific financial auditing guidelines that focus on vital functions and prevent fraud. The facility's financial management operation should be able to sustain an audit from a private accounting firm or government accounting agency. An acceptable internal auditing system should also be in place throughout the budgeting cycle.

Part of maintaining the integrity of a budget is being able to demonstrate the ability to protect and maximize the use of the financial resources entrusted to staff stewardship. This will require internal controls to prevent waste, fraud, and abuse. Internal requirements might include:

- A system for accountability for purchasing and maintaining property and equipment with a high acquisition value
- Policies limiting or eliminating personal use of equipment
- A second level of procurement authority to ensure that contracts and small purchases are competitively priced and available to all eligible contractors and vendors
- Documentation to support the destruction or removal of property that is no longer useful or cannot be repaired

Jail or prison staff must be aware of special funds for which staff have stewardship responsibility. These funds may include inmate accounts, inmate wages, and canteen profit accounts designated by law or internal regulation for special institutional purposes. Inmates can be sensitive about these funds, and they have been the subject of inmate-originated lawsuits. Such accounts also receive, and deserve, close attention from outside auditors. It may be advisable to develop an inmate canteen committee that would allow inmates to have a voice in the use of canteen profits. At least a portion of these funds should be used to benefit the entire inmate population, such as purchasing recreation equipment or augmenting the children's area of the visiting room.

Senior institution staff must also pay particular attention to the expenditure of public money for employee travel and attendance at conferences or special training events. This type of activity can easily generate undesirable attention if such events are not reasonable and appropriate. Per diem expenses of on-duty staff who are working outside the institution offer great potential for abuse and subsequent negative publicity. Management should exercise care in the approval process for training locations; if a conference is in a resort area, it can draw criticism from the public.

Public scrutiny must be an important consideration in all aspects of institution spending. A correctional facility, whether government run or privately operated, is largely funded by the taxpayers. Therefore, management of institution resources must be logical, acceptable to the public conscience, and based on common sense. Institutional staff must develop a keen sensitivity to what people might consider inmate luxuries and avoid spending money on items that are not acceptable to the average citizen. In general, equipment and programs available to inmates should not be better than those available to the average free citizen.

■ Conclusion

Basic training of senior correctional personnel in the use of financial reports and general budgeting principles and techniques helps ensure financial accountability within a prison or jail environment. While most senior managers are not from a business or accounting background, there are many benefits to providing basic financial knowledge to correctional leaders. Most correctional budgets represent multi-million dollar operations, so the return on basic financial training is significant.

The financial operation of a prison or jail is a critical management responsibility. Stewardship of the public's resources requires conservative decision making, a well-developed sense of integrity, and the ability to apply administrative accountability to the overall process. Effective and efficient operations and programs require fiscally responsible budget planning and execution.

DISCUSSION QUESTIONS

1. How does the staffing pattern and roster affect the institution budget?
2. As a senior administrator, what level importance would you assign financial management in your institution and why?
3. How do you think public scrutiny and perception relate to institution expenditures?
4. What methods can be used to ensure fiscal accountability?
5. What challenges might a prison or jail administrator face with regard to financial operations?

ADDITIONAL RESOURCES

R. Phillips and C. McConnell, *The Effective Corrections Manager: Correctional Supervision for the Future* (Sudbury, MA: Jones and Bartlett Publishers, 2005).

Working with the Media

15

Judith Simon Garrett

Chapter Objectives

- Explain the positive and negative aspects of media access to prisons from the perspective of the public and of prison administrators.
- Identify the key elements of an effective media strategy.
- Outline aspects of good media training.

The public is fascinated with crime and criminals, and such fascination does not stop when an offender is hauled off to jail or prison. The interest simply shifts from the crime itself and the workings of the criminal justice system to the prison system and life behind bars. Accordingly, reporters frequently produce stories about inmates and the prisons in which they are housed. Unfortunately for prison and jail administrators, the stories usually center on the plight of the inmates. Some stories focus on the harsh conditions that offenders face, the violence reputed to pervade prison life, or misconduct by staff (ranging from physical abuse of inmates to mismanagement of government funds).

Most stories portray prisons negatively and rarely provide an accurate description of what goes on behind institutional walls and fences. Media stories often fail to mention the many positive things that transpire in institutions, and distort the motives and actions of staff. Prison and jail administrators should work with media representatives to provide the public a more accurate, positive picture of what goes on inside correctional institutions.

Most prison administrators will have to respond to a news media request to interview an inmate or do a story about some aspect of his or her institution's operations. Such a request may involve making an inmate available in the visiting room for 30 minutes, allowing a reporter to shadow a staff member for a week, or anything in between. Therefore, a written policy regarding the handling of media requests is very helpful. Media policies should strike a balance between the burden on prison staff associated with accommodating a media request (including any potential risks to the safety of inmates, staff and the community) and the need for the media to inform the public about prison operations (including prisoners' rights to express themselves through the media). The media does not have the right to enter a correctional institution to complete a story, and inmates do not have the right to unlimited or unrestricted access to media representatives. But appropriate media coverage can help keep the public informed about prison operations and the expenditure of tax dollars.

■ Media Access: Legal Considerations

How much access should the media have to prison and jail operations? Some correctional agencies permit representatives of print and electronic media into their facilities upon request. Other agencies believe such contact with inmates can be disruptive and a threat to the orderly running of the institution. In 1974, in *Pell v. Procunier*, the Supreme Court upheld the California Department of Corrections' prohibition on face-to-face interviews between individual inmates and representatives of the news media.[1] The Court concluded that this restriction was permissible because inmates had other avenues of communicating: They could correspond with the media and receive visits from family and friends. Thus, prisoners had adequate means of expressing their concerns about their conditions of confinement. Furthermore, journalists were permitted tours of the prisons and could ask inmates questions while visiting.

At the same time the Court decided *Procunier*, it decided another media access case involving the Federal Bureau of Prisons (BOP). In *Saxbe v. Washington Post*, the Court ruled that the First Amendment does not provide the press a constitutional right of access to information that is not generally available to the public.[2] The Court upheld the BOP's rationale that giving individual inmates access to the media would create undue attention for specific inmates who may already be notorious (the "big wheel" theory), potentially causing tension among inmates or between inmates and prison personnel. Even though the BOP won this legal battle, the agency does permit media representatives to interview individual inmates under clearly identified policy parameters and on terms set by the agency.

On occasion, state legislatures have stepped in to try to require departments of corrections to alter their policies to ensure that the media are granted appropriate access to prison facilities. In one state, the legislature passed a bill that reversed an agency policy that prohibited members of news agencies from securing interviews with particular inmates.

■ Interview Considerations

Many correctional jurisdictions believe it is important to permit representatives of the media into institutions to interview specific prisoners. In these situations, there are significant questions that should be asked and evaluated before granting approval.

Who Is Making the Request?

- Does this person have press credentials, such as an affiliation with a newspaper, television station, radio station, magazine, or book publisher?
- Does this person have a relationship with the inmate, victim of the inmate's crime(s), or a member of the staff?
- Does this person have a known agenda that is likely to substantially bias the reporting?

What Is the Purpose of the Intended Story?

- Is the inmate who is the focus of the story particularly well known (or notorious)? Did his or her case generate substantial media interest before and during the trial phase?
- Is the author hoping to garner sympathy or support for a particular inmate's case?
- Is the author hoping to encourage a public outcry against the treatment of one or many prisoners held at the facility?
- Are prison operations being targeted as being corrupt or excessively harsh?

What Will Be the Demands on Staff?

- How many people will be interviewed and for how long?
- How many people will be entering the prison in order to complete the story?
- For how long will the media representative be in the institution?

What Will the Effect Be on Prison Operations, Including Security?

- To what extent will the media representative disrupt daily operations? Will the visit prevent scheduled activities from taking place?
- How are the inmates likely to react to the presence of the media representative and to the completed story?
- How are people (including other members of the media) likely to react to the completed story? Are they likely to demand changes to prison operations or make additional media requests?
- To what extent will any pictures that are taken provide members of the public or inmates with information that could be used to fashion an escape or plan other disruptive activities?

There are no right or wrong answers to these questions, nor should the answer to any one question necessarily mandate a particular conclusion, but consideration of these questions will help make the administrator aware of the many possible ramifications of granting or denying media requests.

■ Administrative Concerns

The administrator's foremost responsibility is to maintain the safe and orderly operation of the prison facility. At times, fulfilling this responsibility will conflict with granting a media request, in which case the media representative must accept, if not understand, the reason that the request has been denied. For example, permitting extensive filming of staff training, of prison perimeter fencing, and of control room operations could create security concerns. Allowing substantial or sustained media coverage of a particular inmate has the potential to create difficulties for that inmate (who may become the target of jealousy and anger from other inmates) and for prison administrators (if the inmate decides he or she has special status and wants special treatment).

In cases where the media request can be accommodated with minimal disruption and little chance of a threat to the safe and secure operation of the prison, it may make sense to grant the request. It should always be assumed that denying a media request will give rise to accusations that the institution is attempting to hide something or that the prison administration is silencing the inmate in order to protect itself, an elected official, or some other government representative.

■ Media Access Policies

In general, the more restrictive the access policy is, the more a correctional agency stands at risk of angering the public, representatives of the media, and elected representatives. Severe restrictions on access prevent correctional staff from gaining the public's confidence and support. In the face of the public pressure to rehabilitate criminals and use tax funds wisely, it is important that prison and jail administrators be held accountable for their stewardship of public funds. The best means of disseminating information is welcoming outsiders into the closed environment of the prison or jail. Finally, administrators should be cautious not to routinely grant media requests made by one source and deny requests from another. Doing this would create a perception of favoritism that could result in a variety of problems.

The accreditation process of the American Correctional Association (ACA) requires that a prison or jail facility has a written policy that provides for reasonable access between inmates and media representatives, subject only to limitations necessary to maintain good order and security, as well as to protect an inmate's privacy.[3]

An institution's written policy regarding access by the media should include a variety of key provisions, including the following:

- All media requests should be in written form and should include an acknowledgment by the requestor stating that he or she is familiar with institution, rules and regulations and agrees to comply with such rules.
- The representative must make reasonable attempts to verify all allegations leveled against inmates, staff, or the institution and provide the institution with an opportunity to respond to allegations prior to the publishing of the story.
- The media representative must make an appointment to visit the institution.
- Inmates may not receive compensation for interviews with the media.
- The request for an interview with news media representatives may originate with the representatives or with an inmate; an inmate's request or consent to be interviewed must be in writing.

- The administrator should approve or disapprove media requests in a timely fashion, and all denials shall be provided in writing. Reasons for denial may be based on a variety of factors, including the inmate's medical or mental condition, the threat to the health or safety of the interviewer, the threat to the safety and good order of the institution, the threat to the safety of the inmate, or a court order forbidding news interviews.

Media Representatives

When the prison or jail accommodates a media request, it is the responsibility of the institution's liaison or spokesperson to work with the author or producer to learn as much as possible about what the story will say and to try and minimize distortions of prison operations. The public information officer (PIO) is responsible for managing media requests and all communications with the media. This person should be screened appropriately and given extensive training in how to work effectively with the media. The training can be provided in-house by staff with experience in media relations or the PIO can attend one of the many commercially available classes. This training will help the PIO develop strategies and techniques for communicating effectively with the media.

An effective PIO will shape a response to the media request by providing effective "sound bites" to ensure the most positive portrayal of the institution and its staff. Additionally, an effective PIO makes clear the parameters governing the media's access to inmates, staff, and the institution at the outset. Finally, an effective PIO will ensure that media representatives respect the right to privacy that all inmates and staff enjoy and acknowledge the institution's responsibility to protect this right.

There are occasions when newsworthy events occur at a correctional institution and the administrator or PIO should contact the news media. Examples of such events include escapes, disturbances, and deaths of inmates. When such events occur, administrators should provide information to the media that is considered public, such as an inmate's name, register number, and sentencing data. Similarly, information about staff that is generally considered public includes position title, job assignment, number of years of service, and previous duty stations. Any incident that has the potential to give rise to criminal prosecutions should be discussed only in the most general terms and, to the extent possible, with the advice of legal counsel. As a general rule, it is wise to provide few details at the outset, at least until all relevant facts have been established to a substantial degree of certainty.

Community Coverage

Sometimes media representatives will request access to write or produce a story about a positive aspect of prison operations. On a rare occasion, the media will become interested in a particular program or aspect of prison operations, such as a program to permit women inmates to care for their young children inside prison. It is more likely to be the case that the prison or jail administrator will have to actively solicit media support for worthy programs. The chief executive officer or administrator might occasionally invite representatives of the news media to visit the institution and observe particular programs or operations. It would be inappropriate to extend such invitations on a regular basis because it might give the impression that administrators were more concerned with attracting media attention than with operating the facility.

Another means of obtaining positive media coverage is through members of the community. Garnering support in the community for prison programs and operations is generally an effective strategy to gain positive media coverage. Many prison administrators create a community relations board comprising senior prison staff and members of the community, including representatives from local businesses, elected officials, and others. Through this board, potential conflicts between the prison and the community, such as expansion of the prison or a change in the security level, often can be avoided by facilitating factual discussions rather than emotionally charged debates. Such debates are often played out in the media, and thus it is important to expend considerable resources to avoid conflict at the outset. Additionally, community relations boards often give rise to partnerships between the community and the prison that effectively serve the interests of all participants. For example, prison inmates can assist the community by building homes for underprivileged families or building toys for needy children from scrap materials. The community benefits through free labor, the inmates benefit from the satisfaction gained from helping others, and the prison staff benefit because inmate idleness tends to breed unrest.

■ Conclusion

Granting media representatives access to correctional institutions, staff, and inmates is a sensitive matter that must be considered carefully and managed by experienced personnel. The focus of media coverage undoubtedly will vary; there will be opportunities for positive exposure for the institution and the community, and thus it is essential to maintain positive relationships with representatives of the media.

Clearly, the media play a significant role in shaping public opinion. Correctional administrators must think of communication specialists as information messengers. Through media representatives, administrators can convey a sense of the challenges and complexities of working in and managing correctional institutions.

DISCUSSION QUESTIONS

1. How much access should the media have to prisons and prisoners?
2. What are some key strategies to ensure accurate media coverage from the perspective of prison administrators?
3. What role does the media play in shaping public perception about prison life?
4. What aspects of media relations pose challenges to prison and jail administrators?
5. What are the crucial elements of an effective media strategy?

NOTES

1. *Pell v. Procunier*, 417 U.S. 817(1974).
2. *Saxbe v. Washington Post*, 417 U.S. 843 (1974); C. Cripe and M. Pearlman, *Legal Aspects of Corrections Management* (Sudbury, MA: Jones and Bartlett Publishers, 2005).
3. American Correctional Association, *Foundation/Core Standards for Adult Correctional Institutions* (Washington, DC: St. Mary's Press, 1989).

Community Relations Boards

16

Paul McAlister

Chapter Objectives

- Describe the primary purpose of a community relations board (CRB).
- Outline the goals of successful CRBs.
- Explain the logistics involved with setting up and running CRBs.

Society has become dependent upon correctional institutions. The incredible growth of the number of prisons has made their presence undeniable, yet few people want a correctional facility nearby. A community relations board (CRB) can help a community and a correctional institution live together as neighbors and partners in the effort to deal with the reality of criminal behavior. Institutional personnel can become aware of community concerns. The community can learn that institution staff are members of the community too, and share community concerns. Through ongoing dialogue, communities can learn ways to help strengthen the effectiveness of neighboring institutions, not just endure their presence.

■ Composition of CRBs

There is no prescribed number of CRB members; the number should be determined by each facility. Nominations may be sought from appropriate agencies or individuals. Those individuals may be approved by the board, but final selection should be made by the warden. Each local facility will need to tailor a CRB to fit its specific situation and circumstances; CRB models are highly adaptive.

However, there are several groups of community members who should be represented on a CRB, including:

- People most directly affected by the physical presence of the facility (i.e., immediate neighbors and realtors)
- Local officials with whom the institution will want to build relationships (e.g., police, city council, and county commissioners)
- Outreach members who provide a link to the rest of the community (e.g., educational representatives, clergy, civic groups, charitable organizations, multiethnic organizations, and others)

■ Objectives of CRBs

There are many objectives commonly associated with an institution's CRB, including improved communication, easing reentry, and community involvement.

Improved Communication

First and foremost, a prison or jail administrator should use the CRB to improve communication with the local community. This requires CRB members to become educated in all aspects of institution operations so they can serve as effective information conduits to other citizens. In turn, CRB members can provide honest and worthwhile feedback to senior correctional managers from the community.

Since the early 1980s when CRBs began to develop, their central goal has been to enable the exchange of accurate information between the institution and the community. Honesty is critical to effective institution-community communication. CRB members need to be given accurate information about facility functions and policies and encouraged to ask honest questions. If the institution does not provide honest information, the effectiveness of the CRB will be undermined. Board members must be encouraged to express the concerns of the community openly to the institution.

Through effective and open communication, the CRB can also help to build goodwill and trust between the institution and the community. It can help remind community representatives that staff live in the same neighborhoods that they do and, like them, they want these neighborhoods to be pleasant and safe places to work and live. CRBs can enhance the facility's existing programs by promoting and supporting community volunteer efforts. The board can also suggest ways that the institution can contribute to the community.

While most communities recognize the need for prisons and jails, they typically do not want one nearby. CRBs can help diminish a community's fears by sharing their insight and experience with communities where institutions may be built.

On occasion, incidents occur at correctional facilities that are of heightened interest to the community. Board members should be knowledgeable about institution incidents that would be of public concern. The CRB members can be contacted with accurate information regarding events and can then provide a measure of reassurance and calm to the community. If it takes days or weeks for an incident to be resolved, reassurance may be an ongoing need.

Ease of Reentry

At present, both legislatures and many communities are debating the issue of reentry of inmates. With the ever-growing prison population comes the recognition that many of those currently incarcerated will be released. Faced with this reality, citizens demand good information and reassurance. The issue of reentry is vital to end the cycle of recidivism (colloquially known as the revolving door of corrections). Accountability and support for inmates returning to public life is a vital task. CRBs can serve as important bridges to the community to provide for effective reentry. Community input is indispensable to the development of meaningful reentry programs. If the institution elicits community support for reentry, it should anticipate that the community may ask for additional information, and the institution must respond honestly and appropriately to these requests.

Community Involvement in Intuitional Programming

CRBs also may help enhance community involvement and volunteer participation in institution programs. Inmates will realize that there are people other than correctional staff who honestly care about them. CRBs may also provide options for the inmates to be involved, as appropriate, in community service. CRBs also help the community recognize staff as community members and contributors.

Institutions should provide CRB members with information about trends, programs of the institution, and the difference between facility policy and government-mandated programs or policies. This education can be provided at regularly scheduled CRB meetings or on specific training days. Because policies change regularly, education must be ongoing.

Each board meeting can include a different aspect of the program and introduce staff members involved with that program area. Possible agenda items include emergency preparedness planning, medical and drug programs, reaction to national trends, changes in the institution's mission, and construction or expansion plans. Placing the institution's issues in the context of national trends, legal concerns, and expansion pressures helps enhance CRB members' understanding of their local facility's program needs and mandates.

Educating the community also includes working with educational institutions, civic groups, religious groups, and media representatives. With appropriate concern for safety and inmate privacy, tours also may be an effective aspect of community education. Each institution should provide staff with an internal list of areas that tours would cover so that tours are consistent and key community concerns are always addressed.

The CRB can look for opportunities to suggest that community members become involved with the institution through volunteer programs. The public often fails to realize that inmates generally return to society. Positive changes in inmate behavior are often inspired by dedicated volunteer efforts. The community has a vested interest in inmates' rehabilitation.

The CRB also may help develop programs in which inmates provide the community with needed services. This may be some inmates' first positive experience of giving to someone else. Such opportunities can boost an inmate's self-image and awaken a desire to contribute more to society. Worthwhile projects might involve helping warn young people at risk about the costs of criminal behavior, providing assistance during emergencies, and supporting charitable activities, such as home repair projects for low-income community members.

It must be said that the CRB is not a policy-making entity but a vehicle for the exchange of information between the institution and the community. The CRB may be called on to make recommendations to prison or jail administrators. If the administration has good information to work with concerning community perceptions and concerns, decisionmakers can anticipate community reactions and help community members accept institutional decisions.

It is important for prison staff to be respected by the community and for community members to feel gratitude toward them. Activities that increase public contact with staff should be encouraged. When staff are seen as having a vested interest in maintaining a pleasant and safe community for themselves and their families, it builds trust between the community and the institution.

■ Logistics

CRB members should help decide on meeting times. It is not always easy to find a meeting time that works for everyone. Institution staff should inform prospective members of scheduled times before they are asked to participate. Noon meetings often are preferred, but certain members of the community may have difficulty attending during the day because of employment obligations. Early evening meetings may also work well. Meetings may be monthly, bimonthly, or quarterly. Special meetings can be called if there is a crisis or some decision in which the CRB may be asked to make recommendations. Because a truly representative CRB is the goal, every effort should be made to make full participation as likely as possible.

CRB members may be involved in events outside of regular meetings. To enhance the CRB's understanding of the institution and its staff, CRB members may be encouraged to attend certain social events. During these events, CRB members can learn from staff about what institution programs need enhancement. Board members also may be invited to certain staff meetings such as a staff recall. At such meetings, CRB members will learn more about institutional goals and needs and report what they learned to the community. While confidentiality is necessary in some aspects of institutional operations, CRB members should not feel that the institution is hiding things that the community should know.

Board members' terms will be established by each institution in conjunction with actual community members. There is an advantage to allowing some members to serve for a specified period of time: People can participate and yet not commit themselves too far into the future. A two- or three-year term may be appropriate. It is difficult to provide the education necessary to contribute to the function of the CRB if the term is any shorter. In some cases, members may be willing to commit to an extended period of service. They can provide expertise that is beneficial to all. By allowing automatic reappointments after a term ends, a board can reduce the need for renominating and approving existing members who wish to continue serving. The downside of the extended term is that others who wish to serve are not able to fill a vacated position.

The CRB chair facilitates meetings through handling introductions and working through the meeting agenda. He or she also contributes to the development of meeting agendas. Each agenda should be sensitive to developing issues and address ongoing information and education issues. The chair should be nominated by the board members, who also will decide the chair's term. A term of one year with the possibility of reappointment is common.

Each facility and CRB may draft its own bylaws. The bylaws normally will be structured by the warden or superintendent with input from members and should spell out details concerning membership, terms of service, and functions. Bylaws may also cover issues of security and confidentiality.

There needs to be a public awareness of the CRB and its members. If the community members do not know who the CRB members are, they cannot learn about the institution or give their opinions about the institution to CRB members. An article in the local paper can help community members learn about the CRB, its role, and its members.

■ Conclusion

Community relations have become a very important aspect of institutional planning. A CRB can contribute greatly to an institution and its surrounding community.

Boards are only as effective as their ability to establish open and honest lines of communication. Perfunctory CRBs are doomed to fail. Institutions need to provide honest answers to questions generated by CRB members or communicated by the CRB on behalf of the community. The capacity of the CRB to react to incidents or to prevent confusion within the community can be a great service to all. Wardens can use CRB members as sounding boards. CRBs must carry information and understanding from the facility to the community and from the community to the facility. Both directions of this information channel must remain open for CRBs to be effective. The growing number of people incarcerated and the ongoing need for new institutions mean that the issue of community relations will remain very important for many years to come.

Serving on a CRB is a privilege. CRB members often feel a sense of meaningful investment in developing a partnership between the community and the facility and in helping to ensure a safe and effective approach to criminal justice. Members of a CRB have a significant opportunity to contribute to this vital partnership and make a difference in a local community.

DISCUSSION QUESTIONS

1. How can citizens provide useful insights to institutional administration?
2. What difficulties are there in the use of CRBs?
3. What opportunities do CRBs offer?
4. How can a CRB assist with media coverage?
5. Who should serve on a CRB?

ADDITIONAL RESOURCES

J. Jones, "Community Relations Boards," *Federal Prisons Journal* 2, no. 2 (1991), pp. 19–22.

American Correctional Association, *Helping Hands: A Handbook for Volunteers* (Laurel, MD: ACA Press, 1993).

Political Involvement

Judith Simon Garrett

Chapter Objectives

- Distinguish between a spoils system and career service.
- Outline the importance of correctional administrators working with legislators.
- Identify the types of issues posed by implementing legislation in the correctional environment.

In the 1990s, while the Uniform Crime Reports data indicated that crime rates were declining, the fear of crime remained a primary concern for most Americans.[1] Legislators, eager to be responsive to their constituents, remained steadfast in their "tough on crime" and "war on drugs" approaches that began in the 1980s. As a result, today there are more than 2.3 million adults in prisons and jails in the United States.[2] State and local governments spent nearly $60 billion dollars on corrections in 2001, an increase of 529 percent since 1982. During the same period, spending for police and the judiciary increased only 281 percent.[3]

Following the terrorist attacks of September 11, 2001, the country's priorities concerning justice administration shifted dramatically, focusing almost exclusively on preventing acts of terrorism on American soil and eradicating the forces that allowed them to occur in the first place. Accordingly, the federal budget available for issues other than homeland security and war-related functions became very limited. The public's tolerance for continuing to spend billions of dollars on imprisonment seems to be waning for the first time in many years.

While crime and criminal justice remain largely state and local issues, the legislative and executive branches of the federal government attempt to influence state and local laws and policies through grants and other means. Congress and the president more directly influence federal law through statutes, regulation, and executive orders. Elected representatives and those seeking election also do not hesitate to use public forums and the media to express their increasingly conservative views about crime and criminal justice administration.

■ Political Interest in Prison Operations

As a result of their desire to influence criminal justice policy and demonstrate their commitment to being tough on crime, federal and state legislators attempt to substantially affect correctional operations. The combination of enacting longer sentences and placing new restrictions on early release mechanisms (such as limiting or abolishing parole and good time) has been directly responsible for the tremendous increase in prison and jail populations across the country since the 1980s. At around this same time, legislators forced correctional officials to reduce or eliminate many programs and recreation opportunities in an attempt to create "no-frills" correctional institutions with the belief that a harsher environment will increase the deterrent effect of prisons. For example, the Federal Bureau of Prisons is prohibited from repairing, replacing, or purchasing new weightlifting equipment and musical instruments. Other legislation is introduced regularly that requires inmates to work 50 hours per week or be confined to their cells (for 23 hours per day) if they are not medically able to work.

It is unfortunate that most lawmakers have little or no direct knowledge of prison operations. Many would benefit substantially from an educational tour of a correctional institution. Most corrections managers would disagree with legislators' view that prison programs (including recreation opportunities) and modest amenities such as television rooms are frills that make prisons less of a deterrent. According to most correctional managers, these programs and amenities can be instrumental in ensuring an inmate's successful reintegration into the community following release from prison.

Institution recreational opportunities allow inmates the opportunity to release some of the anxiety and stress that is inherent in prison life and help create an environment with less tension for the staff and inmates. Research has found that prison programs help offenders reintegrate successfully into the community through lower rates of recidivism and higher rates of employment. This supports correctional professionals' belief that institutional programs (e.g., education, vocational training, prison industries, substance abuse treatment) are essential to effective correctional settings. Nearly all prison administrators agree that making prisoners' lives needlessly uncomfortable and unpleasant does little more than make the jobs of staff, particularly correctional officers, more difficult.

Some correctional managers are proponents of the spartan prison existence and help foster the misconception described above; they invite legislators to mandate that all prisons be operated in a no-frills, severe manner. Sheriff Joe Arapaio from Maricopa County, AZ, was a classic example of this philosophy. He required some inmates to sleep in tents, and required sex offenders to wear pink underwear.[4] In contrast, most correctional administrators believe that prisons should be decent places in which to live and should offer reasonable program and recreation opportunities that provide self-improvement activities, offer a safe and healthy environment, and properly control inmate behavior. Institution operations should not seek to embarrass prisoners or create unnecessary resentment.

■ Implementation of Laws

The degree to which politics influence the administration of government agencies has evolved with time. Woodrow Wilson, former U.S. president, is largely responsible for abolishing the spoils system in the federal government. His 1887 essay, "The Study of Administration," presented the philosophy of the Progressive movement, created by reformers unhappy with the spoils system.[5] Under the spoils system, employees of government agencies were fired each time a new political party was elected. Political patrons of the party in office were brought in, and the entire direction of the government entity was subject to major change. Federal patronage had a particularly large influence on government operations. The Progressive movement lobbied to establish a professional civil service that would be neutral on political issues. Over time, the system of political spoils was abolished at the federal level, except for a limited number of high-level positions such as cabinet heads and select senior members of their staff.

For government administrators to run their institutions effectively, they must avoid the strife of politics. As Wilson stated many years ago, the administration of an agency is a function of business and must be removed from the arena of political rhetoric. Professional administrators in all areas of government must focus on the impartial governance of their agencies. Public administration is the systematic and detailed execution of public law; legislators establish policy, and administrators implement their decisions.

In the federal prison system, leadership at all levels has always been career civil servants—individuals who worked their way up the organization. However, in many states, the directors of the departments of corrections are political appointees who serve at the pleasure of the governor. Turnover in these states is generally quite high, posing challenges for the professional correctional administrator. In many states, it has been common for prison wardens to also be political appointees—an individual who had assisted the governor with the election was rewarded with the warden's job in return. These wardens rarely had any relevant experience prior to arriving at the prison. Their ability to effectively manage a prison and lead a staff varied greatly. In many instances, they merely served as a figurehead, and the associate warden or other career staff directed prison operations.

Regardless of whether the wardens are political appointees or career civil servants, it is essential that corrections administrators implement laws in a timely and effective manner, with the least disruption to institution operations. The personal views of the administrators regarding the wisdom of the laws are irrelevant.

Institution administrators should not become involved in the political process. Institution chief executive officers should be seen as impartial, professional administrators who have a responsibility to implement legislation and executive orders fairly.

As experts in corrections, professional administrators sometimes are called upon to share their expertise and facilitate the consideration of proposed changes in the law. While senior institution staff should always welcome the opportunity to educate lawmakers, they must be cautious not to give the appearance that they are trying to influence the legislative process. One approach to educating legislators is to provide educational tours to members of congress and their staff. The Federal Bureau of Prisons demonstrates the realities of institutional management by opening its prison doors to elected officials and the public. Many legislators and, more frequently, their staff take advantage of such opportunities and find the experience interesting and enlightening.

■ Political and Societal Changes Affect Penal Facilities

Political winds shift with the times, sometimes quite rapidly, causing dramatic shifts in prison operations. During the 1970s, prison wardens would proudly show their college campus prisons and well-funded vocational training programs. Today, the public expects prison and jail environments to be more severe and punishing, with special attention paid to preventing radicalization of inmates and the recruitment of terrorists.

It is difficult and dangerous to withdraw privileges or programs from prisoners once these privileges and programs have been given to them; such withdrawals must be planned and implemented carefully. Therefore, when implementing laws that impose new restrictions on prisoners, administrators must communicate with inmates (through staff–inmate interaction, posted memos, and other means) and emphasize that the change in practice must be accepted by everyone— prisoners and staff alike. They must make prisoners understand that misbehaving in response to the planned change will result in swift and certain punishment and will not forestall the intended implementation.

At different times, all aspects of the reasons for confinement (punishment, rehabilitation, general deterrence, and specific deterrence) will be emphasized. Prison administrators must accept these changes and not become publicly or emotionally invested in any particular political approach. Administrators are not precluded from holding their own personal and professional views regarding the most appropriate manner in which to operate a prison, but as civil servants all staff are expected to fulfill the requirements of the law as set forth by their legislative bodies. The public is best served by prison and jail leaders who implement policy established by elected officials effectively. The legislators' newfound interest in the administration of justice is significant and should be considered a positive development.

In a text on prison management, Richard McGee, California's venerable correctional administrator from 1944 to 1967, postulated that officials should fight political pressure.[6] He strongly advised penal leaders to avoid the outside turmoil and maintain a tightly organized and operated environment inside the institutions. Rather than joining in on the political dialogue external to the agency, his advice was to focus on making prisons and jails as safe, manageable, and humane as possible. He affirmed, however, the need to make institution operations accessible to external players who have great influence in the political dialogue so they can understand the facts of life behind bars.

■ Conclusion

Correctional systems demand a substantial portion of state and federal criminal justice budgets. As a result, politicians and the public are increasingly interested in prison operations, and they demand results, such as reduced recidivism. Additionally, the public wants to know that terrorists are not taking advantage of the captive prison audience to recruit warriors to their cause. Correctional administrators often resist the interest of outsiders, viewing it as undue interference with daily operations by those who know very little about the difficult tasks faced by prison and jail staff on a daily basis.

Senior staff of U.S. prisons and jails must exert leadership behind the fences and in the world beyond. It is not appropriate for civil servants to lobby or otherwise directly seek to influence elected representatives of the people. It is, however, acceptable and desirable for those who are experts in managing prisons and jails to educate the judicial and legislative branches of government. Correctional systems will not adapt well to today's turbulent external pressures unless they manage these sources of influence successfully. Leaders must comprehend the political context and appropriately position their agencies to withstand the political winds of change.

Successful leaders of correctional agencies are those who can cultivate the outside support necessary to counter individuals who are intent on wholesale change of the institution regimen. Outside support is necessary in the local, state, and national communities and is generally composed of the media, the judiciary, and the popularly elected legislative bodies. The image of corrections is very much in the hands of chief executive officers of facilities and heads of correctional agencies. Effective leaders in these roles will gather support for their work by being responsive to their constituencies and bringing them inside to show them the realities of institutional management.

As public administrators, the leaders of institutional operations must be responsive to those who pay the bills and operate according to one of the important principles on which our country was established—the rule of law.

DISCUSSION QUESTIONS

1. What are the pros and cons of a spoils system as compared to career civil services?
2. Why should the public and legislators be interested in prison operations?
3. In what ways can corrections administrators develop a positive relationship with legislators?
4. How have expectations of the public and legislators changed regarding prison operations?
5. Why should correctional administrators be concerned about the political process?

NOTES

1. National Crime Prevention Council, *Are We Safe? The 2000 National Crime Prevention Survey* (Washington, DC: U.S. Department of Justice, 2000).
2. P. Harrison and A. Beck, *Prisoners in 2005* (Washington, DC: U.S. Department of Justice, 2006).
3. Bureau of Justice Statistics, *Justice Expenditures and Employment Extracts* (Washington, DC: U.S. Department of Justice, 2004).
4. J. Hill, "Arizona Criminals Find Jail Too in-'Tents'," July 27, 1999, available at http://www.cnn.com/US/9907/27/tough.sheriff/, accessed July 17, 2007.
5. F. Mosher, ed. *Basic Literature of American Public Administration, 1787–1950* (New York: Holmes and Meier, 1981).
6. R. McGee, *Prisons and Politics* (Lexington, MA: Lexington Books, 1981).

Staff Management

III

YOU ARE THE ADMINISTRATOR
Lessons from Abu Ghraib

The mistreatment of Iraqi prisoners held in Abu Ghraib prison in Baghdad was terrible for American military leaders and those responsible for the oversight of American intelligence. The situation at Abu Ghraib involved American soldiers abusing and humiliating Iraqi prisoners, and the media seemed to delight in presenting American hypocrisy—the forces of good caught in an evil act within the military prison.

Questions that arise from such an event are appropriately asked of those in charge of the institution. How could daily activities, as alleged by some, be happening without the knowledge of the supervisory personnel? Senior administrators should be aware of abusive behavior and stop it when it occurs. The leadership structure of a prison or jail facility must know what is going on within the walls and fences of the institution and hold all personnel to a high standard of behavior.

When institutional leadership fails, as Abu Ghraib demonstrates, extremely bad things can happen.

- *How can prison and jail administrators stay in touch with the ongoing issues in their facilities?*
- *How can negative behavior by staff be prevented?*
- *Should senior personnel be held accountable for the behavior of junior staff?*
- *What key lessons about staff management can be learned from the Abu Ghraib scandal?*

Source: "Abuse of Iraqi POWs by GIs Probed," *CBS News*, April 28, 2004, available at http://www.cbsnews.com/stories/2004/04/27/60II/main614063.shtml, accessed March 7, 2007.

Organization and Management

Peter M. Carlson and John J. DiIulio, Jr.

18

Chapter Objectives

- Outline three key factors that contribute to excellence in penal management and leadership and identify the basics of a well-run prison or jail.
- Differentiate between management and leadership.
- Name three primary means of establishing measurement and accountability in a government agency.

■ Introduction

The world of prison and jail management has long been unfairly considered a backwater of public administration, organizational theory, and management behavior. While the academic centers of the nation have been preparing future gurus of business—tomorrow's leaders of city, state, and federal government and assorted other public administrative functions—few have taken on the complex task of developing prison and jail administrators. Almost all of today's penal leaders have entered the field in an entry-level position and earned their way to top management roles. Programs in correctional leadership must be expected to keep up with the national increase in prisons and jails.

Leaders of state and federal correctional systems, chief executive officers of major institutions, and all those who work in developmental assignments under these senior leaders are responsible for millions of dollars of buildings and equipment, thousands of staff members, and many thousands of inmates within each facility or jurisdiction. The field of corrections is a demanding environment in which the leaders and managers of institutions must be bright and capable of governing thousands of staff and inmates. The challenges of prison and jail management have never been greater, with expanding inmate populations, legal complications, more aggressive and dangerous inmates, politicians who want to micromanage institutions, and staff issues that are complex and unending.

The need to manage correctional facilities effectively—to focus staff on the basics of operating safe, secure, and humane institutions—has never been more critical. Prison and jail professionals are responsible for public safety and performing a public service of great significance, but other institutional management tasks are very important as well. A warden must constantly work to ensure staff safety, staff integrity, proper stewardship of government financial resources, and a safe environment for inmates. Corrections leaders are accountable to all constituents: the public, elected representatives, the judiciary, superiors in government, staff, and inmates. Management and leadership skills are essential in this role.

■ Decentralized Organization

In the United States, prisons and jails are decentralized. The federal government, each state, and most counties and cities operate individual correctional networks. Each system tends to operate independently with little or no linkage among agencies. Federal corrections are primarily the responsibility of the Federal Bureau of Prisons (BOP), but other federal agencies are responsible for immigration detention facilities and military prisons. Adult prisons are also operated by private corrections companies that contract to house state and federal offenders. Because so many jurisdictions are involved in prison and jail administration, the administration of justice in the United States is a mixture of various organizational systems and management structures.

Besides the obvious impediment to open communication created by the fragmentation within justice administration, this separation has greatly hindered interagency cooperation in the development of meaningful policy. For instance, many institution administration professionals have accepted that they simply have to receive and process the people the police, prosecutors, and courts send their way. The political process often precludes correctional experts from providing any meaningful input to justice policy-setting circles. Correctional professionals, as a result, have become very myopic, often unable to see beyond the boundaries of their own agency.

Correctional organization, or disorganization, in the various U.S. jurisdictions has created a disarray of services and programs. This complicates any concerted effort to improve corrections systemically. It is nearly impossible to shift financial or other resources among federal, state, and local jurisdictions; to establish common goals and coordinated policy; and to share exceptional procedures and programs between institutions in adjoining states or even among facilities within a single county or city. Simply put, agencies operate independently according to the philosophy of their jurisdiction's governing body—the sheriff, mayor, commissioner of corrections, governor, or federal government.

Oversight

General oversight and control of correctional agencies varies throughout the United States. Some jurisdictions combine penal management with other public service organizations such as law enforcement or mental health departments. Other jurisdictions operate independent agencies ranging from very small city jails up to huge, centralized prison systems. An increasing number of state correctional organizations have evolved into separate departments with the chief executive officer selected by the state's governor. In 2007, 32 states have separate departments of correction reporting to the governor; 11 have departments reporting to boards or commissions; 5 operate under a department of public safety umbrella; and 1 falls under a social services umbrella. Twenty-four of the separate departments have been organized in this manner since 1979.[1]

Clearly, vast differences within the field are created by these separate jurisdictions. Institutions differ with respect to mission; types and amounts of funding; the numbers of staff in roles as psychologists, counselors, case managers, teachers, etc.; and the types of inmate populations served. But these agencies have one unifying characteristic—they each serve the people by confining individuals who have violated the law. Because the laws that govern conduct are established by the political authority of a specific community, city, county, state, or federal jurisdiction, the system of justice administration also varies.

■ Multiple Missions

Prisons and jails generally are expected to accomplish several often conflicting goals in dealing with law violators. They are asked to punish, incapacitate, and rehabilitate offenders, as well as deter others from violating society's rules and regulations. The organization and structure of a penal facility is affected significantly by these different goals. Many people believe the conditions necessary to reform or rehabilitate an inmate conflict with the conditions necessary to punish offenders. Confronted by these contradictory pressures, correctional administrators often try to walk a fine line between opposing missions.

Criminologist Donald Cressey has said that correctional institutions may be placed on a continuum of organizational structures, ranging from a maximum security, old-line penitentiary surrounded by gun towers to a minimum security, program-oriented, unfenced facility.[2] At one extreme is the highly controlled, custody-oriented prison, and on the opposite end is the relaxed, unstructured institution focused on treatment and rehabilitation. However, most prisons and jails attempt to achieve both custody and resocialization goals.

■ Organizational Theory

Organizational theory conceptualizes how authority is distributed within an organization and how it is used to accomplish the agency's mission and goals. A correctional setting's organization is extremely important to staff and inmates. While a public organization may seem impersonal and monolithic, in reality each public entity is a complex mixture of people, personalities, programs, rules, and behaviors. Every organization is composed of people who act individually and collectively to create a culture. These individuals are affected each day by the organizational structure within which they work, so the system of management and control of individual staff members is a key variable in work productivity, morale, and overall agency efficiency.

In the theory and practice of public organization, particularly in correctional administration, efficiency is the point around which everything turns. Many management experts have discussed how best to develop private business or government to produce the most efficient operations. In the early 1930s, business executives James Mooney and Alan Reiley noted several important principles of organizational structure:[3]

- *Unity of command:* Work is coordinated through a hierarchy of leaders; every staff member has only one supervisor with a goal of strong, executive leadership.
- *Scalar principle:* A vertical structure outlines different responsibilities throughout the hierarchy.
- *Separation among departments:* Distinct responsibilities are outlined for different roles and divisions.
- *Relationship between line and staff roles:* Line command flows through the direct chain of command, whereas related staff support offices (personnel, financial management) provide advice and assistance to the chief executive/warden.

This set of principles is an effective means of considering how a penal facility can and should be organized.

■ Correctional Models of Management

Within corrections, three different organizational models exist: an authoritarian model, a bureaucratic model, and a participative model.

The authoritarian model is generally characterized by the presence of a strong leader, very firm control of the prison environment, and the harsh discipline of inmates (or staff) who do not acquiesce to the central authority. This style of institution was prevalent in the United States from colonial days through the mid-1900s.[4] This highly centralized style creates a regimented workplace with consistent application of rules for all. It funnels all decision making to the central power figure, even though some decisions could be better made at a lower level. This model denies all other staff the experience of making decisions and can create an arbitrary and capricious system that may easily become corrupt.

The bureaucratic model also revolves around a strict hierarchical system but is not focused on one dominating personality. Organization control flows through the hierarchy with a strict chain of command and a formal process of communication. Rules and regulations for the correctional institution are written and specific. The facility has a clear set of standard operating procedures. The practical benefit of this model is that a correctional system or institution is not overly dependent upon one or two people and can easily promote or substitute personnel. Additionally, the policy parameters stressed in this management structure are clear for all parties, and staff can be held accountable if they do not comply with the written expectations. On the negative side, written rules do not guarantee consistent enforcement, and they are not helpful in every situation. Bureaucratic processes are slow to respond to change and do not encourage staff to demonstrate new initiative at any level of the organization.

The participative model of management is much more open and democratic than the first two models, though it is not as effective in dealing with fast-moving crisis situations. This method allows and is dependent upon staff input about how the organization should be run. In a few experiments, inmates have given feedback as well. The assumption inherent in this model is that agency and correctional goals are more efficiently accomplished when all staff have participated in reaching a consensus on how to proceed. The participative style gives staff an increased sense of ownership in planning and operations, often resulting in better attitudes toward and support of routine events and new initiatives. Unfortunately, formal and open discussions and negotiations—collective participation in institutional operations—can be time-consuming.

The authoritarian and bureaucratic organizational structures are much more prevalent in correctional administration. These models, unfortunately, do not lend themselves to change; the built-in resistance to new ideas is self-defeating. Yet few successful administrators champion the looseness and lack of structure that can be associated with participatory management.

Many successful agencies have adapted the bureaucratic model to include elements of the participatory style. These correctional leaders decentralize as much of the daily decision making as they can and seek participation from all staff in many avenues. The involvement of midlevel managers and line staff in specific work groups or in overall strategic planning can be extremely beneficial. Staff generally enjoy such activity and, as the individuals closest to the work arena, can make significant contributions. Such forms of representative democracy in the correctional workplace are considered quite effective.

Management Structure

As prescribed by organizational theory, the prevailing management structure in correctional facilities in the United States is hierarchical, centralized, and paramilitary. The bureaucracy of institution management is very controlling and often inflexible, yet it is the most efficient and functional structure for the coordination and control of hundreds of staff members and thousands of prisoners. The critical and dangerous task of running prisons requires uniformity within each specific facility (fairness and equity—the perception that all inmates receive the same treatment) and precision of control (see **Figure 18–1**).

Chief Executive Officer (CEO)
This is the individual responsible for the crucial responsibilities within the jail and prison and may be called warden, superintendent, or administrator. This individual holds the highest executive position and is accountable for all aspects of institution life. The senior administrator establishes policy for the facility and is responsible for personnel, property, programs, and activities. This staff member is also charged with handling the external world of the facility: the public, the media, the politicians, and the courts.

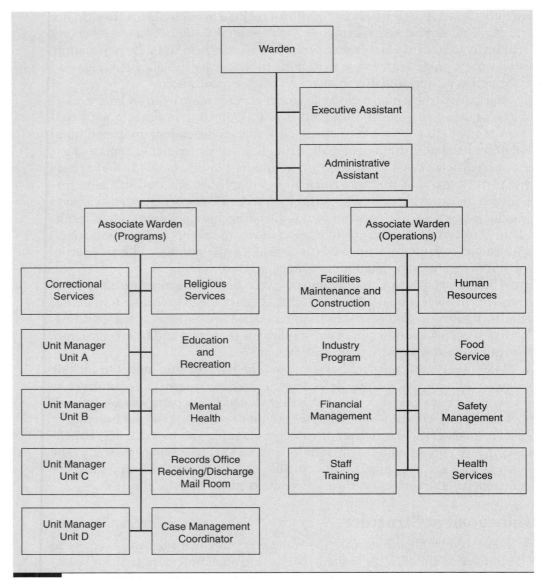

Figure 18–1 A Typical Adult Institution Organizational Chart
Source: Federal Bureau of Prisons

Associate Warden (AW)

Several deputy or associate wardens (AWs) usually head the various organizational structures in large-scale institutions and report to the warden. Often, there is one AW who oversees custody, another who supervises operations, and a third who handles programs. The custody AW is responsible for all security matters and supervises all correctional officers. The operations AW is accountable for all support services within the facility: food and medical service, facilities maintenance and construction, human resource management, and financial management. The programs AW usually heads classification, unit and case management, religious services, mental health programs, education, recreation, and records functions. If a prison has a large industrial operation, there may also be an AW responsible for this production area.

Department Heads

These are generally veteran employees who are experienced in the tasks and skills required within each functional area of the institution. For example, department heads are assigned to areas such as health services, food services, correctional services, or the records office.

Other Supervisors

Other lower-level supervisors may include assistant department heads or first-line supervisors.

Line Staff

Line staff are the individuals who are responsible for the correctional facility on a day-to-day basis. Individuals such as correctional officers or case managers, for example, have face-to-face contact with inmates and are in charge of inmates' daily care.

Decentralized Management

In bureaucratic style, significant decisions within a centralized power structure are made by relatively few individuals, and these decisions are generally consistent and in accordance with the need to operate the penal institution in a safe and secure manner. However, there are also several disadvantages. Department heads and line staff are not empowered and may feel very little ownership of any aspect of institution operations. Additionally, the very bureaucratic process of referring all decisions to senior staff can add great delays, and high-level administrators may not be as familiar with a situation as lower-level staff. If these decisions are not made locally (i.e., the authority rests in a headquarters office rather than the field facility), these negative factors can be greatly magnified.

Decentralized management—the process of dividing and distributing authority and responsibility to administrative personnel—often enhances the effectiveness of administrative operations. When decision-making responsibilities are delegated to staff, staff members gain greater expertise in and ownership of problems and solutions. Critics of decentralized management in the correctional setting believe that dispersing authority and responsibility lessens accountability and does not promote consistent decisions. Decentralizing the process is often considered more expensive, because the process involves training more staff about key issues.

One of the best examples of the successful decentralization of correctional management is unit management, which can be found in many state and all federal institutions. In this model, classification and inmate management authority is delegated to a team of staff members who work in close association with an assigned number of inmates. These staff members (representatives of case management, education, and psychology departments) have offices in the inmate living units and are fully responsible for the day-to-day aspects of the inmates' lives. For example, the unit teams classify inmates for security level and custody needs, make work and program assignments, and handle disciplinary matters for their prisoner caseloads. The staff members report to a department head (referred to as a unit manager) and are held accountable for the overall operation of the unit and management of the inmates.

■ Leadership and Innovation

Correctional institutions work best when administrators stick to the basics of inmate care and custody, as exemplified by wardens who, despite all the competing demands on their time and attention, leave behind the purely administrative tasks and paperwork and practice "management by walking around." The field's most successful executives and managers do not get into routine ways of doing things; rather, they are open to both human resource and technical innovations. Prison and jail leaders who last long enough to innovate are almost invariably individuals who, whatever their personality or ideology, operate as pragmatic professionals and political realists.

Successful prison and jail administrators know from experience that they operate in the context of multiple and competing public objectives (punish, rehabilitate, deter, incapacitate), ever-shifting legislative priorities, small to sweeping judicial interventions, and always incomplete (and often crisis-driven) media renderings of their work. But they wisely seek neither to master nor to withdraw from external demands and pressures. Instead, they and their staffs are responsive to reasonable external demands and pressures, and they engage in the civic life of the surrounding community.

The overarching challenge facing contemporary prison and jail administrators is to remain focused on specific safety, program, and other operational goals and activities. Corrections executives and managers have more or less direct control over these aspects of institutional life. Therefore they shoulder legal, moral, fiscal, and administrative responsibility for these areas, regardless of how the legislative, judicial, community relations, or media winds blow.

Present-day institutional corrections leadership is neither an art nor a science but a craft. Fortunately, the field of institutional corrections, for all its problems, for all its real or perceived failures, has been blessed with a truly extraordinary number and variety of superb craftspeople.

Still, have even the best prison and jail administrators kept pace with their peers—superior public sector entrepreneurs in other arenas of government? The field's present and aspiring leaders must answer this question for and among themselves with respect to at least three sets of issues: boundary spanning, performance management, and public communications.

Boundary Spanning

As Donald F. Kettl has definitively argued, most of contemporary governance and administrative politics involve government by proxy; in other words, no taxpayer-supported bureaucracy operates entirely independently. Rather, a public institution's administration involves working in partnership with other entities. These entities include government agencies, private firms, and nonprofit organizations.[5] Even within prison and jail administration, which some unknowing public management specialists still cite as the paradigm of traditional or direct public administration, government by proxy is commonplace, with all its attendant administrative complications and possibilities. For example, prison and jail administrators interact constantly with the courts and other law enforcement agencies, including agencies representing other jurisdictions or levels of government. They deal with government safety or health inspectors. They hire for-profit and nonprofit consultants. Increasingly, they contract for one or more basic or auxiliary services, from food services to physical plant maintenance.

In fact, in institutional corrections the list of daily and long-term planning functions performed as government by proxy is getting longer all the time. Thus, the work of contemporary prison and jail administrators, like that of most contemporary public sector executives and managers, requires what Kettl terms "boundary spanning," which includes three crucial capacities:

1. The capacity to work productively in a bureaucratic hierarchy and across divisions

2. The capacity to establish constructive professional and administrative ties to other public sector organizations that can affect everything from the agency's daily internal operations to its vulnerability to legal action or its level of legislative support

3. The capacity to enter into cost-effective relationships or contracts with both nonprofit and for-profit service providers

Prison and jail administrators must begin to think strategically about organizational boundary spanning. Accordingly they need to develop innovative and forward-looking preservice and inservice training and staff development programs.

The steps toward boundary spanning follow those outlined by Jameson W. Doig and Erwin C. Hargrove in *Leadership and Innovation:*[6]

1. Identify new missions and programs for the organization.

2. Develop and nourish external constituencies to support the new goals and programs.

3. Create internal constituencies that support the new goals through changes in recruitment systems and key appointments.

4. Enhance the organization's technical expertise.

5. Motivate and provide training for members of the organization that transcends standard or accepted training goals.

6. Systematically scan organizational routines and points of internal and external pressure in order to identify areas of vulnerability to mismanagement, corruption, the loss of leaders' positions, and blows to organizational reputation.

Prison and jail systems administrators who have already taken these preliminary steps to limit vulnerability to mismanagement and corruption should forge ahead with boundary-spanning retooling of their training and staff development programs.

Performance Management

The U.S. Department of Justice has determined several main goals of performance-based management in corrections, each with their own indicators:

- Security—security procedures, drug use, significant incidents, community exposure, freedom of movement, staffing adequacy

- Justice—staff fairness, limited use of force, grievances (number and type, the grievance process), the discipline process, legal resources and access, justice delays
- Safety—safety of inmates, safety of staff, dangerousness of inmates, safety of environment, staffing adequacy
- Conditions—space in living areas, social density and privacy, internal freedom of movement, facilities and maintenance, sanitation, noise, food, commissary, visitation, community access
- Order—inmate misconduct, staff use of force, perceived control, strictness of enforcement
- Management—job satisfaction, stress and burn-out, staff turnover, staff and management relations, staff experience, education, training, salary and overtime, staffing efficiency
- Health—stress and illness, health care, dental care, counseling, staffing for programs and services
- Activity—work and industry, education and training, recreation, religious activities

Yet neither the Federal Bureau of Prisons (BOP) nor any other institutional corrections agency has developed performance-based management systems designed to serve both internal administrative and external constituency-building needs. As Kettl explains:

> The biggest difficulty in thinking through the problems of performance-based management is that reformers and managers far too often consider it simply as a problem of measurement. Committing the government to performance-based management, of course, requires that officials identify and measure results. The more fundamental question, however, is what to do with these measures. Performance-based management is most fundamentally about communication, not measurement. Moreover, this communication occurs within a broader political process, in which players have a wide array of different incentives. Performance based management will have meaning only to the degree to which it shapes and improves incentives. How does what [executives and managers] know about results shape decisions about what programs they adopt? And how does the process of measuring results affect the behavior of political institutions?[7]

Some institutional corrections agencies, like BOP, utilize performance-based management systems and have existing data systems, boundary-spanning training, and staff development programs. Other institutional corrections agencies, however, have yet to begin even the basic measurement phase of performance-based management innovations.

Public Communications

Improvements in correctional institution performance can only occur if practitioners of the correctional administration improve public communications about what they actually do and how they do it, subject to legal and other constraints. In particular, institutional corrections may be the only area of public sector life in which academics, analysts, activists, journalists, judges, lawyers, and lobbyists have turned the field's practitioners into little more than bit players in defining their clients, budgetary and organizational needs, and the likely social costs and consequences of alternative ways of performing their functions.

Prison and jail wardens and superintendents must actively enter these public debates; they are the practitioners and experts in the field. Even fundamental factual matters such as what the criminal records and physical health, mental health, and employment histories of most prisoners look like—the stuff of rap sheets, presentencing investigation reports, and reception and diagnostic admissions forms— are now largely defined by contending camps of nonpractitioner (or ex-practitioner) reformers and experts. The seeming lack of interest and involvement in these issues by current correctional facility administrators is puzzling.

Presumably, prison and jail administrators know a great deal about the official criminal records and troubled life histories of the confined populations they manage and have an organizational stake in sharing their best, most objective understanding of who their day-to-day clients are. They should also be keenly aware of the demands and challenges that they as institution administrators face in providing care and custody to large, transient populations of troubled and troublesome persons. Presumably, they have ideas about which, if any, subpopulations of inmates should be sentenced or treated differently to better serve the public. However, the poor state of public communications about the fundamentals of institutional corrections, the relative silence, is rather deafening. For the field's present and aspiring entrepreneurs in government, this silence is professionally and organizationally self-defeating and ought not to persist.

■ Management versus Leadership— An Unfair Dichotomy

Many exceptional researchers and authors distinguish between leadership and management. Warren Bennis describes the differences between leaders and managers as the differences between those who master the context and those who surrender to it. He believes the manager focuses on systems and structure while the leader stresses the people in the business. Bennis believes managers are more "blue collar" and heavily involved in the practical details of organizational life while leaders are more engaged in the loftier calling of organizational directing.[8]

Tom Peters and Robert Waterman are a little kinder toward management, although they also stress that leaders go beyond the daily grind of basic decision making. Their research reveals that successful people learn from their mistakes, think globally, encourage innovation, and master new knowledge.[9]

Other observers of leadership roles in prisons have softened the line between management and leadership. Kevin Wright describes the merging roles of managers and leaders.[10] While he describes many farsighted responsibilities of leaders, he puts them in the context of day-to-day management activities. A warden must watch the bottom line of the budget but also be able to articulate the necessity of accomplishing longer-term goals. Prison and jail administration requires unique individuals with exceptional people skills. Penology demands institution experience and leadership abilities.

As Chester Barnard said, "It is important to observe . . . that not all work done by persons who occupy executive positions is in connection with the executive functions."[11] In other words, if the warden or superintendent of an institution stops to listen to a complaint from an inmate or staff member, this is not expressly an executive task. As Barnard notes, nearly all high-ranking executives do a considerable amount of non-executive work, and this effort is sometimes more valuable than the executive function. The hands-on managerial tasks are generally associated with maintaining an organization and ensuring that systems of cooperative effort are emphasized and enforced.

It is critical for today's prison and jail chief executive officers to successfully combine the reality of management in the trenches with the ability to lead others toward the future. A warden must stay in close touch with the daily responsibilities of institution management yet set a tone that defines what the facility is and where it is headed. Separating management from leadership requires an artificial division that does not reflect the reality of organizational behavior. Hans Toch notes that penology deals with policy and administration, process, and procedures.[12] Daily oversight is necessary; a warden's job is very much a hands-on experience.

Many factors determine the quality of prison administration, but the most important is management practice. Strong prison and jail governance is key to the attainment of effective and efficient operations, but a chief executive officer with a strong, involved management style may possess many intrinsic leadership qualities.[13] The skills must be merged. The effective, reality-driven manager must model other characteristics of a leader. He or she must be able to empower and delegate to capable staff; demonstrate sensitivity, diplomacy, and vision; and show a willingness to take responsible risks.

■ The Challenge for Institutional Leadership

All wardens or senior-level leaders in any correctional environment want to be known for running a high-quality operation and getting positive results. But what are positive results in a prison or jail environment? An effective prison or jail operation is generally considered to be an institution that is safe, secure, clean, and responsive to the needs of its staff, inmates, and external constituencies. Yet if we gathered a group of correctional leaders and asked them to define these qualities, there would be some disagreement. Which factors are the most important? How do we operationally define successful attainment of each factor? How should staff go about trying to reach each goal, and how do we know when the goal has been attained?

Wardens and senior administrators must establish a vision of a successful correctional operation. They must delineate institutional goals, train staff to ensure that all personnel are aware of the desired outcomes, implement programs in support of the goals, establish a system of feedback on progress toward the goals, and create a means of reinforcing successful accomplishment and good performance. Once this process is in place, accountability and tracking institutional operations become key.

Delegation

Management is often defined as the art of getting things done through other people. A leader must delegate the responsibility for specific tasks to qualified individuals while remaining ultimately responsible for these functions. Today, delegation is often referred to as the act of empowering staff. The senior staff member not only must give the employee responsibility for the task but the authority and the resources to accomplish it. This can be a frightening concept in a prison or jail, where so many things can go wrong, and some errors have extreme consequences. Yet delegation is necessary if a major facility is going to operate effectively 24 hours a day. Staff must be able to take responsibility for work in the facility. Failure to delegate operational supervision to competent staff will guarantee personal burnout, damage the development of subordinate staff, reduce productivity, and harm overall morale.

Delegation is critical at all levels of correctional institutions. While the responsibilities of correctional management are more complex than those in many other lines of work, leadership and management in all organizations requires the sharing of responsibilities. Once leaders do delegate tasks, they must learn how to provide oversight and establish accountability.

Even though delegation is so essential to leadership and management, it is often a difficult skill to develop. Richard Phillips and Charles McConnell believe there are three primary barriers to effective delegation: old habits, lack of faith in subordinates, and the perceived lack of adequate time to train staff.[14] Traditional, comfortable work patterns can be difficult to change. To overcome this inertia, leaders must develop an awareness of this shortfall and practice new behaviors. Leaders must first surround themselves with a quality staff and then mentor them until the leaders become more confident in the staff's skills. These solutions require time; leaders more naturally assume that the job will be done better if they do it themselves. Unfortunately, this assumption is not always true. True leaders invest time in training and supporting others for the sake of personal sanity and for the future of the organization.

Measures of Success

Within the parameters of leadership and institution management, the most important factor in creating successful operations is the establishment of accountability. A superior manager creates high performance expectations, delegates the task, and follows up to evaluate results. Staff cannot and will not be able to comply with a superior's expectations if they are not communicated clearly. Senior managers and executives must share their ideas and concerns with staff. Once the information and desired expectations are placed on the table, it is important to set procedures in place to ensure compliance.

Paul Light highlights three primary avenues to accountability in a government agency:[15]

1. Compliance accountability seeks to measure conformity with written rules and regulations that are defined clearly. Auditing for compliance essentially looks for occurrences of non-compliance and corrects the situation, if possible; staff who erred are corrected with negative sanctions.

2. Performance accountability uses incentives to encourage staff performance. This positive program seeks to encourage appropriate and voluntary compliance with policy at the inception of a task.

3. Capacity-based accountability requires an organization to place necessary resources in positions that are required for a task. Staff, money, and technology must be available for the work to be accomplished in an effective manner.

The proper stewardship of resources is a fundamental responsibility of all correctional agency managers and staff. For example, the federal government, in an effort to support results-oriented management, implemented the Government Performance and Results Act. This Act requires all federal agencies to develop strategic plans, set performance goals, and report annually on actual performance compared to goals. These plans and goals must be integrated into several areas:

- Budget processes
- Operational management of agencies and programs
- Accountability reporting to the public on performance results
- The integrity, efficiency, and effectiveness with which these goals are achieved[16]

Management accountability is defined as the expectation that agency leaders will take responsibility for the quality and timeliness of program performance, control costs, and mitigate adverse aspects of agency operations. Leaders must assure that programs are managed with integrity and in compliance with applicable law. Management controls—organization, policies, and procedures—are tools to help program and financial managers achieve results and safeguard the integrity of their programs.

While there is no single best method to ensure accountability, in government the preferred method is compliance accountability. Auditing generally involves teams of personnel, internal or external to the agency, reviewing processes and procedures in order to compare actual performance with the expectations established by policy.

Historically, accountability in bureaucratic organizations has meant limiting staff discretion by utilizing carefully developed rules and regulations. Prison and jail employees in many situations and certain jurisdictions are treated as untrustworthy adolescents. The command and authority of large-scale organizations is generally very hierarchical, and every staff member's role and responsibility are delineated clearly. If all significant decisions are made by a limited number of senior executives, staff quit caring and pass all decisions up the ladder. Independence and creativity do not mesh well with this system. Ironically, in this inflexible accountability structure the very mechanisms intended to ensure quality sometimes reduce quality. Innovative employee behavior and risk taking can be lost when administrative control is too tight. Yet in corrections, the desire for tight administrative control is understandable. Prisoners of the state are not confined for conforming to laws. Many inmates are very willing to ignore a rule, policy, or general expectations of polite society. In short, they need external control. In order to ensure consistent application of institutional standards of behavior as well as develop a fair and standard way of operating, staff must know and follow correctional policy.

■ Creating Accountability

In prison and jail environments, the first step is to ensure that all staff clearly comprehend the agency's mission. The guiding vision must be established for all. If security and safety are to be primary goals, leaders must spell this out for staff. All personnel must grasp the purpose of the organization. Once this is established, the chief executive must establish the standards for operation that will support the organizational purpose. There are several important means of creating standards for institutional operation and ensuring compliance with them.

Policy

All prison and jail facilities must have written policy and procedures. These documents should establish the philosophy of operation for each program area, identify the outcomes expected, and define what is required of staff and inmates. Policy must accurately reflect current expectation and be expressed in measurable terms. In the prison and jail setting, control of a facility is organized around rules, and consistency is maintained by ensuring that all staff enforce rules in an impartial manner. This requires extensive staff training so all staff understand policy requirements. Ongoing monitoring by first-line supervisors ensures that policies are applied correctly.

Training

It is not enough to plan, write, and publish policy in a business or government setting. Individuals who are expected to comply with rules and regulations, and those expected to enforce policy, must receive appropriate training in the process, program, or procedure. Inservice training is extremely important to ensure that all personnel are fully informed of the strategy, whether it is a new process or a change in the current practice. In general, staff at all levels of the correctional agency will fare better with policy if senior managers take the time to explain why it is important and seek staff input.

Compliance Audits

For policies to have meaning, organizational leaders must assess compliance with written policy directives. Program audits are the best means of assessing staff operations. These routine reviews enable senior management to gauge program performance, determine the degree of risk, test the adequacy of internal control, and make midstream adjustments to operations to help achieve the desired results. It adds validity to the process if the reviewers know the program area but are organizationally independent.

Another aspect of auditing that can be very helpful is the requirement that program staff perform an internal review of their operations at scheduled intervals. Honest self-monitoring is truly the best means of keeping an organization on track and effective.

Benchmarking

It is always useful to compare one's operation to other similar programs. Key data points of institution management are identified easily (i.e., number of escapes, homicides, suicides, assaults, disciplinary reports, inmate grievances). Once this information is gathered, it is easy to compare data longitudinally at the same facility or for the same periods in similar institutions. Such benchmarking can provide managers with helpful data. For instance, does a positive "hit rate" of 13 percent of all urinalyses indicate an unusually high use of illegal and contraband substances? By comparing this percentage with the percentage at that institution one year earlier or with the percentage at another facility of similar security, leaders can better answer that question.

Accreditation

The American Correctional Association (ACA) has established standards for adult correctional institutions and separate standards for adult detention facilities. The ACA will, for an established fee, help correctional jurisdictions develop a local accreditation process and then provide a team of auditors to assess compliance with the nationwide standards on a pre-established schedule every three years. This excellent system provides external oversight of local implementation of nationally recognized standards. Such feedback can be very helpful in assessing institution operations, defending against legal challenges, and providing ongoing comparisons of significant aspects of facility management for line staff and senior managers.

Identification of Corruption

All correctional jurisdictions must have an internal affairs office to ensure that agency resources are used in a way that is consistent with the mission and government resources are protected from waste, fraud, and abuse. Staff and inmates must be accountable for their behavior and should understand clearly the expected standard of conduct. A crucial piece of management accountability is the obligation to ensure that institution operations are conducted with integrity and in compliance with the law. Inmate and staff allegations of impropriety must be investigated promptly and all individuals held accountable for their actions. Federal agencies are subject to the Inspector General Act and the Chief Financial Officers Act; state and local jurisdictions generally have similar watchdog legislation that puts teeth into management control.

Strategic Planning

Core values are critical to an organization, and they are rooted firmly in the mission of the agency. A warden must have a well-developed and well-publicized vision of what an organization is and what it is striving to become. Effective leadership does not require a sheriff, warden, or senior administrator who is highly charismatic and larger than life, whether feared or revered. Indeed, some of the great leaders in the private world of business avoid the limelight, focus on creating an organization that handles basic functions well, and keep their eyes on goals to which staff are deeply committed.[17]

Staff at all levels must be involved in this process of preparing the institution for tomorrow's trials and tribulations. Once a strategic plan is developed, it should serve as a guideline for moving the facility through the future issues that develop. One word of caution: Management accountability depends partly on the chief executive's ability to select appropriate organizational responses to new challenges and to ensure that staff stay on track with well-conceived goals. Planning commitments are very powerful; leaders can approve actions or plans that, while helpful in the short run, create lasting constraints on the organization. The best managers know when to make commitments . . . and when to break them.[18]

■ Large-Scale Management Structures

Governments often have huge bureaucracies to ensure that personnel do what they are paid to do. In the past, this has meant many people supervising other people perform routine tasks in a mediocre manner. Government leaders have expended great effort in reinventing how government employees lead and manage, and this has led to some recognition that less can be more.[19] It is important to create an atmosphere where staff believe in what they are doing, have input to how work is structured, and are permitted to exercise judgment in day-to-day tasks. While some prison and jail managers argue that top-down management is better management, this is not necessarily true. These opposing philosophies are not incompatible. While it is critical to have policy that establishes broad parameters and requires ethical enforcement, it is not necessary to require mindless compliance with little room for individual innovation. Government reformers are absolutely correct in their quest to put common sense back into government and to encourage individual enthusiasm. Too much emphasis on policy compliance can drive innovation out of an organization, but a sense of reasonableness should prevail. The establishment of management accountability is the heart and soul of being a high-quality leader within the fast-paced correctional environment.

■ Conclusion

Prisons and jails throughout the United States operate under many organizational and management structures—some clearly more effective than others. The more progressive systems have integrated their operations into cohesive systems that work effectively to meet the societal goals of confining inmates and preparing them for their eventual release back to their communities. In these few systems, the courts, corrections, parole, and probation staff seamlessly work with felons and then pass them from one stage of the correctional process to the next in a goal-focused manner. Unfortunately, most systems do not operate this way. In these institutions, the agencies work on their own disparate goals and typically are involved with prisoners only from their isolated perspectives. They then pass inmates on with little continuity of care and no ownership in the success, or failure, of the overall process. This disorganization affects many correctional systems and is the biggest weakness in the system of justice administration in the United States.

DISCUSSION QUESTIONS

1. Do you believe there is a difference between leadership and management?
2. Explain the importance of accountability and performance measurement in the correctional environment.
3. Is there a unique requirement in prisons and jails for this type of assessment that is different from other types of government, non-profit, or for-profit organizations?
4. How does a correctional leader establish an organizational culture of accountability and integrity?
5. Why is that culture important to prison and jail operations?

ADDITIONAL RESOURCES

G. Gaes, S. Camp, J. Nelson, and W. Saylor, *Measuring Prison Performance: Government Privatization and Accountability* (Walnut Creek, CA: Alta Mira Press, 2004).

J. DiIulio, D. Kettl, and G. Garvey, *Improving Government Performance* (Washington, DC: Brookings Institute Press, 1993).

R. Freeman, *Correctional Organization and Management: Public Policy Challenges, Behavior, and Structure* (Philadelphia: Elsevier Science and Technology Books, 1999).

R. Phillips and C. McConnell, *The Effective Corrections Manager: Correctional Supervision for the Future* (Sudbury, MA: Jones and Bartlett, 2005).

W. Wilson, "The Study of Administration" (1886), available at http://teaching americanhistory.org/library/index.asp?document=465, accessed August 30, 2007.

NOTES

1. C. Riveland, "The Correctional Leader and Public Policy Skills," *Corrections Management Quarterly* 1, no. 3 (1997), pp. 22–25.
2. D. Cressey, "Prison Organizations," in *Handbook of Organizations*, J. March, ed. (New York: Rand McNally, 1965).
3. J. Mooney and A. Reiley, *The Principles of Organization* (New York: Harper and Row, 1939).
4. I. Barak-Glantz, "Toward a Conceptual Scheme of Prison Management Styles," *The Prison Journal* 61 (1986), pp. 42–60.
5. D. Kettl, *Government by Proxy: (Mis)Managing Federal Programs* (Washington, DC: Congressional Quarterly Press, 1988), p. 50.
6. J.W. Doig and E.C. Hargrove, eds. *Leadership and Innovation: Entrepreneurs in Government* (Baltimore, MD: The Johns Hopkins University Press, 1990).

7. D. Kettl, "Building Lasting Reform," in *Inside the Reinvention Machine: Assessing Governmental Reform*, J. DiIulio, Jr. and D. Kettl, eds. (Washington, DC: The Brookings Institution, 1994).

8. W. Bennis, *On Becoming a Leader* (New York: Addison-Wesley Publishing, 1989), p. 37.

9. T. Peters and R. Waterman, *In Search of Excellence: Lessons from America's Best-Run Companies* (New York: Warner Books, 1982), p.118.

10. K. Wright, *Effective Prison Leadership* (Binghamton, NY: William Neil Publishing, 1994), p. 3.

11. C. Barnard, "The Executive Functions," in *Classic Readings in Organizational Behavior*, J. Ott, ed. (Belmont, CA: Wadsworth Publishing Company, 1989), pp. 265–275.

12. H. Toch, *Corrections: A Humanistic Approach* (Guilderland, NY: Harrow and Heston Publishers, 1997), p. xiv.

13. J. DiIulio, Jr., *Governing Prisons: A Comparative Study of Correctional Management* (New York: The Free Press, 1987), pp. 6–7.

14. R. Phillips and C. McConnell, *The Effective Corrections Manager* (Gaithersburg, MD: Aspen Publishers, Inc., 1996), p. 60.

15. P. Light, *Monitoring Government: Inspectors General and the Search for Accountability* (Washington, DC: The Brookings Institution, 1992), p. 3.

16. United States Government, Office of Management and Budget, *Government Performance and Review Act,* revised June 1995 (Washington, DC: U.S. Government Printing Office, 1995).

17. J. Collins and J. Porras, *Built To Last: Successful Habits of Visionary Companies* (New York: HarperCollins Publishers, 1997), p. 8.

18. D. Sull, "Managing By Commitments," *Harvard Business Review,* June 2003, pp. 82-91.

19. A. Gore, *The Best-Kept Secrets in Government* (Washington, DC: U.S. Government Printing Office, 1996), pp. 15–16.

Leadership: Executive Excellence

19

Harley G. Lappin

Chapter Objectives

- Explain the importance of leadership in the field of corrections.
- Outline the fundamental principles of an effective organization.
- Describe the concept of forward thinking and several of the relevant future trends.

▪ Introduction

The United States currently imprisons more people for longer periods than at any other time in its history. According to the Bureau of Justice Statistics, at year end 2005 the total number of prisoners under the authority of federal and state correctional jurisdictions exceeded 2.3 million persons.[1] The number of inmates returning to the community also has increased substantially over the years. The rates of incarceration and releases to the community reinforce the necessity for effective leadership of correctional systems, which is critical to an agency's operations and to achieving its mission of enhancing public safety.

Leadership in the field of corrections starts with the recognition that there are two core purposes of correctional systems:

1. To protect society by incapacitating criminals (preventing escapes and providing a safe, secure, humane environment within the prisons for all staff and inmates)

2. To reduce recidivism (the rate at which inmates who have served their sentences commit new crimes and are returned to the corrections system)

The rate of recidivism nationally is very high for all types of offenses and is influenced by many factors, including employment opportunities, family support, and peer group influence.[2] Correctional systems can help, but sound leadership is required across agency levels (from direct line supervisors to the agency's executive staff) to accomplish both components of corrections' core purpose.

■ A Context to Leading: Fundamentals

Since 2003, the Federal Bureau of Prisons (BOP) has made a concerted effort to reinvigorate staff focus on its core principles or values. There is no question that establishing and maintaining core ideologies is fundamental to the success of the BOP. The heart of the BOP's continuity has been its core ideologies; they are integral to both the agency's expectations for its workforce and its interaction with the outside environment.

The emphasis on core values has been particularly critical as the agency has had to undergo myriad major changes in order to reduce costs and live within its budget. BOP leaders have sought to maintain the essence of its operations while at the same time making substantial improvements and gaining efficiencies. Experience has demonstrated that the following principles are necessary to effectively lead an agency and position it to succeed under future leaders:

- Leaders must enable staff to think for themselves and develop and communicate their own innovative ideas.
- Organizational processes must be in place that can be learned by and communicated to others within the agency so the organization can function regardless of its designated leader.
- An organization must adhere to a firm core ideology and still be able to adapt to environmental changes; it must be both visionary and still have exceptional daily execution of the "nuts and bolts."

Core ideologies have allowed the agency to maintain consistency, even when faced with shifting political currents and other external demands. They have facilitated the management of the significant expansion of the federal inmate population and the BOP's transformation into the largest correctional system within the United States with a current total population of almost 199,000 inmates.[3] For the prison system, core values have kept resources applied to what is essential and have anchored the agency's national strategic plans. They make it easy for staff to understand the importance of what they do on a daily basis and understand how practices may change even though core values do not.

The following ideas reflect the importance of core ideologies to the successful accomplishment of the BOP's mission:

1. A safe environment for both staff and inmates
2. Secure institutions to confine offenders and protect the public
3. Skills-building programs to offer inmates the opportunity to prepare to live crime-free lives upon release
4. Service and stewardship to the public and a continued tradition of excellence
5. Staff who are ethical, professional, well-trained, and diverse

The first three ideas are specified within the BOP's mission. The fourth concept (service and stewardship) is a requirement based upon the agency's role as a steward of the public's trust and taxpayer funds. And finally, the fifth item recognizes the BOP's most important resource and the qualities it desires in its workforce—without quality staff, nothing else is possible.

While laws establish minimum standards of care to which all inmates are entitled, the BOP has always worked to achieve the highest of standards: to manage inmates effectively, to establish and maintain its reputation of excellence and outstanding public service, and to continue its leadership role in the field of corrections. Staff are largely the reason for this success due to three qualities that consistently have been present in those individuals who have attained significant professional and personal success: excellence, respect, and integrity. Striving for excellence and having respect for oneself and others are not only key to achieving success both in and out of work, but are also very important values, particularly for those who work in institutions, to model to the inmate population.

The current environment requires having the most capable, qualified, talented, hard working, committed, professional leaders possible. Leading is an enormous responsibility that carries with it high expectations. Individuals selected for advancement are acknowledged, formally, through the promotion itself, and informally, through the recognition of their hard work, dedication, and professionalism. And while they have already demonstrated their commitment to excellence on some level, with advancement their responsibilities expand significantly, as do the BOP's expectations of them.

No single individual has a magic formula for leading or managing, and one's philosophy should evolve with experience; however, review of the BOP's history indicates that leaders tend to be far more successful than others if they have certain qualities:

- Integrity—high integrity and ethics; conduct on and off the job must be beyond reproach
- Effective communication—ability to establish and communicate high performance standards and expectations, listen and communicate effectively with staff, have a positive attitude, and be a good role model
- Respect—recognition that each employee is an individual and that expectations may not be the same for everyone
- Self-awareness—knowing oneself and being aware of one's strengths and weaknesses
- Resourcefulness—ability to recognize and use wisely the extensive expertise and resources available, which are designed to help prison facilities succeed, and appropriately acknowledge assistance received
- Ability to analyze—ability to assess a given situation using the skills of the head (technical knowledge, delegation, strategic thinking) and skills of the heart (interpersonal relations, emotional intelligence, humility, acting as a team player) and knowing when these skills apply
- Loyalty—unwavering loyalty to the agency while effectively balancing doing what is right for staff and what is in the best interest of the agency

- Understanding of the big picture—an understanding of the climate and culture of the location and external factors that affect the agency (e.g., legislation, prosecutorial initiatives); knowledge of the agency's history and of lessons learned
- Problem-solving capabilities—the ability to zero in on causes and on what is important and overcome obstacles; the ability to make informed, sound decisions, even under stress, including the tough ones

Additionally, effective leaders must have a desire to continue growing and learning. If corrections professionals wish to expand their scope of impact by pursuing higher level leadership roles, then their performance and conduct must be exemplary, and these individuals must continue to enhance those skills necessary to succeed (e.g., by taking advantage of mentoring opportunities). No one is above learning, and each person is responsible for expanding his or her own horizons. It is critical that a corrections professional keep learning and stay on top of developments in the field. Stagnation is extremely counterproductive, because the agency, with or without a particular individual, will continue to adapt to its environment to ensure readiness to meet future demands. Thus, an individual must commit to ongoing professional development for himself or herself and for the agency.

Integrity

Staff integrity is key to accomplishing the BOP's mission and dealing with its very broad range of constituents effectively, including:

- Inmates and their families
- Government and court officials
- Law enforcement
- The public
- Advocacy groups
- Oversight entities
- The media

Integrity affects the agency's credibility; the level of trust and confidence garnered from its constituency groups depend on it. It affects how the BOP's message is received and the support that is returned. At the most basic level, integrity affects the agency's ability to secure funding and select sites for prisons.

As stewards of the taxpayers' trust, corrections professionals should not expect notice or praise from their constituents when they do the right thing, conduct themselves professionally, or excel at performing their duties. Doing the right thing is expected—in fact, it is what all correctional staff, management and line staff alike should expect of themselves, and those wishing to lead must model that, first for their peers, then for their staff.

Michael Josephson and the Government Ethics Center Commission (part of the California-based Josephson Institute of Ethics) offer five principles designed as a guide to public service ethics.[4]

1. Public office is a trust; use it only to advance public interests, not personal gain.
2. Make decisions on the merits, free from partiality, prejudice, or conflicts of interest.

3. Conduct government "openly, efficiently, equitably, and honorably" so the public can make informed judgments and hold public officials accountable.

4. Honor and respect democratic principles; observe the letter and spirit of laws.

5. Safeguard public confidence in the integrity of government by avoiding appearances of impropriety and conduct unbefitting a public official.

It is important to remember that the public views each BOP staff member as representing the agency, and corrections systems go as their staff go. Corrections professionals must conduct themselves in a manner consistent with the highest of ideals in all aspects of their lives. Leaders must model this behavior and demand it of their staff.

Understanding the Big Picture

Effective leadership requires a solid understanding of the larger context within which the agency operates. A leader must keep in mind that his or her institution is one small part of the agency, which in turn, in the case of the BOP or state departments of corrections, is just a small part of government. That larger context and factors external to the agency can directly or indirectly affect the correctional system: Using the budget as an example, if more is funneled to one department, less is available to the others, because the government (whether federal, state, or local) must live within certain total overall dollar limits.

As a large correctional system, the BOP deals with a variety of complicated issues and subpopulations that by necessity require a broad range of programs and services. Several challenges have driven much of the agency's decision making and planning over the past few years. But to grasp the enormity of these challenges and their impact, leaders must first look closely at the agency's history over the last several decades.

The BOP's inmate population growth was gradual through its early history until the late 1980s, when the Comprehensive Crime Control Act became effective. This legislation abolished federal parole and established sentencing guidelines. Several other major factors have affected the federal prison system, including legislation establishing mandatory minimum drug and firearms sentences; the National Capital Revitalization and Self-Government Improvement Act of 1997 (which required the Bureau to absorb the entire felony population of Washington, DC); and prosecutorial initiatives targeting gangs, firearms, immigration, and drug offenses.

The events of September 11, 2001, resulted in a pronounced shift in this nation's resource allocation to those departments or agencies responsible for homeland security and counterterrorism activities (and later to the military in support of the war in Iraq). The end-result to other domestic agencies, including the federal prison system, was several years of very tight budgetary constraints, further exacerbated by the country's response to the various natural disasters that followed, most notably Hurricane Katrina. Again, if one department receives more, less is available to the others from the finite budget, so the BOP has had to manage growth and crowding during this period of diminishing resources.

Extensive restructuring and streamlining initiatives had to be implemented to reduce costs and allow the agency to live within its means. Cost reduction initiatives cut across all agency levels, including:

- Closing four independent prison camps that were determined to be too costly to update and operate
- Discontinuing the intensive confinement program that had proven to be no more effective than regular camps at reducing recidivism (even though they were more labor intensive)
- Creating centralized designation, sentence computation, and classification processes
- Implementing a medical classification system (modifying how medical services are provided by identifying institution care levels based on availability of resources and designating inmates to those matching their specific medical needs)
- Consolidating two training sites

Some initiatives are still ongoing, including mission changes at a limited number of institutions (e.g., security levels) to better manage the inmate population and a restructuring of Federal Prison Industries, otherwise known as UNICOR.

These initiatives resulted in displacement of a large number of staff as positions were abolished, but it was extremely important for the agency to retain its highly skilled, trained, and experienced workforce. A vacancy clearinghouse process was established to help place displaced staff; the majority of displaced staff chose to remain with the BOP.

Since beginning the streamlining process, the result of these changes has been the elimination of thousands of positions and a cost avoidance of tens of millions of dollars, all accomplished without compromising the safety and security of the BOP's institutions. The flexibility, creativity, and dedication of BOP staff have contributed significantly to these achievements. By establishing and promoting a staff feedback mechanism on the agency intranet, the agency's executive staff was able to solicit staff input that identified many areas of potential cost savings. There is no question that open, two-way communication has been key to the success of these initiatives.

To optimally achieve the core purpose of corrections, leaders must ensure the BOP continually refines its processes and services. For example, in each of the past several years, the agency has released an average of more than 43,000 federal inmates per year back to U.S. communities.[5] Most inmates who go to prison also leave prison and return to the community. Consequently, BOP leaders must emphasize the importance of the release preparation portion of the mission, actively demonstrate their support in meeting this objective, and demand staff performance in a manner consistent with this goal (e.g., role modeling prosocial behaviors and appropriate interaction).

■ Federal Prison Industries (FPI)

Federal Prison Industries (FPI) is one of the agency's most important correctional work programs. This program is critical to the safety and security of agency institutions and helps it occupy about 18 percent of the work-eligible inmate population. Research has confirmed that inmates who work in FPI are 24 percent less likely to recidivate.[6]

Recent legislative changes significantly limited FPI's mandatory source status, making it necessary for FPI to compete for virtually all its sales. This adversely affected FPI's office furniture program, causing sales to drop and some factory closures. Congress continues to consider bills that would, if passed, significantly affect how FPI operates. Consequently, FPI is focusing its efforts on expanding the services, fleet management, and recycling business areas, which operate without any mandatory source preference and compete against private vendors (in some instances, even against low-cost providers from outside the United States). Again, by FPI leaders making a proactive decision to change based on their awareness of the political climate and public sentiment, they have been able to control their own future to the maximum extent possible.

■ Interagency Collaboration

Providing effective programming in prison is just one component of a successful reentry strategy. Corrections and treatment research demonstrate that treatment support to offenders entering the community under continued criminal justice supervision (i.e., transfer to a halfway house, probation and parole) reduces recidivism. Thus it is incumbent upon the agency to ensure continuity of appropriate programs, care, services, and support tailored to the specific needs of inmates, particularly for those with greater needs.

As partners in the criminal justice system consider the unique challenges related to community reentry, leaders must recognize some basic truths about what will be needed to succeed. No single agency or individual can do it alone; many agencies share some responsibility for ex-offenders. Leaders must ensure effective collaboration and communication during the entire incarceration process and well before release, involving the right people—in this case, involving all parties with a stake in the outcome. Agencies must make full use of available technology and automation to improve data flow within and across agencies and to reduce redundancies.

The National Offender Workforce Development Partnership (OWDP) includes representatives from several different agencies and government departments. It emphasizes collaboration, information sharing, and resource development with the purpose of improving reentry success by increasing career-oriented employment opportunities for ex-offenders. The bureau's involvement in this partnership ties in perfectly to its Inmates Skills Development (ISD) initiative. BOP leaders understand that if ex-offenders have viable, reliable jobs that provide a reasonable living wage, they are more likely to stay crime-free. In fact, work programs have been a key component of the Bureau since 1934 and have provided training to develop and enhance inmates' marketable skills since that time.

The BOP makes sure its employees understand that viable employment reduces the potential for rearrest and promotes a climate that encourages active support towards this objective. As change agents, leaders must share that awareness with staff, actively solicit and support innovation on their part that enhances collaborative efforts, and direct the implementation of those ideas that show promise or value. Various agency changes or activities have followed the decision to participate in the National OWDP, including:

- Guidance has been provided to all wardens on implementing partnerships at the institution level.
- The National Institute of Corrections (NIC) created and hosts a website that provides detailed information and points of contact for partner agencies.
- Joint Offender Workforce Development training is being provided by the NIC, the agency's ISD Branch, and Office of Probation and Pretrial Services for field staff and local community partners at targeted cost-effective geographic locations.

As one can see, a single decision (in this case, to be a member of the national OWDP) easily can trigger multiple consequences.

■ Forward Thinking

The seeds of the future are always planted in the present; the trick is to identify which will germinate. It is impossible to forecast some circumstances, such as what the consequences for the criminal justice system might be if there is another major successful terrorist attack, or a budget crisis, or a change in society's outlook regarding the criminal justice system's responsibilities regarding the drug trade and drug use. But it is a leader's responsibility to encourage innovative, future-oriented thinking and consider organizational enhancements to operational practices and culture.

The BOP strives to be its own best critic and to preserve a strong risk assessment capability. Critical self-examination readies the agency to meet future demands, as does the concept of forward thinking that was introduced in the agency by former Director Kathleen Hawk Sawyer in April 2001.

Many agencies have a forward-thinking process in place. In starting the agency down this path, Sawyer explained that the agency historically had done a great job strategically planning three to five years out. However, she recognized the need to plan strategically for the long-term (e.g., 20–25 years into the future), a process that would involve framing what the future might look like and enhancing the agency's flexibility. In order to best position itself for the future, the BOP has focused on incorporating proactive thinking into its entire culture, still an ongoing process. Ultimately, this approach will spread across the agency and get everyone thinking of the big picture and future possibilities. The essence of this approach is creating a new way of thinking for the agency's future leaders.

In the years since Sawyer posed this challenge, BOP leaders have been immersed in the process of forward thinking, researching and analyzing trends that may affect it, scanning the environment, and creating scenarios that will help track and plan for the effects of various outside influences on the prison system— all to ensure the agency's continued success. To that end, a Forward Thinking Review Team was established that oversees ongoing training to agency staff, dissemination of information related to this specific initiative, continued environmental scanning and trend identification and analysis, and the development of mechanisms to monitor scenarios. It also ensures the inclusion of a forward thinking element in all new agency initiatives.

Research was conducted on a global scale in the categories of environment, justice, technology, and workforce to identify emerging trends. Commonalities in trends were evaluated to determine the likelihood of having an impact on the agency in 20 to 25 years. The Forward Thinking Review Team further consolidated these trends and identified the top 30 trends potentially having an impact on the future of the agency. Information on these trends serves as a resource in the development of policy, recommendations for pilot programs, and preparation of executive staff papers. It provides a framework for discussing how the BOP is preparing to meet anticipated demands and changes, as well as provide some questions to challenge leaders or those aspiring to leadership positions.

Prevention, Rehabilitation, and Reentry

The review team found that partnerships with correctional facilities and their communities needed to focus on reentry programs for all offenders and that this should remain a primary goal of all correctional institutions in an effort to reduce recidivism. They issued a challenge to leadership: "What are correctional systems doing now, collectively or as individual agencies, that will effectively prepare the agencies for this future scenario?"

Inmate Work and Vocational Training Programs

The review team also recognized that inmate work and vocational training programs should be designed in partnership with community business leaders to support local economies and help ease the transition of inmates into the community's workforce. Towards this end, leaders must seek to design programming efforts both in partnership with and in support of the local economy and not in competition against it. Inmate labor is valuable to both correctional institutions and their communities, providing both valuable work experience and service while reducing the likelihood of inmate misconduct. As such, vocational training and educational programs are crucial to correctional facilities. The review team challenged leadership in this arena as well by asking, "How can wardens obtain buy-in from local community business leaders for such partnerships?"

Resource Allocation

A shift in national priorities means that resources are moving towards terrorism prevention and public education and away from corrections. As such, the review team challenged correctional leaders to "identify changes that may need to occur to enable corrections and supervision agencies to successfully accomplish their missions in the future if more resources are pushed to other areas."

Workforce Demographics

Trends in the workforce indicate that advancing technology creates the need for more educated and skilled workers. As such, the role of corrections professionals is expanding, and they will be required to perform more advanced skills. For example, the Health Services' Electronic Medical Record (EMR) initiative aims to enable all agency health care practitioners to access the health care records of all inmates housed in its facilities. The challenge in this area is assessing how technology will affect the 'nuts and bolts' of corrections.

■ Conclusion

Many opportunities and challenges exist for individuals seeking leadership roles in the field of corrections. Each individual is entirely responsible for his or her own future, both personally and professionally. To have an impact, particularly in leadership roles, individuals must strive for and demand personal excellence of themselves. The field of corrections absolutely demands it. Adopting this strategy while maintaining one's sense of humanity will yield success, but more importantly, the greatest sense of fulfillment.

DISCUSSION QUESTIONS

1. Why is strong leadership important in the field of corrections?
2. What are the fundamental principles of an effective organization?
3. What is forward-thinking, and how is it relevant to the field of corrections?
4. What forward-thinking trends have been identified by the Federal Bureau of Prisons?
5. In what ways can the leadership of the Bureau of Prisons respond to the challenges of forward thinking?

ADDITIONAL RESOURCES

Federal Bureau of Prisons, http://www.bop.gov
National Institute of Corrections, http://www.nicic.org
UNICOR, Federal Prison Industries, Inc., http://www.unicor.gov

NOTES

1. P.M. Harrison and A.J. Beck, *Prisoners in 2005*, January 18, 2007 ed. (Washington, DC: U.S. Department of Justice, Office of Justice Programs, 2006).

2. P. Langan and D. Levin, *Recidivism of Prisoners Released in 1994*, Bureau of Justice Statistics Special Report (Washington, DC: U.S. Department of Justice, Office of Justice Programs, 2002).

3. Bureau of Prisons Website, *Federal Inmate Population*, available at http://www.bop.gov, accessed June 21, 2007.

4. M. Josephson and the Josephson Institute of Ethics, Government Ethics Center Commission. *Preserving the Public Trust: The Five Principles of Public Service Ethics* (Los Angeles: Josephson Institute of Ethics, 2005).

5. Federal Bureau of Prisons Office of Research and Evaluation, internal agency unpublished documents.

6. W.G. Saylor and G.G. Gaes, "PREP: Training Inmates Through Industrial Work Participation and Vocational and Apprenticeship Instruction," *Corrections Management Quarterly* 1(2) (1997).

Governing: Personnel Management

20

Robert L. Wright

Chapter Objectives

- Outline the desired characteristics of correctional staff.
- Explain leadership selection in a healthy and effective organization.
- Explore the role of professionalism in running a penal institution.

With most correctional facilities spending at least 80 percent of their appropriated funds on personnel services for staff salaries and other payroll expenses, most prison administrators recognize that staff are the heart of their programs.[1] Staff are involved in every aspect of a correctional setting. They serve as ambassadors, peacekeepers, and role models. They lead, supervise, and control the activities of the men and women who serve time. Staff ensure that a facility is safe, humane, and efficient—a place where meaningful change may occur.

■ Staff Opportunity

People who have been successful in corrections are honest, hard working, and dependable. Their performance at previous jobs has been above average or better. They are valued members of their communities, participating in local organizations and helping to build strong, wholesome places to raise families. Corrections work provides an excellent career opportunity for people who have these distinguishing qualities.

Corrections has been one of the leading growth industries in the United States, and corrections continues to be a field where an employee may begin at the entry level—as a correctional officer, budget technician, teacher, recreation leader, or counselor, for example—and end up as a top-level administrator. With the growth in the industry, opportunities abound. The only limiting factors may be the employee's interest, education, and initiative. Many administrators recognize that the best wardens and superintendents have risen through the ranks and supplemented their correctional experience with criminal justice or human services education and specialized correctional education. Often, this education is offered by the American Correctional Association (ACA), the National Institute of Corrections (NIC), and local criminal justice academies. As correctional leaders recruit, train, and promote new employees, they recognize that they are preparing the next generation of correctional leaders.

■ Correctional Standards and Laws

The ACA publishes *Standards for Adult Correctional Institutions*, which establishes expectations for correctional administration and personnel management in at least 10 areas of critical importance:

1. Written policy and procedures
2. Staffing
3. Affirmative action and diversity
4. Selection
5. Probation
6. Physical fitness
7. Compensation and benefits
8. Ethics
9. Personnel records
10. Employee counseling

Other standards address staff training and the minimum guidelines for adult correctional facilities.

Correctional professionals regard the development of these standards as the most significant accomplishment made during the 20th century in corrections. Correctional administrators are encouraged to become fully conversant in these guidelines in their respective areas. Personnel and training are 2 of the 32 areas of correctional administration addressed by the 463 standards for adult correctional facilities.

Correctional administrators must also understand the law and statutes of their area; there are statutes, administrative codes, executive orders, court decisions, and personnel and civil service merit system rules that are specific to each jurisdiction. A working knowledge of these laws is central to enlightened, effective management. Legal counsel and personnel specialists should guide correctional leaders in the specifics of these legal concerns.

■ Workplace Behavior and Professionalism

There are other areas that corrections leaders must understand as well. Some wardens and superintendents recognize that certain staff may focus on the technical aspects of the business (such as shakedowns, searches, and counts), while paying less attention to issues concerning workplace behavior and professionalism. However, staff members must also pay attention to current expectations about workplace behavior and professionalism. Individual employees may be subject matter experts or may have made significant contributions, but if they lack self-discipline, disregard published departmental expectations, or neglect to treat fellow employees with dignity and respect, their corrections careers may be short lived.

Each correctional employee should be informed of the department's or agency's expectations. Employees at all levels should expect from and provide to each other:

- Decent, civil speech, without profanity
- Common courtesy
- Respectful conduct
- Cooperation and teamwork

These simple expectations are important. Where they are observed, all employees can expect a relatively pleasant, professional, and fair workplace. That workplace will be free from harassment, rude behavior, retaliation, disparate treatment, intimidation, and discrimination. Illegal discrimination includes any discrimination based on age, sex, marital status, race, religious or political beliefs, creed, color, or national origin, as well as the presence of any sensory, mental, or physical disability.

To encourage helpful, healthy working relationships, it is essential for correctional management to articulate these expectations about professionalism. And agency managers, without exception, must not only talk about the desired behavior but model it in every interaction. These understandings should be included in an employee handbook that is presented to each employee, emphasized during recruitment and selection interviews, discussed and modeled in new employee orientation, and reviewed in annual inservice training.

Meeting the important objectives of professionalism and teamwork—with correctional employees conducting themselves in a manner that is courteous, respectful, and businesslike—is essential for the agency and its mission. These minimum expectations are also essential for ensuring that correctional employees meet their personal and career goals.

Correctional staff must manage the lives of others—individuals who, when left on their own, may be out of control. To achieve good results with others, correctional staff must first manage themselves. As agents of the city, county, state, or federal government, employees can expect that there will be inquiries about performance. Everything correctional staff do is subject to scrutiny. The correctional agency must have well-established procedures—published and available to all staff—that outline the agency's practices for conducting business and handling inquiries into allegations about staff performance and conduct. These reviews of employee conduct must ensure that staff have an opportunity to review allegations. The agency conducts an investigation, the results are shared with the employee, and the employee has a hearing and opportunity to respond.

When employees conduct their day-to-day business in accordance with agency policy, administrative codes, and the law, they have the full support of the agency and the attorney general and other jurisdictional legal counsel. When staff at any level operate outside the published agency guidelines, they are on their own, individually responsible for their behavior and decisions.

Employees should always:

- Be mindful—while at work, stick to business.
- Be polite—work with others with courtesy and respect.
- Respect procedures and follow them.
- Respect honesty and deal factually with fellow employees and supervisors.
- Avoid gossip—staff should not pass along any information that is not factual.
- Contribute to a healthy workplace.
- Let things go—no one in public sector employment has any authority or license to participate in retaliation or intimidation behavior intended to frighten, belittle, discourage, threaten, menace, harm, get even, or "pay back."
- Honor coworkers and expect the best.
- Understand sexual harassment—harassment is behavior that is unwelcomed by another; it is not determined by the intentions of the instigator but by the recipient.
- Respect the rights, feelings, and opinions of others—rude behavior is unacceptable.

All employees in the corrections agency need to understand that there is no higher priority than the workplace behavior of employees. Every employee must remember that each employee is responsible for the healthfulness of the corrections workplace. Fellow employees must be respected, and concerns must be conveyed factually and with courtesy, never in a derogatory or mean-spirited way.

Correctional staff are challenged to be professional—to be honest, hardworking public servants dedicated to the public good. The public constantly evaluates correctional staff at all levels. The correctional employer and the agency should welcome inquiry and inspection. Whenever correctional staff have disagreements, they should remember their role as peacemakers. Employees should go directly to the individual with whom they have a concern and discuss the issue, seeking resolution. If that fails to resolve the situation, they are encouraged to seek assistance from their supervisor. There is enough negative activity in the corrections workplace that has been generated through the lives of the inmates serving time. It is destructive for corrections employees to be sidetracked or manipulated into negative behavior.

■ Leadership Development

Quality institutions with effective, determined, and inspired leaders that meet challenges effectively and professionally will foster many future leaders—unit supervisors, program managers, lieutenants, assistant superintendents, and superintendents.

Some of the most successful superintendents and wardens have learned from an effective, progressive leader. They have had the opportunity to observe the organization and the leader up close. Approaches and issues have been discussed, with the rationale for decisions examined and alternatives reviewed. Astute, attentive, aspiring corrections executives are able to build on and apply the lessons they learn from their organization. Interaction with their peers provides opportunities for them to work together effectively. As these individuals participate in the business of the organization, modeling the behavior of the leaders, they are also providing opportunities for evaluation of their own leadership qualities and integrity.

Healthy and effective organizations will foster healthy and effective leaders. Correctional administrators should expect surprises among the talent pool in the effective and healthy organization and select leaders solely on the basis of merit. It is not beneficial to attempt to identify future leaders too early. Rather than expend special training resources on only selected individuals, it is best to create a rewarding and rich environment where excellence is the order of the day, performance is evaluated, competition is healthy, and the level of performance of all employees is raised. In the healthy organization, people understand that any entry-level employee could end up running the place. The only limit is their interest, their education and experience, and their ability to apply their knowledge effectively and work with others.

These views represent the opinions of wardens who have served in healthy and effective organizations run by enlightened leaders. Truly great leaders often credit their work in a vital organization as the key to their success. There are few satisfactory substitutes for strong mentoring. However, many forums exist that are designed to assist and prepare future leaders. The ACA and NIC offer training, and professional corrections associations meetings can provide substantial opportunities for corrections practitioners to meet and share strategies and experiences. Also, the opportunity to read and study the accounts of correctional leaders can stimulate the insight, reflection, and assessments critical to the development of effective correctional leaders.

■ Setting the Tone

It has been said that corrections management is an art, not a science. Some people have degrees and substantial credentials, but they can lack common sense and the ability to understand key issues and significantly influence the organization. An effective corrections manager is never satisfied with the status quo. Goals and expectations are set and communicated, and the organization moves toward those goals. In large part, the effectiveness of the organization depends on the example that the leaders set. There is a direct correlation between the expectation of the leader and the results. When leaders or managers tolerate mediocrity, the result is mediocrity. When excellence and professionalism are the expectations, the momentum of the organization increases, and employees perceive the ways they may serve as helpful contributors.

Corrections is a people business. The staff of any correctional organization are common people; working together, they can achieve an exceptional result. Honest, hard working, and dependable, they follow instructions and treat others with dignity and respect. They show integrity and compassion, and they are effective team members. All staff must understand that they must follow agency guidelines and be consistent in performing their duties and enforcing rules. There is no room for free agents or mavericks; they create chaos and make the facility a dangerous place for others to work. When dealing with inmates, staff must remember to be firm, fair, and consistent. Staff should not show favoritism toward certain inmates and should hold inmates accountable for their actions. Staff in a healthy corrections organization realize that one of the greatest opportunities they have is to serve as an example and influence the inmates in their care by their professionalism.

Leaders must inform staff at all levels of the agency's position concerning rehabilitation. For example, they should remind staff that there is no quick fix for crime. While the agency provides opportunities for those in custody to make wholesome use of their time, staff should recognize that if change occurs, it comes from within the mind and heart of the individual. Change is possible, and many correctional employees are motivated by the hope that change will occur. A healthy institution is filled with hope, meaningful work and education opportunities, and significant self-help and leisure activities that challenge and motivate inmates.

Staff should understand that they set the tone in their facility. They establish the community and the quality of life. When staff understand their impact, facilities can offer life-enriching experiences for inmates serving time and a healthy and safe place for people to live, where staff are proud and honored to serve. Lives are affected positively while the interests of the public are served. Employees are truly role models—for other staff members and for the inmates of a prison or jail.

■ Corrective and Disciplinary Action

In a strong, successful organization, staff are selected on the basis of merit and everyone continues to learn. Therefore, it is in the agency's best interest to provide the training and assistance to help all employees be successful—to help them handle the challenges of corrections. The agency has a duty to outline what each position requires, prepare staff for changing roles, develop operational procedures, and show employees how to follow those procedures. This means senior management must provide appropriate opportunities for staff to become informed and proficient.

Training and orientation in a correctional institution should cover rules and regulations and their rationale—not only for the inmates but also for the staff. Training should also cover the consequences of not following rules and regulations. Every organization must have discipline in order to achieve its objectives. Supervisors, to be effective, must know the rules and administer them fairly.

The communication between supervisor and subordinate is critical. Supervisors must be mindful that agencies must train and improve staff. Positive reinforcement helps develop personnel, but when an order or rule is disobeyed, supervisors cannot ignore it. It is critical that any corrective or disciplinary action is conducted in private. It is embarrassing for anyone to be corrected in a public setting, and it makes for a much more positive encounter if these discussions are handled in an office that affords a confidential conversation. In nearly every situation, public employees have considerable rights to privacy.

The most common method of correcting unacceptable behavior is a simple warning. A minor disciplinary action such as a warning must be instructional and constructive. It must help the employee understand the importance of and reason for the corrective procedure being followed and the logic behind the facility's roles about how a task should be performed. Employees must understand that consistency—having all staff carry out their duties and procedures in a uniform manner—is essential for the institution.

Effective supervisors are thorough and thoughtful in addressing performance issues. When an event occurs, they find out what happened; who was involved; and where, when, and why it occurred. They address the issue with the employee privately in a calm and factual manner, without anger or debasing remarks. The warning should present the facts and be appropriate for the individual and the situation. It is often helpful to try and find out why the employee took the action in question. For example, asking, "I see you did it this way. Is there a reason for that?" The answer can help establish the motivation of the staff member. Once the employee realizes and admits to the error, and understands how to behave in the future, the warning process should be ended gracefully.

It is generally not helpful to threaten a staff member about what would happen if future mistakes were made. The supervisor should instead make sure that the employee understands that the "air is clear" and the situation is history. There are few situations for a supervisor that require more tact, good judgment, common sense, and fairness than the handling of corrective action with a staff member. Every supervisor must understand that the objective of corrective action is to help the employee be successful. Often, the manner in which corrective action is handled matters more than what particular issues are addressed.

Warnings, well-intended as they may be, are not always enough. When repeated rule or procedural violations occur, a supervisor should follow up with the staff member. In these situations, the supervisor should follow the same guidelines as in preparing for a warning but also document the facts of the situation, the conversation with the employee, and the expectations for the staff member in writing. It is helpful to have the employee sign this document to acknowledge that the facts are correct. This written document can be used in a memorandum of instruction or a letter of reprimand in the future.

The document should be reviewed with and presented to the staff member in private, and a copy should be provided to the individual. If performance issues persist, it will become necessary to take progressive corrective or disciplinary action.

Normally, corrective or disciplinary action for successive infractions will follow a progression in the disciplinary process:

1. Verbal instruction
2. Verbal warning

3. Written memo of instruction
4. Letter of reprimand
5. Disciplinary reduction in pay
6. Disciplinary suspension without pay
7. Demotion
8. Discharge from employment

Warnings and letters of reprimand are generally understood to be corrective action. Disciplinary action, on the other hand, is more severe and formal. Corrective action is usually taken by the supervisor, and disciplinary action is taken by the agency.

There are events or incidents that result in or require immediate separation from service and for which the principle of progressive discipline is not necessary. These include:

- Conviction in a civil court for domestic violence
- Crimes of moral turpitude (rape, child molestation)
- Reporting to work under the influence of alcohol or drugs
- Trafficking institutional contraband
- Sexual or inappropriate relationships with inmates
- Establishing personal relationships with families of inmates
- Bringing weapons into the institution

These and similar offenses may be expected to result in the employee's immediate suspension from duty, an investigation of the allegations, and a hearing as prescribed by the policy of the agency.

■ Conclusion

The chief executive officer, the warden or jail administrator, should strive to create a correctional facility where staff are recognized as the greatest asset, dignity and respect prevail, employees are valued and allowed to grow, and inmates and staff have a wholesome place to live and work. These personnel management duties cannot be outsourced. They are the central business of the agency, and they require the personal direction of the agency's top administrator. Strong central leaders will influence the people who live and work in these prison communities positively.

DISCUSSION QUESTIONS

1. Why is senior management expected to deal carefully with unacceptable staff behavior in a corrective manner?
2. Do you believe the concept of progressive discipline is helpful to the institution?
3. Is progressive discipline helpful to the individual?
4. How are employees in a correctional facility role models?
5. What forms of disciplinary action are available to management?

ADDITIONAL RESOURCES

E. Berman, et al., *Human Resource Management in Public Service. Paradoxes, Processes, and Problems* (Thousand Oaks, CA: Sage Publications, Inc., 2001).

P. Block, *The Empowered Manager: Positive Political Skills at Work* (San Francisco: Jossey-Bass, 1987).

R. Fisher, W. Ury, and B. Patton, *Getting to Yes: Negotiating Agreement Without Giving In*, 2nd ed. (Boston: Houghton Mifflin, 1991).

The Society for Human Resources Management, available at http://www.shrm.org/hrlinks/, accessed August 2, 2007.

K. Wright, *Effective Prison Leadership* (Binghamton, NY: William Neil Publishing, 1994).

NOTES

1. American Correctional Association. *The 2005 Directory of Juvenile and Adult Correctional Departments, Institutions and Other Agencies* (College Park, MD: American Correctional Association, 2005).

A Day in the Life of the Warden

James A. Meko

21

Chapter Objectives

- Identify the personal skills that a warden needs to be successful on the job and explain what staff and inmates look for in an effective warden.
- Determine the four areas of institutional operations that require the warden's close attention.
- Explain how the events at Abu Ghraib prison contributed to the field of correctional leadership in the United States.

There are few mundane workdays for a warden—the chief executive officer (CEO) of a jail or prison. Even on slow days, these leaders must be ready for emergencies or unusual events. Accordingly, it is difficult to describe a typical day for senior administrators of today's jails and prisons.

■ Necessary Skills

The effective CEO must either have, or actively develop, specific "soft" personal skills to be effective on the job. These include a positive work ethic, the ability to screen information to identify problems as well as opportunities, very good oral and written communication skills, and the ability to interact positively with racially and socially diverse staff and inmates. Additionally, the ability to analyze complex situations and to make timely effective decisions are also necessary to be effective.

A prison or jail CEO must be able to respond quickly to challenges that arise both inside and outside an institution. The best warden is a leader, not just a manager. He or she is proactive, anticipating and preparing for most correctional situations. An institution leader who is reactive is often overwhelmed by circumstances and may be slow to respond. Each day, strong leaders accomplish tasks that reflect their values and those of the people they serve.

A warden's day is not always limited to normal work hours, Monday through Friday. A proactive warden is available to staff at all hours, night and day. Such access to staff ensures that the CEO is always aware of important developments at the institution. The proactive warden ensures that communication moves along the chain of command in both directions. A lieutenant or department head should not hesitate to call senior staff on a question related to an important issue at any time.

■ Before the Official Work Day

On a normal day, the warden should review the local newspaper and television and radio broadcasts in preparation for discussing with staff any significant outside events that could influence the institution. Examples of such events include legislation affecting staff or inmates, a local politician's negative remarks about the facility, or an incident at another facility that could affect local operations.

Once in the institution, the CEO will review the shift commanders' logs of the previous day's or weekend's activities. These logs typically summarize events in the institution and identify both the staff and inmates involved in any incident. The warden must be satisfied that these occurrences were handled appropriately. If they were not, immediate corrective action should be taken.

■ Early Morning

A normal institution day often begins with a meeting involving the warden, the associate wardens, the warden's executive assistant, the head of institution security, and any other relevant staff. Generally, the security chief reviews significant events that have occurred during the preceding day or weekend, and each participant is to feel free to contribute new items or discuss plans for the day.

This meeting is important for several reasons. First, it keeps vital lines of communication open. Honest feedback, in both directions, ensures high-quality institutional operations. Such feedback also ensures that key players understand what is important to the warden and the agency, and empowers them to act accordingly. Second, the meeting enables the warden to give direction to one or all members of the group with the benefit that all hear the direction and are able to question the rationale for the direction. Such a system promotes mutual respect, shared values, and camaraderie.

The morning meetings with key staff enable the warden to share such strategies with decision makers at the institution and to ensure that all are on board. These meetings should never be longer than one hour, because each principal is responsible for a vital area of institutional operations; they best serve the agency on the job and not in a meeting. At the close of the meeting, individual members should feel free to seek some private time with the CEO to discuss significant personnel issues not related to the group.

Wardens should never forget that discussions about sensitive issues such as discipline, personnel, tactics, and strategy should be discussed with staff on a need-to-know basis. This means that only individuals with a real need to know should be told specific sensitive information. Just like military operations, corrections tactics and strategy, for example, are more likely to be effective if only those who implement them know the information in advance. If all staff know every detail of these plans, the chance of inmates becoming aware of them also increases, thus hindering success in operations.

The departure of the group gives the warden time to review incoming mail, identify individuals to prepare responses, and establish dates when the response is due. Staff who prepare mail responses must pay special attention to mail from particular individuals, such as those at the agency's headquarters, key legislators, judges, local officials, or local media representatives. The effective warden maintains a personal and continuing dialogue with all these individuals to keep them informed of noteworthy developments at the institution. This demonstrates respect for these individuals' positions and shows them that as a leader he or she is professional and responsive.

Mail reviewed or signed by the warden should be routed through the appropriate associate warden and department head for two reasons. First, it keeps them in the loop and ensures that they will feel a sense of ownership of the finished product. Second, it prepares each for positions of increased responsibility in the future.

■ Midmorning

Time must routinely be set aside to meet with key staff on important matters such as budget, personnel, facilities, industries, and strategic planning. It is important, however, that meetings not consume all of the CEO's time. The prison or jail administrator should establish guidelines for staff conducting meetings so that time is not wasted on extraneous, irrelevant issues. These guidelines should require a clear agenda, shared with participants in advance, and tied to articulated meeting objectives. Guidelines should also specify the preparation required of participants, establish a time limit, state that discussion will be encouraged and dissent tolerated, require that data presented be accompanied by visual aids to facilitate the presentation, and explain that no one will be permitted to monopolize the agenda. The individual chairing the meeting should ensure that each person's participation is monitored in a firm and dispassionately fair manner. These guidelines will ensure that the warden will have the time necessary to manage the institution by walking around.

Management by walking around, first identified by Tom Peters and Nancy Austin, is a trait of administrators who are in touch with both staff and inmates.[1] These senior managers know what is happening in their institutions. Such a philosophy encourages the warden to visit all areas of the facility, ask questions, listen to answers, assess the morale of both staff and inmates, and identify problems to be resolved.

Successful prison leaders visit special housing units, the food preparation area, the health services unit, and staff training venues at least once a week to assess the quality of operations in each area, measure staff performance, and listen to staff and inmate concerns. Other areas of the confinement facility should be visited on a regular basis, but not as frequently. Experience has shown that problems in these four underscored areas are good predictors of more serious issues.

While touring a jail or prison, administrators must really see what is happening. They must ensure that staff are communicating with inmates—not dictating to or haphazardly confronting inmates. If staff members are permitted to deal aggressively with inmates routinely, they quickly develop a confrontation mentality when speaking with inmates. Frank and direct communication demonstrates respect and tolerance. A domination mentality begets fear, intimidation, and hostility. Truly exemplary staff members care and control, exerting both compassion and authority as needed.

A warden must insist that staff be responsive to legitimate inmate concerns and complaints. Showers without hot water, broken heating systems, stopped-up plumbing, and poorly prepared food are small nuisances for inmates, but if these little problems are ignored, they can quickly become the basis of negative, collective inmate hostility. A warden who permits such inmate complaints to be ignored by staff will undoubtedly face major problems.

While walking and talking, the prison administrator must ensure that staff are following agency policies and that, even more importantly, they are respecting basic standards of propriety and decency. With the growing numbers of young and inexperienced staff members, it is often the warden's responsibility to ensure that these staff members know what is expected of them as well as why certain rules and regulations are in place. For example, security concerns dictate that inmates be strip searched upon arrival at a secure institution. However, an inmate's personal dignity can be respected in a search; the search should be performed in an area that protects an inmate's privacy.

Management by walking around also enables the warden to establish and maintain multiple channels of communication. In this way, the warden can take the pulse of the facility; firsthand knowledge of operations is an absolute necessity. It is a big mistake for a prison CEO to sit in his or her office and wait for subordinate staff to bring in news about what is happening. Too often, what the warden hears is what the reporting staff member thinks the warden wants to hear or what makes the staff member look good. Multiple channels of communication provide multiple sources of information to be analyzed and evaluated.

■ Early Afternoon

The afternoon begins with a daily stint "standing mainline" for the lunch meal. A proactive warden requires associate wardens and department heads to attend the serving of one meal each day. This provides additional opportunity to interact with the inmate population and staff. Questions or concerns can be responded to, or written down and responded to later. This is yet another opportunity to assess the atmosphere of the correctional facility. Positive interaction between senior staff members and inmates in the dining room is a very visible way for senior staff to demonstrate care, compassion, and fair management for inmates, as well as an opportunity to be responsive to legitimate inmate concerns.

The CEO thus sends an important message to all inmates: Those who administer the institution are approachable and responsive. The underlying purpose of meeting with inmates is to be responsive to each and every inmate regardless of economic status, race, ethnicity, crime, criminal history, or sexual orientation. It is critical to demonstrate to staff at all levels the importance of caring for each inmate's needs. Communication is always preferable to confrontation. Never does the warden want to send the message that he or she talks to only certain inmate leaders or specific inmate groups; inmates must not be given the message that they must deal through other offenders to get things done.

It is equally important to ensure that all associate wardens and the warden are not in the dining room simultaneously. One of the senior staff members should always be outside the security envelope of the institution in the event of a hostage taking. It is also good practice to regularly eat a meal in the inmate dining room. This allows senior administrators to evaluate the food, demonstrate concern for inmate welfare, and show confidence in the food service staff. Another good practice is to carry a food thermometer to test the temperature of food items on the serving line. This keeps staff on their toes and shows inmates that administrators care about what inmates are served. After mainline, it is worthwhile to visit the staff dining room to relax, visit with staff, field questions, and assess morale.

■ Midafternoon

Midafternoon meetings with various department heads enables the warden to monitor areas of facility operations. This can also be the time for close-outs with agency auditors who have been in the institution assessing the quality of some specific operation (e.g., business office, personnel, case management, etc.). These should be looked upon as learning opportunities. When audit teams critique a department's operation, it is important that staff not be defensive or challenge an auditor's integrity. This is a perfect time to learn about necessary changes and evaluate the performance of individual staff members.

At this time of day, a CEO might also greet tour groups as they enter or leave the facility. These encounters provide the CEO with a great opportunity to personally offer insight into facility operations and present the agency in the best light possible. Such tour groups are often composed of college students, members of the media, or local citizens.

■ Late Afternoon

Before calling it a day, a CEO must set aside time for staff development. The proactive leader sees staff development as a personal responsibility. This entails meeting with new and existing employees to outline the agency's, as well as the warden's, standards for behavior and treatment of inmates. No training or inadequate training sets the stage for problems both in the institution and in the courts. It is important to visit staff training personally as it occurs to assess its quality and to interact with participants and teachers. The effective CEO conducts annual refresher sessions on subjects such as staff integrity and career planning and meets with employees at all-staff assemblies to present issues of the day and anticipated changes in operations. A CEO must endeavor to be available to staff at all levels. Interaction with staff often provides wardens information not available through normal channels. Such conversations also help to identify outstanding employees for recognition or promotion and demonstrate to all the importance of two-way communication. Staff respect CEOs who not only lecture but who participate as students in required agency training. An institution leader must also make time to meet with individual staff members who need career advice and seek mentoring.

The warden should check out with associate wardens before leaving the institution to determine if there are any new developments that require his or her attention. Wardens benefit greatly by keeping associate wardens informed and by providing them with daily feedback about their performance. Administrators must evaluate all the information staff provide carefully in light of information received from personal observations and other sources.

The CEO must never permit a situation to develop in which teamwork is adversely affected by unhealthy competition between associate wardens. Such an atmosphere is dangerous and debilitating. In a jail or prison setting, all associate wardens must be capable of serving in the place of the warden for short periods of time. Given that requirement, each should understand that he or she may have to cover for the CEO tomorrow.

True leadership mandates that the institution warden give the associate wardens the necessary experiences and opportunities for personal development to assist them in achieving their career goals. This must include education, outside management courses, daily decision making, and agency training. In this manner, the leadership team at the institution builds on its strengths and overcomes its limitations.

On the way home or even on the way into work, the warden should consider stopping by the gym and working out. Stresses of the day have less impact if one is able to exercise vigorously for a period of time. The warden must endeavor to maintain balance in his or her life. Once at home, the prison or jail administrator should try to forget the day's events and focus on loved ones. One should never let a job be all-consuming. Whenever possible, the warden should be involved in community activities such as youth groups, church groups, or civic organizations. This helps the administrator maintain a positive perspective and make contributions to the community outside of the job.

■ Recent Lessons on Correctional Leadership from Abu Ghraib Prison

In closing, one should consider an extremely poor example of correctional leadership that occurred in Iraq in 2003 and 2004 at the Abu Ghraib prison under the jurisdiction of U.S. forces. There, Iraqi prisoners were tortured and humiliated by their keepers. Leaders should learn from bad experiences.

In testimony before the U.S. Senate Committee on Armed Services, the chief Army investigator, Major General Antonio Taguba, was asked by Senator John Warner to explain, in simple language, how prisoner abuse arose. The General responded, "failure in leadership sir, from the brigade commander on down, lack of discipline, no training whatsoever, and no supervision. Supervisory omission was rampant."[2]

The effective leader of a penal facility is informed, involved, and prevents such things from happening. The leader must be able to establish and maintain good relationships with people at all levels. Staff and inmates want neither a tyrant nor a pushover as their warden. They want a fair, firm, and consistent administrator who creates a positive interpersonal working environment. Everyone wants an individual who is friendly and cheerful, who listens more than he or she speaks, who keeps an open mind, and who is calm and considerate. A successful warden is one who is involved, praises as well as criticizes, avoids discouraging comments and gossip, is sensitive and considerate regarding others' shortcomings, and holds others in high esteem.

■ Conclusion

Correctional management is complex and demands a high level of energy, fairness, and integrity. Leadership of an institutional community requires a leader in the largest sense of the word; a warden really must be all things to all people. Wardens serve and protect the staff and inmates of the institution as well as the people in the community.

DISCUSSION QUESTIONS

1. In what way does management by walking around make a prison environment better?
2. Why is quality staff training so important in prison and jail operations?
3. What are some examples of effective leadership skills in correctional environments?
4. What aspects of a warden's day are generally predictable?
5. How can a warden ensure maintaining balance in his or her life, given the responsibilities that come with the position?

ADDITIONAL RESOURCES

W.W. Bennett and K. Hess, *Management and Supervision in Law Enforcement* (Belmont, CA: Wadsworth/Thompson Learning, 2004).

E. Buice, "Going the Extra Mile with the Media," *Law and Order*, April 2002, p. 16.

R. Giuliani, *Leadership* (New York: Miramax Books, 2002).

L. Kokkelenberg, "Real Leadership Is More than Just a Walk in the Park," *Law Enforcement News*, February 14, 2001, p. 9.

NOTES

1. T. Peters and N. Austin, *A Passion for Excellence* (New York: Random House, 1985), p. 123.
2. U.S. Senate Committee on Armed Services, *Hearings on Iraqi Prisoner Abuse,* May 11, 2004, transcribed by Media Millwork Incorporated.

Diversity of Correctional Officers

22

Peter M. Carlson

Chapter Objectives

- Explore the position of correctional officers today and explain how the role of correctional officers has evolved over time.
- Describe the importance of a diverse workforce and explain why women have typically had a difficult time being accepted in correctional positions in male facilities.
- Discuss the performance level of female correctional officers in varying corrections environments.

The Role of the Correctional Officer

Correctional officers generally make up the majority of the staff of a penal facility and are charged with directly supervising prisoners. Often, staff begin their careers in corrections as correctional officers. Many state systems and the Federal Bureau of Prisons frequently promote officers to other departments and jobs.

Correctional officers, also known as guards or hacks, have various shifts and responsibilities in the facilities. Correctional officers primarily are responsible for supervising inmates, maintaining order and discipline, and serving as informal counselors and mentors. They also oversee and control inmate housing, common areas throughout the institution, work areas, and the dining room. They assist in transporting prisoners to medical or other correctional facilities and to perimeter security.

Salaries for these security positions have improved greatly since 2000, and annual starting pay now ranges from a low of $21,119 in Mississippi to a high of $43,523 in New Jersey.[1] The national average annual starting salary is approximately $24,840.[2] In some states, security staff who choose to stay in the correctional officer position can earn above $45,000; lieutenants and captains can earn up to $67,000.[3]

■ Cultural Diversity

In the past, prisons were in rural areas, and security staff were usually hired from the local area. The demographics of correctional employees in the United States have changed drastically over the last 20 years, a shift that has paralleled the transformation that has occurred in other workplaces.

Previously, women and minorities were seen as a threat to the cohesiveness of the correctional workforce, and those officers were often victims of discrimination or abuse. Many African American, Hispanic, and women staff reported that Caucasian, male coworkers were very slow to accept them and were viewed as too supportive of the prisoners or not trustworthy.[4] Today, minorities and women are an integral part of the correctional workforce, as both correctional officers and senior managers. **Figure 22–1** illustrates the overall rise in the percentages of non-Caucasian and female correctional officers.

Administrators have tried to include people of all races and both genders and have become convinced that institutional management depends on the ability to relate to and communicate with inmates. Positive control and accountability within a penal facility cannot be viewed as the ability to respond to negative behavior with force. A well-run institution is clearly one that has open communication among staff members and between staff and prisoners—not tension.

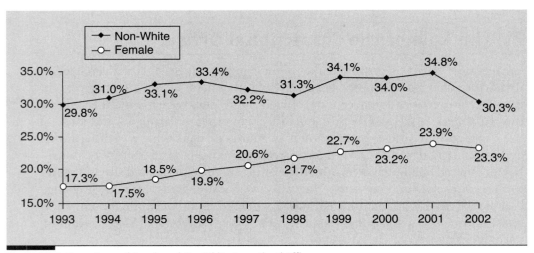

Figure 22–1 Percentages of Female and Non-White Correctional Officers
Courtesy of Camille and George Camp of the Criminal Justice Institute Inc., Middletown, CT.

Major riots and disturbances have often resulted from a lack of understanding and open communication. One of the most notorious rampages in correctional history was the riot at the New York State Correctional Facility in Attica in 1971. This revolt occurred because of many factors, but the commission that investigated the disturbance concluded that the predominantly Caucasian and rural correctional force could not understand or adequately relate to the African American and Puerto Rican inmates who were young and unwilling to accept the authoritarian attitude of the correctional staff.[5]

Attica and similar riots have provided some hard-learned insight into managing prisoners. Administrators have learned that fairness and reasonable treatment are critical factors in any well-operated prison or jail, and that staff members must be able to relate to the inmate population. The staff must "look like" the inmates; the ideal is to have the same proportions of Caucasians, African Americans, and Hispanics in staff and inmate ranks. This rule is a logical part of good institutional management. It makes sense to have a staff that can more easily relate to all inmates. With the continuing growth of inmate populations and the number of newly constructed correctional facilities, prisons and jails must adapt to the changes in the job market because they need more employees.

■ Female Correctional Officers

Stereotyping of occupational roles by sex has deep roots in American culture, including a "sex-linked ethos" in which specific occupational groupings are associated with one sex or the other.[6] The culture and ideologies surrounding the qualifications and pursuit of these employment roles tend to define both the qualified labor pool from which the occupations draw and the sought-after qualities and attributes of the people in that pool. While the number of women in the general labor force has expanded greatly in recent years, women have not been assimilated fully into professional, technical, or management roles, despite the fact that they have performed extremely well in all occupational areas.

Inside the criminal justice arena, the Joint Commission on Correctional Manpower and Training noted that in 1969 while women made up 40 percent of the general workforce they accounted for only 12 percent of the correctional workforce.[7] Women have been associated with prison work from its earliest days, but historically they have been employed for tasks associated with clerical duties, teaching roles, support services, or guarding female offenders. It was not until the 1970s that women correctional officers were placed in prisons with male inmates.[8]

In 1985, women accounted for approximately 13 percent of correctional officers in prisons; by 2002, that number had increased to 22.7 percent.[9] While this larger proportion is significant, it is still not as high as the proportion of women in the workforce in general. Explanations for the historical and contemporary employment bias in corrections cover a broad spectrum of arguments. Many males dispute the ability of women to maintain order and control in an adult men's prison environment. The conviction that prison tasks are "men's work" has been based in the belief that physical strength as well as bravery are requisites, and that women lack these qualities. Men have argued that the isolation and harsh working conditions of prison life are factors working against the entrance of women into corrections; others share the concern that the use of female officers violates the right of privacy of male offenders. Still other male officers and administrators believe that women create management problems by becoming romantically entangled with inmates or staff members.[10]

Resistance to the employment of women as correctional officers goes beyond simple personal bias. This attitude is often embedded in the organizational structure and culture of the prison. As new employees are placed in the correctional environment, they must quickly accept the customs, traditions, values, and other criteria of conduct that are part of the institution's environment if they are to assimilate successfully into their new work culture.

Employees' expectations and internal belief systems often determine how much success they have at work. Researchers note that both sexes behave differently in work organizations because men have more real power and greater promotional opportunity.[11] Individuals of both sexes, when placed in situations in which they are powerless and have limited promotion potential, respond by lowering their goals and developing differing patterns of behavior in comparison with those who see greater potential for opportunity and power. Researchers have found that women in law enforcement lack the aggressive social skills that men bring to the job: Women traditionally have had less experience with the aggressive behavior associated with organized sports and less experience with teamwork or asserting authority.[12] These skills have to be learned on the job and require new patterns of behavior and new body language, including facial expressions that project authority, rather than a pleasant demeanor or subservience.

Women in corrections report that even when they have the skills and commitment to the work, they are often at a major disadvantage. A glass ceiling, defined as an actual or perceived obstacle to organizational advancement for women or other minorities, is partially created by not including women employees in informal social circles, not providing mentoring assistance on the job, and holding women to a different standard of behavior. Paternalism or efforts to protect women often prevent them from working all posts in normal institution roster assignments; this places women at an experiential disadvantage and creates a competitive edge for others at the time of promotion consideration.

Numerous studies have reported great hostility and resistance from male staff when women have attempted to enter the correctional workforce as equals.[13] Criminal justice occupations have been often associated with machismo and masculinity, and the presence of women seems to throw doubt on this association. Some believe the jobs become devalued when women take them.

Additionally, the problem of sexual harassment affects the workplace in very negative ways. Harassment includes behaviors such as swearing, touching, intimidation, or even inappropriate humor. This type of conduct is often prevalent in work environments dominated by men and creates additional stress for women. In the prison and jail setting, men staff and inmates can present challenges even to the most competent women and establish an atmosphere that demeans and blocks women staff from achieving their potential. In its most favorable light, sexist and demeaning behavior creates an unpleasant work environment; at worst, such attitudes and behavior lead to high turnover rates and diminish the ranks of skilled women employees in the correctional facility. Some men staff members fervently believe that women employees working as correctional officers jeopardize the safety of men officers by being more susceptible to rape. Men staff members believe this requires them to place themselves in danger by having to respond to such explosive situations.[14] This concern seems to confirm the belief that the security of the institution, and therefore men's safety, is placed at risk by women's presence.

Female Performance

Women in correctional uniform have performed very well in women's and men's facilities at all security levels. While women have had difficulty finding peer acceptance as correctional officers in prisons, they have adapted reasonably well to this work environment. In the jail setting, studies have found mixed results. As predicted, women officers were perceived as being less effective in breaking up fighting inmates or controlling larger, more aggressive inmates. However, men staff members believe that women officers were very impressive in calming angry inmates and inmates who are mentally or emotionally disturbed.[15] Studies have found that inmates and men staff members have reacted positively to the presence of women personnel; men have controlled their language, acted more politely toward women officers, and exercised more care with their appearance.

Some evidence indicates that women working in male prisons conduct themselves differently as they perform the responsibilities of a correctional officer. Research has noted that some women adopt a more service-oriented demeanor than their male counterparts and take on a less confrontational style of dealing with offenders.[16] Many believe this style is an asset in the prison and jail setting.

A 1983 study of correctional officers in the California Department of Corrections found that men and women staff performed their jobs equally well. However, major differences were noted among the three job groups sampled: male officers, female officers, and male inmates. Both groups of men felt that women were less effective in tasks requiring physical strength and in violent emergency situations. Female officers scored well in all other tasks evaluated, including supervisory performance evaluations, number of commendations and reprimands, and use of sick leave.[17] In general, female officers were found to have established a great deal of self-confidence on the job.

Female officers in the Federal Bureau of Prisons fared well in a more recent study.[18] Both studies of female correctional officers noted many reports of hostility from male staff and male inmates, and examples of sexual harassment of female staff. This study found no difference in job satisfaction between male and female officers.

Many male correctional officers in high-security U.S. penitentiaries expressed their surprise at how effective female officers have become at their jobs—particularly in the cellblocks. Men working in this environment highlight women's ability to relate positively to male offenders. This skill becomes especially useful during tense situations, and many examples have been cited of female officers calming angry inmates. The majority of male staff, while initially opposed to hiring female officers in federal penitentiaries, changed their opinion after observing women on the job.[19]

Women have been integrated successfully into prisons and jails for men of-fenders in the last 25 years. While this public policy shift has not been easily accepted by the staff—mostly men—many institution administrators and line staff members believe women have helped increase the level of fairness and the quality of operations.

■ Conclusion

Equal employment opportunity and workplace diversity are important features of American society and institution administration. Caucasian male correctional officers have had to adjust to the changing workforce. The new personnel have brought many outstanding skills and abilities to the correctional environment, and institutional operations have benefited from this diversity. Minorities and women have become successful correctional officers, supervisors, and senior administrators in all areas of the American criminal justice system.

DISCUSSION QUESTIONS

1. Now that the correctional workforce is more diverse, how have women performed in this unique career environment?
2. Do you believe that the physical size of correctional personnel is an important factor when hiring new staff?
3. What are the benefits of a racially diverse correctional staff to prison management?
4. What are the benefits of a culturally diverse correctional staff to prison management?
5. Should male correctional facilities employee female correctional officers and vice versa?

ADDITIONAL RESOURCES

S. Camp, W. Saylor, and K. Wright, *Racial Diversity of Correctional Workers and Inmates: Organizational Commitment, Teamwork and Worker Efficacy in Prisons*, (Washington, DC: Federal Bureau of Prisons, Office of Research and Evaluation, 2000), available at http://www.bop.gov/news/research_projects/published_reports/equity_diversity/oreprcamp_jq2.pdf, accessed August 11, 2007.

T. Ellis, C. Tedstone, and D. Curry, *Improving Race Relations in Prisons. What Works?* Home Office: Online Report, December 2004, available at http://www.monitoring-group.co.uk/News%20and%20Campaigns/research%20material/Prisons/racism_in_prions_what_works.pdf, accessed August 11, 2007.

K. Scarborough and P. Collins, *Women in Public and Private Law Enforcement* (Boston: Butterworth Heinemann, 2002).

NOTES

1. Mississippi Department of Corrections, "MDOC Job Opportunities," available at http://www.mdoc.state.ms.us/mdoc%20job.htm, accessed June 12, 2007.
2. New Jersey Department of Corrections, "Careers in Corrections," available at http://www.state.nj.us/corrections/careers.html, accessed June 12, 2007.
3. C. Camp and G. Camp, *Corrections Yearbook 2002* (Middletown, CT: Criminal Justice Institute, Inc., 2003), pp. 168–169.
4. J. Irwin, "The Changing Social Structure of the Men's Correctional Prison," in *Corrections and Punishment*, D. Greenberg, ed. (Beverly Hills, CA: Sage Publications, 1977).
5. New York State Special Commission on Attica, *Attica: The Official Report of the New York State Commission* (New York: Bantam Books, 1972).

Chapter Resources

6. R. Kanter, *Men and Women of the Corporation* (New York: Basic Books, 1977).

7. C. Kehoe, "Addressing the Challenges of the Correctional Work Force," *Corrections Today,* August 2004, available at http://www.aca.org/pastpresentfuture/archivemessages.asp, accessed September 17, 2007.

8. J. Pollock, *Sex and Supervision: Guarding Male and Female Inmates* (New York: Greenwood Press, 1986).

9. Camp and Camp, *The Corrections Yearbook*, p. 157.

10. C. Feinman, *Women in the Criminal Justice System* (New York: Praeger Publishing, 1986).

11. Kanter, *Men and Women of the Corporation.*

12. S. Martin, *Breaking and Entering: Policewomen on Patrol* (Berkeley, CA: University of California Press, 1980).

13. P. Carlson, *Assignment of Female Correctional Officers to United States Penitentiaries: Implementation in the Federal Bureau of Prisons* (University of Southern California, doctoral dissertation, 1996), p. 12.

14. G. Ingram, "The Role of Women in Male Federal Correctional Institutions," in *Proceedings of the 110th Congress of Corrections* (San Diego, CA: American Correctional Association, 1981).

15. P. Kissel and J. Seidel, *The Management and Impact of Female Corrections Officers at Jail Facilities Housing Male Inmates* (Boulder, CO: National Institute of Corrections, 1980).

16. G. Alpert, "The Needs of the Judiciary and Misapplication of Social Research," *Criminology* 22 (November 1984), pp. 441–456.

17. H. Holeman and B. Krepps-Hess, *Women Correctional Officers in the California Department of Corrections* (Sacramento, CA: California Department of Corrections Research Unit, 1983).

18. K. Wright and W. Saylor, "Male and Female Employees' Perceptions of Prison Work: Is There a Difference?" *Justice Quarterly* 8 (1991).

19. P. Carlson, *Assignment of Female Correctional Officers to United States Penitentiaries: Implementation in the Federal Bureau of Prisons*, pp. 162–163.

Labor Relations

23

Michael H. Jaime and Armand R. Burruel

Chapter Objectives

- Provide examples of typical and unusual provisions found in collective bargaining agreements covering the corrections jurisdictions.
- Explain why bargaining toward a master agreement in a corrections jurisdiction can be prolonged.
- Outline the components of a sound management and supervisor training program on labor relations and staff management.

Among the many demands that prison and jail administrators face, one of the more difficult aspects of institution administration involves the interaction between management and labor organizations. This complex process is referred to as labor management relations.

All corrections administrators, whether employed by a prison or jail, must become involved in the plethora of labor management relations activities. They must learn the vocabulary of labor relations: grievance, unfair labor practice, job steward, collective bargaining agreement, mediation, arbitration, and so forth. But labor management relations need not be daunting; every organization has a management professional who knows how to deal with labor management relations. Administrators should remain calm and handle any labor relations matter professionally, not personally.

■ Relevant Acts and Laws

In 1935, Congress adopted the Wagner Act, or the National Labor Relations Act (NLRA), which formally recognized employees' rights to form and join labor organizations and to participate in collective bargaining. The NLRA thrust upon the nation new rules that would be codified in later years—in whole and in part—by federal, state, and local jurisdictions. These rules provided rights to employees and labor organizations to organize, bargain collectively, and represent members before management and created definitions of unfair labor practices.

In 1947, Congress further amended the NLRA with the Taft-Hartley Act. The various amendments prohibited unions and labor organizations from engaging in certain activities. Further acts and amendments established other prohibitions and directions for labor and management.[1]

States, in turn, adopted their own laws governing collective bargaining. For example, in 1967 New York State adopted the Taylor Law, which was used as a model by other states. This new legislation took several important steps in labor relations by:

- Granting employees rights to organize and be represented by employee organizations of their choice
- Requiring public employers to negotiate and enter into agreements with public employee organizations regarding their employees' terms and conditions of employment
- Establishing impasse procedures for the resolution of collective bargaining disputes
- Defining and prohibiting improper practices by public employers and public employee organizations
- Prohibiting strikes by public employees
- Establishing a state agency (the Public Employment Relations Board or PERB) to administer the Taylor Law[2]

In 1977, California passed the State Employer Employee Relations Act, later to be called the Ralph C. Dills Act. The Dills Act is a comprehensive labor law that governs the collective bargaining process of California's state employees. The Dills Act, like New York's Taylor Law, took several important steps by:

- Providing employees the right to organize and be represented
- Requiring the governor or a designated representative to meet and confer in good faith on matters within the scope of representation
- Providing for a mediation process (impasse process) in the event the parties fail to reach agreement
- Defining unlawful practices for the state or an employee organization to engage in (unfair labor practices)
- Establishing a state agency (Public Employment Relations Board, or PERB) that would be responsible for the administration of the Dills Act[3]

Collective bargaining laws such as these empowered corrections employees to form and join the labor organization of their choice, via an election process, and to select the organization that would be their exclusive representative in labor matters before management (see **Figure 23–1**).

■ Issues that Are Driven by the Union

Corrections administrators should research issues important to their respective labor organizations. These issues may vary from area to area; however, there are four basic vehicles used by the union to deliver messages to management.

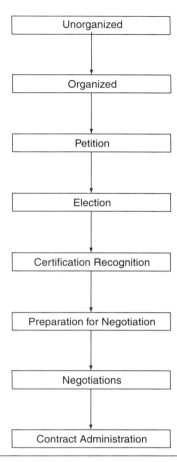

Figure 23–1 The Collective Bargaining Process

1. Master bargaining table issues or local bargaining issues
2. Grievances
3. Unfair labor practice charges
4. Issues of importance that the union discusses with management on a regular basis

Bargaining Issues

The concept of management meeting with those employee organizations that have achieved exclusive representative status is well-established in law and is a routine practice in most jurisdictions of state government and in the federal government. The basic and most important right in any collective bargaining law is the exclusive representative's ability to negotiate terms and conditions of employment through the "meet and confer" process on behalf of the rank and file. Supervisory employees in some local and state governments (e.g., police, fire fighter, corrections, and other first responder entities) may also be afforded the right to be represented by an exclusive representative and the right to engage in the meet and confer process.

"Meet and confer" is a phrase commonly used in labor relations that essentially means to bargain. Management and the exclusive representative are obligated to personally meet and confer promptly for a reasonable period of time upon request by either side. In these meetings the parties exchange information, opinions, and proposals and endeavor to reach agreement. The product of this meet and confer process is a master agreement. "Master agreement" is a phrase describing the main collective bargaining agreement or contract to which other later agreements or subordinate agreements may be appended. Master agreements are considered a binding contract under the enabling jurisdiction's collective bargaining laws and are perceived by state courts and federal courts as such.

Depending upon the provisions of each jurisdiction's bargaining law, items outside of the master agreement, such as a management change in policy or practice that is to occur after the master agreement is settled, may have to be negotiated with the employee organization. This commonly requires that management provide a written notice to the exclusive representative stating an intention to negotiate the item that is subject to the change in policy or practice and allow time for a written response. The impact of the change that is being made—not the decision to make the change—is generally subject to the meet and confer process. However, depending upon each jurisdiction's collective bargaining law, in the absence of a master agreement that provides controlling language, management may be obligated to bargain with the exclusive representative over the decision to make the change as well.

Grievances

A grievance is a dispute between the employee organization and the employer, or a dispute of one or more employees against the employer, involving the interpretation, application, or enforcement of the provisions of the master agreement. The grievance process typically will be specified within the collective bargaining agreement. Those jurisdictions that do not have such an agreement may have a grievance or a complaint process spelled out within some form of law. The exclusive representative will use the grievance process to challenge management's administration of the collective bargaining agreement. If not satisfied with management's response to the grievance, the union may ultimately elevate the grievance to a binding arbitration process. It is at this step in the grievance process that management and the union must abide by the final decision made by a neutral third party, the arbitrator. Normally, only contract grievances can be arbitrated.[4]

Once a grievance has been filed properly, management should review the practice that is creating the concern promptly. Each institution should have a clearly articulated grievance procedure and grievance investigation process to help ease this labor management relations process.

Unfair Labor Practice Charges

The unfair labor practice typically proscribed under an enabling collective bargaining law is an action or decision by an employer or a representative of the employer that directly or indirectly interferes with the organizational rights of employees or employee organizations.

Under most collective bargaining laws, actions or decisions that may be unlawful for the employer to engage in include:

- Imposing or threatening to impose reprisals on employees
- Discriminating or threatening to discriminate against employees
- Interfering with, restraining, or coercing employees because of the exercise of their guaranteed rights under a collective bargaining law
- Refusing to meet and confer in good faith with a recognized employee organization

Similarly, actions or decisions that may be unlawful for the union to engage in include:

- Causing or attempting to cause the employer to violate a collective bargaining law
- Imposing or threatening to impose reprisals on employees
- Discriminating or threatening to discriminate against employees
- Interfering with, restraining, or coercing employees because of their guaranteed rights under a collective bargaining law
- Refusing or failing to meet and confer in good faith with the employer
- Refusing to participate in the mediation process[5]

Collective bargaining laws typically include provisions stating that management may not dominate or interfere with the formation or administration of any employee organization, contribute financial or other support to such an organization, in any way encourage employees to join any organization instead of another, or refuse to participate in the mediation procedure. This portion of the collective bargaining laws typically are considered applicable during those periods of time, either specified in law or by the administering agency such as a PERB, when two or more competing organizations are petitioning and soliciting employees to vote for the competing organizations to be their exclusive representative.

An unfair labor practice is adjudicated before the administrative body, such as a PERB, that is empowered by legislation to oversee the appropriate administration of the collective bargaining law. For instance, under the NLRA, unfair labor practice charges are brought before the National Labor Relations Board.

Issues of Importance

Employee organizations or their representatives frequently bring up other matters of concern to the organization and the membership within the context of other forums. These forums may include an informal or formal meeting with the agency director, warden, sheriff, chief of police, or administrators of a jail or prison. Although the issue or item being referenced by the employee organization may either be a topic in current or previous negotiations or a grievance, arbitration, or unfair labor practice case, the union nonetheless may take every available opportunity to redirect its efforts in dealing with the issue. All management staff must be aware of what is important to the corrections or law enforcement organization. If the issue that is raised appears to be a significant item to the union, it would be very helpful for the manager to communicate this to the labor relations or employee relations office and the administration. It is best to coordinate a uniform response to the employee organization when dealing with matters of policy or significant practice that concern the employee organization.

■ Successful Labor Relations

Each corrections organization generally has a professional or group of professionals who are trained in dealing with labor organizations and labor issues. Typically, in any prison or jail operation, this will be the employee relations or the labor relations office. In an effective corrections organization, the labor relations office provides resources to its departmental management staff. Investing administrative resources (e.g., time, staff, and training) will help create a coordinated program effort that will meet the challenges presented by the labor organization. Typically, a labor relations office or program within any governmental entity will be charged with representing management in all areas of labor management relations, including contract negotiations and administration, resolution of employee grievances and complaints, arbitration cases, unfair labor practice charges, and related court litigation.

Training is the key to the success of an effective labor relations program. Training should be conducted in areas such as contract administration, grievance handling, the manager's and supervisor's conduct with the union, the bargaining agreements process, and basic interaction with job stewards or union representatives. For example, the California Department of Corrections and Rehabilitation offers basic and advanced training in labor relations to supervisors and managers. The training also includes a management development training module on how to negotiate.

In California's basic labor relations training for supervisors, first-line supervisors learn why the agency has collective bargaining agreements with employee organizations, how to interact with the job steward, how to investigate and respond to a first-line grievance, how to identify the basic properties of an unfair labor practice, and how to avoid unfair labor practices. The training that addresses advanced labor relations for supervisors instructs second-line supervisors on the detailed background of and case law concerning the negotiation process. Second-line supervisors are given instructions on how to bargain, and an overview of what makes a management bargaining team, the importance of taking notes at a bargaining table, and how to implement a bargaining agreement. When second-line supervisors are promoted to the management ranks, they must attend a three-week management development training program. This includes a hands-on course on how to conduct local bargaining. Participants hear lectures on the negotiations process and become involved in a mock bargaining scenario. This training scenario allows the participants an opportunity to respond to hypothetical union proposals on staffing, health and safety issues, as well as other routine issues that regularly confront correctional managers.

Training sessions such as these will not make the average supervisor or manager a highly skilled labor relations professional. However, they do allow supervisory and management staff to become familiar with labor relations and gain confidence in addressing the issues presented by employee organizations.

Working with an employee organization representative on various issues does not mean always reaching agreements and concessions. It does require treating the representative with integrity and respect. It is always beneficial for management to resolve problems at the lowest level, working with the representative as a problem-solving team. Management should strive to make handling grievances a positive experience rather than one filled with conflict and hostility. Whether the representative becomes a link or a barrier between management and employees depends on the way management deals with day-to-day situations.

Taking care of problems is the key to effective relations with the labor organization. All issues, large or small, can be equally important to the union. A small, unattended problem can eventually blow up into a huge administrative nightmare. By dealing with issues promptly and professionally as they are brought forward, prison and jail administrators will be communicating a very important message to the union: Management is responsive and responsible. Administrators should not worry if the response they must provide is negative. "No" is a valid answer. Sometimes, management's response to the union will be affirmative. Management must know how to separate frivolous issues from bona fide ones.

Management's response should be deliberate and well thought out. It should initiate a course of action to take care of the problem. Management should avoid the "you can trust me" approach with no clear plan of action for follow through, which leads to the "you can't trust me at all" syndrome. If a management official loses credibility on one issue, it can affect all subsequent discussions about other problems.

As mentioned earlier, grievances in a specific department should indicate the problem areas being encountered in the interpretation of the bargaining agreement. When in doubt about how to respond appropriately to any union concern or grievance, managers should contact the employee relations or labor relations office.

■ Developing the Management Team

Communication helps keep the management team cohesive and effective. Corrections administrators should share the routine as well as the unusual occurrences involving the union with the department's labor relations staff. Administrators are not expected to know everything about labor management relations. If managers have questions, they should contact professionals who will be able to help. Training programs for administrators, managers, and supervisors on labor management relations develops understanding, clarifies expectations, and improves the abilities and confidence in the management team members to effectively address the labor organization while representing management's interests.

The management team consists of all administrators, managers, and supervisors within the correctional facility or agency. Management staff involved in any specific issue may vary depending on the issue. For example, during master bargaining table negotiations, the management team will be led by the agency director, who make the decisions. Wardens, jail commanders, other corrections administrators, specialists in labor relations and human resources management, and other administrative staff may participate as the main body of the negotiating team. It is not advisable to include top decision makers directly at the bargaining table. Similarly, at the local level, the warden or other senior administrator serves as the decision maker, and subordinate institution administrators, custody management, and administrative staff typically will make up the main body of management's bargaining team that develops and researches proposals. Again, if the management team members have received appropriate training, their familiarity with the labor relations process will make them stronger management team members.

■ Ethics and Labor Relations

Ethics refers to standards of conduct—standards that indicate how one should behave based on moral duties and virtues, which themselves are derived from principles of right and wrong.[6] The same principles used in evaluating all ethical behavior can be applied to ethics within the labor relations process. The following six "pillars of character" developed by Michael Josephson model ethical behavior in this process:

1. *Trustworthiness.* When interacting with the union or employee organization, will the supervisor be looked upon as honest in communication and conduct? A supervisor should be perceived as a reliable person who gets to the bottom of an issue and can interact in good faith with the union.

2. *Loyalty.* As a manager, being loyal to the corrections organization is the top priority, even though this is often somewhat difficult. As long as a manager is straightforward with the union and not thought to be playing games, the union will respect his or her loyalty. The union representatives will argue a decision, but they will come back again and again if a supervisor is perceived as fair to the union and loyal to the corrections organization.

3. *Respect.* One can command the respect of others only if one gives others respect. While a union representative or job steward on the line may be a subordinate, when that representative is discussing labor relations issues with a manager, the representative should be treated as an equal. A manager should not be expected to take abuse from the representative. Civility, decency, and tolerance will promote a good relationship between both representatives.

4. *Responsibility.* Being in charge of an institution, a housing unit, or other corrections area is a big responsibility that demands accountability, a pursuit of excellence, perseverance, and self-restraint.

5. *Fairness.* Probably the most important ethical characteristic that labor organizations expect from management is fairness. If a manager is resolving a grievance or an informal complaint, the job steward will always compare the manager's actions with actions taken in the past. Did management interpret the contract uniformly? Was any favoritism involved in the action? If a manager's actions are perceived to be unfair, the union will make life miserable for all concerned.

6. *Caring.* Having a caring attitude does not mean giving everything away. However, if a manager is perceived as not having a caring attitude, the goodwill between labor representatives and management can erode quickly. Then, instead of being considered an ethical corrections

administrator who is loyal to the institution, he or she may be perceived as lacking integrity.[7]

Politics in Public Sector Labor Relations

As money makes political access a reality to various organizations, so too, public sector labor unions use money to influence their access to the different strata of politics. The state-level corrections and local-level jail environment is not different in this regard. Political oversight of correctional facilities, whether at the federal, state, or local level, brings opportunities for politically sensitive public sector labor organizations. In recent years, some public sector labor unions have successfully received dues increases through a membership vote. Some funds received by the labor organizations from the dues increases were plainly advertised for the purpose of influencing voters during elections at local and state level offices and influencing elected officials and providing future election support.

This reality of political access must be understood by correctional workers, whether line workers, supervisors, or managers. Political interest in the correctional environment may be heightened for a variety of reasons. Labor unions or employee associations with political access can draw public attention to the many operating practices of the correctional organization. Local governments are not immune to similar political complexities and challenges and also have labor unions representing corrections, law enforcement, fire fighters, and other first responders.

In "right to work" states, where there is a restricted or nonexisting right for public sector employees to organize for collective bargaining purposes, employee associations may still form under other protections found in federal or state laws. In these states, the associating practices of persons with a shared community of interest can, theoretically, also lead to a powerful money engine that may be used for political access. However, these types of associations typically present a less compelling political voice by comparison to those associations found in collective bargaining states.

These differences, based on the political fabric and the role of associations and labor unions, necessitate that corrections administrators, managers, and supervisors pay close attention to the earlier mentioned principles of effective management practices, including the call for ethical decisionmaking.

Conclusion

Labor relations issues in any correctional environment are always affected by the perceptions of both parties. Management must deal with all labor management relations issues in a professional, honest, and straightforward manner. Emotions should never enter the arena; once an issue becomes personal, it will be extremely difficult to resolve. All prison and jail administrators have a network within the local agency and across jurisdictions. When a particularly tough issue arises, administrators should call other administrators or wardens and find out how they have handled similar issues. When in doubt, one should request assistance from labor relations staff.

While labor management relations can be contentious, both sides should remember that they are working for the same agency and share the goal of operating safe and effective correctional facilities.

DISCUSSION QUESTIONS

1. Should public sector labor unions have limitations on how much political access they can exercise on matters related to corrections administration?

2. Should supervisors or managers in a corrections jurisdiction have collective bargaining rights?

3. If ethics and ethical decisionmaking is important to the corrections environment, what should an ethical relationship between a corrections labor organization and a corrections administration look like?

4. How does a corrections administration ensure its ability to manage operations in order to ensure public safety and public service?

5. How would a collaborative partnership to reduce crime and victimization that involves community-based organizations, law enforcement, corrections, parole, and faith-based organizations be affected by the bargaining rights and political influence of labor unions?

NOTES

1. L. Kahn, "The Law of Labor Relations: An Overview," in *Primer of Labor Relations*, 25th ed. (Washington, DC: Bureau of National Affairs, 1994).

2. New York State Governor's Office of Employee Relations, "New York State Public Employees Fair Employment Act—The Taylor Law." Available at http://www.goer.state.ny.us/CNA/bucenter/taylor.html, accessed August 22, 2007.

3. State of California Government Code, sec. 3512–3524.

4. *Agreement between the State of California and California Correctional Peace Officers Association, Bargaining Unit 6*, September 18, 1992 through June 30, 1995, article VI, Grievance and Arbitration Procedure, sec. 602, def. (a).

5. California Department of Corrections, *Basic Supervision Lesson Plan, Labor Relations*.

6. M. Josephson, *Making Ethical Decisions*, 4th ed. (Marina del Rey, CA: Josephson Institute of Ethics, 1996), p. 2.

7. Josephson, *Making Ethical Decisions*, pp. 9–17.

Preventing Corruption

24

Sam S. Souryal

Chapter Objectives

- Describe public service corruption and differentiate between public corruption and prison corruption.
- Determine characteristics of professionalism.
- Define acts of misfeasance and distinguish acts of malfeasance from acts of nonfeasance.

Corruption of prison personnel traditionally has been thought to be part of the broader spectrum of public corruption. Prison corruption and public corruption are similar: Both are acts of betrayal committed by persons entrusted with preserving fundamental interests of society.

■ Public Service Corruption

When public officials display professionalism, they encourage trust. When they behave corruptly, they betray the public's trust. Corruption by public officials has been considered much more sinister than corruption by private sector employees, for several reasons:

1. Citizens have no choice but to use the available public services (e.g., to drive a car, to run a business, to pay taxes, to petition for a license), whereas they can choose among the services offered by different organizations in the private sector.
2. Public officials take an oath to execute the laws of the land faithfully and to serve society, making their failures, especially when unjustified, seem more "sinful."

3. Because of their sovereignty, public agencies can inflict greater damage on unsuspecting citizens than can officials in the private sector. When corruption is discovered, citizens can lose faith in their political system, their elected officials, and themselves.

Especially in a democracy, people expect their public servants to be efficient and civil (hence the term "civil service") and to consider their duties sacred obligations. While people should expect public officials to demonstrate a higher level of integrity than the average person, private officials and contractors must adhere to the rules and practices good faith management.

Alert and conscientious managers can estimate the extent of corruption fairly by monitoring several indicators, including:

1. Formal and informal complaints filed against employees by dissatisfied customers, supervisors, or other employees

2. Disciplinary actions taken against employees for violating agency rules and regulations

3. Patterns of depressive episodes or questionable behavior by workers such as involvement in alcoholism, drug use, or domestic violence

4. Erratic behaviors by workers such as more out-of-town trips than are customary, radical changes in lifestyle that might indicate the sudden acquisition or loss of wealth, or unexpected requests for reassignment or resignation

5. Graffiti on the walls and inside bathrooms

Based on these indicators, management should be able to determine when to intervene.

■ Prison Corruption

Prison and jail corruption differ slightly from public corruption. These differences are functions of the uniqueness of the environment, occupational opportunities, and patterns of social relationships that develop inside correctional institutions.

Prison officials not only serve in environments that are relatively closed to public scrutiny (making it easier to carry out corrupt acts or to suppress evidence of such acts), they are also engaged in unusually stressful jobs for much longer periods of time. In most instances, they are given the difficult task of controlling a reluctant, resistant, and sometimes hostile inmate population whose welfare may seem better served by corruption than by honest compliance with prison rules and regulations.

Because prisons and jails have played a major part in maintaining order in society, their operation has become a massive industry. More than two million Americans are now behind bars and another five million are under some sort of correctional supervision.[1] When huge numbers of inmates are confined to small spaces, prisonization increases. A culture of manipulation, violence, and—at times—barbarism may ensue. This can wear down the professional fiber of correctional officers, especially those assigned to highly stressful tasks. As a result, prison personnel may experience more resentment and cynicism than their counterparts in other public agencies.[2]

Prison and jail operations also have become complicated and expensive. This can increase the opportunity for economic corruption, especially if the workers are not quite professional and supervision is lax. As the complexity of the operation and the amount of capital involved grow, corruption tends to increase. The demographics of confined inmates today may be more conducive to another kind of corruption: racial cruelty and racial oppression. Prison and jail populations do not proportionately represent the general population of Americans. This disparity has grown more pronounced in the last 30 years. While African American males make up less than 12 percent of the U.S. population, they compose almost half of prison and jail populations, while the majority of their officers are Caucasian.[3] This composition of institution populations and the disparity between the racial distribution among inmates and staff may contribute to staff violence against minorities.[4]

Correctional management has undergone a series of changes more radical than those confronted by any other public institution. In the last 20 years alone, an avalanche of new rules emerged concerning overcrowding, judicial review, parole conditions, acquired immune deficiency syndrome (AIDS), gang members, drug use, the aging of inmates, and the use of tobacco products by inmates and correctional officers. These rapid changes can cause serious managerial problems, complicating the maintenance of discipline inside a correctional facility. As a result, institution officials—generally more attached to security issues than social issues—may feel hesitant to enforce the new rules, which they may consider vague, confusing, politically motivated, or possibly dangerous. Furthermore, new rules have prompted the hiring of unprecedented numbers of correctional officers, creating yet another difficulty: inadequacies in screening and training. Ironically, this chronic state of institutional uncertainty that has engulfed prison operations must be navigated every day by the least experienced personnel.

Lastly, given the relatively low pay of correctional officers, especially of those at the lower levels, the potential gains from corrupt behavior may be too attractive to resist. Correctional officers may, over time, become dependent on inmates for the completion of some tasks or the smooth management of the tier. In return, they may overlook inmate infractions and supervise with some favoritism.[5] This behavior may be more common among young and inexperienced personnel.[6]

■ Standpoints on Prison Corruption

From a sociological standpoint, corruption in prisons and jails may be considered an abuse of power, because the term denotes the use of power to achieve a purpose other than that for which it is granted. An officer might hire or promote a less-qualified worker because he or she is a relative of a superior or because that is the warden's desire; treat inmates preferentially because the inmates serve as house trustee; or deny civil rights to a group of inmates because of their faith or religion.

From a legal standpoint, corruption can involve the use of oppression or the use of extralegal methods to suppress the will of others. An officer might write up inmates for violations they did not commit because of their race or ethnicity, amplify charges against inmates (or other officers) because they are viewed as troublemakers, or permit physical abuses to be inflicted upon inmates (or other officers) because they are gay or lesbian.

From a moral standpoint, corruption may be the failure of staff to demonstrate compassion or to keep a promise; public officials are morally obligated to care for the needs of those in their custody or under their supervision. An officer might ignore an inmate's cries for medical attention, believing that his or her shift is understaffed; trick an inmate into giving information regarding illicit or illegal activities of other inmates with the promise of better treatment that then is not rendered; or, in a parole hearing, withhold helpful testimony regarding the good behavior of an inmate because of his or her refusal to respond to the officer's sexual advances.

From an economic standpoint, corruption in the correctional environment could mean abusing authority for personal gain; public workers are forbidden to take bribes, kickbacks, or any unauthorized payments for discharging regular duties. An officer might write up a procurement contract that fits a specific vendor who had promised to pay a kickback if selected; exploit inmates by threatening to "make their lives miserable" unless their families pay a bribe; bring or sell contraband to inmates; or use prison equipment without authorization.

Corruption among correctional workers occurs at all levels and in many different forms. In some cases it is very limited in scope, such as a minor conflict with a staff member. In other cases, correctional officers have been terminated for smuggling contraband or for having inappropriate relations with an inmate. Other corruption cases involve more elaborate schemes and involve substantially more money. In 1998, the Kansas Court of Appeals overturned the conviction of a county jail administrator who took money from an inmate account (an account holding inmates' personal funds, abandoned inmate funds, and jail telephone commission profits) and put it into an interest-bearing bank account. The transferred money was used to buy equipment for the jail and pay jail commissary bills. While the money was not used for the personal benefit of the administrator, the Court of Appeals ruled that the transfer was improper because it violated the terms of the trust and held the administrator, as custodian, responsible. However, because the money was neither county nor state property, the charge of "misuse of public funds" was deemed inappropriate.[7]

In another case, the Secretary of the Florida Department of Corrections, James Crosby, was forced to resign in February, 2006, following a large-scale state and federal investigation into the Florida Department of Corrections. The investigation uncovered widespread corruption, including problems with contract vendor accounts and abuse of the department's athletic program. Allegedly, non-employee "ringers" were furnished with department security identification cards to be able to participate in department softball games and were encouraged to use steroids. As a result, several high-ranking officials in Florida's Department of Corrections were fired, including two regional directors, four wardens, and three assistant wardens.[8]

Because of the enclosed environment of prisons, corruption can take the form of use-of-force violations, including punitive and excessive use of force. For example, in 1997, Wayne Garner, Georgia Commissioner of Corrections, was implicated in an alleged mass beating of inmates in Georgia's Hays State Prison. Garner allegedly watched while the inmates, some in restraints, were beaten until blood covered the walls.[9]

It is noteworthy that the use of force in prisons and jails need not cause serious bodily harm to be excessive. In *Hudson v. McMillian*, the Supreme Court held that inmates can claim cruel and unusual punishment under the Eighth Amendment even if a correctional officer's use of force did not result in serious bodily injury. In this case, Keith Hudson, a Louisiana inmate, was punched in the face and stomach while the correctional supervisor looked on and told the officers only "not to have too much fun." Hudson suffered minor bruises and swelling, and some loosened teeth in the beating. The Court determined the malicious and sadistic nature of the use of force violated the inmate's rights.[10]

As a result of these rulings, a new discipline of correctional law was born. This new discipline articulated matters of civil liability. Accordingly, inmates can now file lawsuits against prison officials seeking a change in prison conditions, policies or procedures, as well as monetary damages. Federal civil rights lawsuits can now be filed by inmates under U.S.C. Title 42 Section 1983. The right of inmates to file such lawsuits (against correctional officials) was first endorsed by the U.S. Supreme Court in *Cooper v. Pate*.[11] Historically, under the doctrine of sovereign immunity governments could not be sued unless they consented to being sued. The Supreme Court later further clarified the application of section 1983 lawsuits against government officials in *Monell v. Department of Social Services of the City of New York*.[12] The Supreme Court holds local governments, counties, and municipalities liable, because they are considered "persons" and, as such, can be sued under Section 1983. Yet, regardless of how it is defined, prison corruption inevitably falls into one of the three following categories of acts:

1. *Misfeasance*. These are deviant acts that an official is supposed to know how to do legally (through education, expertise, and training) but are willingly committed illegally for personal gain. Misfeasance is most likely to be committed by high-ranking officials in the prison hierarchy or by others associated with the correctional facility through a political or a professional appointment. For example, if a member of the oversight board stretched the limits of his or her discretion allowing for indiscretions by contractors that undermined the public interest and benefited the board member personally, this would be considered misfeasance.

2. *Malfeasance*. These are criminal acts or acts of misconduct committed by institution officials in violation of the criminal laws of the state or agency regulations. Such violations are usually committed by officials at the lower or middle management levels. Examples of malfeasance include theft, embezzlement, trafficking in contraband, extortion, official oppression, or the exploitation of inmates or their families for money, goods, or services.[13]

3. *Nonfeasance.* These acts constitute failures to act in accordance with one's administrative responsibilities. They are basically acts of omission or avoidance by an official. Acts of nonfeasance are committed across the board, regardless of people's positions in the agency's hierarchy. Because of their subtle nature, acts of nonfeasance may be more responsible for corrupting correctional officers than acts of misfeasance or malfeasance.[14] Two types of acts are common in this category: (a) selectively ignoring inmate violations of institutional rules, such as looking the other way when marijuana or other drugs are smuggled into the facility by inmates or visitors in return for payment; and (b) failing to report another employee involved in misconduct out of loyalty or as a repayment for a previous favor.[15]

■ Prevention of Prison Misconduct

Official corruption cannot be prevented; it can only be minimized. Because workers are not born professional and cannot be counted upon to police themselves—especially when faced with acute moral dilemmas—direction, guidance, and leadership must be provided. Ideally, a manager should serve as a role model, an arbitrator, a disciplinarian, and the conscience for all workers and inmates. However, this may be more easily said than done.

To have any chance at success, management must first be credible. Senior staff must create a work environment that is conducive to honesty, fidelity, and obligation. Honesty means telling the truth at all times unless concealment is justified for a higher good. Fidelity is keeping all promises made to workers, inmates, and to any other group associated with the prison enterprise. If management claims that it treats its officers fairly, fairness must be provided at each step of the officer's career, including assignments, promotion, demotion, and discipline. Obligation is treating each rule, policy, or directive seriously and not acting in bad faith. Therefore, if management declares that it will enforce a rule by which every officer is to be searched at a point of entry, everyone, including the warden, must be searched faithfully.

Ethical institution leaders must also be consistent, reasonable, and sympathetic to the needs of officers and inmates. The behavior of institution leaders—in public as well as in private—must be above reproach and their managerial decisions borne by moral reasoning, regardless of who is to win or lose. The use of manipulation and hidden agendas must be shunned, because it can add substantially to the resentment and cynicism of prison personnel. Management should exercise its leadership methodically, rather than through casual remarks made at commencement ceremonies or staff meetings. If correctional officers note that their leaders do not truly care, they stop caring themselves, leading them to pursue personal interests that may be much more profitable.

To minimize prison corruption, management should articulate its position on corruption and corrupters and develop and implement an anticorruption policy. Management should prepare a policy statement outlining its position regarding corruption and corrupters and distribute it to every official. The message must be perfectly clear: Professionalism counts, corruption will not be tolerated, and all employees will be held accountable for absolute integrity in everything they do. By publicizing such a policy statement, managers help raise the consciousness of workers about the depravity of corruption and make it clear that the agency supports high ethical conduct and will identify and deal with policy violators—in short, that no one will be exempt from compliance with the agency's professional standards.

The Role of Policy Statements

An ideal policy statement should articulate the activities generally accepted as being corrupt by prison or jail officials, including criminal acts such as theft, assault, forgery, bringing in contraband, maintaining an illegal sexual relationship with an inmate or another officer, and falsification of evidence (by a member of internal affairs). It should aslo specify the investigatory procedures and penalties to be meted out in each of the previous categories.

As a matter of policy, the internal affairs division or an independent office should conduct investigatory procedures, and such procedures should not undermine the constitutional rights of the accused (e.g., the presumption of innocence, due process, and easy access to legal defense). Penalties should be fair and reasonable and may include suspension (with or without pay), termination, reassignment, payment of a fine, or in more serious cases, a judicial sentence to be imposed during a court trial.

After identifying and publicizing the agency's anticorruption policy, management must establish an anticorruption action program. Wardens may be reluctant to pursue an overtly aggressive anticorruption program because of concern about reactions from the correctional officers union, workers' morale, and an unfair or a vindictive media response.

Anticorruption Programs

Prison administrators should design their anticorruption programs to fit their specific function, culture, and resources. Regardless of how such programs are designed, four strategies should be included. First, upgrade the quality of correctional personnel. A natural place to begin developing agency defenses against corruption is the recruitment office door. There are two significant obstacles to pursuing this endeavor: low entry-level pay for correctional officers, resulting in relaxed educational requirements, and a higher turnover rate than in other public agencies. The national turnover rate among correctional officers in 1997 was 20 percent before completion of their probationary period and 80 percent after completing the probation period.[16] In Texas in 2006, 92 percent of newly hired correctional officers were separated from employment during their first year of employment.[17]

Managers of correctional institutions should take several steps to fight corruption, including the following:

Utilize Hiring Standards to Keep Out High-Risk Applicants

Careful attention should be given to conducting background investigations and reference checks during the screening process. Advanced psychological testing should be utilized to check the character of those who make the final cut, and a mandatory interview by a hiring board should be a routine procedure prior to appointment.

Establish Quality-Based Supervisory Techniques

Traditional supervision in correctional facilities focused on quantitative standards, such as the classification of 25 inmates or the preparation of 150 meals. These should be replaced with a quality-based supervisory system that focuses on how well and to what extent the tasks are performed. Supervisors should be trained to overlook trivial policy, but should report and aggressively investigate serious transgressions—regardless of who the perpetrators might be. In this respect, well-trained and quality-oriented supervisors are expected to possess the professional wisdom to know which is which—without being told. At the outset of the training initiative, management may have to face substantial employee resentment over changes to the status quo, and perhaps sabotage by some, but the eventual outcome should be worth the investment.

Strengthen Fiscal Controls

Most acts of official corruption involve the illegal acquisition of money. An effective tool to check corruption in correctional institutions is the proper design and administration of pre-audit and post-audit controls within the agency. Internal auditors can determine if bidding procedures have been followed, expenditure ceilings observed, and vouchers issued only for actual expenditures. Toward that end, the American Institute of Certified Public Accountants produced three volumes of comprehensive accounting and auditing standards, and the responsibility of internal auditors has been expanded to include the investigation of all aspects of fraud, waste, and abuse. But controls by internal auditors are obviously not foolproof. These auditors may be deceived when superior officials collect checks for services that are not rendered, bribes are paid for negotiated contracts, overtime pay is collected by workers who are on vacation, travel expenses are absurdly padded, and institution equipment in good working condition is sold as scrap metal.

Furthermore, it is possible that the auditors themselves are involved in the corruption. In such cases, the challenging question would be "Who then watches the watchdogs?" To establish accountability in correctional facilities, the prison director, the regional director, or the warden must ensure that internal auditors are honest, are experts in the latest advances in the accounting field, and are willing to check out every business transaction, regardless of how simple or complex. Internal auditors also must be autonomous in their decisions, save only for scrutiny by state auditors and members of the General Accounting Office. Advanced methods of control now include the establishment of a telephone hotline where whistleblowers can pass on tips about misconduct they may have observed. This brings about another important observation: The tendency of correctional agencies to accept and appreciate the practice of whistleblowing (rather than frantically fighting it) confirms their eagerness to cultivate a healthy ethical culture. In professional agencies, employees should be encouraged, rather than discouraged, to report misconduct, and managers should not be disturbed by such practices because they should have nothing to hide.

Emphasize True Ethical Training

Correctional institutions have recently been involved in serious training at all levels—basic, managerial, professional, executive, and so forth. The American Correctional Association has been responsible for determining the minimum amount of training for all prison and jail systems, and the Commission on Accreditation for Corrections has been offering far more for the accredited institutions. Some of the more popular courses offered have been in cultural diversity, sexual harassment, stress reduction, classification techniques, and job satisfaction. Ironically, one of the least popular courses has been about ethics in corrections. While ethical training can make certain individuals feel guilty, it can make more people feel confident about themselves, their values, and the benevolence of their careers. Leaders of professional institutions therefore should, make every effort to increase ethical training, both inhouse and at national and regional conferences.

Leadership in this area should be visible and active. Leaders should subscribe to ethics journals and learn the arts of moral reasoning. They should participate in panel discussions debating what constitutes right and wrong behaviors, what distinguishes rational from irrational decisions, and how to promote a healthy ethical culture in their institutions. They should not shy away from facing their subordinates, engaging them in question and answer sessions, and guiding them in the pursuit of true professionalism. Anything short of this would defeat the purpose of establishing an anticorruption program.

■ Conclusion

Both professionalism and corruption are important concepts in institution management. Professionalism is an ideal toward which correctional personnel should strive, and corruption is a shameful reality they should eradicate in themselves and in the institutions they serve. As long as corrections is part of the mandate to "establish justice and ensure domestic tranquility," correctional managers must ensure justice by stamping out corruption.

DISCUSSION QUESTIONS

1. Why does the closed nature of correctional facilities increase concerns around corruption?
2. Why is public corruption so detrimental to society?
3. What are some examples of corruption by prison staff?
4. What steps can administrators take to minimize or limit corruption by prison staff?
5. What is professionalism and how is it demonstrated by prison personnel?

ADDITIONAL RESOURCES

American Correctional Association, "The State of Corrections: Proceedings," Annual Conference (1990).

T. Gray, *Exploring Corrections* (Boston, MA: Allyn and Bacon, 2002).

D. Farrington and C. Nutall, "Prison Size, Overcrowding, Prison Violence, and Recidivism," *Journal of Criminal Justice,* vol. 8 no. 4 (1980), pp. 221–231.

J. Warren and T. Reiterman, "Overview on Crowding in CA Prisons," *LA Times,* March 13, 2005, available at http://realcostofprisons.org/blog/archives/2005/03/la_times_overvi.html, accessed August 22, 2007.

R. Muraskin and M. Muraskin, *Morality and the Law* (Upper Saddle River, NJ: Prentice Hall, 2001).

J. Pollock, *Prisons Today and Tomorrow* (Sudbury, MA: Jones and Bartlett Publishers, 1997).

S. Souryal, *Ethics of Criminal Justice: In Search of the Truth* (Cincinnati, OH: Anderson Publishing, 2003).

NOTES

1. D. Gilliard and A. Beck, *Prisoners in 1997* (Washington, DC: U.S. Department of Justice, 1998).
2. M. Braswell, B.R. McCarthy, and B.J. McCarthy, *Justice, Crime, and Ethics* (Cincinnati, OH: Anderson Publishing Company, 1984); J.M. Pollock, *Ethics in Crime and Justice: Dilemmas and Decisions*, 2nd ed. (Belmont, CA: Wadsworth Publishing Company, 1994).
3. *The Real War on Crime: The Report of the National Criminal Justice Commission*, S.R. Donziger, ed. (New York: Harper Collins, 1996).
4. Pollock, *Ethics in Crime and Justice: Dilemmas and Decisions.*
5. *Ibid.*
6. Braswell et al., *Justice, Crime, and Ethics.*
7. *Estate of McDonald v. Unified Government of Wyandotte County*, No. 92,699, 2006 WL 1170101, Kansas Court of Appeals, (2006).

8. M. Galnor, "Prison System Purge Goes On," *Florida Times-Union*, March 16, 2006, A1.

9. R. Cook, "Guard Recalls Beatings as Payback Time," *Atlanta Constitution*, June 29, 1997, 1C.

10. *Hudson v. McMillian*, 503 U.S. 1 (1992).

11. *Cooper v. Pate*, 378 U.S. 546 (1964).

12. *Monell v. Department of Social Services of the City of New York*, 436 U.S. 658 (1978).

13. Braswell et al., *Justice, Crime, and Ethics*.

14. *Ibid.*

15. *Ibid.*

16. C. Camp and G. Camp, *The Corrections Yearbook* (South Salem, NY: Criminal Justice Institute, 1998), p. 150.

17. Texas Department of Criminal Justice Training Manual (2006), p. 5.

Chapter Resources

Sexual Misconduct

25

Anadora Moss

Chapter Objectives

- Identify some of the factors that can contribute to sexual abuse of inmates by correctional staff.
- Outline the parameters that define staff sexual misconduct and permissible behaviors between staff and inmates.
- Describe some of the approaches taken by correctional administrators and legislators to minimize sexual misconduct.

Sexual violence and abuse within U.S. correctional facilities have become the focus of an unprecedented public policy discussion. This heightened national focus in the area of sexual misconduct has resulted in more research and created a greater sense of urgency within correctional institutions to develop and enforce effective misconduct policies and practices. It has also resulted in the passage of important legislation.

■ The Prison Rape Elimination Act

On September 4, 2003, President George W. Bush signed the Prison Rape Elimination Act (PREA) into law. This legislation sends a clear message of intolerance of any deliberate negligence by correctional leaders. While sexual misconduct in prison culture historically has been discussed only rarely, PREA addresses staff and inmates in all settings within the criminal justice system, including juvenile settings. The act speaks to staff and inmate sexual abuse and promotes a zero-tolerance approach to all prison rape and sexual assault. It also focuses on sexual abuse occurring within the offender population and assigns specific responsibilities to a number of federal agencies. For example:

1. The Bureau of Justice Statistics (BJS) collects, reviews, and analyzes the incidence and effects of prison rape.

2. A review panel conducts hearings on prison rape.

3. The National Institute of Corrections (NIC) offers training and technical assistance, and provides a national clearinghouse for information.

4. The U.S. Attorney General's office publishes national standards for the detection, prevention, reduction, and punishment of prison rape.[1]

■ Defining Staff Sexual Misconduct

Definitions of sexual misconduct range widely in state law and policy. There are, however, forces that increase the likelihood of greater consistency in defining staff sexual misconduct in corrections.

Initially, definitions emerged from settlements and consent and court decrees reached in early lawsuits in women's prisons. One policy regarding sexual misconduct of staff and inmates developed under court order defines sexual misconduct as "sexual behavior directed toward inmates, including sexual abuse, sexual assault, sexual harassment, physical conduct of a sexual nature, sexual obscenity, invasion of privacy, and conversations or correspondence of a romantic or intimate nature."[2]

A second set of definitions developed in the context of state laws that prohibit the sexual abuse of persons in custody. These definitions reflect what was achievable within the political arena when the legislation was written.

The policies of state corrections agencies provide a third set of definitions of sexual misconduct among staff and inmates. These definitions initially mirrored federal sexual harassment policies. For example, Title VII of the 1964 Civil Rights Act defines sexual harassment as a form of sex discrimination. Likewise, the Equal Employment Opportunity Commission (EEOC) policies provide that:

> *Unwelcome sexual advances, requests for sexual favors, and other verbal or physical conduct of a sexual nature constitute sexual harassment when:*

1. Submission to such conduct is made either explicitly or implicitly a term or condition of an individual's employment,

2. Submission to, or rejection of such conduct by an individual is used as the basis of employment decisions affecting such individual, or

3. Such conduct has the purpose or effect of unreasonably interfering with an individual's work performance or creating an intimidating, hostile, or offensive working environment.[3]

Though sexual harassment is defined clearly as sexual misconduct, the case and statutory law related to staff–inmate relationships acknowledges that the difference in power between staff and offenders fundamentally distinguishes those interactions from those of employees to their coworkers or employers. In particular, people in custody cannot leave the presence of those who have authority over their day-to-day existence. Therefore, the concept of a "welcome" advance cannot be applied to staff sexual misconduct involving inmates.

The NIC defines staff sexual misconduct as:

> *Any behavior or act of a sexual nature by an employee, contractor, service provider, volunteer, or any persons or entities acting on behalf of a correctional agency to:*
>
> - A person under the care or custody of the department or agency
> - Any collateral contact of an offender not related to correctional purposes
> - Victims or their families
> - Any other persons who have official contact with the department or agency on behalf of offenders, such as lawyers or social workers
>
> *Staff sexual misconduct includes, but is not limited to, acts or attempts to commit acts such as:*
>
> - Sexual assault, abuse, and harassment
> - Actions designed for sexual gratification of any party
> - Conduct of a sexual nature or implication
> - Obscenity
> - Unreasonable invasion of privacy
> - Inappropriate viewing
> - Conversations or correspondence which suggests a romantic or sexual relationship[4]

Over the years, a greater understanding of staff sexual misconduct and the inherent difference in power between staff and inmates has led most correctional leaders, experts, and legislators to the conclusion that there can be no consensual sex between staff and inmates. Case law has further clarified that consent is not a defense to criminal prosecution for staff sexual interaction with offenders.[5]

The Bureau of Justice Statistics defines specific behavior comprising staff sexual misconduct as "any behavior or act of a sexual nature directed toward an inmate by an employee, volunteer, official visitor, or agency representative." This includes romantic relationships, intentional touching of sexual areas of the body, sexual acts, indecent exposure, and invasion of privacy or voyeurism for sexual gratification.[6] Further, BJS defines staff sexual harassment as "repeated verbal statements or comments of a sexual nature to an inmate by an employee, volunteer, official visitor, or agency representative, including demeaning references to gender or derogatory comments about body or clothing; or profane or obscene language or gestures."[7]

Definitions of staff sexual misconduct have continued to evolve. With the development of standards from accrediting organizations and the anticipated standards being developed by the National Prison Rape Elimination Commission, greater consistency and operational impact is anticipated by many professionals working in this area.

■ Legal Issues and Evolution of State Laws

Just as very visible cases in the media highlighted domestic violence and brought the issue to the public, highly visible cases of staff sexual misconduct in women's prisons have raised awareness of the issues of correctional staff engaged in sexual misconduct with inmates. This trend began emerging in the 1990s when allegations of widespread sexual abuse of women in Georgia's Milledgeville State Prison were unveiled. Corrections staff were found to have engaged in rape, sexual assault, inappropriate viewing, and verbal degradation of female prisoners in their custody.[8] The court orders and remedies in these cases provided models of policy and practice that began to establish a correctional management framework for responding to an issue that affects all institutions. The monetary awards resulting from individual inmate cases increased the concern in the public sector as well as among correctional professionals, legislators, and advocacy groups.[9]

Sexual misconduct is an issue in both men's and women's correctional facilities. In fact, recent data from the BJS reports that the highest incidence of staff sexual misconduct, as reported from administrative records, is that of female staff engaging in staff sexual misconduct with male offenders.[10]

Today, every state has enacted legislation prohibiting sexual contact between correctional staff and inmates. While this progress is laudable, gaps still remain in criminal law protections for offenders from staff sexual abuse.[11] These gaps result in a lack of coverage when an incident involves particular types of agency personnel, agencies other than prisons and jails, and certain types of conduct. Criminal laws also have failed to address other issues consistently, including staff's duty to report misconduct, retaliation and coercion against staff and offenders who report misconduct, and the appropriateness of sanctions for false reports by staff and offenders.[12]

■ Determining Prevalence of Sexual Violence in Correctional Facilities

Incidents of staff becoming sexually involved with inmates has always been a part of the dark side of prison and jail culture, and for many years the accepted response has been based on an assumption that all professions have a few bad apples. News headlines frequently report abuses of power and inappropriate sexual involvement in the military, classrooms, religious communities, and other institutions. Although the leaders of these agencies or organizations are concerned about the consequences from such incidents, they often claim that fault lies with a few unscrupulous individuals. The PREA raises the urgency of the corrections field's response to these issues by viewing this problem as a widespread phenomenon. A careful reflection on the history of correctional institutions yields many anecdotal and documented cases of staff and inmate sexual liaisons, including documented cases among otherwise dedicated correctional staff.

The understanding of prevalence of sexual misconduct in institutions continues to grow with the enforcement of new laws, policies, and reporting mechanisms. Like domestic violence and other arenas where physical or sexual assault occurs, it is difficult to determine the prevalence of such behaviors. Historically, the occurrence of incidents of sexual misconduct in correctional settings has been difficult to determine for several reasons:

- Data on investigations have often been documented under more general categories not specific to sexual misconduct, such as assault or drug investigations.
- The degree to which sexual misconduct is reported parallels the historical underreporting of other forms of sexual assault.
- Prisoners may not report sexual misconduct for fear of reprisal, fear that they will not be believed, or because the relationship meets their needs in some way.
- Sexual misconduct is difficult to investigate, and investigative techniques that corroborate or add information beyond the word of the inmate are generally necessary to substantiate a claim.
- Staff may not recognize the signs indicating a potential problem.
- A code of silence may prevail either within the inmate population or among staff, creating a significant barrier to reporting.
- Effective reporting mechanisms often are not in place.
- Investigations often are not perceived to be objective.

Recent work funded by the NIC gathered staff perspectives on barriers to reporting that complicate the task of defining prevalence of misconduct.[13] This work suggests the existence of the following barriers to accurate reporting:

- Problems in inmate reporting—changing stories and lack of cooperation
- Difficulties in obtaining physical evidence
- Inappropriate and uncompassionate approach to inmate interviews
- Inadequate protocols training on investigation for all staff
- Codes of silence
- Lack of resources
- Lack of information on the progress of an investigation
- Use of outside investigators with limited or nonexistent correctional expertise
- Lack of coordination among various parties involved in the process
- Problems with maintaining confidentiality
- Lack of support by leadership and relevant outside agencies

Overall, staff at all levels and in all types of facilities see that responding to sexual assault is part of good correctional practice. It is critical that correctional staff engage the issues of staff sexual misconduct and recognize their individual responsibility to prevent, respond, and report such behavior. Staff engagement through a well-documented process ensures that the prevalence and types of sexual misconduct are understood better. If staff are committed to addressing staff sexual misconduct, then timely and accurate documentation of its prevalence should follow.

Unfortunately, an effective reporting process can only occur in a culture that supports it. No research published to date outlines successful change of correctional staff attitudes from a culture of silence to one that supports responsible response and reporting of incidents of misconduct. There is promising work emerging on correctional culture and sexual violence, but it is still in its infancy.

Determining Prevalence

PREA requires the federal government, through the BJS, to study the prevalence of sexual violence in correctional facilities nationally. The mandate under the law requires the BJS to develop new national data collections on the incidence and prevalence of sexual violence within correctional facilities. Their national survey reports allegations of sexual assault in prisons, jails, and other adult correctional facilities (see **Figure 25–1**).

Because the survey collected only reported data, it is important to note that at present there are no reliable estimates of the extent of unreported sexual victimization among prison and jail inmates. BJS is developing and testing self-administered surveys for measuring the incidence of sexual violence more fully.

To better understand the prevalence of sexual violence in correctional facilities, BJS is developing and testing methods that would wholly measure the incidence of this problem. Data collection will be based on victim reports of sexual violence in a survey of current and former inmates to permit reliable comparisons that overcome the limitations of administrative records.

Current research, legislation, policy, and management practice are used to define and combat staff–inmate sexual misconduct. The evolution of this environment from one in which mere assumptions guided dealings with misconduct has come about because of a number of factors, both internal and external to the correctional environment. With this evolution and the congressional mandate that BJS conduct multiyear surveys addressing prevalence, the field of corrections is developing a better understanding of the occurrence of sexual violence and abuse in our correctional systems.

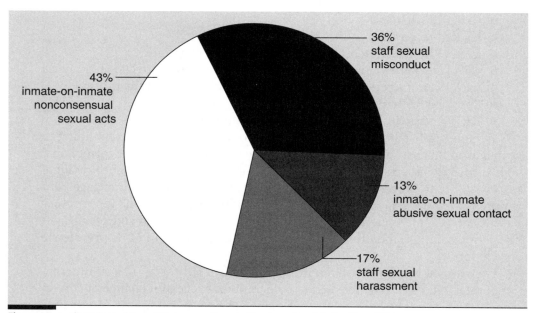

Figure 25–1 Allegations of Sexual Violence as Reported by Correctional Authorities, 2006
Source: P. Harrison and A. Beck, *Sexual Violence Reported by Correctional Authorities,* 2006 (Washington, DC: U.S. Department of Justice, Bureau of Justice Statistics, 2006), available at http://www.ojp.usdoj.gov/bjs/abstract/svrca06.htm, accessed September 4, 2007.

■ Influencing Factors

Correctional management's increased awareness and concern about staff sexual involvement with offenders has evolved gradually. A number of factors have contributed to an increased emphasis on responding to this issue in a way that goes beyond the few bad apples theory. By recognizing some of these key factors, practitioners can learn from observing the ways in which the public, nongovernmental organizations, legislators, the faith community, litigators, and correctional leadership are powerfully interrelated players in the development of public policy.

Growth of Corrections

The prison population continues to grow each year, and for the correctional manager, the reality of building new facilities, hiring large numbers of staff, and managing prisons during such extreme periods of growth has had a tremendous impact on day-to-day operations. With the growth of the offender population and prisons, more supervisors have been needed to operate all aspects of prison facilities. Promotions have occurred with personnel spending far less time in subordinate positions than in earlier decades. Supervisors themselves may feel uncomfortable confronting sensitive issues such as sexual misconduct because they might have been working as peers with the people they are supervising and were moved quickly to a supervisory position. Moreover, prison and jail crowding, staff shortages, and design flaws have contributed to sexual violence, as officers may feel unable to protect inmates.

Cultural Collision

With the increase in the construction of correctional facilities, inmates are often far removed from their homes. The backgrounds of the staff and offenders may be very different. Also, the demographic profile of offender populations in some parts of the country has changed markedly. Today's offenders are much poorer, less educated, less skilled, and more urban than those in previous prison and jail populations.

Cultural differences reflected in the day-to-day life of a prison can create a collision of perceptions, communication styles, life experiences, and values. The diversity of staff continues to increase as well. Cultural collisions within institutions are more likely under these conditions, creating difficulty in relationships among staff and offenders, both between and within groups. An increased loyalty among subcultures within the facility, including both with staff and offenders, may contribute to what is understood as the code of silence. The code of silence refers to an unwritten code among groups that reporting infractions, abuse, or misconduct is not acceptable within the group. In fact, breaking the code may have serious consequences for the member stepping out of line.

Advocacy Groups

Groups and individuals advocating for inmates' rights are justifiably concerned that correctional leadership, law, and policies should address abuse of any kind in institutional settings. Several advocacy groups have issued reports highlighting the issues of staff sexual misconduct in women's prisons. The U.S. General Accounting Office, NIC, and Office of the Inspector General have published reports on staff sexual misconduct in women's prisons with recommendations, and these reports have raised awareness in the corrections community of this important issue.[14]

Other national and state groups have served as watchdogs, observing prison officials and addressing concerns in subsequently written reports or hearings. Some of these groups have also worked with officials from departments of corrections to develop remedies to train inmates and staff how to address sexual misconduct.[15] These groups have played a role in constructive reform. Most notably, an unlikely coalition of groups played a substantial role in the passage of PREA.

Cross-Gender Supervision

When staff members of one sex supervise inmates of the opposite sex, it is called cross-gender supervision. Though cross-gender supervision may increase the likelihood of heterosexual sexual misconduct, it is important not to identify cross-gender supervision as the key reason that sexual misconduct exists. This limited view would negate the importance of appropriate training, supervision, and adherence to strong professional boundaries that provide the tools for opposite-sex supervision. However, the need to manage ratios carefully within facilities that employ both genders, to assess appropriate assignments and procedures in areas with high levels of privacy needs carefully, and to stay abreast of current practice emerging through case law remain critical to all correctional administrators.

The competing concerns of privacy for offenders and gender-neutral hiring practices remain a tense policy area for many administrators. While Title VII of the Civil Rights Act of 1964 requires gender-neutral hiring, historically many women's prisons have had a much greater percentage of women as officers than they are currently employing. Likewise, women have gained greater access to positions in men's institutions. Case examples in state departments of corrections and the Federal Bureau of Prisons indicate that sexual misconduct is predominantly heterosexual. However, there are documented cases of same-sex misconduct as well.

Public Interest and Increased Awareness of Abuse of Power

Costly lawsuits combined with the general public's frustration with crime have contributed to an increased interest in corrections. Highly visible cases in the military, the academic community, churches, civic organizations, and the private and public sectors have raised the awareness of abuse of power in various settings. These conditions have led to a continued focus on issues related to sexual violence in the corrections profession.

The growing awareness of sexual violence in prisons parallels increasing awareness of domestic violence. Any social or cultural change begins by naming or acknowledging the issue. Today, the public has an even greater awareness of the existence of the abuse of power. Domestic violence is more visible than in the past and provides an example of the dynamics of one individual having power over another. Examples from all settings in which figures have authority over other individuals have only affirmed the importance of addressing issues of sexual violence and abuse in correctional settings.

▪ Dynamics of Staff–Inmate Sexual Misconduct

People may wonder why staff do not understand the simple directive not to have sex with inmates. However, most staff do understand this directive and would never intentionally violate their professional duties, yet, the boundaries between staff and inmate can become blurred. If, for example, staff members feel isolated or verbally abused themselves, then they may come to see themselves as having more in common with inmates than with other staff or administrators. This alienation can create a vulnerability to identifying more closely with a sympathetic inmate. Staff feeling particularly vulnerable in their personal lives may have an increased vulnerability to crossing lines of appropriate behavior, particularly if they have had minimum training and support in understanding the importance of professional boundaries.

Similarly, prisoners do not leave their emotional needs or needs for the basic comforts of life in the courtroom. These needs will become apparent to staff, particularly staff who have come to identify with inmates. Unlike training for clinical professionals such as psychologists, social workers, and clinical chaplains, training for correctional staff typically does not include discussions of the critical boundaries between professionals and patients.

Correctional staff are in continual close contact with those they supervise, but very few correctional training programs address the feelings and emotional dilemmas of officers or staff members that occur when inmates become attached to them personally, or when the reverse is true. Some correctional staff and inmates feel that they have fallen in love, or sex may be used to obtain certain favors. While some situations may be obviously coercive or violent, each instance of misconduct is an abuse of power. Even cases in which staff member and offender enter into a willing relationship often involve abuse, particularly when the relationship ends.

Staff and inmate interactions must always be understood within the context of an environment based on a paramilitary structure. If the culture of the institution does not support the objective reporting of and response to other nonsexual infractions, then the atmosphere for reporting staff sexual misconduct is greatly hampered. Correctional professionals should continue to learn about the profile of prisoners in both men's and women's facilities. Understanding more about the effects of childhood abuse and its impact on adult behavior patterns can also be helpful in identifying ways to respond more effectively to interactions between staff and inmates.

Research shows that many male and female inmates have experienced significant childhood abuse. A study by Chesney-Lind suggests that there are gender differences in the dynamics that follow an abused child into adulthood. For instance, girls are much more likely than boys to be victims of sexual abuse; these same girls often fall into abusive relationships in adulthood.[16] This may suggest that female staff who were abused children may continue to be victims when interacting with male inmates with aggressive tendencies. Understanding these pathways to offending can provide rich insight into the dynamics of staff sexual misconduct.[17] This is particularly true in understanding the dynamics of staff–inmate relations in women's facilities.

People often believe that sexual misconduct in prisons is about men officers abusing women inmates. Although accurate data on the rate of occurrence is yet to be determined, research shows that sexual misconduct is prevalent among female correctional staff and male inmates. Nonetheless, it is important to know that cases of sexual misconduct have involved all levels of staff. Some preliminary work has been done to develop a clinical profile of inmates and staff who become involved, but these samples are small. Case examples used for training purposes demonstrate that even staff members who might be thought least likely to become involved in misconduct may sometimes cross the line.

The insights of first responders, clinicians, investigators, and administrators into sexual violence in correctional facilities can increase administrators understanding of post traumatic stress in correctional settings. Models of effective response to victims should be implemented in all correctional settings.

Any discussion of staff–inmate dynamics would be incomplete without the acknowledgement of staff concerns of false allegations from offenders. This difficult topic creates one of the major barriers to gaining staff confidence in addressing sexual misconduct in correctional settings. Many staff members still fear that inmates will use false allegations to manipulate them or other staff. Many experts believe that while false allegations are a reality, ultimately there is no real benefit for the inmate. Invisible retaliation from staff and inmates usually greets the false accuser. Some correctional staff members may worry that greater attention to the issue of sexual misconduct will create a spike in allegations. However, experienced practitioners recognize that each allegation must be investigated fully. If false allegations are a continuing problem in a facility, then other problems in the supervision of inmates may exist as well.

■ Approaches to Prevent and Address Staff Sexual Misconduct

In order for organizations to respond effectively to staff sexual misconduct, a comprehensive strategy is needed that addresses more than the issue of sex. Staff sexual misconduct cases must be viewed in the context of an organization's culture and practices. In other words, such incidents are often symptoms of other problems in the correctional setting in which they arise. Even well-run facilities can be vulnerable to incidents of sexual assault and misconduct; therefore, careful attention must be given to the organizational culture and practice. **Table 25–1** lists components that can affect the effective management of correctional organizations dramatically.

During the last decade, federal, state, and local correctional professionals have become increasingly aware of the need to create strong policy and practice that

Table 25–1	Areas of Consideration to Reduce Incidents of Sexual Misconduct in Correctional Institutions
Issue	**Considerations**
Law	All states prohibit staff sexual misconduct. Know the law. Improve the law if needed. Train all staff on the law.
Policy	The policy must be strong and unambiguous, with clear definitions and descriptions of prohibited behavior. The emphasis should be on zero tolerance: that all cases will be investigated and, if founded, will result in an emphatic response up to and including termination. Policy speaks to such considerations as mandatory reporting by staff, multiple channels for inmates to report, protection, and prohibition on retaliation, and investigative protocols that include clear channels of authorization for action. The policy also prescribes training for staff and orientation of inmates, connections to medical and mental health services, and relationships with outside authorities such as law enforcement personnel and community hospitals.
Leadership	Support from the top is essential to success. The agency's leaders must demonstrate their personal commitment to addressing the problem of sexual misconduct. A program for addressing staff sexual misconduct will only be credible in action, when staff and offenders alike perceive that the policies will be followed and appropriate, proportional steps will be taken if a violation occurs. Correctional administrators should be reminded that they are vulnerable to lawsuits and liability if they fail to have effective strategies to deal with staff sexual misconduct. Agency leadership must insist upon healthy institutional cultures that do not tolerate abuse. All allegations of sexual misconduct must be investigated appropriately.
Staff Training	Developing an understanding of agency policy and state and federal law through training is a fundamental management tool. This training should center on events that create safe opportunities for discussion of professional roles and the anticipated dilemmas created in the correctional environment. A dynamic training approach should be based on the realities of institutional life for staff and offenders. Effective training includes designs that offer opportunities for small group discussion, role play, and case examples. The ability of the trainer to facilitate discussion and respond comfortably to sensitive questions from participants is critical. Credibility and concern on the part of the trainer for the subject area must be established in the beginning of the training experience and be demonstrated throughout the delivery of training. Training must clearly define, explain, and improve the institution's response to sexual violence by skill-based training.
Inmate Orientation	The inmate orientation should include education on the following issues: 1. Institutional policy and corresponding sanctions (including agency commitment to addressing sexual assault) 2. Awareness of sexual violence, including red flags for potential victimization and inmate prevention/protection strategies such as boundaries, coping skills, and the grooming and coercion process 3. The health and safety risks of consensual sexual relations 4. Staff readiness to assist inmates and avenues for reporting both fears and incidents 5. The investigation process (including timeliness and evidence preservation) 6. Sanctions for false reporting 7. Warnings to predatory and sexually aggressive inmates 8. Availability of counseling and other treatment services

continued

Table 25-1	Areas of Consideration to Reduce Incidents of Sexual Misconduct in Correctional Institutions, continued
Investigative Process	The training must encompass a wide variety of considerations: 1. The gender of the investigators 2. Their office location and placement in the organization 3. The physical evidence and circumstances involved; techniques for interviewing victims 4. The feasibility and legality of such techniques as conducting polygraphs 5. Monitoring mail and phone calls 6. Secret tape recording A staff sexual misconduct investigation requires a skilled interviewer who can ask very explicit questions without alienating or traumatizing a complainant. Investigators must be trained to investigate institutional assaults effectively, adequate resources should be allocated to pursue allegations, and coordination and collaboration among other departments should be cultivated.
Operational Issues	These considerations encompass topics ranging from facility design and the placement of cameras to the way in which overtime is allocated and inmate jobs are assigned. The following are some strategies for the prevention and response to sexual violence in correctional facilities: 1. Direct supervision 2. Single cells 3. Creating treatment programs for victims of assault 4. Use of cameras and other emerging technologies 5. Controlled data collection methods 6. Data collection management 7. A classification system that identifies and separates vulnerable and predatory inmates 8. Assessing isolated areas Rotation of staff assignments, awareness of facility design, careful training of investigators, clear reporting mechanisms, and strong relationships with volunteers and the community all promote a work environment where abuse is less probable.
Media	Managing media is critical. Provided policy, training, and protocols have been put into place, the agency is in a position to respond effectively to the event by emphasizing its zero-tolerance approach and the fact that an aggressive investigation with the consequences is occurring.
Physical and Mental Healthcare	Specialized training for medical and mental health staff emphasizes that they are part of the response team in ways that they may not be if the issue were a more traditional security breach such as conveyance of contraband or the existence of a gambling ring. They must have the appropriate treatment and reporting protocols in place and be prepared to share information with all key professionals essential to the process. This approach calls for each agency to engage in a critical policy discussion, including consideration of resources and use of outside medical resources.
Aftermath	An after-action review answers the questions of how the incident happened and such specifics as whom, when, and where. By collecting data and developing a matrix of incidents, the affected organization can identify patterns and take preventive measures.

address staff sexual misconduct. A dramatic increase in state laws, more defined policy, and improved operational practices within facilities contribute to greater awareness around this sensitive issue. The opportunity to improve policy and practice continues to look promising.

The evolving body of research on this topic developed by the NIC describes the benefits of developing preventative strategies to counter sexual misconduct. These strategies employ leadership and management practices to ensure an increased level of awareness by staff and an active response to address staff sexual misconduct by leadership. Many federal, state, and local practitioners have attended training programs and developed preventative strategies based on the work of the NIC. Correctional leaders in many systems have implemented a zero-tolerance policy statement concerning staff sexual misconduct. With the availability of programmatic and research grants from the federal government under the PREA, an increased focus on staff sexual misconduct is occurring.

Professional organizations such as the American Correctional Association, the Association of State Correctional Administrators, the American Jail Association, the American Probation and Parole Association, and the National Sheriffs Association have all adopted resolutions condemning staff sexual misconduct as an egregious violation of professional boundaries and one that compromises the field's core mission of public safety. Many agencies have frankly and forthrightly identified this behavior as a serious problem and are dedicated to eliminating misconduct. Similarly, institutions have acknowledged that inmate-on-inmate sexual assault occurs in correctional institutions and are striving towards its prevention, intervention, and elimination.

■ Conclusion

Within the last decade, sexual misconduct among staff and inmates has come to the forefront of the agendas of correctional administrators and the criminal justice community as awareness of sexual misconduct has increased. The relationships of correctional personnel who work on the frontline with prisoners often create challenges for both correctional staff and offenders. An understanding of the importance of professional boundaries and the role of correctional staff in an often emotional environment requires knowledge of inmate dynamics and institutional culture maintained by the example of institutional leaders and policies.

State and federal laws provide a structural framework for identifying sexual misconduct and recommending consequences for inappropriate behavior. It is important that the issue of sexual misconduct be discussed openly with staff and inmates and that everyone understands and supports the effort to end staff sexual misconduct. From highly visible lawsuits to national policy, the issue of staff sexual misconduct is clearly at the forefront of contemporary corrections.

DISCUSSION QUESTIONS

1. What is staff sexual misconduct with inmates? What sorts of behaviors are included and why are they prohibited?
2. What are some of the reasons that staff engage in misconduct?
3. What can correctional systems do to prevent this type of behavior?
4. Give some examples of actions that have been taken over the past few years to respond to this problem in prisons and jails around the country.
5. What has been the impact of the Prison Rape Elimination Act?

ADDITIONAL RESOURCES

National Prison Rape Elimination Commission, http://www.nprec.us
Office of the Inspector General, "Deterring Staff Sexual Abuse of Federal Inmates," http://www.usdoj.gov/oig/special/0504/index.htm
Stop Prisoner Rape, http://www.spr.org
Vera Institute of Justice, http://www.vera.org
A. Moss and A. Wall, "Addressing the Challenge of Inmate Rape," http://spr.org/pdf/soc/AddressingtheChallengeofInmateRape.pdf

NOTES

1. U.S. Department of Justice, *Annual Report to Congress* (Washington, DC: National Institute of Corrections, 2004).
2. U.S. Department of Justice, *Addressing Staff Sexual Misconduct with Offenders* (Washington, DC: National Institute of Corrections, 2004).
3. Equal Employment Opportunity Commission, "Guidelines on Discrimination Because of Sex," Title 29, Volume 4, Chapter XIV (Washington, DC: U.S. Government Printing Office, 2006), available at http://www.access.gpo.gov/nara/cfr/waisidx_06/29cfr1604_06.html, accessed September 4, 2007.
4. U.S. Department of Justice, *Addressing Staff Sexual Misconduct with Offenders*.
5. B. Smith, *Fifty State Survey of State Criminal Laws Prohibiting Sexual Abuse of Prisoners* (Washington, DC: National Women's Law Center, 2005).
6. A. Beck and T. Hughes, *Sexual Violence Reported by Correctional Authorities, 2004* (Washington, DC: U.S. Department of Justice, 2005).
7. *Ibid.*
8. *Cason v. Seckinger*, 231 F.3d 777, 784 (11th Cir. 2000); *Women Prisoners of the District of Columbia Department of Corrections v. District of Columbia*, 877 F. Supp. 634 (D.D.C. 1994).

9. *Daskalea v. District of Columbia*, 227 F.3d 433, 441 (D.C. Cir. 2000).

10. Beck and Hughes, *Sexual Violence Reported by Correctional Authorities, 2004.*

11. B. Smith, *Fifty State Survey of State Criminal Laws Prohibiting Sexual Abuse of Prisoners.*

12. J. Yarussi, N. Simonian, and B. Smith, *Legal Responses to Staff Sexual Misconduct with Individuals Under Correctional Supervision*, National Institute of Corrections, unpublished draft.

13. U.S. Department of Justice, "Staff Perspectives: Sexual Violence in Adult Prisons and Jails," *Prison Rape Elimination Act*, June 2006, vol. 1 (Washington, DC: National Institute of Corrections and The Moss Group, Inc., 2006) available at http://nicic.org/Downloads/PDF/Library/021619.pdf, accessed September 4, 2007.

14. Human Rights Watch, *Women in Prison*, available at http://www.amnestyusa.org/women/pdf/womeninprison.pdf, accessed September 4, 2007.

15. B. Smith, *An End to Silence. Prisoner's Handbook on Identifying and Addressing Sexual Misconduct,* 2nd edition (Washington, DC: American University, Washington College of Law, 2002).

16. M. Chesney-Lind and L. Pasko, *The Female Offender* (Thousand Oaks, CA: Sage Publications, 2004).

17. B. Owen, *In the Mix: Struggle and Survival in Women's Prisons.* (Albany, NY: State University of New York Press, 1998).

Volveering

26

Richard L. Stalder

Chapter Objectives

- Explain the role of volunteer work within a correctional environment.
- Outline the best mechanism to recruit and secure volunteers.
- Determine methods to ensure long-term success of volunteer programs.

Administrators of correctional programs must meet expanding service requirements with a resource base that is shrinking on a real-dollar-per-capita basis, yet long-term solutions to crime require meaningful programs that provide an opportunity for change—an opportunity that helps inmates develop basic skills and new ways of thinking. The only solution is to marshal the help of volunteers to compensate for inadequate financial resources. In the future, volunteers will be the cornerstone of correctional programs that fully meet the demands of society and the needs of the offender. Volunteers will no longer be peripheral; they will be essential to fulfilling the mission of correctional facilities.

Types of Volunteers

There are two primary classifications of volunteers in a correctional program: direct service and indirect service volunteers.

Direct Service Volunteers

Direct service volunteers generally provide onsite service to the program, its staff, or offenders. The scope of services may range from infrequent brief participation with a large group of volunteers to daily or weekly involvement as individuals. This category of participation typically demands the greatest sacrifice of time from a participant and may involve additional risk, depending upon the area in which an individual is serving. Direct service volunteers are the core of most successful programs. They interact with staff and offenders and, in a properly administered volunteer initiative, can be very effective in contributing to the success of agency goals.

Indirect Service Volunteers

Indirect service volunteers are typically not involved onsite and generally do not have contact with the primary receivers of service. They can, however, offer valuable assistance. Indirect service volunteers participate in a variety of tasks. Tasks may include fundraising; developing policy, procedure, and training manuals; offering technical assistance in the review of budgeting, accounting, and financial audit issues; and donating materials and supplies, among others.

Many programs have come to rely on the fundraising assistance of outside organizations to enable them to provide services that are not included in their core budget. Indirect service volunteers often provide or coordinate access to facility beautification funds that are not otherwise available. Individuals who sacrifice their time to raise money are real volunteers who have a real impact.

Volunteers may be professionals in administrative and training areas and offer advice and assistance in their areas of expertise. Volunteers can contribute by developing or reviewing core policy documents and training curricula or manuals. Many professionals will donate time to assist correctional institutions in areas for which budgeted funds are not available.

Often overlooked are those volunteers who solicit and donate materials and supplies for various projects. Important elements of community restoration are often not possible without this type of involvement. In Louisiana, for example, appropriated monies were not available to provide supplies to an inmate artist who wanted to paint and donate pictures for patient rooms in a local hospital; volunteers convinced individuals and organizations to donate the needed materials. Other inmate groups were interested in making toys for needy families at Christmas. Again, budgeted funds were not available for wood, glue, tools, and the many other supplies needed to make the project a success. Volunteers came to the rescue and provided everything necessary to build thousands of toys.

■ Sources of Volunteers

Sources of both direct and indirect volunteer services include:
- Individuals with particular skills known by key agency or facility staff
- Civic clubs
- Religious organizations and churches
- Fraternal organizations
- Crime victims
- Special purpose organizations

Perhaps the best mechanism to meet a defined volunteer need is to identify and solicit, through key staff, individuals who can provide the specific service. All staff must help develop a resource bank of individuals who can be called upon and who will provide assistance. In an ideal world, citizenship through service would be a part of everyone's agenda, and prison, jail, and field service staff would simply call upon individuals when they were needed. Regrettably, this ideal is rarely met. Therefore, the likelihood of securing individual participation may improve if potential volunteers are contacted by someone they know who is familiar with the program.

The support of civic, religious, fraternal, and special purpose organizations is critical to any volunteer initiative. Civic clubs (such as Lions, Rotary, Optimist, Toastmasters, and Jaycees Clubs) consider service to their community one of their primary functions. They do not discriminate against corrections and are underutilized only when they are not called upon. Their service can be both direct and indirect. Many individuals from these organizations are community leaders. Their volunteer service helps them become better informed, more effective advocates for corrections.

Religious and fraternal organizations generally have outreach programs that are related to corrections initiatives. While the scope or type of service may be defined by the organization and not the agency, the ultimate goals typically coincide. Many mainstream religious denominations, as well as the Volunteers of America and the Salvation Army, have core service commitments to criminal justice. They provide strategic assistance to offenders and their families in areas where there would otherwise be a void.

Crime victims have been used effectively in some correctional jurisdictions. This group offers a unique perspective in terms of their ability to reflect and focus inmates on how crime has affected their lives in significant ways. Such programs have been developed and utilized in community-based correctional organizations.[1]

Special purpose organizations also enable and enrich the delivery of critical services. Alcoholics Anonymous and Narcotics Anonymous best typify this category of organization. Broad-based programs of substance abuse education and counseling would simply not be available in many correctional institutions and work release centers or to probationers and parolees without these groups. There is little question of their importance since the majority of the individuals under correctional supervision have substance abuse experiences that relate to their criminality. Without Alcoholics Anonymous and Narcotics Anonymous, many inmates would not change their substance abuse behaviors and, therefore, would be less likely to succeed after release.

■ Components of an Effective Volunteer Program

The components of an effective volunteer program include initial development and organization, recruitment, selection, orientation, training, and recognition.

Development and Organization

The initial development and organization of the volunteer program in a correctional institution will help ensure its ultimate effectiveness. Initially, policies and procedures must be drafted and incorporated into the formal structure of the departmental program that utilizes the volunteers. This will minimize the possibility of disruption and provide consistency and direction to volunteers and staff alike. Inherent in this effort should be the establishment of clear lines of authority and the appointment of a volunteer coordinator. Whether full-time or part-time, this important position will improve communication among volunteers and staff and allow for the clear definition of expectations.

Recruitment

Recruitment strategies must be well-defined. Failure to enlist the support of enough volunteers will endanger the program's success. Too many people can be overwhelming and result in chaos with reduced productivity. Recruitment strategies must aim for the right number of the right kind of volunteers. The volunteer coordinator must use the personal contacts of staff and the reputation of established organizations to ensure that recruitment is successful. If well-known organizations have integrated their volunteers in the institution successfully, others will be encouraged to do so.

Selection

Selection follows recruitment. Improper volunteer selection will weaken a program. Neither the agency nor the volunteer nor the community benefits from informal selection practices. Unfortunately, in volunteering, one size does not fit all. Senior citizens who were very successful in hospital programs may find that the stress of a foster grandparent assignment in a secure juvenile correctional facility renders them ineffective. Interviews, education about agency expectations, and discussion of what the potential volunteer wants to accomplish are essential elements of proper selection practices. Because there are so many areas in which volunteers participate, a motivated individual can almost always be utilized in some capacity.

Background checks are important, especially in direct service. Current or recent clients of the criminal justice system are generally not the best candidates for facility volunteer assignments. Anyone who visits an inmate in a secure facility must be restricted from participating as a volunteer. Relationships with offenders in direct volunteer service must be limited to the boundaries of the volunteer duties. Formal recruitment and selection procedures are the best first line of defense to ensure that a volunteer program does not jeopardize the security of the facility. Programs involving juveniles must use extra scrutiny in volunteer selection.

Orientation

Prior to beginning their duties, volunteers should participate in a structured orientation program that covers at least the following topics:

- The basic mission and goals of the criminal justice system and the agency in which services will be provided
- The facility or field service area and the specific division in which services will be provided

- The basic security procedures appropriate to the program security level (including definitions and control of contraband and the importance of maintaining professional relationships with offenders)
- Safety and emergency procedures
- Cultural diversity awareness

Training

It is often feasible to use existing staff to train volunteers. The curricula and the length of training will vary in accordance with the scope and frequency of services provided. Initial orientation training—including a system of registration and identification that will track the volunteer throughout the period of active service—should be completed before any volunteer activity begins. Short in-service training segments can often be combined with appreciation banquets or similar functions.

Recognition

Perhaps one of the most important elements of ensuring the long-term success of a volunteer program is the formal recognition of volunteers' contributions. Plaques and letters of commendation become important mementos of service rendered. The simple act of saying "thank you" can help foster goodwill and commitment.

■ Volunteer Programming Ideas

The impact of a comprehensive program of volunteer services can be very positive. The result of volunteers' efforts may be seen in many areas, including religious services, recreation, staff training, social services, substance abuse treatment, and pre- or postrelease programs. These are only examples of areas in which volunteers can assist. Each institution will have unique needs and unique opportunities. For example, volunteers may also be used in institutional library settings, administrative support roles, legal service programs for staff and offenders, and health care.

Religious Services

Perhaps the most traditional and largest area of volunteer service in a jail or prison is religious programming. In fact, this is one of the oldest types of volunteer programs in American corrections; members of the clergy first became involved in prison programs in the 1700s and 1800s. Their primary purpose was to "help offenders repent."[2] The staff chaplain often handles the overall coordination of volunteers. Facilities typically are unable to provide assistance to all faith groups without the help of volunteers. Under federal law, institutions must ensure access to (often obscure) religious faiths and practices. Volunteers may make the difference in whether compliance can be achieved and maintained.

Recreation

Another important area of service is recreation. The community is an important ingredient in sports programming in a facility. Leadership and sportsmanship can be taught very effectively by volunteers who compete as individuals or teams, participate in training officials, or serve as coaches or fans. This organizational component of most institutions is often unsupported by appropriation and therefore depends on volunteer involvement.

Staff Training

Staff training represents an often overlooked area where volunteers can make a contribution. Many institutions do not have sufficient staff or funds for outside contractors to provide training in areas such as cultural diversity, employment law, management and supervision practices, and technical medical and mental health issues. The professional development of staff is a good way to involve the community in the institution.

Social Services

Offender education programs require volunteer support to provide service to any significant number of inmates who need to develop academic and vocational skills. Volunteers can work in literacy, adult basic education, and job skills development programs. Education is a cornerstone of any program to reduce recidivism by improving opportunities for legitimate employment after release. Education also is traditionally underfunded. Therefore, it should be a primary focus of the volunteer initiative.

Substance Abuse

As previously mentioned, without volunteers, the ability of facilities to provide broad-based programs of substance abuse education would be severely diminished. Alcoholics Anonymous and Narcotics Anonymous have a long, widely heralded record of accomplishment in correctional institutions. The commitment of individual volunteers, many of whom have previously conquered substance abuse problems, is a significant factor in this success.

Pre- and Postrelease

Prerelease and postrelease programs are other examples of critical areas that receive too little budgetary support given their importance in meeting correctional goals. Volunteers fill this gap (see **Table 26–1**). They provide service—not just enrich that which is already provided. Community support to the released offender is frequently the only support available. Involvement begins at the institution with education in a prerelease setting about how to access services. The involvement continues after release, with volunteers serving as mentors and providing guidance and support.

■ Conclusion

Everyone is a winner in a properly organized and administered volunteer program. Inmates receive services they would otherwise not receive; this can only improve the odds for successful adjustment in and out of the institution. Staff receive the benefit of community expertise, allowing them to sharpen their professional skills and be more effective in their jobs. The agency and facility are opened to the community, helping erase traditional misconceptions about jails and prisons as external support increases. The most important player, the volunteer, is also a winner; citizenship through service is its own reward.

Table 26–1	**Suggestions for Starting and Maintaining a Volunteer Program**

Evaluate the need: After determining what tasks are not getting done or are overextending staff, decide whether these tasks could be handled effectively by volunteers.

Develop goals and job descriptions: Write up the goals of the volunteer program, as well as job descriptions for volunteers, so that administrators, staff, and the volunteers themselves know what volunteer positions entail.

Involve staff: Be sure to include staff (especially staff who will work directly with volunteers) in all planning and implementation of volunteer programs. If their input is included, staff will have a greater desire for the program to succeed because they will share a sense of program ownership.

Actively recruit volunteers: There are many organizations one can contact to find volunteers. Churches, civic groups, retirement organizations, and colleges and universities are all good choices for volunteer recruitment.

Explain security needs to volunteers: Instruct volunteers on the institution's security policies and procedures, and explain why they are needed. Otherwise, volunteers may resist institution security precautions simply because they do not understand their purpose.

Give volunteers the big picture: Teach volunteers about the institution's mission and services so they have a sense of how their contributions are a part of the facility's overall operations.

Evaluate program effectiveness: Once a volunteer program is in place, it is crucial to know how well it works. All volunteer activities must be carefully documented so program and volunteer effectiveness can be evaluated. Once the program has been established, it also should be formally evaluated by staff, inmates, and volunteers. With this information, one can make sure the program's purpose is being served.

Recognize the volunteer's contribution: Volunteers, like all of us, need to be recognized for their work and accomplishments. A pat on the back can go a long way, particularly in a demanding field like corrections. Recognizing volunteers for their contributions can help keep them motivated and involved.

Educate volunteers about inmates: Before inmate contact begins, caution volunteers on the pitfalls that await those who are not familiar with the inmate culture and who may be easy prey for manipulative inmates. This is necessary to protect volunteers from being used and to maintain the institutions security.

Source: Reprinted with permission from Kevin R. Ogburn, "Volunteer Program Guide," *Corrections Today* 55, no. 5 (1993), p. 66, American Correctional Association, Alexandria, VA.

DISCUSSION QUESTIONS

1. How do direct and indirect volunteer service providers differ?

2. Why should volunteers be required to undergo an orientation process before working with offenders in an institutional setting?

3. What types of information about inmates should volunteers be exposed to and why?

4. What are the advantages of a properly organized and administered volunteer program?

5. What types of pre- and postrelease programs can help an offender make a successful reentry into his or her home community?

ADDITIONAL RESOURCES

American Correctional Association, *Helping Hands: A Handbook for Volunteers* (Laurel, MD: American Correctional Association, 1993).

B. Love, "Volunteers Make a Big Difference Inside a Maximum Security Prison," *Corrections Today,* vol. 55, no. 5, pp. 76–79.

Prison Literacy Project, *Prison Literacy Project Handbook* (Philadelphia, PA: Prison Literacy Project, 1993).

R. Sigler and K. Leenhouts, *Management of Volunteer Programs in Criminal Justice* (Denver, CO: Yellowfire Press, 1985).

P. Weston, *Volunteers in Justice* (Washington, DC: National Association of Volunteers in Criminal Justice, 1977).

NOTES

1. J. Costa and A. Seymour, "Crime Victims, Former Offenders Contribute a Unique Perspective," *Corrections Today* (Laurel, MD: American Correctional Association, August 1993).

2. J. Coleman, "Chaplains: God's Partners in Prison," *Corrections Today*, December 2003, available at http://www.correctionalchaplains.org/gods%20partners/gods%20partners.html, accessed September 10, 2007.

Inmate Management and Programming

YOU ARE THE ADMINISTRATOR
Legal Oversight

Fictional inmate Phil McKenzie was determined to get out of the segregation unit of the prison at all costs. He cut his arms and legs and tried to hang himself with the tube from his breathing machine (provided to address his sleep apnea). When that didn't work, he broke off a sharp piece of metal from the machine, which he first used to slice his neck and then swallowed, in hopes of causing internal bleeding. Corrections staff managed to save McKenzie and released him from isolation to get him much needed medical and mental health treatment. McKenzie had a long history of depression and when he returned to segregation five weeks later, he hanged himself, thereby becoming the 13th inmate in the state to commit suicide in less than two and a half years.

Inmate advocates have filed suit in federal court claiming that McKenzie and 18 other inmates who committed suicide or attempted suicide were driven to harm themselves by the conditions they endured in isolation units in prisons around the country. Most states operate segregation units, though some have recently reacted to lawsuits by no longer putting mentally ill inmates in such units. Others have implemented more frequent monitoring of the inmates, increased training of staff, and removed fixtures that could be used for hangings.

Around the country, department of corrections staff are struggling with how to treat violent inmates who are out of control and need to be segregated from others for their own safety, as well as the safety of other inmates and staff. However, for the mentally ill inmates who fall into this category, placement in segregation (which often results in being locked in a cell for 23 hours a day with just one hour for shower and recreation in an outdoor cage) can amount to a death sentence.

In the case of McKenzie, the expert retained by the state to examine the case concluded that "confining suicidal inmates to their cell for 24 hours a day only enhances isolation and is antitherapeutic." The Department of Corrections vowed to adopt all of the resulting recommendations, including better inmate assessments, better supervision and monitoring of inmates, and better officer training.

- How can prison and jail administrators deal with mentally ill and special needs offenders?
- Are segregation units ethical and safe?
- How can specialized inmate programming help offenders like McKenzie?
- What standard should the courts employ to determine cases like this one?

Source: P. Belluck, "Mentally Ill Inmates at Risk in Isolation, Lawsuit Says," *New York Times*, March 9, 2007, available at http://www.nytimes.com/2007/03/09/us/09prison.html?ex=1331096400&en= 7b9093960e12d25d&ei=5124&partner=permalink&exprod=permalink, accessed October 10, 2007.

Disciplinary Procedures

27

Clair A. Cripe

Chapter Objectives

- Describe the goals of an inmate discipline policy and outline the essential elements of a good disciplinary program.
- Explain the constitutional provision that governs inmate discipline procedures.
- Name major legal decisions governing prison discipline proceedings.

Social control of inmate behavior is critical to the successful governance of a correctional institution. To oversee inmates appropriately, positive discipline is necessary. An inmate discipline policy establishes the institution program that regulates inmate conduct, attempting to keep that conduct within the limits of acceptable standards of institutional behavior. Good inmate behavior helps ensure the orderly and safe running of any prison or jail. A functional and well-implemented inmate disciplinary policy will instill respect for authority. It is hoped that the good behavior and respect for authority will persist after the offender's release.

◼ Importance of Inmate Discipline

Those who have worked in prisons or jails accept without question that an inmate disciplinary process is essential. Those who are new to corrections may wonder why it is necessary to have additional discipline when these persons are already locked away from society. What are correctional staff and administrators trying to achieve with their disciplinary procedures? Moreover, are those goals clear, and do their actions in disciplining inmates bring them closer to those goals?

Inmate discipline should achieve several agreed-upon goals:

1. Making inmate conduct conform to a standard of behavior that ensures a safe and orderly living environment

2. Instilling respect for authority

3. Teaching values and respectful behavior (in a group of people who, by definition, have not displayed good values and behavior) that inmates may continue to use once they reenter community

Society has adopted rules of behavior for its own protection and wellbeing. These rules are called criminal laws. They are enforced by law enforcement officers. From the patrols and investigations of police officers, through prosecutions and criminal trials, into the correctional facilities that carry out the sentencing orders of the courts, the criminal justice system attempts to achieve the same kinds of goals that the inmate disciplinary system does within institutions: making behavior conform to accepted norms, protecting the safety and property of all, and instilling respect for authority.

When people visit a prison for the first time, most of them have the same reaction: "I am amazed at how much movement there is. At many times of the day, inmates appear to be moving in many directions. What is the purpose? How can all that activity be supervised and tolerated?" That reaction is nearly universal for two reasons. First, ideas about prisons and jail often have been formed by misleading sources such as movies, television shows, and media reports that portray correctional facilities as closely regimented, locked-down (that is, with inmates in their cells all the time) facilities. Indeed, there are some locked-down, tightly regimented facilities, but there are only a few of them in the whole country. Second, most corrections facilities emphasize free movement of inmates during daytime hours and evening activities. This free movement allows for various programs such as work, education, recreation, etc. Inmate movement lessens tensions and normalizes day-to-day living, thus benefiting both inmates and staff.

However, the movement of inmates—and the large variety of workplaces, programs, and activities in which inmates are involved—also increase the need for discipline. Most inmates appreciate the opportunity to join a variety of activities, and they adjust to the requirements of living in a prison environment. To assist those inmates in their good adjustment to prison life and to help staff in their primary mission of maintaining a secure and safe institution, it is necessary to punish those who break the rules. For those reasons, inmate disciplinary procedures are necessary.

■ Essentials of Inmate Discipline

Keeping in mind the goals of disciplinary actions, there are three aspects of discipline essential to an effective inmate disciplinary program.

1. There should be a written set of rules defining expected inmate behavior and procedures for handling misconduct. Most institutions will have a misconduct code—a list of offenses that are subject to punish-

ment in the prison or jail. An adjunct of that list of offenses should concern the types of sanctions that may be imposed if the rules are broken.

2. The rules for discipline must be communicated carefully and thoroughly. All staff members must be taught how the inmate disciplinary program works, in general and specific terms, because every employee may be involved (as a witness, reporting official, or in some other capacity) in the disciplinary system. Inmates must also be given the details of the system. They must know the kind of responsible behavior that is expected of them, and they should learn the penalties for misbehavior. All staff and inmates should be given a written statement of the policy and hear an oral presentation explaining how the policy works, with the opportunity to have their questions answered.

3. The disciplinary policy must specify clearly how inmates will be notified of suspected misconduct, how sanctions will be imposed, and what their rights to be heard are (along with their rights to appeal, if any).

It is extremely important to have an inmate discipline program that is supported by written policies and specifies precise procedures. These procedures must be applied by staff who understand the importance of the disciplinary program and its procedural requirements. It is also important to apply the policy consistently to engender respect for the rules among staff as well as inmates.

■ Informal Resolution of Misconduct

It must be noted that informal handling of many kinds of misconduct occurs and is essential to the smooth running of any correctional system. Officers in correctional facilities are given authority to use their good judgment, and they will handle some misconduct informally. This does not mean that the inmate behavior is excused if it violates the rules. It means that there is official recognition that the goals of the disciplinary system (achieving a safer and more secure facility and instilling respect for authority) may be achieved in many cases without processing misconduct through the official system.

Informal handling is usually used for less serious misconduct. For example, an inmate, especially one who has just arrived, may violate a rule without knowing that the conduct is considered misbehavior. In these cases, the officer's reaction is more instructional than punitive: Taking the inmate aside and explaining the proper way to do things communicates the expected behavior for the institution, and it also may open the way for good relations between the officer and the inmate. When to handle such an incident in this way will obviously be a matter of judgment. The officer must assess the likelihood that the inmate did not understand what was expected. The officer's assessment should always take into account other factors, such as the seriousness of the behavior, the sophistication level of the inmate, and the effect this handling of the incident will have on other inmates. For example, being out of bounds, that is, in an unauthorized place at an unauthorized time, may be relatively innocent behavior for a naïve new inmate, but it could be extremely serious behavior for a sophisticated inmate.

As the behavior becomes more serious, or if an officer is not certain whether to have an informal discussion with the inmate or to write up charges, the matter can be referred to a supervisor who will make that decision. In addition to the officer on the scene, who may choose to refrain from filing certain charges, the correctional supervisor should be given authority to dispose of charges informally, rather than pursuing them in a formal disciplinary procedure. Again, these informal dispositions are to be used at the lower end of the spectrum (obviously, assaults, drug offenses, and escape attempts must be pursued with formal charges), but more than half of the offenses on the misconduct list can probably be handled informally in the right circumstances.

In a more sensitive type of informal disposition, the staff member is certain that the inmate has violated a rule, but a sanction is given without charges being filed. Sometimes, this is done by negotiating with the inmate ("I will drop the charges, if you agree to . . ."). More often, and preferably, staff may use certain minor and agreed-upon sanctions in these situations (e.g., extra work, restrictions in movement, or even changes in work assignments, recreation activities, or housing assignments). There need not be a written code for informal sanctioning, because these sanctions should consider an inmate's current and past behavior, though sanctions should be applied consistently. To that end, facility management and supervisory correctional staff must specify how informal sanctions may be used. Supervisors should check that the power given to staff to punish inmates informally is not abused. When power is abused, inmates lose respect for authority.

There are benefits of informal resolutions of inmate misconduct. The most obvious is that they help prioritize cases within the disciplinary process, reserving the procedural hearings for the more serious offenses. Inmates will have greater respect for staff who handle matters fairly but informally.

■ Due Process Requirements

The Fifth Amendment grants that no person may be "deprived of life, liberty, or property, without due process of law," and the Fourteenth Amendment grants that "nor shall any State deprive any person of life, liberty, or property, without due process of law; nor deny to any person within its jurisdiction the equal protection of the laws." Whenever a claim is made (for example, by inmates claiming violation of their rights because of disciplinary action), there is first the question of whether there has been any deprivation of life, liberty, or property. Not every type of government action that a person dislikes raises a due process question. Generally, due process is considered a set of procedures that ensure that the action taken is fair, given the circumstances. Again, as a general rule, the more serious the action taken by the government, the more procedural protections (due process) will be required. The U. S. Supreme Court has looked at disciplinary actions in prisons and established some relevant constitutional standards based on a variety of cases.

Sandin v. Conner (1995)

The Supreme Court gave new guidance as to the circumstances under which due process is required in prison actions in the case of *Sandin v. Conner*.[1] The Court greatly simplified the constitutional standard to be used to decide when prison action amounts to a deprivation of liberty that requires due process. The Court recognized in *Sandin v. Conner* that the purpose of prison disciplinary action is to achieve good prison management and prisoner rehabilitative goals. As long as disciplinary action is in pursuit of those goals, and does not add on to the sentence already being served or go beyond the conditions contemplated in the sentence being served, the disciplinary action does not create a liberty interest. The Court held that Conner's discipline in segregated confinement did not present the type of atypical, significant deprivation in which a state might conceivably create a liberty interest. The regime to which he was subjected as a result of the misconduct hearing was within the range of confinement to be normally expected for one serving an indeterminate sentence of 30 years to life.

Thus, the question is not whether the inmate is punished, or even punished severely. The question is whether the punishment is within the range of conditions, restrictions, and sanctions that are contemplated for the type and length of sentence that the particular inmate is serving. This substantially reduces the number of cases where due process protections are constitutionally required because the action taken by prison officials deprives inmates of liberty. Lower courts have been busy interpreting the scope of the *Sandin v. Conner* ruling. From the facts of the case and the Court's language, segregated confinement (or other housing restrictions or the withdrawal of privileges) would not trigger due process requirements, but actions that might extend the time to be served (such as good time awards being taken away or parole release being affected) would require a due process hearing. Some correctional agencies, as a matter of caution as well as fairness, continue to require certain procedural protections for any serious act of inmate misconduct.

Wolff v. McDonnell (1974)

Wolff v. McDonnell is the leading case in the Supreme Court on inmate discipline.[2] The Supreme Court recognized the special nature of prison disciplinary proceedings and specifically rejected claims for procedures (such as representation of inmate by lawyers) that would "encase the disciplinary procedures in an inflexible constitutional straitjacket that would necessarily call for adversary proceedings typical of the criminal trial." In other words, prison hearings are "administrative" and call for much less procedural protection than court proceedings.

There are five minimum due process standards for a prison disciplinary hearing:

1. There must be advance written notice to the inmate of the claimed violation.
2. The hearing should be held at least 24 hours after the notice, so that the inmate has time to prepare for his or her appearance.
3. The inmate should be allowed to call witnesses and to present documentary evidence, unless permitting him or her to do so would be unduly hazardous.
4. A representative (who may be an inmate or a staff member) should be allowed to assist the inmate if the inmate is illiterate or if there are complex issues involved.

5. There should be a statement by an impartial disciplinary committee of the evidence relied on to support the fact findings and the reasons for the disciplinary action taken.

Two requirements that had been strenuously sought for inmate hearings were specifically denied in *Wolff v. McDonald*: There was no requirement for the inmate to be permitted to confront and cross-examine adverse witnesses (because of the special hazards those actions could present in a prison setting) and there was no requirement for allowing inmates to have legal counsel.

Other Relevant Cases

Other Supreme Court cases added additional guidelines for prison disciplinary hearings.

In *Baxter v. Palmigiano* (1976), the Court reiterated its ruling that inmates are not entitled to either retained or appointed counsel in disciplinary hearings.[3] A new issue was the inmate's right to remain silent in the disciplinary proceeding. The Court said that it was permissible for officials to tell inmates that they could remain silent, but that their silence could be used against them (to draw an adverse inference) at their hearing. However, the inmate's silence by itself would be insufficient to support a decision of guilt by the disciplinary committee. In addition, if the inmates were compelled (that is, ordered) to furnish testimony that might incriminate them later in a criminal proceeding, that testimony could not be used against them in the criminal trial.

In the case of *Superintendent v. Hill* (1985), the Court ruled on the amount of evidence that was required, constitutionally, to support a prison board's findings.[4] The Court said that due process required only that there be "some evidence" to support the findings of the disciplinary board—about the lowest standard of proof that could be devised.

■ Use of Informants

There are some issues of inmate disciplinary proceedings that have not been addressed by the Supreme Court. One of the most troublesome is the use of evidence that comes from confidential informants or develops as a result of the information given by such informants. Much information for the safer control of correctional facilities is supplied by inmate informants. Many correctional supervisors and officers rely on this information to keep track of sensitive inmate misconduct. In fact, officers often develop regular sources of such confidential information and even provide rewards to the informants. (Again, this practice has its counterparts in the community's law enforcement activities.) Because a central goal of prison management is the maintenance of security and safety, there is a need to have such suppliers of information. However, to ensure the fairness of the disciplinary process (to have some due process in prison hearings), there must be protection against inmates fabricating information or providing false accusations, which they may do for a variety of reasons.

Although the U.S. Supreme Court has not addressed this issue, lower courts (and correctional officials and lawyers) have considered it and have concluded that there must be some way to ascertain the reliability of informants' stories. Informants, to protect their safety, cannot be called as witnesses at hearings. By the same token, their identities cannot be revealed to the accused inmates. In the policy of most agencies, the disciplinary committee makes inquiry, apart from the accused, as to the reliability of informant information that is present in the case. A conclusion that the informant is reliable is entered into the record. That conclusion may be based on the circumstances of the case, in which the informant's assertions have been corroborated by other facts; on a report from an investigating officer who gives reasons for believing the informant; on other situations where the informant gave reliable information; or on detailed information that could be known only by someone who was present to observe the facts that otherwise proved true.

Until the Supreme Court provides firm guidance in this area, correctional staff will have to look to any rulings that have been made by courts in their jurisdiction and follow those rulings. Staff responsible for drafting policy must make certain that the inmate disciplinary policy reflects any court rulings on use of informant information (and any other area where local courts may have special requirements, outside of or beyond those given by the Supreme Court). Line staff implementing the disciplinary policy must be able to rely on its accordance with any court rulings that apply in the local jurisdiction. With the Supreme Court's ruling in *Sandin v. Conner*, the exposure of staff to individual liability for constitutional violations in disciplinary proceedings has been greatly reduced but not eliminated.

■ Inmate Appeals and Grievances

Appeals from disciplinary decisions have not been required by the courts as a necessary component of inmate discipline procedures. For the more serious levels of misconduct, an appeal system is common under administrative policy. There may be one or two levels of appeal of disciplinary action. In some systems, appeals would be made to the disciplinary hearing officer or committee or to the warden. In agencies with regional organization, there may be a level of appeal to the regional office, and many agencies do allow appeal to the headquarters level.

Because inmates are often unhappy with the results of the disciplinary proceedings, it is no surprise that disciplinary matters are appealed frequently. In some agencies, inmate disciplinary actions may be appealed through the inmate grievance system, in the same way other prison actions or conditions may be grieved. Many agencies do not allow disciplinary matters to be taken into the regular grievance system; in these agencies, there will be a separate procedure for review or appeal of inmate disciplinary actions. In those agencies where discipline actions are grievable, it is common for appeals concerning discipline to account for the greatest number of inmate complaints. Similarly, when inmates go to court, disciplinary actions are, in most jurisdictions, the most frequent category of matters taken into court.

Review of disciplinary actions on appeal is typically limited to procedural review. The records (the discipline offense report, the investigation report, and the written report of the hearing officer or hearing committee) are examined, to ensure that the procedures required by the agency's disciplinary policy have been followed. The facts of the case and the sanction imposed are summarily reviewed. Some evidence must exist to support the findings of the disciplinary authority. The reviewer ensures that the sanction imposed is within the range of punishments authorized for that offense. There is a legal requirement that the hearing officer or committee record the evidence relied on to support the conclusion reached, and that the reasons for the sanction(s) imposed be given.

There are other significant legal requirements for inmate disciplinary proceedings, and corrections staff should be aware of them. As noted previously, the inmate disciplinary program attempts to establish respect for authority. For the inmate disciplinary program to receive respect, it must be administered fairly and according to rules of due process. Apart from being the law, these rules are important for maintaining a fair and humane living environment in correctional facilities.

■ Personal Liability

Legally, it is also essential that staff follow rules to avoid exposure to personal liability. In litigating inmate complaints over the last 40 years, the most frequently used legal action has been under the Civil Rights Act of 1871, particularly what is usually called a Section 1983 lawsuit.[5] That federal statute provides that:

> *Every person who, under color of any statute, ordinance, regulation, custom, or usage, of any State or Territory, subjects, or causes to be subject, any citizen of the United States or other person within the jurisdiction thereof to the deprivation of any rights, privileges, or immunities secured by the Constitution and laws, shall be liable to the party injured in an action at law, suit in equity, or other proper proceeding for redress.*

This law allows people who claim their constitutional rights have been abridged to go into federal court for legal relief. (While Section 1983 only applies to state actions and state employees, its standards have been applied equally to federal actions and employees by the Supreme Court.[6]) Section 1983 has been popular for inmates to use for several reasons. In most states, federal courts were seen as being more liberal and more receptive to inmate complaints than state courts. Legal rulings were obtained in one part of the country, which could then be used as leverage to get favorable rulings in federal courts in another part of the country.

Section 1983 could be used to obtain two kinds of relief. If constitutional violations were proved, the federal courts could give injunctive relief and also monetary damages against the offending officials. Injunctive relief could require an agency to stop doing something it had been doing, to change its procedures, or to start doing something that was constitutionally required. Many changes were required by courts in injunctive orders under Section 1983. More threatening to the individual official is the risk of being ordered to pay monetary damages because of constitutional violations.

A corrections worker may be held personally liable if he or she does not follow constitutional requirements that have been established by court rulings at various levels. A ruling by the U.S. Supreme Court is the most authoritative and must be followed throughout the country. There may also be rulings in lower federal courts or in state courts that govern a particular type of activity covered by the Constitution or state statutes. It is the responsibility of lawyers for each correctional agency to make sure that employees are aware of any such local rulings and that agency policy reflects the current constitutional requirements given in court rulings.

Correctional staff can protect themselves by receiving staff training, which points out legal standards and constitutional requirements, and following agency policy, which must be most carefully done in those areas of possible constitutional liability where the courts have ruled. All staff must be careful to follow those rules established by the courts with regard to disciplinary actions.

■ Conclusion

Disciplinary policy must be written carefully to ensure fairness and to guarantee basic due process standards. Staff must follow that policy carefully to protect themselves from liability and maintain the integrity of the correctional environment.

DISCUSSION QUESTIONS

1. Why is it necessary to have a prison's discipline policy spelled out in writing?
2. What are the benefits and risks of handling inmate misconduct informally?
3. Would having an attorney present at a prison discipline hearing to represent the inmate be a good or a bad thing?
4. What are the benefits and risks, in using inmate informants to take disciplinary actions against other inmates?
5. To what extent is an inmate discipline process similar to the criminal law process, and how is it different?

ADDITIONAL RESOURCES

C. Camp and G. Camp, *The 2002 Corrections Yearbook—Adult Corrections* (Middletown, CT: Criminal Justice Institute, 2003).

C. Cripe and M. Pearlman, *Legal Aspects of Corrections Management,* 2nd edition (Sudbury, MA: Jones and Bartlett Publishers, 2005).

NOTES

1. *Sandin v. Conner,* 585 U.S. 472 (1995).
2. *Wolff v. McDonnell,* 418 U.S. 539 (1974).
3. *Baxter v. Palmigiano,* 425 U.S. 308 (1976).
4. *Superintendent v. Hill,* 472 U.S. 445 (1985).
5. *Civil Rights Act of 1871,* U.S. Code vol. 42, sec. 1983 (1871).
6. *Bivens v. Six Unknown Federal Narcotics Agents,* 403 U.S. 388 (1971).

Grievance Procedures

28

Lisa Hutchinson Wallace, Kevin I. Minor, and James Stephen Parson

Chapter Objectives

- Describe the history of inmate grievance procedures, including when and why they were developed.
- Explore the principles of model grievance systems and the application to actual systems.
- Detail the core elements of an inmate grievance system, as outlined by the Civil Rights of Institutionalized Persons Act, and the potential benefits of inmate grievance systems.

An inmate grievance system is a structured, institutional process that provides a forum for inmates to seek redress for issues or complaints surrounding their incarceration. The system comprises a set of established, written rules detailing the issues that may be grieved, timeframes associated with the grievance process, and specific steps to be taken when filing a grievance. Such a system allows inmates the opportunity to lodge a formal complaint regarding issues that they have been unable to resolve informally.

To fully understand the span of an inmate grievance system, it becomes necessary to comprehend the underlying intent of the term grievance. Historically, this term has been rather ambiguous. This has resulted in a lack of clarity regarding the issues that may be grieved, as well as the proper procedures for addressing such issues. Due in part to this ambiguity, the definition of this term has been established formally. The most recent legislation that delineates written inmate grievance systems, the Civil Rights of Institutionalized Persons Act (CRIPA), defines a grievance as a "written complaint by an inmate on the inmate's own behalf regarding a policy applicable within an institution, a condition in an institution, an action involving an inmate of an institution, or an incident occurring within an institution."[1]

■ History of Inmate Grievance Procedures

Inmate grievance processes have been an integral part of correctional processes for many decades. While the current systems are formalized, those utilized in the past typically have been very informal and lacking in common characteristics. During the 1970s, however, a number of common mechanisms for reviewing and responding to inmate complaints emerged, although most were still informal in structure. Some of these systems provided for some form of relief to inmates, although many fell short of their intended purpose of resolving inmate complaints effectively.

One of the most common mechanisms for dealing with inmate grievances utilized during this time period was the ombudsman system. Typically, this system involved bringing in someone from outside the institution to investigate and review complaints. The individual overseeing the hearing was able to make recommendations for corrective action but was prohibited from recommending any other form of relief.

Another common system involved the use of an inmate advisory group to review grievances. In this type of system, a group comprised solely of inmates reviewed all complaints and made recommendations regarding appropriate actions. As with the ombudsman system, recommendations were limited to those entailing corrective action.

The third type of system most commonly utilized during this time period was hearing panels. Under this system, a committee of staff members reviewed all complaints. While these panels were also limited to remedying the situation through corrective action, they were also often empowered to order, rather than simply recommend, such action.

Although the previously described mechanisms were intended to provide a forum for complaint resolution, many of them employed disorganized, informal processes that produced inconsistent and contentious results. Section 1983 of the Civil Rights Act of 1871 allows inmates to allege that state officials (including correctional officers) have violated their constitutional rights.[2] The lack of a formal process to resolve inmate complaints provided the basis for numerous inmate lawsuits under this section of the Act. While the Civil Rights Act of 1871 has been in existence for quite some time, it was not until the 1970s that inmates began to use the act as a means to access the courts. As previously noted, this timeframe was also when the majority of the grievance procedures were informal and lacking in consistent guidelines, providing a breeding ground for numerous inmate lawsuits.

In 1941, the U.S. Supreme Court's decision *Ex Parte Hull* opened the floodgate for inmate lawsuits, as it recognized for the first time that inmates had a constitutional right to access the courts.[3] The number of inmate lawsuits continued to increase well into the late 1970s, when the frivolous nature of some lawsuits and the undue burden they placed on the system became unsettling to the public. In response to the increased number of lawsuits that were filed, citizens began to press legislators to find ways to curtail these suits. Nowhere was the need to identify alternatives to lawsuits more apparent than in the inmate grievance procedures, which served as the impetus for many of these lawsuits. Therefore, as the number of inmate lawsuits and the public's cry to curtail such suits grew, the need for a more formalized process to resolve inmate grievances was realized.

Establishment of Written Inmate Grievance Systems

The need to address the increased number of inmate lawsuits was met with the passage of the Civil Rights of Institutionalized Persons Act in 1980, which formalized the procedures comprising inmate grievance systems.[4] This act required the Attorney General to establish the minimum standards to be utilized in inmate grievance procedures. Additionally, this act provided for the establishment of a certification process by which states could submit their inmate grievance systems for certification based on the specified criteria. As a result, the Office of Inmate Grievance Procedure Certification was established to oversee the certification process. The Office has the responsibility of certifying all inmate grievance systems proposed by the states. Once a process has been certified, a written copy of the process, along with materials explaining it, must be distributed to all employees and inmates in the institution. Additionally, all employees and inmates should receive an oral explanation of the process.

Core Elements of a Grievance System

In order to receive a certification from the Office of Inmate Grievance Procedure Certification, systems must include a number of core components. The specific requirements within each of these components varies widely among states and institutions, but the act specifies the issues to be included under each component as well as the minimum threshold that the process must incorporate.

Specification of Written Grievance Procedure

The inmate grievance system must be recorded formally. This procedure must specify the institutions to be covered by the system, as well as the process by which adherence to the procedure will be documented. In addition, both inmates and employees should have an opportunity to fundamentally participate in the formation and implementation of the grievance process. The requirement for inmate and employment participation was revised in 1992 to provide clarification regarding the role of inmates in the adjudication of claims. The amendment specifies that inmate participation does not imply that inmates must sit on panels that adjudicate grievances, merely that they have an opportunity to be involved in the formation and implementation of the system in some manner.

Communication of Process

All inmates and employees in the institutions who will be covered by the grievance process must be notified in writing of that process. CRIPA also requires that materials explaining the process must accompany the written documentation.

Further, these materials must be presented in all languages that are spoken by a significant segment of the population. Finally, oral notification of the process is required upon entrance into the institution and periodically thereafter.

Specification of Accessibility to Process

All inmates must have access to the inmate grievance process. Restrictions based on disciplinary issues, classification reasons, or administrative designations may not prohibit individuals from accessing the process.

Applicability of Process to Complaints

CRIPA requires that the grievance processes apply to a wide variety of complaints. While the language of CRIPA does not detail the nature of each complaint, the act does require each state to establish a written grievance process that clearly specifies the complaints covered under its auspices. CRIPA further specifies a range of complaints to be covered; specifically, inmates must be allowed to file grievances regarding issues of policy or conditions within the institution. Additionally, they must be allowed to file grievances regarding actions of employees or the institution. Finally, if an action within the institution affects them personally, then they are allowed to file a grievance. Because CRIPA allows for a broad continuum of complaints to be included in the grievance process, it is not surprising that certification requirements vary significantly among states.

Remedies Available Under the Process

Inmate grievance processes must provide meaningful remedies. The range of remedies varies as widely among states as does the range of allowable complaints. Today, remedies are not limited solely to corrective actions; they can include restitution, monetary reimbursements, actions against personnel, change in classification level, changes in institutional policy, modification of record, and others. Regardless of the type of remedy chosen, all remedies must be formally recorded in the written grievance system.

Appropriate Protection Against Reprisals

The grievance process must provide safeguards to guarantee protection for inmates who choose to utilize the process. All inmates who participate in the grievance process must be assured that they will be protected from any reprisals. A reprisal policy must explicitly state the preventive actions that will be taken to guarantee such protection along with the penalties for any policy violations.

Provisions for Emergency Grievances

Inmate grievance systems must also specify the exclusive actions to be taken under emergency grievances. Prior to specifying those actions, each system must first define those situations it will deem emergencies. At a minimum, any situations in which the adherence to the fixed time limits of the grievance system might result in personal injury or other serious harm must be classified as emergency situations. Next, all systems are required to identify the process to be utilized when an emergency grievance is filed. CRIPA further specifies that upon classification as an emergency grievance, the grievance must be automatically forwarded to the level of action authorized to provide corrective action. No substantive review is required prior to the forwarding of an emergency grievance.

Records

CRIPA requires the documentation of the grievance throughout the system. Initially, all grievances must be initiated through the process of completing a simple standard form. Such forms must be readily available to inmates and should not contain technical or cumbersome language that would impede an inmate's ability to complete it. Additionally, all institutions are responsible for systematically collecting data at the aggregate level regarding the number of grievances filed, the nature of the grievances, and the outcome. All files regarding inmate grievances must be kept for three years after the final resolution of the complaint. Finally, the act requires that all records regarding inmate grievances be kept confidential. Such records must not be disclosed to other inmates or staff with the exception of their being handled by other staff for clerical purposes.

Evaluation

All inmate grievance systems must describe the evaluation strategy to be used to evaluate and modify the system should it be warranted. At a minimum, the evaluation report must be submitted to the Attorney General on an annual basis. CRIPA specifies the basic information that must be reported in the evaluation, which includes:

- Number of complaints filed
- Types of grievances
- Frequency and type of emergency grievances
- Resolution of grievances
- Average lengths of time between the filing and resolution

Other Applicable Requirements

In addition to these core requirements set forth by CRIPA, several procedural recommendations for the operation of inmate grievance systems have been established.

First, grievance systems must specify the time limits that govern the written replies to grievances and include an explanation of the decision in such a reply. Specific time limits must be established for each level of decision making involved in the process. The entire grievance process must be completed within 90 days of the filing of a grievance, unless the inmate provides written agreement to an extension. If the time limits expire during any stage of the process, the inmate's grievance is automatically moved to the next level of review.

Second, the grievance investigation process must be explicitly stated within the system. CRIPA prohibits the involvement of any inmate or employee allegedly involved in the circumstances surrounding the grievance to be involved in the investigation.

Third, a written response to the grievance must be provided at each level of review. The written response should include a description of the reasons behind the decision, outline the inmate's rights to further review (should such review be appropriate), and provide information regarding the steps to be taken to pursue such a review.

Finally, the last core requirement of an inmate grievance system involves the issue of independent review. Inmates are entitled to a review of decisions made as a part of the grievance process, although the review must be made by an individual or entity that does not fall under the direct supervision or institutional control. When the grievable issues involve an institutional policy or practice, the agency must give serious consideration to the recommendations produced under an independent review but need not adhere to them.

■ Rationale Behind the Establishment of Written Grievance Systems

The primary impetus behind the establishment of formal inmate grievance processes was the desire to establish a systematic, unbiased procedure for resolving inmate complaints. The idea was that if inmates were provided an organized process in which they could lodge their complaints, then they would be less likely to file frivolous or unnecessary lawsuits. Ultimately, the goal of an inmate grievance system is to resolve or alleviate tension within the institution to the extent possible.

Grievance systems provide a number of benefits. One of the most important benefits is that it provides both practical and legal protection for inmates and correctional staff. From a practical standpoint, the mere act of allowing an inmate the ability to lodge a complaint formally and have that complaint reviewed in a systematic and unbiased manner helps alleviate the tension and deprivations suffered in prison. In researching the social structures created and maintained by inmates, Gresham Sykes identified certain deprivations that inmates suffer when incarcerated in prisons. Sykes is careful to point out that these "pains" involve a psychological reaction to the deprivations experienced during incarceration.[5] The result of this deprivation is the formation of a solidarity among inmates opposing the correctional staff. Inmates form this cohesive unit as a way to mitigate the effects of the "pains of imprisonment" and to exercise some form of control over their situation.[6] The simple act of offering inmates a formal system where they are allowed to lodge a complaint regarding some aspect of their confinement helps to alleviate the pains of imprisonment and thus preclude the formation of solidarity that directly opposes the staff.

With respect to the legal benefits of inmate grievance systems, a formalized system benefits both inmates and staff because it generally reduces the number of lawsuits filed. When inmates have an opportunity, to have their complaints heard independent of the court system, they are less likely to file a lawsuit. After the implementation of the formal grievance system in 1980, the federal government began to place more responsibility and authority in the decisions resulting from such systems. In 1996 Congress passed the Prison Litigation Reform Act (PLRA). Section 803 of this act amended portions of CRIPA, specifically by requiring the exhaustion of all remedies through the inmate grievance systems prior to the filing of a lawsuit.[7] This exhaustion requirement has been upheld by the U.S. Supreme Court as a necessary precursor to the filing of a lawsuit, even if the remedies available under the inmate grievance system do not provide for the specific remedy being sought by the inmate.[8]

Therefore, regardless of remedy sought, inmates must exhaust the remedies available under the institution's inmate grievance system prior to filing a lawsuit, thereby significantly reducing the number of lawsuits filed in the courts. The U.S. Supreme Court has also upheld the exhaustion requirement for all cases including those involving allegations of excessive force. Because CRIPA requires that provisions for emergency grievances (such as issues of excessive force) be specified in the grievance system, the Court felt it appropriate to expand the exhaustion requirement to cover all aspects of inmate life.[9]

Not only has the use of these internal systems become an obligatory precursor to the filing of a lawsuit, these systems have also become a good source of information if a lawsuit is filed. Because the formal systems require thorough and appropriate documentation of the nature of the grievance and the responses to the grievance at the various levels of the system, accurate records exist when lawsuits reach the court.

Inmate grievance systems also provide correctional staff with a wealth of information they can utilize to help improve the secure and orderly operation of the institution. For instance, the statistics regarding the number of grievances filed paint a picture of the specific areas within the institution or the individuals that need to be addressed. The identification of potential issues allows correctional administrators to address issues without court intervention. Additionally, statistics regarding the number and types of grievances filed provides an overall glimpse into the climate that currently exists within the institution. Finally, a documented process for resolving complaints opens the line of communication with both staff and inmates.

■ Weaknesses of Inmate Grievance Systems

Although there are many benefits to a formalized system, the mere act of establishing a binding grievance process does not come without limitations. The promise of justice inherent in a formal grievance system can sometimes lead to false expectations. When resolutions do not turn out the way that inmates expected, their frustration may be greater than if their resolution was never heard. Additionally, if inmates misuse the system and file frivolous complaints, the system can quickly become overburdened. Such a situation increases the workload of the correctional staff and other involved persons who must process and hear these complaints.

Additionally, a formal system inherently produces an increased amount of paperwork. Initially, the grievance process must be well-defined, well-recorded, and well-promoted to staff and inmates. When grievances are filed, each step of the process must be documented thoroughly. Additionally, the evaluation component required by CRIPA specifies that annual reports regarding the types and numbers of inmate grievances as well as the resolution of those grievances must be aggregated and reported to the U.S. Attorney General. All of these requirements place an increased burden on the correctional staff and administration.

These systems must also function in the manner proscribed and approved by the Attorney General so that maximum benefits may be realized. If the ultimate goal is to alleviate tension, then the inmates must perceive the systems and the systems' administrators to be fair. Regardless of their perceived effectiveness, formal grievance systems have gained popularity since the passage of CRIPA and their use has increased substantially.

■ Conclusion

Formalized inmate grievance systems have become integral to the institutional operations of many facilities. Since the establishment of these systems through the passage of the Civil Rights of Institutionalized Persons Act, the use of formal inmate grievance systems has increased. Additionally, the passage of the Prison Litigation Reform Act and subsequent rulings by the U.S. Supreme Court upholding the exhaustion portion of this act have enhanced the impact of grievance systems on inmate access to the courts. It is imperative that these systems be administered effectively to ensure that their intended benefits are realized.

DISCUSSION QUESTIONS

1. When did inmates grievance procedures first start to evolve in American corrections systems?
2. What are some of the reasons why these procedures were created?
3. What are some of the principles of effective inmate grievance procedures?
4. What types of challenges do inmate grievance coordinators face?
5. What potential benefits do inmate grievance systems offer?

ADDITIONAL RESOURCES

United States Department of Justice, Civil Rights Division, http://www.usdoj.gov/crt

NOTES

1. *Code of Federal Regulations,* vol. 28, sec. 40.1 (d).
2. *Civil Rights Act of 1871*, U.S. Code vol. 42, sec. 1983 (1871).
3. *Ex Parte Hull*, 312 U.S. 546 (1941).
4. *Civil Rights of Institutionalized Persons Act*, U.S. Code, vol. 42, sec. 40.1 (a) (1980).
5. G. Sykes, *The Society of Captives: A Study of a Maximum Prison* (Princeton, NJ: Princeton University Press, 1958).
6. G. Sykes and S. Messinger, "The Inmate Social System," in R.A. Cloward, D.R. Cressey, G.H. Grosser, R. McCleery, L.E. Ohlin, G.M. Sykes, and S.L. Messinger, eds., *Theoretical Studies in the Social Organization of the Prison* (New York: Social Science Research Council, 1960), pp. 5–19.
7. *Prison Litigation Reform Act of 1995*, U.S. Code, vol. 42, sec. 803 (a) (1996).
8. *Booth v. Churner*, 532 U.S. 731 (2001).
9. *Porter v. Nussle*, 534 U.S. 516 (2002).

Protective Custody

29

Kevin I. Minor, Lisa Hutchinson Wallace, and James Stephen Parson

Chapter Objectives

- Define protective custody and its forms.
- Differentiate between disciplinary status and protective custody status and explain the stigma of protective custody.
- Understand protective custody as a component of the wider prison subculture.

This chapter addresses several considerations related to protective custody (PC). It defines the term, estimates the number of PC prisoners, and addresses procedural and operational considerations, including entry to, life in, and transition out of PC. It also discusses the effects of segregation and the major legal considerations involved with its use. Finally, this chapter explores PC as a component of the wider prison subculture, acknowledging the need for greater attention to clarifying legal issues and safely transitioning prisoners out of protective status.

■ Defining Protective Custody

Protective custody was first introduced in the 1960s. It is defined by the American Correctional Association (ACA) as "a form of separation from the general population for inmates requesting or requiring protection from other inmates for reasons of health or safety."[1] It is most often manifested as specialized, segregated housing for several different types of inmates. Formal definitions of protective custody are applied by prison authorities, and many jurisdictions include an official definition of PC in their statutes or policy documents. Informal applications of the term also may be applied to inmates and staff within the prison population.

Table 29–1 outlines elements of protective custody as well as similar types of segregation such as disciplinary segregation (whereby those prisoners believed to have violated rules are isolated from the general population as discipline or punishment) and administrative segregation (whereby continued presence of the

Table 29–1	Types of Segregation		
Question	Disciplinary Segregation	Administrative Segregation	Protective Custody
Who typically assigns the prisoner to segregation?	A standing disciplinary committee or hearing officer	An ad hoc committee or single official with the authority to do so	A standing classification committee
What is the process?	Relatively open, formal hearing with rules of evidence, witnesses, etc. and the prisoner present; focus on specific action or occurrence alleged to have taken place; result is a finding of fact or determination of guilt; availability of a standard appeals process	Relatively closed informal hearing or discussion often without the prisoner present; focus on ongoing or emerging general circumstances; result is a finding of likelihood or probability of what might take place if segregation is not imposed; appeals process is nonstandard or nonexistent	Prisoners may request PC placement and have the request reviewed by prison officials; alternatively officials may recommend PC status, but prisoners often have an option to sign a waiver and opt for non-PC status
What is the length of segregation?	Definite or set period	Indefinite period	Indefinite period
Where is the prisoner segregated?	The disciplinary segregation unit, or in prison argot, "the hole"	Any of several secure units or facilities (e.g., separate administrative segregation unit, disciplinary segregation unit, protective custody unit, county jail, out-of-jurisdiction unit, etc.)	A separate protective custody unit or within a larger segregation unit housing non-PC segregated prisoners
Why is the prisoner segregated?	As punishment for a specific rule infraction to which the prisoner has admitted or for which he or she has been convicted; actual outcome is specific (i.e., intended deprivations and punishment)	To disrupt or prevent an ongoing or potential security threat; actual outcomes are variable and general (e.g., prevention of security threats, unintended or intended punishment of select prisoners)	For protection against victimization; actual outcomes include possible temporary protection, unintended deprivation, short- and long-term stigmatization in the eyes of both staff and other prisoners, as well as a potentially higher likelihood of future victimization

inmate in the general population might pose a serious threat to security). While terminology varies across jurisdictions, this threefold distinction is representative of most prison systems. PC sometimes is treated as a subset of administrative segregation, but the two are actually quite distinct. Administrative segregation is a generalized tool of management and control that many prison officials readily embrace and many prisoners perceive as enhancing their status. PC is a strategy targeted specifically around protection that officials and prisoners perceive in more distasteful terms.

Application

Although prison staff officially determine who is placed in formal PC, the prison subculture exerts considerable influence over PC assignments by fomenting aggressive and other negative behaviors toward certain types of prisoners that can culminate in PC placement. Moreover, the subculture negatively skews the official definition of PC by attaching a stigma to the status itself and to those prisoners (and even staff) assigned to PC.

Informal PC status is more subjective, and less readily defined and identified, than formal protective custody. In short, PC is much more than official transfers to a PC unit. The key distinction between informal PC and other protective activities is that the former involves stigmatized custodial change. Unofficial PC is achieved via some stigmatizing change in the custody of an individual prisoner other than placement in a formal PC unit or program. Though it may or may not be regarded as a formal custody change by staff, the effect of the change is real because the lifestyle and movement patterns of the prisoner are altered significantly.

Self-Imposed Protective Custody

Not every act taken by staff or prisoners to promote protection qualifies as PC. Dick Franklin argues that many prisoners, instead of seeking official PC status, take alternative protective actions within the prison subculture—a "self-imposed protective custody."[2] This includes situations such as:

- Intentionally breaking institutional rules to receive administrative or disciplinary segregation (commonly known as "taking a fall")
- Organizing activities to avoid interaction with the person or group believed to be a threat (commonly known as "PC-ing up")
- Establishing patterns of close affiliation with staff, as in the case of certain trustees (commonly known as "hangin' on the man's leg")
- Establishing affiliation with other individual prisoners or groups of prisoners, such as a gang (commonly known as "cliquing up")
- Seeking to protect themselves through aggression directed at the perceived threat (commonly known as "spinning out")

Stigma

Prison subculture applies a kind of normative, universal stigmatization to protective custody status, which includes a lack of social acceptance and diminished respect from others. Prisoners who formally or informally adopt the PC role are perceived as subhuman and treated accordingly in the subculture. The "PC punk" label, as it is commonly perceived, may lay the stage for the PC prisoner to be the target of extreme degradation and violence.

Table 29-2	Illustrative Typology of Protective Activities		
Initiating Party	Formal PC	Informal PC	Non-PC Protective Activities
Prisoner	Seeking out PC status on one's own accord	Disengaging adaptations to threats (e.g., avoiding certain areas and/or people, skipping meals or recreation, etc.)	Engaging adaptations to perceived threats (e.g., physically attacking the source of the threat)
Staff	Recommending PC during classification activities	Managing individual prisoners to promote protection through disengagement (e.g., individual housing reassignments or transfers for vulnerable prisoners)	Managing prisoners in the aggregate to promote protection (e.g., increased surveillance of particular areas)

While the stigma initially derives from the reason the prisoner originally acquired PC status (e.g., having entered prison upon conviction of child molestation), an added stigma results from the sheer act of opting for the PC label, rendering PC prisoners vulnerable to forms of victimization at the hands of other prisoners and staff.[3] It is important to note that many protective acts do not entail disengagement of prisoners from the general population and carry no stigma in the prison subculture (see **Table 29-2**).

Estimating the Number of PC Prisoners

Estimating the prevalence of PC is difficult since it is impossible to know the number of informal PC cases and terminology differs across jurisdictions. However, estimates range from 6000 to 8000 persons in PC nationwide, or approximately 1 percent of the total prison population.[4] While data are lacking on unofficial PC and non-PC protective activities, it is safe to conclude that both are more common in the prison environment than official PC designations and operations.[5]

Formal PC Procedures and Operations

At a most basic level, prisoners acquire formal protective custody status through one of two procedures.

1. Inmates may request the status and then have it granted at the discretion of institutional staff, frequently after having served the staff in the capacity of informant or "snitch" to justify PC placement.

2. Staff members working in the area of classification may decide that a particular prisoner should be housed in PC and classify the prisoner accordingly, although prisoners typically can waive PC status.

There are issues surrounding both of these procedures. PC consumes money and staff time because of the special provisions it entails. Even when a separate PC housing unit is not made available, special arrangements have to be made within whatever segregated housing unit does exist, and many PC prisoners require single cells. Although PC prisoners often live under restrictive and deprived conditions, the same services and programs provided for the general population should be accessible to them, because in theory they are being segregated for protection rather than punishment.

As such, depending on a jurisdiction's laws and policies, as well as space availability, officials often screen requests, under the assumption that a substantiated rationale for PC needs to exist. It is clear that some prisoners who might benefit from PC are reluctant to seek it. They may see the designation of PC status as incapable of guaranteeing protection—as actually increasing the risk of harm through the stigma of the label—or they may see it as incompatible with their sense of who they are and the norms of the prison subculture.[6]

What may be more difficult is to determine if all prisoners who request PC belong there. For example, there may not be a genuine and serious threat against the person making the request, the requestor may be attempting to achieve some outcome besides safety (e.g., a one-person cell and greater solitude or a transfer to another facility), or the requestor may even represent a threat to prisoners in the PC population. On the other hand, prison officials have traditionally been concerned that denial of requests for protection increases their legal liability.

The challenge for staff is to balance liability concerns against the need to weed out illegitimate requests. But this is not easily accomplished because genuine threats are sometimes difficult to discern. In some situations, for instance, certain prisoners are identified as "marks" and convicts then wager bets on who can drive their mark to check into PC first, making it difficult to distinguish genuine and serious threats from manipulations posing little real or long-term threat to safety. In addition, staff can actually heighten a prisoner's vulnerability to victimization, and potentially their own legal liability, by insisting that prisoners who ask to enter PC first act as informants about certain illicit activities of other prisoners or even divulge everything they know about illicit activities in the subculture (snitch) as a means of establishing the threat. In such instances, some prisoners will exaggerate, embellish, or outright fabricate information to make their case for a threat and to be taken seriously, especially when staff press to know more and more.

Equally important issues are associated with a decision by staff to classify a prisoner as PC. This decision can produce enormous stress and anxiety for prisoners who view the status of PC as stigmatizing them and rendering them more vulnerable to harm over the long term. These prisoners may take the fact that staff have recommended them for PC as proof of their vulnerability.[7] People who object strongly to a PC housing assignment can become difficult to manage, so in most places, prisoners can waive PC status. Even if a prisoner has no angst or objection to a PC assignment, there is still the difficulty of accurately determining if he or she should be classified as PC, as there are few objective, empirically established guidelines to follow. As Austin observes, external classification systems are much more advanced than internal ones, and PC is an internal classification matter (i.e., it involves deciding where a prisoner will be housed within a specific institution rather than deciding to which institution across the jurisdiction the person will be sent).[8] Finally, there is the possibility of staff abusing the PC status. For example, in an effort to gain some kind of needed cooperation from a convict in the general population, staff may threaten the convict with application of the PC label (and dissemination of its stigma) in hopes of coercing the cooperation.

Given the difficulties of internal classification and the issues discussed above, most institutions with PC units end up with a diverse population in those units. For example, a unit may contain:

- Persons vulnerable to assault or death in the general population for any number of reasons (e.g., unpaid gambling debts)
- Individuals hoping to avoid sexual assault
- Persons hoping to avoid disciplinary infractions that would add to their sentences
- Individuals who have served as snitches to gain PC status
- Prisoners hoping to manipulate special privileges
- Persons who are mentally unstable[9]

The result of such diversity, of course, can pose serious management challenges within the PC unit itself. The problem may be compounded when PC prisoners are housed in or in close proximity to a larger administrative or disciplinary segregation unit that houses predatory prisoners.

Limited research has been conducted on the distinguishing characteristics of those entering PC.[10] A Missouri study compared PC and non-PC prisoners and demonstrated that PC prisoners exhibited greater deficiencies on measures of physical, psychological, and social functioning. PC prisoners were found to be serving longer sentences, and a disproportionately high number had been convicted of sex crimes and homicide.[11] However, a Canadian study by Wormith et al. uncovered more similarities than differences between these groups.[12] Wormith et al. did find more evidence of substance abuse and mental disorder among those in PC, and sex offenses were also more common. Further, PC prisoners had been exposed to more aggression in the general population, were significantly more fearful, and held more negative perceptions of treatment received from staff. A 2001 Canadian study by Zinger et al. also uncovered greater psychological problems among segregated prisoners.[13]

Living in PC

The conditions under which PC prisoners live vary widely across jurisdictions (and even within them). An institution may have a separate PC unit or may house

PC prisoners as part of a larger segregation unit that contains administrative and/or disciplinary cases. Where separate PC units exist, they are typically small, averaging fewer than 50 persons. Either dormitory or cellblock living space may be provided.[14]

PC prisoners need maximum-security protection irrespective of their own individual custody levels assigned at classification.[15] For this reason, a low-custody-grade PC prisoner who presents little threat to anyone, but is nonetheless at high risk of being victimized, may live under conditions similar to those for a prisoner in segregation with a very high custody grade who poses significant threats to other prisoners or staff. As a general principle, ACA standards (which many institutions attempt to follow for accreditation purposes) hold that "inmates in PC should be allowed to participate in as many as possible of the programs afforded the general population, providing such participation does not threaten institutional security."[16] In addition to having access to programs like mental health, religion, and education, PC prisoners generally should receive services and privileges (e.g., food, hygiene, medical, phone, mail, recreation, etc.) that, although not necessarily identical to those provided the general population, are as comparable as possible, absent a compelling security rationale.

The reality is that enormous variations in conditions exist across institutions. Like other policies and procedures, the ACA standards leave much leeway for staff to justify PC conditions that differ appreciably from conditions found in the general population. Moreover, despite some exceptions, the standards tend to treat all segregated populations as a single category for purposes of specifying conditions, meaning that there is minimal room for distinguishing PC prisoners from those in disciplinary segregation. This contributes to the fact that PC prisoners have been reported to live under conditions that are often far more restrictive and depriving than those found in the general population.[17]

On a related note, some researchers have reported that PC populations experience negative stereotyping and treatment, not only at the hands of other prisoners, but also staff.[18] Where negative attitudes and treatment are found on the part of staff, these may arise from a variety of issues:

- Stigmas attached to the types of crimes for which many PC prisoners have been incarcerated
- The problem behaviors of these prisoners in the general population that culminated in their seeking or being assigned to PC
- A perception that some prisoners are in PC in order to manipulate special privileges
- A general demeanor toward PC prisoners as weak and inferior
- The ongoing challenges posed by daily interactions and work with PC populations

In summary, PC prisoners may experience a double dose of negative bias—one from fellow prisoners and the other from staff—with one compounding and reinforcing the other.

This is not to say that life in PC is entirely or universally negative. Some prisoners derive genuine protection from such a placement, and evidence has been reported that many individuals feel safer and experience less stress in PC.[19] In addition, some prefer the relative solitude and tranquility.[20] At the same time, however, it is clear that some (but certainly not all) PC units experience a disproportionately high number of problems. These can include psychological breakdown among prisoners (resulting in such things as suicide attempts, self-mutilations, and psychotic symptoms), physical and verbal attacks directed toward other prisoners and staff, group disturbances, fire setting and property destruction, as well as sexual assaults and pressure for sexual favors.[21] In part, such problems derive from the peculiar mix of prisoners found in PC, but they can also derive from excessive idleness and a paucity of services and programming.[22]

Transition from PC

Stays in PC range from very brief to permanent. Prison classification and casework officials make decisions about the length of stay. Prisoners may transition from PC to the outside world, to another institution or community-based facility, or to the general population within the same prison.

Although prison officials normally hope to transition as many prisoners as possible out of PC due to the drain on resources, such transitions are sometimes inadvisable. Regardless of the setting to which the person is transitioning, an important consideration is the extent to which threats to the person's safety exist or persist in that setting. These may be threats that have carried over from the time preceding PC placement or new ones that have emerged for reasons at least partly related to the PC stay. The fact is that many PC prisoners have exhibited specific behaviors deemed highly offensive in the prison subculture (e.g., sex crimes on the outside, failing to pay debts, snitching, etc.), and the mere act of entering PC is regarded as offensive as well.

Contempt for persons in PC can be so intense that members of the general population who do not capitalize on an opportunity to victimize a PC inmate may end up endangering themselves. Despite the potential gravity of this issue, few jurisdictions have been found to have systematic programs for transitioning PC prisoners.[23] Instead, transition efforts seem to be unsystematic and grafted onto such extant practices as counseling, classification reviews, security investigations, and prisoner transfers.[24] Far more attention appears to have been directed to deciding who belongs in PC than to deciding how to transition prisoners to a safe environment following PC.

Yet, if the goal is really to provide protection, neglecting transition is a serious mistake, especially given the rather elaborate informal systems of prisoner communication that exist, not just within and across institutions, but also with the outside world. It is obvious enough how prisoners doing time in the same institution could easily learn of a prisoner's impending return from PC to the general population. But prisoners hoping to victimize those coming out of PC (or hoping to have them victimized by others) can communicate with friends and associates housed in facilities to which former PC inmates will be transitioning. They can also communicate with potential victimizers on the streets. Elaborate gang networks are hardly a prerequisite for all this to take place. Underestimation of the prisoner subculture in this respect can result in death or serious injury to a person transitioning from PC to another environment.

■ Effects of PC

In principle, it is difficult to disagree with the assertion that a properly adminis- tered protection unit can contribute to a more orderly, stable, and humane insti- tutional environment. Some positive effects of protective segregation include protection from victimization, feelings of greater safety, and lowered stress and environmental stimulation. However, this section focuses on some of the nega- tive, often unintended, consequences of PC, especially with regards to its asso- ciated stigma, because stigmatization lies at the crux of many of the negative outcomes associated with PC.

In broad terms, the net negative effects of PC range from the most infamous and extreme to the subtle and sometimes unrecognized. For instance, at the in- famous end are the events that transpired in February 1980 at the Penitentiary of New Mexico (La Pinta) at Santa Fe, when an enraged mob of drugged prison- ers forced their way into the protection unit and executed a "hit list" of segre- gated inmates with such objects as metal rods and acetylene torches.[25] At the more subtle, far less publicized end are prisoners who jeopardize parole release by entering PC, in part because their PC status is interpreted as evidence by of- ficials that they are incapable of adjusting to prison.[26] In short, although protec- tive segregation can help shield a prisoner from physical, psychological, economic, and social forms of victimization, it can also contribute to such victimization and, indeed, can lead these forms of victimization to interact with and feed off one another.

Some research has addressed the sociopsychological effects of PC. For example, Brodsky and Scogin uncovered a variety of "psychopathological con- sequences" (e.g., extreme anxiety, sleep dysfunctions, psychosomatic symp- toms, delusional thought, depression, etc.) among two thirds of the PC prisoners at two of the institutions studied; PC prisoners in these institutions had lit- tle, if any, access to programs and spent most of their time in their cells. However, the researchers found no comparable consequences in a prison that permit- ted PC prisoners more mobility and access to programs. This suggests that the negative psychological effects of PC are situational and contingent on the particular segregation environment.[27] Zinger et al. found that although those entering segregation displayed more psychological problems than nonsegre- gated prisoners, there was no evidence of deteriorated psychological func- tioning after two months in segregation.[28] So although certainly a possibility, measurable negative psychological effects are by no means an inevitable out- come of PC.

Owing to their disrespected status, PC prisoners become tagged as easy and deserving marks for general population prisoners seeking to enhance their own status as convicts or unleash pent-up hostilities. Additionally, in many institu- tions, a general population prisoner who passes an opportunity to intimidate or otherwise victimize a PC prisoner can face serious repercussions from other mem- bers of the general population.

According to Erving Goffman, PC prisoners may respond to disqualification and stigmatization in several, sometimes overlapping ways. They may

- Try to correct the source of the stigma, such as by arranging to pay off a gambling debt
- Adopt the unconventional identity implied by the stigma (e.g., admitting that one is weak)
- Use the stigma as an excuse for shortcomings or to gain favors
- Come to view the stigma in beneficial terms and, correspondingly, direct focus to the negative characteristics of those not possessing or approving of it
- Come to avoid contact with peers who lack the stigma
- Approach interaction with non-stigmatized persons with much uncertainty, apprehension, and anticipation of how to act, constantly looking for signs and making adjustments during the course of daily interaction
- Feel that "minor failings or incidental impropriety" are being interpreted by others as further evidence of the stigma; in other words, the stigma can begin to overshadow normal ways of relating to others, resulting in exaggerated acts of "cowering" or "bravado"[29]

Goffman's insights provide a sense of how everyday social interaction patterns and individual identities are shaped around the stigma associated with the PC status. In addition to adversely affecting the way the PC prisoner is perceived by others, PC status can have deleterious effects on the perception of self and accompanying behaviors. Lemert described how initial acts of deviation from a particular set of norms (what he called primary deviance) can, when followed by stigmatizing reactions from others, result in the person gradually slipping into secondary deviance. A person who has undergone such slippage begins to regard him- or herself as deviant and reorganizes life activities accordingly, including the possible commission of further deviance.[30] The process of mortification of self that Goffman sees as typically transpiring in closed institutional settings is likely to be exacerbated for PC prisoners, because they confront double stigmatization—stigmatization from the prison society as well as from the outside society owing to their criminal status.[31] The effect can be a radical and long-term alteration of self-perception.

It is worth noting that efforts to undo the stigma of PC status will ordinarily have little, if any, success in prison. To understand why, it is useful to extend Lemert's distinction between primary and secondary stigmatization. Applied to PC, primary stigmatization attaches to whatever reason(s) a prisoner incurred PC status. But in the prison subculture, PC status in and of itself (whether formal or informal) carries a stigma above and beyond any "transgression" that initially gave rise to it. In short, stigma can act as both cause and effect of the PC status.

■ Legal Considerations

While a considerable amount of case law exists regarding both staff liability and prisoners' rights, very little has dealt specifically with PC. Prison officials sometimes do not distinguish PC from administrative segregation, and although this distinction is a drastic one in the prisoner subculture, the courts have not recognized a clear separation. Without a body of law specifically addressing PC, it becomes difficult to determine the legal rights of PC prisoners and corresponding staff liability, exacerbating the already tenuous responsibility of managing this population.

Of the case law that exists, two domains of legal implications have emerged. One involves issues of liability for correctional officials, and the other involves violations of prisoners' constitutional rights, specifically those associated with due process and conditions of confinement.[32] To fully understand these domains, brief mention of the mechanisms by which prisoners bring about legal action is warranted.

Prisoners bring legal action through torts, habeas corpus actions, and the federal Civil Rights Act, U.S.C. Title 42, Section 1983. The issue of liability for correctional staff is typically brought before the courts through tort claims, which allow prisoners to recover monetary damages. Claims of violation of constitutional rights usually involve Section 1983 actions, which allow both monetary damage and injunctive relief. For PC prisoners suing over constitutional violations, a Section 1983 action would potentially provide the most relief due to its ability to establish liability as well as to require prison officials to either stop a practice found to be in violation of prisoners' rights (restrictive injunction) or compel them to provide a service not previously offered (mandatory injunction).

With respect to staff liability, a consensus has emerged from courts that correctional officials have a basic duty to take reasonable precautions to protect individuals under their care. Such responsibility, however, does not expressly require officials to ensure safety unless a jurisdiction's statute so specifies.[33] Additionally, requests for PC cannot be ignored, but officials are not bound to provide protection for all who ask.[34] Finally, officials are required to maintain conditions within the institution that "offer a reasonable level of control" over prisoners who pose a potential danger to others, but the measures designed to control such inmates vary widely across jurisdictions.[35]

The courts have ruled that the key to whether correctional officials should be held liable for their actions is the determination that an inherent duty to protect exists. If such a duty is found to exist and is subsequently breached, then courts have generally found liability to attach. With respect to the general liability of correctional officers, the U.S. Supreme Court has ruled that prison officials are not civilly liable for their actions provided they did not ignore a clearly established right held by the prisoner.[36] The most significant exposure to liability occurs when a prisoner seeks and is denied PC status and is subsequently harmed. Given the Court's ruling that liability does not exist unless violation of a clearly established right has occurred, an important issue for correctional officials is the delineation of clearly established rights for PC prisoners.

The issue of clearly established rights raises questions regarding the constitutional rights of PC prisoners per se. While some issues have been addressed, as yet there is no clear consensus about these rights. A substantial portion of existing case law on this question has resulted from Section 1983 litigation, which requires a claim that a violation of federal rights has occurred. The most common claims of constitutional violations for individuals in PC involve due process and conditions of confinement issues. For issues relating to PC the key question is whether the deprivation associated with being placed in protection is significant enough to trigger the due process clause. The manner in which a prisoner is placed into PC determines the amount of due process to which he or she is entitled. The courts have ruled that those who request PC placement are not entitled to due process protections such as hearings or periodic review of their status, while those who are involuntarily placed into PC are automatically entitled to basic due process rights.[37]

The second constitutional right of PC prisoners that has been recognized by the courts involves the Eighth Amendment prohibition of cruel and unusual punishment. Most of the issues brought before the courts involve claims associated with the conditions of confinement, specifically defining what constitutes cruel and unusual punishment for PC prisoners. As previously discussed, there is much variation with respect to the conditions under which PC prisoners live, variation that is responsible for much of the legal action surrounding PC. Most courts require prison officials to ensure that services and facilities for persons in PC are similar to those provided to the general population; some courts have placed a heavy burden on prison officials to ensure similarity.[38] The courts have also found that prison officials may be held liable for Eighth Amendment violations if "deliberate indifference" is evident, defined as existing when officials are aware of the risk of serious harm to an individual, yet fail to take reasonable precautions to protect the individual.[39]

The lack of legal consensus regarding the rights of PC prisoners has made the assignment of liability an arduous, difficult, and seldom-accomplished task. When directly addressing the liability of correctional officers with respect to PC prisoners, the courts have ruled that when officials demonstrate reasonable care in the protection of such persons, there is generally no liability for harm incurred.[40] Additionally, the courts have found that mere negligence on the part of prison officials does not equate to a constitutional violation and therefore does not invoke liability, even if injury results.[41] However, they have been willing to hold prison officials liable when deliberate indifference exists. With respect to PCs, deliberate indifference is found to exist when prison officials are subjectively aware of a risk of serious harm and fail to take reasonable steps to decrease it.[42]

The lack of consensus regarding the rights of PC prisoners, and the high standard for institutional liability, which requires proof of deliberate indifference, are both factors that further render PC status untenable for many prisoners. The reluctance of courts to address the legal issues surrounding PC, combined with the high standards necessary for relief, would only exacerbate prisoner reluctance to seek PC status due to the stigma it entails. Ironically, it seems that since the Supreme Court has not addressed the specific rights associated with PC adequately, but has given considerable attention to administrative segregation, at least some individuals from whom officials are attempting to protect PC prisoners have more due process rights than the individuals being protected.

■ Conclusion

Three main directions for future policy emerge from this material. First, protective custody status is a unique one in the prison and jail environment, and the courts need to clarify the major issues surrounding it. At minimum, attention should be directed to clarifying the rights of PC prisoners vis-à-vis those living in the general population, administrative segregation, or disciplinary segregation. Administrators must also consider liability if prisoners who requested and were denied PC status are subsequently harmed. Further, while there is minimal legal guidance concerning formal PC, there is less guidance still regarding staff liability and informal PC.

Second, far greater attention needs to be given to systematically and safely transitioning prisoners from official PC status to other settings. The ethical obligation to provide protection does not end when a person leaves the protective unit. Indeed, based on what is known about prisoner communication and networking systems, the obligation may only be beginning. Disproportionate attention has been given to the issue of screening people entering PC, to the neglect of those leaving it. Far more is needed in this area by way of research, development of effective procedures, dissemination of information about effective practices through training and publications, and clarification of legal guidelines.

Third, prison subculture must be considered when exploring PC. Treating PC as detached from subcultural norms, values, and practices can promote a false sense of security or protection, both within facilities and across them. The mere fact that a protective unit is needed at all in an institution is symptomatic of problems endemic to the prison environment, especially as regards violent, intimidating, and predatory behavior. When persons are sentenced to institutional environments where safety is so lacking in the general population that they must live in constant fear, it is unreasonable to suppose that a safe haven can exist within those environments, effectively protecting those who need it indefinitely.

Humane imprisonment involves undertaking long-range, and at times expensive or politically unpopular, efforts to change the underlying sources and correlates of prison victimization, to the point that formalized PC becomes largely superfluous.[43] This means finding ways to address issues that are deeply embedded in the prison and its social organization—crowding and the associated competition over resources; problematic race relations; gang formations; illicit sub rosa activities and the control thereof through institutionalized snitch systems; norms and traditions condoning intimidation, retaliation, and predation—to cite some interrelated examples.

Although there is certainly a place for PC in contemporary prison management, ultimately PC is a band-aid approach to symptoms that reflect the prison subculture and, indeed, the totality of prison environments in the United States. Penologists should avoid becoming so focused on the technical aspects of PC management that they lose sight of the conditions needed to curtail or eliminate its use.

DISCUSSION QUESTIONS

1. Why do confined offenders need protective custody?
2. Should correctional personnel encourage PC status or encourage inmates to try to survive in general population?
3. How would you suggest that correctional officials head off the negative effects of PC or disciplinary status housing?
4. What distinguishes formal protective custody from informal protective custody?
5. What steps should be taken to assist inmates transitioning out of protective custody?

ADDITIONAL RESOURCES

J. Henderson and R.L. Phillips, *Protective Custody Management in Adult Correctional Facilities: A Discussion of Causes, Conditions, Attitudes, and Alternatives* (Washington, DC: National Institute of Corrections, 1990).

NOTES

1. American Correctional Association, *2004 Standards Supplement* (Lanham, MD: American Correctional Association, 2004), p. 318.
2. D. Franklin, "Protective Custody: A Window to Institution Culture" in *Contemporary Issues in Prison Management: Additional Readings*, available at http://www.nicic.org/Library/015778, accessed September 10, 2007.
3. L.H. Bowker, *Prison Victimization* (New York: Elsevier, 1980).
4. J. Austin and K. McGinnis, *Classification of High-Risk and Special Management Prisoners: A National Assessment of Current Practices* (Washington, DC: National Institute of Corrections, 2004).
5. K. Carriere, "Protective Custody in Canada: A Review of Research and Policy Responses," *Canadian Criminology Forum* 10 (1989), pp. 17–25.
6. H. Toch, *Living in Prison: The Ecology of Survival* (New York: Free Press, 1977).
7. H. Toch, *Men in Crisis: Human Breakdown in Prison* (Chicago, IL: Aldine, 1975).
8. J. Austin, *Findings in Prison Classification and Risk Assessment: Prison Division—Issues in Brief* (Washington, DC: National Institute of Corrections, 2003).
9. R. Angelone, "Protective Custody Inmates," in P. Carlson and J. Garrett, eds., *Prison and Jail Administration: Practice and Theory* (Sudbury, MA: Jones and Bartlett Publishers, 1999), pp. 226–231.

Chapter Resources

10. A. Perez and M. Hageman, "Dilemma in Protective Custody: Some Notes," *Journal of Offender Counseling, Services, and Rehabilitation* 7 (1982), pp. 69–78.

11. T. Pierson, "Social and Psychological Correlates of Protective Custody (PC) Status: A Comparison of PCs and Non-PCs," *Journal of Offender Counseling, Services, and Rehabilitation* 14 (1989), pp. 97–120.

12. J. Wormith, M. Tellier, and P. Gendreau, "Characteristics of Protective Custody Offenders in a Provincial Correctional Centre," *Canadian Journal of Criminology* 30 (1988), pp. 39–58.

13. I. Zinger, C. Wichmann, and D. Andrews, "The Psychological Effects of 60 Days in Administrative Segregation," *Canadian Journal of Criminology* (2001), pp. 47–83.

14. J. Henderson, *Protective Custody Management in Adult Correctional Facilities* (Washington, DC: National Institute of Corrections, 1991).

15. P. Gendreau, M.C. Tellier, and J.S. Wormith, "Protective Custody: The Emerging Crisis Within Our Prisons?" *Federal Probation* 49 (1985): 55–63.

16. American Correctional Association, *Standards for Adult Correctional Institutions*, 3rd ed. (Laurel MD: American Correctional Association, 1990).

17. S. Brodsky, and F. Scogin, "Inmates in Protective Custody: First Data on Emotional Effects," *Forensic Reports*, 1 (1988), pp. 267–280.

18. P. Priestley, *Community of Scapegoats: The Segregation of Sex Offenders and Informers in Prisons* (New York: Pergamon, 1980).

19. L. Alarid, "Sexual Orientation Perspectives of Incarcerated Bisexual and Gay Men: The County Jail Protective Custody Experience," *The Prison Journal* 80 (2000), pp. 80–95.

20. Toch, *Living in Prison*; Wormith et al., "Characteristics of Protective Custody Offenders in a Provincial Correctional Center."

21. Alarid, "Sexual Orientation Perspectives of Incarcerated Bisexual and Gay Men"; Perez and Hageman, "Dilemma in Protective Custody"; and Toch, *Living in Prison*.

22. Brodsky and Scogin, "Inmates in Protective Custody."

23. Austin and McGinnis, *Classification of High-Risk and Special Management Prisoners*.

24. Henderson, *Protective Custody Management in Adult Correctional Facilities*.

25. I. Hirliman, *The Hate Factory: The Story of the New Mexico Penitentiary Riot* (Agoura, CA: Paisano Publications, 1982).

26. Perez and Hageman, "Dilemma in Protective Custody."

27. Brodsky and Scogin, "Inmates in Protective Custody."

28. Zinger et al., "The Psychological Effects of 60 Days in Administrative Segregation."

29. E. Goffman, *Stigma: Notes on the Management of Spoiled Identity* (New York: Simon and Schuster, Inc., 1963).

Chapter Resources

30. E. Lemert, *Human Deviance, Social Problems, and Social Control* (Englewood Cliffs, NJ: Prentice-Hall, 1967).

31. E. Goffman, *Asylums: Essays on the Social Situation of Mental Patients and Other Inmates* (New York: Doubleday, 1961).

32. Henderson, *Protective Custody Management in Adult Correctional Facilities.*

33. *Fleishour v United States*, 244 F.Supp. 762. 767 (N.D. Ill., 1965).

34. *West v Rowe*, 448 F. Supp. 58 (E.D. Ill. 1978); *Hall v. Unknown Agents of NY Department of Corrections*, 825 F.2d 644 (2nd Circuit, 1987).

35. Henderson, *Protective Custody Management in Adult Correctional Facilities*, p. 13.

36. *Harlow v. Fitzgerald*, 457 U.S. 800 (1982).

37. *Sweet v. South Carolina Dept. of Corrections*, 529 F.2d 854 (4th Circuit 1975).

38. *Williams v. Lane*, 851 F.2d 867 (7th Circuit 1988); Remedial Order Feb. 28, 1989 81 C 355.

39. *Farmer v. Brennan*, 511 U.S. 825 (1994).

40. Henderson, *Protective Custody Management in Adult Correctional Facilities.*

41. *Daniels v. Williams*, 474 U.S. 327 (1986); *Parratt v. Taylor*, 451 U.S. 527 (1981).

42. *Farmer v. Brennan* (1994).

43. M. Silberman, *A World of Violence: Corrections in America* (Belmont, CA: Wadsworth, 1995).

Suicide

30

Daniel W. Phillips, III

Chapter Objectives

- Outline the prevalence of suicide in the United States and its prison populations.
- Explain the correlation between social and individual characteristics and suicide.
- Describe the steps needed to construct a proper correctional suicide prevention and treatment program.

Suicide is a major problem in the United States, claiming more than 30,000 lives per year. Prison suicide rates are only slightly higher, but jails have a suicide rate more than four times that of the U.S. population.[1] Fortunately, jail and prison suicide rates have been dropping significantly for the last 20 years. The decrease in suicide rates has not been unique to correctional settings. The US suicide rate decreased by 15 percent from the late 1980s through 2005, perhaps due to advancements in pharmacology and a new class of antidepressants. Despite this reduction, suicide still remains a major concern for correctional institutions.

■ Social and Individual Considerations

Research indicates that the suicide rate is correlated with certain social factors (e.g., race, gender, and age) and with individual factors (e.g., mental illness and substance abuse).[2] Sociologists believe that certain demographic factors, such as gender, cause people to experience life differently, to encounter a different number of stressors, and to develop different coping skills. For example, boys may be encouraged more to use guns than girls are and a boy is therefore more likely to use a gun to end his life than his female counterparts. Girls may invest in interpersonal relationships more than boys do and may therefore have bigger networks of relationships to fall back on when they experience stress, thus decreasing the likelihood that they will commit suicide.[3]

In the United States, males constitute 80 percent of all completed suicides.[4] In fact suicide is the eighth leading cause of death for males.[5] Despite the fact that men are four times more likely than women to complete suicides, women self-report that they attempt suicide three times more often than men.[6] One reason for this great differential may be the specific methods preferred by each gender. Men are more likely to use a firearm, a method of suicide that is more than 90 percent successful, when they commit suicide.[7]

Race and age are also strongly correlated with completed suicides. Caucasians are twice as likely as African Americans to complete a suicide. Young people (those aged 15–24) have very low rates of death, but suicide is the third leading killer in people of this age group.

In addition, there are many individual suicide risk factors, including:

- History of mental illness
- Substance abuse
- Previous suicide attempts
- Isolation
- Relationship loss
- Feelings of hopelessness
- Physical illness

Individuals may benefit from several protective factors including:

- Family and community support
- Interpersonal coping skills
- Cultural or religious beliefs that suicide is wrong[8]

■ Stigma

Religious organizations have differing attitudes toward suicide. Some religions state that suicide leads to eternal damnation while others are careful to treat the deceased and his or her family with the utmost respect. Socially, Americans are still likely to punish the family of a suicide victim. People sometimes shun the victim's family or imply that the family is at fault for the suicide. Sometimes people do not attend the victim's funeral or grieve with the family. Research indicates that some families carry around the grief and guilt of a suicide for generations.[9]

Those who attempt suicide may also be stigmatized. Because the public already has negative ideas about inmates in general, they may look at suicidal inmates as a group not worth saving. Pubic perceptions are important because even a well-designed suicide prevention program will not work if the corrections officers who are supposed to implement it devalue the suicidal person. Suicide prevention programs must focus on suicide stigma reduction and a revaluation of suicidal inmates.

The stigma of suicide combined with the stigma of being incarcerated leaves the suicidal inmate as one highly stigmatized individual. This may explain why saving suicidal inmates' lives have not been a priority in the past. Correctional staff must learn to value inmates' lives so that they can save those who are suicidal.

■ Correlates of Suicide

Past research by the Department of Justice has revealed many correlates of jail and prison suicide. Recent studies indicated that small jails (those with fewer than 50 detainees) had a suicide rate five times that of the largest U.S. jails. Males and white people were most likely to kill themselves, as is true in the general U.S. population. Jail suicide rates increased significantly with age, although 18-year-olds also had a high rate of suicide in correctional settings. While male and female prisoners had similar suicide rates, male jail inmates had a suicide rate 50 percent greater than female inmates. Jail suicide was concentrated in the first week, but two thirds of prison suicides took place after the first year.[10] Hanging was the most common suicide method used in jail and prison.[11]

■ Legal Considerations

The legal responsibility of jail and prison staff with respect to inmate suicide depends on the concept of deliberate indifference. This concept comes from the U.S.C. Title 42, Section 1983, which allows incarcerated people to sue jail and prison workers and officials for deliberately not providing medical treatment such as the kind needed to prevent a suicide. Lawsuits filed under Section 1983 have to meet the deliberate indifference standard, which means that corrections workers must have known that a person was suicidal (possibly from a record of previous suicide attempts or knowledge of a person's serious mental health problems) and did nothing about it. This failure to act or provide treatment then must result in real damage such as a death by suicide or serious impairment from a suicide attempt. Deliberate indifference is a higher standard to meet than mere negligence or even gross negligence. Simply making a mistake is not enough to be sued successfully under Section 1983.[12]

■ Attitudes and Training

In the past, people have viewed suicides as inevitable and therefore unpreventable. However, most people who attempt suicide will never complete it. Sometimes a prisoner attempts suicide in an effort to gain attention but ends up killing himself or herself.

Since the early 1980s jails and prisons have invested money and time to reduce the number of suicides. Staff members have been trained, and procedures have changed. Staff members ask incoming inmates about suicidal risk factors such as previous suicide attempts and mental health or substance abuse problems. Staff may prevent at-risk inmates from being alone (because more than 80 percent of all jail and prison inmates who kill themselves do so in their cells) and may refer such inmates to a mental health or substance abuse professional.[13] Smaller jails have partnered with community mental health facilities to provide mental health care, while prisons and large jails have hired full-time mental health staff.

Lindsay Hayes of the National Center on Institutions and Alternatives (NCIA) developed a six-point framework for developing a written suicide prevention policy in institutions:[14]

1. **Training**—Jail and prison staff members need to receive training on suicide prevention (as well as CPR and first aid) as they are being trained to become correctional officers. Officers need to understand why it is important to save the life of a prison who is suicidal. This needs to be followed up by regular in-service training. Training should include information about suicide and its underlying causes, how to best prevent suicide, and how to intervene should the occasion present itself.

2. **Identification and Assessment**—There must be a way of identifying and assessing people who may be suicidal. Corrections officers interview inmates with these surveys and score them based on the number of suicide risk factors. Suicidal persons may be evaluated by someone with clinical training such as a psychiatric nurse or licensed clinical social worker.

3. **Housing**—Solitary confinement, segregation from the general population, and placement in mental health or medical units should be seriously considered for suicidal inmates.

4. **Levels of Supervision**—The more suicidal an inmate, the more often he or she should be observed. Some must be observed constantly, others once every few minutes, while others are observed once every 15 minutes. Still controversial is the use of fellow inmates to watch those who may be suicidal in the general population. Suicidal inmates do not do best by themselves. If the inmate has to be removed from the general population, it is still best if he or she is not alone.

5. **Intervention**—Each facility should have a plan to deal with a suicide attempt before an incident occurs. Officers have found themselves in the unenviable situation of watching from outside a cell as an inmate inside a cell dies from self-inflicted wounds. The officer was following rules that stated that no officer should enter a cell alone. While technically the officer would have been abiding by institution rules, this sort of scenario would be very uncomfortable for jail or prison management to recount during a civil trial.

6. **Follow-up/Administrative Review**—Every aspect of the suicide incident should be reviewed. All involved personnel, such as the medical staff, mental health staff, corrections officers, and management, should be queried about the incident. Policy should be reevaluated, and the training of those involved in the incident should be examined. Changes should be made based on the results of the review.

■ Program Example: Jail Mental Health Crisis Network

The high rate of suicide in Kentucky jails prompted an investigation in 2002. This crisis was the impetus for needed change. Mental health training is now required for all jail personnel, and legislation funds an innovative program offering a system of care to the state jails. The Jail Mental Health Crisis Network, developed and operated by the Bluegrass Regional MH-MR Board, provides jails the following services:

1. Training is provided for jail personnel on the signs and symptoms of suicide and mental illness.
2. Screening instruments identify the risk and needs of people who have a current or past history of mental illness or suicide risk.
3. Assessment by a licensed mental health professional in a toll-free telephonic triage system identifies four risk levels for suicide or mental illness.
4. Management protocols are defined for each risk level, including protocols for housing, supervision, and management.
5. Mental health followup offers face-to-face crisis counseling, consultation, and diversion by a local mental health professional.

Since its inception, the Kentucky Jail Mental Health Crisis Network has reduced the rate of suicide in participating jails by 84 percent. This program saves lives and provides an unprecedented handshake between jails and mental health service providers.[15]

■ Community Health Workers

A study of Kentucky jails revealed that in rural states, many of the mental health services are not provided by in-house mental health staff but rather by private psychiatrists, psychologists, and social workers. Other jail systems contract for mental health services from their local community mental health centers.[16]

Community mental health centers employ several different types of workers, including psychiatrists, psychologists, or social workers. Their services can be paid for by the jail or prison. Often correctional facilities have standing contracts with community mental health centers to provide services. There is, however, concern that with two agencies coordinating activities a mental health worker could not be located in an emergency.

■ Conclusion

Since the 1980s, suicide in correctional settings has decreased significantly more than in the general U.S. population. Yet, suicide still happens in correctional settings, and much more so in jails than in prisons. Research has uncovered correlates of suicide so that correctional management can better prevent suicide in jails and prisons. Suicide prevention programs have been advanced by knowing that half of all jail suicides occur during the first week of incarceration and that those with mental health and substance abuse problems are more likely to kill themselves. Correctional management must take suicide seriously, and correctional staff should be well-trained to detect and deal with suicide. Additionally, jails and prisons should employ mental health professionals to treat suicidal inmates.

DISCUSSION QUESTIONS

1. What is the best allocation of resources to help prevent inmate suicide?
2. Should jails be expected to have full-time mental health staff to deal with persons with mental illness or should they contract with local community mental health centers or doctors in private practice in the community?
3. Why are the rates of suicide in jails higher than the rates in prisons and society at large?
4. What is the correlation between race and gender and suicide?
5. Why is this correlation so strong?

ADDITIONAL RESOURCES

Criminal Justice/Mental Health Consensus Project, http://consensusproject.org
Jail Suicide/Mental Health Update, http://66.165.94.98/cjjsl.cfm
National Center for Injury Prevention and Control, http://www.cdc.gov/ncipc/factsheets/ suifacts.htm
National Commission on Correctional Health Care, http://www.ncchc.org

NOTES

1. C. Mumola, *Suicide and Homicide in State Prisons and Local Jails* (Washington, DC: Office of Justice Programs, 2005).

2. Centers for Disease Control and Prevention, National Center for Injury Prevention and Control, *Suicide: Fact Sheet, 2006*, available at http://www.cdc.gov/ncipc/factsheets/suifacts.htm, accessed March 6, 2007.

3. L. Butler and S. Nolen-Hoeksema, "Gender Differences in Responses to a Depressed Mood in a College Sample," *Sex Roles* 30 (1994), pp. 331–346.

4. Centers for Disease Control and Prevention, National Center for Injury Prevention and Control, *Web-Based Injury Statistics Query and Reporting System*, available at http://www.cdc.gov/ncipc/wisqars/default.htm, accessed March 7, 2007.

5. R. Anderson and B. Smith, "Deaths: Leading Causes for 2001," *National Vital Statistics Report*, vol. 52, no. 9 (2003), pp. 1–86.

6. E. Krug et al., *World Report on Violence and Health: Summary* (Geneva, Switzerland: World Health Organization, 2002).

7. Physicians for Social Responsibility, *Firearms and Suicide* (Washington, DC: Physicians for Social Responsibility, 2003).

8. Health and Human Services, *The Surgeon General's Call to Action to Prevent Suicide* (Washington, DC: Department of Health and Human Services, 1999).

Chapter Resources

9. E. J. Dunne, J. L. McIntosh, and K. Dunne-Maxim, eds., *Suicide and Its Aftermath: Understanding and Counseling the Survivors* (New York: W.W. Norton and Company, 1987).

10. Mumola, *Suicide and Homicide in State Prisons and Local Jails.*

11. J. Goss et al., "Characteristics of Suicide Attempts in a Large Urban Jail System with an Established Suicide Prevention Program," *Psychiatric Services* 53 (2002), pp. 574–579.

12. M. Middleton, *Typical Section 1983 Claims*, FindLaw for Legal Professionals, available at http://library.findlaw.com/1999/Jan/1/126485.html, accessed March 7, 2007.

13. Mumola, *Suicide and Homicide in State Prisons and Local Jails.*

14. L. Hayes, *Prison Suicide: An Overview and Guide to Prevention* (Washington, DC: National Institute of Corrections, 1995).

15. J. Adams and S. Shipley, "Locked in Suffering: Kentucky Jails and the Mentally Ill," *Louisville Courier Journal*, four-part series, February 24–March 3, 2002.

16. D. Phillips and C. Mercke, "Mental Health Services in Kentucky Jails: A Self-Report by Jail Administrators," *Journal of Correctional Health Care*, vol. 10, no. 1 (2003), pp. 59–74.

The Death Penalty

31

Julie C. Eng

Chapter Objectives

- Describe the role of religious and activist groups in the capital punishment debate.
- Explain why the death penalty has become such an emotional topic in the administration of justice in the United States.
- Examine how states differ in their capital punishment laws.

Capital punishment is a highly controversial topic and raises many important questions: Is the death penalty morally wrong? Is it an effective punishment? For prison and jail administrators, capital punishment raises many other questions about the details of housing and caring for inmates sentenced to death.

■ A Brief History

The history of capital punishment goes back to the earliest human cultures, when methods of execution were extremely cruel. The goal of the punishment was to create a painful experience; stoning, whipping, and boiling were common. With the passage of time and the evolution of ideas, punishments have changed. Early in the 20th century, death-sentenced inmates in the United States were executed by hanging, electrocution, or firing squad. Now, these practices are more limited (see **Table 31–1**). New execution methods allow the government to execute criminals in the most civilized and humane manner, with the least amount of suffering. States that use more than one method of execution allow sentenced inmates to choose their method of death.[1]

Table 31–1	Methods of Execution
Method	**Number of States Permitting**
Lethal Injection	37
Electrocution	9
Lethal gas	4
Hanging	3
Firing Squad	3
Total number of states in 38 and the Federal Government where the death penalty is legal.	

Source: T. Snell, "Capital Punishment, 2005," Bureau of Justice Statistics (Washington, DC: U.S. Department of Justice, 2006).

Particular problems noted with the electric chair have encouraged the development of new methods of executions. Several executions have been documented where the electricity administered was not enough to kill the condemned person. The most infamous example of this occurred in Alabama in 1983, during the execution of John Louis Evans. Three separate attempts of 1900 volts of electricity were required to complete Evans' punishment. In Florida, during an electrocution on March 25, 1997, the electricity created flames that erupted from convicted murderer Pedro Medina's masked head during the execution. An international outcry resulted, causing many states to review of this method of execution.

Though some states (such as Michigan, Rhode Island, Wisconsin, Minnesota, and Maine) abolished the death penalty in the early 1900s, other states abolished the death penalty only to reinstate it later.[2] Most states, however, have maintained their capital punishment laws with little change, except for court-required restrictions.

After the Supreme Court articulated a new standard in the 1970s for the use of the death penalty, states returned to a greater use of the death penalty as their ultimate sanction. From 1930 through the 1950s, those who were sentenced to die generally were executed promptly. Nearly 4000 men and women were executed in the United States during this time. In the 1960s, executions slowed dramatically as courts became more involved in this arena, and death row inmates often exercised lengthy appeals. This trend culminated in the U.S. Supreme Court's ruling in *Furman v. Georgia* (1972) that capital punishment was cruel and unusual.[3]

In *Furman*, the Court held that the decision to execute as a punishment was not applied fairly to all defendants. As a result of this ruling, the death penalty was annulled in 39 states. Many jurisdictions then embarked on an effort to rewrite their sentencing laws to deal with the issue of arbitrariness. In order to combat the Supreme Court's charges, state legislators mandated capital punishment for certain offenses and created specific guidelines for the judicial system to follow when deciding whether to implement the death penalty. In 1976, the first test of the new laws came before the Court in *Gregg v. Georgia*.[4] Although the Supreme Court ruled against mandating the death penalty for certain crimes, the guidelines for juries to follow were approved. Accordingly, the Georgia version of guidelines became the model for many other states.

■ Arguments Against the Death Penalty

As the death penalty became more common within the United States, so did the controversy surrounding it. Critics have raised several complaints, including:

Long Waits on Death Row

The greatest argument against capital punishment seems to be that few individuals are actually executed. On January 1, 2007, about 3350 inmates were awaiting execution. The state with the highest number of total executions—Texas—has a current backlog exceeding 390 people. In California, 660 inmates are currently on death row. Between 1973 and 1995, only 5 percent of inmates on death row nationwide were executed.[5] Despite these statistics, judges and juries continue to hand out an average of 2300 death sentences each year.[6]

Unclear Qualifications

The lack of clarity around who qualifies for the death penalty has contributed to the debate over its use. Federal and state prisons have an immense backlog of death row inmates simply because the laws addressing this controversial issue are nebulous. Although the goal of the moratorium imposed by the *Furman* decision was to reduce capricious sentencing, inmates today are able to find myriad loopholes within the laws to prolong their time on death row.

Determined lawyers manage to contest capital punishment rulings in creative ways, often because of the lack of consistent sentencing guidelines across state and federal jurisdictions. This inconsistency stems from ambiguous standards within the law. Most death penalty sentencing guidelines are based on seven criteria:

1. Murder committed in the commission of a felony (e.g., robbery, rape, or kidnapping)
2. Multiple murders
3. Murder of a police or correctional officer acting in the line of duty
4. Especially cruel or heinous murder
5. Murder for financial gain
6. Murder by an offender having a prior conviction for a violent crime
7. Causing or directing another to commit murder

Because these guidelines are defined loosely, interpretations may differ radically among judges and juries in various state and federal jurisdictions. "Multiple murders" could refer to hundreds of deaths resulting from a bomb exploding in a large office building or the shooting of two individuals during a robbery. Another judge could believe that a single murder is particularly cruel. Thus, when the courtroom deliberation finally results in a decision on how to apply the law—a process that can be extremely lengthy—this decision can become the basis of many appeals.

Complexity of Appeals

All three branches of government are concerned about the time delays and complexity of death penalty appeals. The Anti-Terrorism Effective Death Penalty Act (1996) was intended to shorten the lengthy and cumbersome state and federal appeal process. It limits the appellate process for those under sentence of death to one federal appeal; if the individual seeks additional appeals, he or she must first receive approval from a three-judge panel.

With the current death penalty laws, an estimated 40 percent of capital punishment decisions are reversed.[7] The inmate is then resentenced, which starts a new cycle of retrying cases and appealing decisions. Taxpayers are left paying for keeping men and women on death row in prison for years while their cases await judicial review. Even if the current rate of execution were to double, it would take nearly half a century for the backlog of sentenced inmates to be executed.

High Financial Cost

Another major shortcoming of capital punishment is the immense cost. Many Americans support capital punishment because they believe it to be less expensive than a life sentence in prison without parole. Contrary to this conventional wisdom, capital punishment is actually the more costly option. Considering the average age of incarceration and typical life expectancy, the estimated cost ranges from $750,000 to $1 million per execution.[8] Some states even estimate the cost to be much higher. These costs are not affected by the method of execution but by the cost of extended legal review. Retaining attorneys and expert witnesses and conducting the investigation create significant expenses. In Florida, the average cost of trying and executing one person is $3.2 million. This is almost three times the cost of a single Florida inmate's imprisonment for 40 years.[9]

Racial Discrimination

When debating the death penalty, one cannot avoid the argument that the process discriminates against some races and ethnicities. Dr. David Baldus, professor of law at the University of Iowa, points out that, "About half of all the people who are murdered each year in the United States are black. Yet since 1977, the overwhelming majority of people who have been executed—more than 80 percent—had killed a white person. Only 11 percent had killed a black person."[10] These statistics suggest that the judicial system appears to value the lives of Caucasian citizens more than the lives of individuals of other races or ethnicities. Baldus also completed a study in Georgia that revealed that convicted offenders with Caucasian victims received the death penalty 4.3 times more often than those with African American victims.

Other studies also reveal bias with respect to the death sentence. In 2006, 53 inmates in 14 states were executed. All were men; 61 percent were Caucasian, and 39 percent were African American. If the statistics mirrored the U.S. population, only about 22 percent of the death row population would be African American, though in 2005, 42 percent of inmates on death row were African American.[11] Sociologist Michael Radelet has concluded from his studies that those of lower social class and poor economic status are also disproportionately given death sentences.[12]

Limited Effect on Deterrence

Many argue that the infrequency with which capital punishment is imposed can hardly be expected to deter others from crime. In the study of capital punishment in Georgia, Baldus determined that a mere 23 percent of death penalty-eligible criminals received that punishment.[13] The irregularity and inconsistency of dispensing the death penalty may be its greatest weakness.

Failure of Legal Representation

The American Bar Association (ABA) voted in February 1997 to seek a halt to the death penalty. Its primary concern stemmed from the inconsistent quality of legal representation provided to criminal defendants facing the ultimate sanction. Some ABA attorneys argued that these critical judicial decisions turn not on the nature of the crime but on the quality of representation for the accused, citing examples of lawyers who were inadequately paid or incompetent.

Unfair Application Based on Location of the Crime

Locations of crimes also have a large affect on whether or not a criminal is assigned the death penalty. The University of Maryland and the Nebraska Crime Commission have each released studies that show prosecutors in urban areas seek the death penalty as much as five times more frequently than those in rural areas.[14]

Wrongful Execution

Opponents of the death penalty also argue that the possibility of killing an innocent human being is a weakness of capital punishment. They believe the chance that an innocent person may be wrongly killed is enough of a reason to ban the death penalty altogether. In the past 30 years, 115 people were found to be innocent and released from death row; more than a dozen of these were exonerated due to DNA testing.[15]

■ Support for Capital Punishment

Even with the negative publicity concerning its costs and delays, many people still favor the death penalty, based on some of the following arguments. First, death penalty advocates argue that while every life should be valued and every American is guaranteed due process of law by the Fourteenth Amendment, this right does not confer the power to enjoy, as Raymond Paternoster calls it, "super due process."[16] While the Supreme Court has repeatedly claimed that the death sentence requires a more in-depth case study, the time-consuming appeals process has become extreme. If capital trials are not extensively prolonged and the appeals process is simplified, costs would be significantly lower in capital punishment cases than life imprisonment ones. Only when the judicial aspect of the death penalty is accelerated does capital punishment become substantially more economical than life imprisonment.

Others contend that the death penalty is valid because it deters would-be criminals from committing violations. Because death is the ultimate sanction a court can deliver, some believe that those criminals who consider the consequences of committing certain acts will certainly refrain rather than become a member of the death row population. Paul Rubin coauthored a 2001 study that found that the death penalty deterred 18 crimes per execution, and polls find that 40 percent of people believe that capital punishment deters crime.[17]

Perhaps the greatest reason that people support the death penalty is that it offers merited punishment, known as just deserts. According to this argument, a person who takes the life of another has forfeited the right to his or her own existence, and the moral fiber of a community is strengthened when it takes the life of a murderer because it can express its outrage. Even if little crime reduction or deterrence results from the death penalty, the public continues to embrace the deserved punishment theory.

Walter Berns readily explains this argument:

> *We surely don't expect to rehabilitate [murderers], and it would be foolish to think that by punishing them we might thereby deter others. The answer is clear: We want to punish them in order to pay them back. We think that they must be made to pay for their crimes with their lives, and we think that we, the survivors of the world they violated, may legitimately exact that payment because we, too, are their victims.*[18]

Berns' argument allows the public the right to be angry with criminals and to act on that anger.

The overwhelming amount of support the death penalty shows how deeply the concept of retribution is rooted in the United States. A belief in justice—the creed of "an eye for an eye" (in this case, a life for a life)—holds that retribution is the justification for capital punishment. People need absolute assurance that a convicted murderer will never again have the opportunity to become a repeat offender, and only the death penalty can offer this comfort.

Lastly, some people contend that the death penalty comforts the families of victims, giving them a sense of relief and greater peace of mind. On the other hand, the long appeals process can be particularly frustrating to families of victims as they spend hours inside courtrooms during repeated testimonies during the inmate's appeal process. Finally, some victim's families oppose the death penalty.

■ Death Row Operations

Condemned inmates in U.S. penal facilities spend many years on death row. The average inmate spends 147 months (12.25 years) on death row before being executed.[19] Operating institutions that house those sentenced to death is a unique challenge. Each correctional system must consider how to house and treat these inmates. Prison administrators must decide if they will place these individuals with the general population or operate a separate death row housing area.

Several states have elected to mainstream their death-sentenced inmates and allow them to participate fully in general population work, education, recreation, and other program opportunities. These jurisdictions do not separate the inmate from others until the individual's date of execution is imminent. However, most state correctional agencies have developed separate death row operations. Generally, prison administrators have chosen to run these cellblocks as segregation units, highly controlled custodial environments that offer a high degree of accountability for these inmates who are deemed to present the most extreme threat to society. These individuals spend the majority of their time in secure cells and are out solely for minimal periods for recreation and showers. Other death row operations keep this group apart from general population but permit relatively free interaction among the individuals under sentence of death. These units generally offer communal dining, recreation, and work experiences. Inmate programming (such as religious or educational programming) is often offered to these inmates within boundaries of the unit.

■ The Emotional Ordeal of an Execution

Correctional administrators have expressed concern about the emotional toll that carrying out the death penalty takes on the staff.[20] Despite the harsh rhetoric engaged in by proponents of capital punishment, even by correctional staff, it is extremely difficult to participate in an execution without experiencing personal trauma. Participating staff should be selected carefully.

Many wardens responsible for an execution personally pick the staff who will participate in the death watch during the two or three days prior to an execution and use only the most mature and experienced staff for the actual execution. This selection process often involves staff who volunteer for the task, with final selection made based on their experience and ability. Some agencies will not utilize staff who volunteer; instead, they ask certain individuals if they would be willing to participate. Many institutions carefully protect the identities of staff who are involved.

The use of medical staff in an execution is another very sensitive matter. Many physicians will not participate in the actual execution and limit their involvement to the pronouncement of death. The American Medical Association sternly speaks out against physicians' participation and has stated it is a violation of their professional ethics and the Hippocratic Oath. This can present a dilemma for those states that use lethal injection as the method of execution, because it is a medical procedure. Accordingly, most executioners are nurses or emergency medical technicians who are skilled at inserting a needle into an individual's arm. All jurisdictions offer complete anonymity for these individuals.

After an execution, it is important to debrief all those who participated. This emotional time is the most difficult for the execution team, and senior staff work very hard at providing necessary social services support.

■ Special Exemptions: Juveniles and People with Mental Retardation

Two groups of people are now exempt from capital punishment: juveniles and people with retardation.

Juveniles are defined differently according to particular states. Most states have designated 17 or 18 minimum years of age, with a few states making this designation at 16 years of age. Arguments in favor of juvenile capital punishment tend to focus on this issue of crime prevention, citing statistics that homicide committed by juveniles is worse in the United States than in most other countries. Public fear of juvenile homicide is extremely high, and some argue that juvenile offenders seem to be nonresponsive to other, nonviolent types of crime prevention efforts.[21]

Table 31–2	Relevant Supreme Court Cases

Furman v. Georgia (1972)—The death penalty does not violate the Constitution, but the way that states assign it is considered cruel and unusual based on the prevalence of racial discrimination in sentencing.

Gregg v. Georgia (1976)—The sentencing phase of capital trials must be separate from the case trial.

McClesky v. Kemp (1987)—Georgia state death penalty is constitutional even though there is racial bias in the application of sentencing and executions.

Thompson v. Oklahoma (1988)—The Court ruled that offenders under the age of 16 at the time they committed the crime cannot be executed.

Stanford v. Kentucky (1989)—Executions for offenders who were 16 or 17 when they committed their crimes are constitutional.

Atkins v. Virginia (2002)—Executing mentally retarded criminals violates the Constitution's ban on cruel and unusual punishment.

Roper v. Simmons (2005)—Offenders who were younger than 18 when they committed their crimes may not be executed.

The argument against the death penalty for juveniles contends that youth offenders do not have fully matured brain impulse control, which usually develops in the late teens or early twenties. Juvenile offenders may have endured terrible childhoods, and they may lack a realistic understanding of death, thus strongly diminishing the deterrence factor associated with the death penalty.

While scholars may continue to debate the merits of and objections to juvenile executions, *Roper v. Simmons* (2005) effectively has ended the legal discussion. In this case, the U.S. Supreme Court ruled that offenders under the age of 18 when they committed their crimes may not be put to death, because this violates the Eighth Amendment ban on cruel and unusual punishment. The Supreme Court also decided in *Atkins v. Virginia* (2002) that it is considered cruel and unusual punishment to execute mentally handicapped persons (see **Table 31–2**).[22]

■ Conclusion

The government-sanctioned taking of a life is a contentious and emotionally charged issue. Nearly every argument for or against capital punishment can be refuted by opponents. Whether one invokes economic, moral, social, or legal reasoning, it is not enough to convince the other side of the rightfulness or wrongfulness of the law. The debate over the death penalty has not been resolved by logic. Clearly, emotional arguments are more significant and influential than arguments of reason. As long as most Americans believe in the concept of just deserts, it seems clear that the death penalty will remain a viable force in criminal justice for years to come.

DISCUSSION QUESTIONS

1. Do you support or oppose the idea of capital punishment?
2. Why does the American public consider some offenders to be worthy of capital punishment and not others?
3. Do you feel that the execution of juvenile or mentally handicapped offenders constitutes cruel and unusual punishment?
4. How can the disparity of capital punishment based on location and age be resolved?
5. How should death penalty laws be framed to consider these concerns?

ADDITIONAL RESOURCES

Death Penalty Information Center, http://www.deathpenaltyinfo.org

F. Zimring, *The Contradictions of American Capital Punishment* (Oxford, UK: University Press, 2003).

E. Mandery, *Capital Punishment: A Balanced Examination* (Sudbury, MA: Jones and Bartlett Publishers, 2005).

National Center for Policy Analysis, www.napa.com

NOTES

1. T. Snell, *Capital Punishment, 2005*, Bureau of Justice Statistics (Washington, DC: U.S. Department of Justice, 2006).

2. P. Keve, *Corrections* (New York: John Wiley and Sons, Inc., 1981), p. 468.

3. *Furman v. Georgia,* 408 U.S. 238 (1972).

4. *Gregg v. Georgia,* 428 U.S. 153 (1976).

5. Death Penalty Information Center, *Death Row Inmates by State and Size of Death Row by Year*, http://www.deathpenaltyinfo.org/article.php?scid=9&did=188#year, accessed September 17, 2007; J. Leibman, *A Broken System: Error Rates in Capital Cases, 1973–1995*, June 2000, available at http://www2.law.columbia.edu/instructionalservices/liebman, accessed September 17, 2007.

6. NAACP Legal Defense Fund and Educational Fund, *Death Row USA*, available at http://www.naacpldf.org/content.aspx?article=297, accessed September 17, 2007.

7. L. Bartle, *Current Death Penalty Statistics* (Washington, DC: Death Penalty Information Center, 1996).

8. M. Costanzo and L. White, "An Overview of the Death Penalty and Capital Trials: History, Current Status, Legal Procedures, and Cost," *Journal of Social Issues* 50 (1994), pp. 1–18.

Chapter Resources

9. Costanzo and White, "An Overview of the Death Penalty and Capital Trials: History, Current Status, Legal Procedures, and Cost," p.10.

10. E. Eckholm, "Studies Find the Death Penalty Tied to Race of Victim," *New York Times,* February 23, 1995, p. 1.

11. Death Penalty Information Center, available at http://www.deathpenaltyinfo.org, accessed September 17, 2007; Bureau of Justice Statistics, *Correctional Trends*, available at http://ojp.usdoj.gov/bjs/glance.htm#cptrends, accessed September 17, 2007.

12. M. Radelet, *Facing the Death Penalty* (Philadelphia: Temple University Press, 1989), p. 10.

13. Eckholm, "Studies Find the Death Penalty Tied to Race of Victim," p. 1.

14. American Civil Liberties Union (ACLU), *Scattered Justice: Geographic Disparities of the Death Penalty*, March 5, 2004, available at http://www.aclu.org/capital/unequal/10532pub20040305.html, accessed September 17, 2007.

15. The Justice Project, available at http://www.thejusticeproject.org/, accessed September 17, 2007.

16. R. Paternoster, *Capital Punishment in America* (New York: Lexington Books, 1991).

17. P. Rubin, *Does Capital Punishment Have a Deterrent Effect? New Evidence from Post-moratorium Panel Data,* Emory University Economics Working Paper No. 01-01, available at http://ssrn.com/abstract=259538, accessed August 14, 2004.

18. W. Berns, *For Capital Punishment: Crime and the Morality of the Death Penalty* (New York: Basic Books, 1979), p. 152.

19. Death Penalty Information Center, *Time on Death Row*, available at http://www.deathpenaltyinfo.org/article.php?&did=1397, accessed September 17, 2007.

20. Robert Watson, personal interview by the author, Smyrna, Delaware, October 13, 1994.

21. V. Streib, *The Juvenile Death Penalty Today: Death Sentences and Executions for Juvenile Crimes, January 1, 1973 – April 30, 2004*, May 4, 2004, available at www.internationaljusticeproject.org/pdfs/JuvDeathApril2004.pdf, accessed September 17, 2007.

22. *Roper v. Simmons* 543 US 551 (2005); *Atkins v. Virginia* 536 U.S. 304 (2002).

Gang Management

32

Mark S. Fleisher

Chapter Objectives

- Describe which criteria distinguish prison and street gangs in a correctional institution.
- Examine management dilemmas that prison and street gangs pose to the safe and orderly operation of prisons.
- Outline how unmonitored communication networks can create prison security problems.

■ Defining a Gang

Generally speaking, the term "gang" has been applied to prisoner social groups that show some level of organization.[1] Institutions recognize two main types of gangs: prison gangs and street gangs that operate within the prison walls. Prison gangs and street gangs are distinguished from each other mainly by the presence or absence of an internal inmate hierarchy of members and leadership style—in other words, who gives orders and how these orders are enforced. Styles range from face-to-face coercion with reciprocity ("You do this for me, and I'll do that for you") to sheer application of power ("You do this, or I'll kill you").

Traditional prison gangs of the 1960s and 1970s consisted of adult criminals who were organized into hierarchies with rank differentiation. This meant that a prison gang was an efficient criminal organization that could smuggle drugs into prison and engage in crime to support such rackets. Violence and threats of violence strengthened crime pursuits and were directed toward both opposing gang enemies and deadbeat inmates who did not pay drug or gambling debts as well as toward the gangs' own members. Street gangs, on the other hand, were social groups whose members were younger inmates, had egalitarian or relatively simple hierarchies, less (or no) status differentiation, and less (or no) defined leadership styles. Street gangs were less effective criminal organizations.[2]

Traditional prison gangs originated in state and federal correctional facilities in the 1960s and 1970s. Prison gangs include, but are not limited to:

- Aryan Brotherhood
- Mexican Mafia
- *La Nuestra Familia*
- Black Guerrilla Family
- Texas Syndicate
- Dirty White Boys
- *Mexikanemi*

Classic examples of street gangs include the Gangster Disciples and Latin Kings.

Many years of improved effective prison management have blurred distinctions between prison gangs and street gangs. Today, the term security threat group (STG) is used more often. This term applies to any inmate group in which the following characteristics are present:

- Affiliation is based on race, ethnicity, geography, ideology, or a combination of these and other factors.
- Members seek one another's protection.
- There is an economic objective, such as drug distribution, that often is linked to violence or threats of violence.

The internal structure and leadership style of security threat groups differ.

Scholarly research is limited on prison gangs, street gangs in prison, and the control and management of prison and street gangs in correctional facilities. Additionally, access is limited to evaluation literature on prison gang intervention and prevention practices.[3] However, research does demonstrate that gang influence on general prison populations has waned in recent years. Inmates in major institutions often report the absence of gang violence and gang-run rackets. Due to improved prison management, information management systems, and threats of new criminal charges, the influence and power of STGs have been stifled.[4]

■ Gang Management Program

The availability of inmate programs affects long-term gang management. On the street and in prison, the most effective crime intervention is prevention. A gang management program should be one of a number of major programs in an institution's management plan, not an afterthought. The success of gang control plans is measured by management control data such as incident reports as well as the institutional climate.

Gang management requires a comprehensive and holistic policy that specifies legal precedents, procedures, and guidelines, including classification and verification of gang members. A strong policy should also include:

- A statement of an institution's philosophy on gang management
- An operational definition of an STG that fits an institution's management plan, culture, and climate
- An established list of ground rules in gang investigations and prosecutions

However, a well-planned management program does more than gang classification and intelligence gathering. It links STG management to both institution security and inmate programs. This dual link should be emphasized. Removing those gang members who distribute drugs and engage in strongarm activities from the general population may offer temporary control, but other inmates will replace those locked up, the network will persist, and the crime problems will recur. A gang management program must be able to guide an institution with an increasing number of criminals with street gang ties.

Other relevant strategic issues include:

- Identifying strategies to prevent newly committed inmates from participating in gang activities
- Allowing current, albeit marginal, gang members a way out of STGs
- Developing an organizational culture that exhibits zero tolerance for gang activity while also encouraging inmates' participation in work and other programs

The Role of Prison Staff

The technical side of gang management—the paperwork and procedures necessary to classify gang members and the strategies for investigating gang crimes—is straightforward. The human resource side of gang management is more difficult. If an institution does not have strong staff professionalism, high inmate and staff morale, open communication among staffers and between inmates and staffers, and, most importantly, an ample number of productive jobs for inmates, then STG management alone may be insufficient to create a positive institution.

Generally speaking, the role of staff in this effort is equivalent to that of gang detectives on the street. They have a sensitive job with significant implications for institution safety. Selecting qualified personnel is the single most important step to establish and maintain a gang-control program. A careful screening of potential applicants is essential. It should not come as a surprise to find inmates' siblings, cousins, girlfriends, former wives, street companions, and lovers applying for employment as correctional staff. Examining STG members' presentence investigations may determine if relatives of gang members are already employed under different surnames.

Curbing Communication Networks

Gang management personnel should keep logs of STG members' visitors, monitor telephone calls and mail, and stay in touch with local police agencies. Additionally, requiring inmates to obtain permission to send letters to inmates in different correctional institutions may help control gang communications.

The Internet has helped opened prison gates by substantially broadening prisoners' information flow and networks. Prior to the development of the Internet, inmates sustained rackets through phone calls and visits with street contacts. These social ties often quickly weakened and disappeared. However, the advent of the Internet has made communication with the outside relatively easy. This means that inmates can search for information about their fellow prisoners and even track prison transfers through inmate locators if Internet use is not monitored carefully.

Social Control

Research has shown that small but dysfunctional aspects of the prison environment may affect larger prison culture.[5] Not attending to items or programs that need development or repair (e.g., broken showers, weak educational programs, poorly structured work assignments) may create an environment well-suited for the development and expansion of STGs. Therefore, it is the responsibility of an institution's senior leadership to ensure that these seemingly minor factors are addressed appropriately.

To be sure, hard-core gang members will not likely discard weapons or stop drug distribution in exchange for new recreation equipment and better food; however, research shows that a prison's social and physical environment does influence inmates' behavior. This means that correctional administrators and managers can influence inmates' behavior through organizational development, strong management, and proactive planning.

Improving the prison environment may seem counterintuitive; it might seem that infractions would be controlled best by removing privileges and locking inmates down over long periods in high-security institutions.[6] However, withdrawing inmate incentives and using lockdowns is a financial and social gamble. If inmates do not behave well after receiving harsh punishments, administrators may lose control of their institutions.

Corrections administrators have been effective in their use of super-maximum security prisons as housing for the most violent and disruptive gang members. Removing a violent leader and his or her close companions may weaken a gang. Beyond the veil of crime, prison gangs afford some degree of social security. However, followers may be affiliated with a criminal organization out of fear and they may behave how they believe others want them to behave rather than how they want to behave themselves.

Prison management must achieve a fine balance. Pressuring organized disruptive groups too hard with behavior control mechanisms and offering inmates too few incentives to behave well may facilitate violence and other disruptive behavior. Ignoring STGs (standing back and not interfering in gang activities) may cause the same result. Somewhere between taking moderate action and tightly squeezing gangs is a strategy to fit each institution.

Housing can also play an important role in controlling STGs. Putting gang members together for housing or work recreates street gangs' geographic isolation (e.g., one neighborhood, one gang), reinforces the ties among gang-affiliated inmates, and may create animosity about who has the best cells or best jobs. Allowing gangs to run cell houses, "sell" cell residences, or "own" territory on the yard has proven to be disruptive to institution security.

Intelligence and Data Collection

Today's tightened prison management, improved surveillance, improved layout, and sophisticated intelligence-gathering strategies have curtailed prison gangs' rackets. A key to comprehensive gang intelligence is having valid data that is updated continuously. Gang members move about, so staff should take steps to ensure that they know about local, regional, and national gangs, including:

- Developing strong nationwide ties to gang units in police departments
- Participating in national correctional conferences on gang intelligence
- Maintaining collegial relations with fellow STG management personnel in state and federal institutions
- Establishing contacts in police agencies and state's attorneys' offices

The responsibility of collecting gang information should be spread across prison staff in all departments. A gang management program may have only a few specialists, but a prison has hundreds of employees. STG management should be the job of all employees in a correctional facility.

Gangs are social groups; their members want face-to-face communication. They walk together, hang out in the yard, and sit together in the dining hall. Case managers, correctional counselors, leisure time activities specialists, food service employees, industry and work supervisors, and other staff must be familiar with gang tattoos, graffiti, and symbols that demonstrate an inmate's gang affiliation. At a minimum, all staff should be trained in observational methods of data collection, and institutions should have clear procedures for reporting information to the institution's intelligence personnel.

Gang data, however, is worthless if it is not well-organized and carefully analyzed. Today's intelligence staff should be trained on databases and have the ability to create custom databases to accomplish specific tasks. An example of this would be linking visitors to inmate incident reports on offenses such as the possession and use of drugs. If management personnel cannot write software for specific facility needs, then staff should know their specific programming needs and be able to describe those needs to someone who can accomplish these tasks.

Conclusion

Prison STGs, like gangs on the street, seem to have become the passion of young inmates. "Old heads" have aged out of prison gangs or have been killed or locked down. Talking to inmate oldtimers about prison and street gangs in prison suggests that prison gangs are not as active and as threatening as they were 10 to 20 years ago. Nevertheless, proactive planning will keep staff alert to such problems reemerging.

A proactive defense against gang expansion will include the development of a comprehensive institution management system to encourage inmates' cooperation and participation in programs, including jobs with pay scales relative to inmates' skill and motivation level. Good inmate–staff rapport is essential. Inmates should trust staff to protect them from predatory inmates, whether or not these predatory inmates are gang members. In the end, the challenge of curbing STG growth will best be met with modem crime intelligence techniques and good planning.

Chapter Resources

DISCUSSION QUESTIONS

1. What distinguishes prison gangs from street gangs operating within prisons?
2. How could prison program budgets for education and recreation affect gang management?
3. What social functions do prison and street gangs have?
4. How can staff take steps toward prison gang management?
5. How have prison and street gangs changed in recent years?

ADDITIONAL RESOURCES

G. Knox, *National Gangs Resource Handbook. An Encyclopedic Reference* (Bristol, IN: Wyndham Hall Press, 1994).

National Alliance of Gang Investigators Associations, http://www.nagia.org

National Criminal Justice Reference Service, http://www.ncjrs.org

National Youth Gang Center, http://www.iir.com/nygc

J. Turley and R. Petersen, *Understanding Contemporary Gangs in America: An Interdisciplinary Approach* (Upper Saddle River, NJ: Prentice Hall Publishers, 2003).

NOTES

1. R. Fong, "The Organizational Structure of Prison Gangs: A Texas Case Study," *Federal Probation* 59, no.1 (1990), pp. 36–43.
2. J. Jacobs, "Street Gangs behind Bars," *Social Problems* 21, no. 3 (1974), pp. 395–409.
3. A. Goldstein and C. Huff, *The Gang Intervention Handbook* (Champaign, IL: Research Press, 1992).
4. J. Irwin, *The Warehouse Prison* (Los Angeles: Roxbury Publishing Company, 1995), pp. 94–95.
5. J. DiIulio, Jr., *Governing Prisons* (New York: The Free Press, 1987), p. 215.
6. G. Marx and C. Parsons, "Dangers of the Front Line," *Chicago Tribune* 9, no. 1 (1996), pp. 8–9.

Special Needs Offenders

33

Judy C. Anderson

Chapter Objectives

- Describe the variety and prevalence of physical and mental health issues present in a prison population.
- Outline the challenges posed by inmates with disabilities or special needs.
- Identify some of the measures that institutions must take to manage inmates with special needs effectively.

Special needs offenders are those incarcerated men and women with unusual or unique requirements stemming from their physical or mental age or other disabilities such as physical impairment, terminal illnesses, chronic medical conditions, mental illness, and mental retardation. Special needs can encompass many types of conditions, and practitioners have had to find ways of handling such offenders. These approaches have ranged from complete lack of acknowledgment to the provision of specialized services and units.

Special Needs Classification

The Americans with Disabilities Act (ADA), passed by Congress in 1990, defines a disabled person as one who has "a physical or mental impairment that substantially limits one or more major life activities, has a record of such an impairment, or is regarded as having such an impairment." This definition covers all persons, including inmates. Physical impairments are defined as severe mobility, visual, hearing, and speech limitations. People with mobility impairments include those who use wheelchairs or ambulate with assistive devices such as walkers, crutches, or canes. Chronic medical conditions that usually result in hospitalizations or medical segregation include:[1]

- Cardiovascular conditions
- End-stage renal disease
- Respiratory conditions
- Seizure disorders
- Tuberculosis
- Acquired immune deficiency syndrome (AIDS)

Mentally ill inmates include those with any diagnosed disorder defined in the *Diagnostic and Statistical Manual of Mental Disorders*, published by the American Psychiatric Association. Mental illness causes severe disturbances in thought, emotions, and ability to cope with the demands of daily life. Mental illness, like other illnesses, can be acute, chronic, or under control and in remission. Mental health services can range from intake screening, crisis intervention, inpatient admission, long-term therapy, or outpatient treatment.

Mental retardation is defined as having less than normal intellectual competence, characterized by an intelligence quotient (IQ) of 70 or less. Retardation, or developmental disability, usually results in impairments in adaptive behaviors such as personal independence and social responsibility.[2]

Terminally ill offenders are usually defined as having a fatal disease and less than six months to live.

Geriatric offenders are usually categorized as those offenders over the age of 65; however, preventative health care dictates that an earlier age should be used as a guideline. Many correctional systems have adopted 50 years of age as the defining age for geriatric offenders based on their socioeconomic status, access to medical care, and lifestyle.[3]

■ Prevalence

The number of older offenders is rising due to longer and mandatory sentencing. A 2006 Bureau of Justice Statistics special report noted that at midyear 2005, more than 50 percent of state and federal inmates suffered from a mental health problem. Moreover, many of these offenders also suffered from substance dependence or abuse (42 percent of state inmates, nearly 30 percent of federal inmates, and nearly 50 percent of jail inmates).[4]

It is estimated that offenders with mental retardation are over-represented in the entire correctional system, especially in prison. A 1998 study suggested that the number of mentally retarded persons is two to three times greater in prison than in the community. It should be noted that many offenders in custody are considered to have borderline retardation, thus increasing the absolute numbers.[5]

■ Identifying Those with Special Needs

During intake (the process of evaluating inmates as they enter the correctional system) any special needs must be identified as soon as possible in order to provide appropriate services, both during intake and throughout the period of incarceration. Intake is often conducted at a separate facility or unit within a reception and diagnostic center.

Within the first few hours of admittance, a medical examination or at least a medical screening should be conducted. While a thorough medical examination is preferable, it may be impossible to provide if intake volume is high. The medical screening, which serves as a mini-triage, should identify any medical or mental health concerns that need immediate attention; these cases should be assessed immediately by medical staff.

Protocols for Care

Each correctional facility should develop written policies and procedures that are consistent with the ADA. When evaluating services, the institution must ascertain if there are policies, procedures, and practices that would prevent inmates with disabilities from participating. If there are, then reasonable modifications might be indicated to avoid potential discrimination. For example, library or law books might be brought to a wheelchair-bound inmate if the library is located in a building without wheelchair access. The ADA does provide for exclusion from a program if a disabled inmate presents a direct threat to the health and safety of others. For instance, a mentally ill inmate who hears a voice telling him or her to kill another person would need not be included in general population activities.

Segregation or Mainstreaming?

After acknowledging that special needs offenders may be in the prison population, the first major task is to decide how to handle these groups of inmates—should they be segregated or mainstreamed? Reasons cited for separating special needs offenders from the general population include:

- *Cost containment.* It is a more efficient use of funding if special needs inmates are housed and treated as a group.
- *Managed care.* More effective care can be focused on a specialized unit
- *Concentration of resources.* Relevant staff and resources can be more concentrated if special needs inmates are housed in one location.

Mainstreaming (the integration of a person with a disability into the normal prison population) is a basic premise of the ADA, which requires that disabled inmates have complete access to prison programs and services. However, as previously noted, the ADA does allow correctional administrators to exclude or remove an offender with a disability from the general population if he or she is a direct threat to the health and safety of others.

Consequently, a combination of mainstreaming and segregation (with emphasis on the former) should provide services for inmates with disabilities as well as follow the law outlined in the ADA. This approach would also be consistent with the way persons with disabilities are handled outside correctional facilities.

For example, a disabled person who can no longer live alone safely may hire an aide to come into his or her home, move in with someone else, or move to an assisted living, intermediate, or long-term care facility. Changes in status typically represent a progression rather than a jump, for example, from independent living to long-term care, or could involve movement back and forth among various levels of assistance. Within the correctional system, options mirror those found in the community.[6]

Facilities should emphasize mainstreaming and segregate as a last resort, except for the severely mentally ill and retarded and the acutely ill. Even then, services should be provided so that persons with disabilities or illness have an equal opportunity to benefit from programming.

■ Access and Communication

Program access, which usually refers to architectural or design barriers, is also a significant concern when dealing with special needs inmates. Inaccessibility of facilities is not justification for denying programs, services, or activities. While the ADA does require accessibility in new construction and alterations to existing buildings, it does not require that all existing facilities be modified to the new standard. Alternate methods of program delivery can satisfy the law's requirements in many cases if the program, service, or activity can be brought to the offender.

Universal design should be considered for all new construction so that ADA and applicable standards for persons with disabilities can be met. Universal design means that the construction is built to meet ADA standards so that structural changes will not be needed later so that the inmate can age in the same location. For the community, universal design means that the building design employs features that let the space grow and contract with changes in needs and lifestyles. Examples of universal design include wider doors, ramps instead of stairs, higher commode seats, flapper levers on sinks instead of knobs, and heavier building material so that grab bars or railings might be installed later if needed. Universal design allows far greater flexibility in all aspects of programming in the present and the future. While universal design is initially more expensive, it costs less over time as modifications are easier and not as expensive to implement.

Communication is a factor in program access. Communications with special needs offenders should be as effective as it is with other inmates. Examples of ways to enhance communications include auxiliary aids for inmates who are deaf or have hearing or speech impairments include telecommunication devices for the deaf (TDDs), communication boards, and assistive listening devices. Written notes could also be effective for short or routine communications. Certified sign language interpreters should be accessible either on a volunteer or contract basis. Staff members could also be trained as interpreters.

Audio tapes and books in Braille should be available for inmates with vision impairments. Signs in Braille throughout the facility would enhance communications and eliminate confusion. Other adaptations could be large-print memoranda, talking directly to the offender at eye level (if not a security risk), and lowering pitch rather than talking louder.

Program and Activity Availability

Whether an inmate is assigned to a specialized unit or a modified living space in the general population, programs, services, and activities must be available. Again, the program, service, or activity may need to be brought to the inmate rather than the inmate going to the specific location. For instance, certain prison industry tasks, such as folding or packaging items, might be performed on the unit rather than at the prison industries location. If other inmates are offered work assignments involving pay or the possibility of sentence reduction, then the same must be available to inmates with special needs. If boot camps and work release programs are offered, modified or alternate programs should be designed for special needs offenders. If the education area is inaccessible, then a tutor could come to the living unit. Staff must learn to think "outside the box," while still keeping security foremost in the design of individualized programs. Staff also should consider involving special needs inmates in the design process, as these inmates best know their own capabilities.

Partnerships with other federal, state, and local agencies should be developed to provide better and more suitable programs and activities. Because reentry starts the day the person comes into the institution, correctional agencies need to be proactive to ensure an inmate's successful return to the community.

Depending upon the severity of their illness, mentally ill inmates may require segregation in a specialized unit. Where the unit is under the auspices of medical, psychiatric, or corrections professionals, the institution should provide a liaison to ensure that sound correctional practices are followed. Frequently, a medical or human services professional and a correctional staff member will manage the unit together.

The major focus in such units is therapy, whether medical, psychological, or a combination of both. The emphasis is on returning the inmate to a state of wellness. As inmates progress, they may be returned to the general population either by participating in programs and work activities or by moving into a unit that provides transitional care.

Mentally retarded or developmentally disabled inmates may also require placement in a specialized unit. Again, specialized staff will provide programming in concert with correctional staff. Because a major focus must be on the ability to function independently, daily living or survival skills will be central, resulting in activities that may seem out of place in a correctional environment. Such activities might include cooking, shopping, or doing laundry. Based upon the level of functioning, inmates might stay in specialized units until their release from the correctional facility. As with all special needs offenders, continuity of services must be provided as the inmates move back into the community. Partnerships with community agencies for mentally ill or retarded offenders will provide needed services in the institution as well as a seamless delivery of services for offenders transitioning into the community.

Classification Considerations

During classification, the dangerousness and mental stability of inmates are evaluated and, usually, persons with similar characteristics are sent to similar locations. Most classification systems consider risk factors such as prior convictions, current convictions, escapes or attempted escapes, length of sentence, and institutional adjustment. Age, educational level, history of substance abuse, and history of violence may also be considered.

A provision for override allows staff to factor in information such as medical and mental health conditions and to change the custody and security level of the inmate to accommodate these conditions. These override decisions should be made on a case-by-case basis and should place the inmate in the least restrictive custody possible. For example, inmates in wheelchairs should not automatically be placed in a medical unit but should be placed where they could best function, such as a regular housing unit on the ground floor with full access to all programs and services in the facility. An inmate who is suicidal or psychotic should be placed in an appropriate specialized unit for treatment until it is determined that he or she can return to general population or step-down unit. Classification decisions must not be made solely on the basis of a disability or special need but it should be among the many factors considered.

Housing Accommodation

Many special needs offenders will need special housing and programming. Inmates with mental illnesses or mental retardation, like those with acute medical concerns, may need to be admitted to a specialized unit and the intake process completed there or deferred until a later time. Special needs inmates retained in intake should be housed on ground level near the officers' station or in a monitored observation cell. Some type of identification (such as red reflective tape) should be placed on the cell door or at eye level and about a foot from the floor to denote that assistance is required should the facility need to be evacuated. Medical or human services staff should be assigned to monitor these inmates and provide necessary services.

Alternate types of aids may need to be provided. For example, a white cane used by an inmate with a visual impairment is considered a weapon within a correctional facility and could be replaced with a collapsible one if it is consistent with the security level of the facility. To provide for individuals with physical limitations, the institutions should make sure wheelchairs are accessible throughout the unit or have an alternate plan for providing various services.

Special Facilities

Physical plant modifications will probably be necessary for older construction. Modifications include lowered storage lockers, booster seats on commodes, handheld shower heads, bath chairs, and tables that allow a person in a wheelchair to roll close to the table. Blackboards can be hung vertically instead of horizontally to allow accessibility by inmates in wheelchairs. Lower-placed water coolers, bulletin boards, door levers, in addition to telephones with volume controls, specialized door closure systems or automatic doors, and roll-in showers help provide independence as well as accessibility. Air conditioning may also be needed as many special needs inmates are on psychotropic medications, and overheating can cause medical complications.

For the visually impaired, colors need to be distinct so that a person can tell the difference between the wall and the floor, the steps and the riser, and the wall and the door. Rooms, cubicles, and beds should be marked with reflective tape to denote assistance needed for evacuation.

Scheduling can also improve access. A separate meal shift can be established for special needs inmates who either need more time or assistance to eat or who need to be segregated from the rest of the population to prevent victimization. The same rationale should hold true for the amount of time allowed for moving from one location to another, such as from a living unit to a work location or to medical services, or for completing a task such as cleaning a living area.

Prior to locking up a special needs inmate, medical staff should be consulted about what, if any, requirements must be met. These requirements may range from no restrictions to an order for no placement in a segregation unit. If no segregation is ordered, alternate arrangements such as placing a deadlock on the present cell with appropriate, documented visual checks should suffice.

◼ Special Support

Assistance may be required to complete activities of daily living or, in some instances, the task will have to be completed by other inmates or staff. Other inmates can work as caregivers or assistants. These jobs should have clear descriptions specifying that caregivers do not have authority over the special needs inmates. Inmate caregivers should be chosen as carefully as staff. Training should be provided prior to beginning work with special needs offenders and should be offered on a scheduled basis. All inmates who assist should be directed by staff and should consult with specified staff on a daily basis.

◼ Hospice and Palliative Care

Just as hospice and palliative services are becoming more common in the community, the same is happening in the correctional system. Usually terminally ill inmates are placed in a facility with the highest level of medical care available. Treatment of the terminally ill concentrates on palliative care—keeping the patient comfortable and pain-free—instead of curing the disease.

Hospice programs utilizing other offenders as aides focus on providing spiritual, emotional, and supportive care to the persons receiving end of life care. Careful screening, training, and monitoring of hospice workers is vital to the success of the program. This interdisciplinary program involves commitment and support from all level of the institution to be effective.

■ Reentry

Reentry to the community begins the day the offender enters the correctional system. Efforts should focus on returning a productive person to the community. This is difficult to accomplish for any offender and will probably be more difficult for a special needs offender. The focus for the special needs population is ensuring that their needs are cared for adequately.

As outlined in the National Institute of Correction's *Effective Prison Mental Health Services*, discharge planning should center on housing, medical and mental health services, employment, family support, and enrollment or reinstatement of benefits including Medicare, Medicaid, and veterans' benefits.[7] Whenever possible, community agencies should serve as partners to provide services during incarceration and then become the primary service provider upon release. Detailed planning for provision of services should start at least six to eight months before the special needs offender is released.

■ Staff Development

Education, training, and staff development are an integral part of corrections. The National Congress on Penitentiary and Reformatory Discipline held in Cincinnati, Ohio, in 1870 adopted a Declaration of Principles that proclaimed that "special training as well as high quality of head and heart, is required to make a good prison or reformatory officer."[8] While pre-employment and annual training are required for correctional staff, little training focuses on working with special needs offenders. In order to work effectively with this group, all involved staff should be provided with relevant training and development. Training components should include, but not be limited to the following areas as they relate to special needs:

- Familiarization and sensitization techniques
- Educational and medical information on various special needs
- Techniques, tips, and strategies for managing special needs inmates more effectively

The professionally trained staff who work with special needs offenders (particularly with persons with mental illness or mental retardation) have mandated training hours in order to maintain licensure and certification in addition to the agency, accreditation, and statutory requirements that all correctional staff must meet. These professionally trained staff are a good source for providing training for the other correctional staff. Security staff should be able to learn techniques for managing special needs inmates, such as using belly chains instead of handcuffing behind the back for heart patients, transferring an inmate from a wheelchair to a bed, and communicating with inmates with hearing or sight impairments.

Other resources for training include state agencies, medical facilities, advocacy and special interest groups, and other providers of services. Additionally, training plans from other correctional facilities might be available through the National Institute of Corrections' Resource Library or from the facilities themselves. Staff should be encouraged to participate in professional and special interest groups in the community for education and professional growth as well as for networking opportunities.

Correctional staff should also inform the community about special needs offenders. Contact and partnerships should be developed with the appropriate providers of services to provide services during incarceration and after reentry and ensure that information is provided to both inmates and service providers. Intake forms in corrections should be revised to collect the information required by community service providers, especially those managed by state and federal agencies.

◼ Tips for Working with Special Needs Offenders

- Move more slowly than normal. Sudden movements can frighten special needs inmates or make them think they are being attacked.
- Talk directly to the inmates. Those with hearing or speech disabilities may need to read lips.
- Address conversation to the inmates. Do not talk around them as though they are children.
- Talk at face level. If a longer conversation is indicated, either sit or stoop down to eye level with inmates in wheelchairs.
- Speak clearly, in a low tone. Do not talk loudly or in a shrill voice.
- Use terms that the inmates can understand.
- Simplify instructions. Give one direction at a time to avoid confusion. Sometimes it helps to put instructions in writing.
- Establish and maintain a familiar routine. Do things in the same sequence—get up, straighten cell, have breakfast, go to pill line, and so on.
- Talk in positive terms. Talk about what can be done, not what cannot be done.
- Be patient. Allow extra time to complete tasks.
- Be flexible and creative when providing programs, services, and activities.
- Use large type so the font is easier to read.
- Utilize the public address system for announcing changes and to read memoranda concerning changes.
- Ensure that inmates eat properly. Poor nutrition can lead to other problems.

◼ Conclusion

The growing number of special needs inmates and the regulations outlined in the ADA are having a great impact on the U.S. correctional system. In addition to making physical plant modifications or building new construction, correctional administrators will need to lead their staff and facilities toward compliance with all provisions relevant to federal and state laws and ADA and American Correctional

Association standards. Policies, procedures, and practices may need to be altered to provide appropriate programs, services, and activities to special needs offenders.

Resources are available through local, state, and federal organizations and agencies. Staff should partner with local providers so that necessary services can be offered to special needs persons while incarcerated and upon reentry. Line staff must be selected carefully as not everyone is suited for working with special needs offenders. These staff members must be provided the specialized training required to provide care for these inmates. A successful correctional program for special needs offenders takes commitment from all levels—administrative, program services, and security, as well as the community at large.

DISCUSSION QUESTIONS

1. What types of physical and mental disabilities are present in the inmate population and to what extent?

2. In general terms, what does the Americans with Disabilities Act (ADA) require in terms of managing special needs offenders?

3. What measures can be taken to effectively manage inmates with special needs?

4. What are the advantages and disadvantages of separating or mainstreaming special needs offenders?

5. What issues arise when special needs offenders face reentry into the community?

ADDITIONAL RESOURCES

American Correctional Association, *Standards for Adult Correctional Institutions*, 4th ed. (Lanham, MD: American Correctional Association, 2003).

B. Anno et al., *Correctional Health Care: Addressing the Needs of Elderly, Chronically Ill, and Terminally Ill Inmates* (Washington DC: National Institute of Corrections, 2004).

H. Hills, C. Siegfried, and A. Ickowitz, *Effective Prison Mental Health Services: Guidelines to Expand and Improve Treatment* (Washington, DC: National Institute of Corrections, 2004).

NOTES

1. B. Anno, *Prison Health Care: Guidelines for the Management of an Adequate Delivery System* (Washington, DC: National Institute of Corrections, 1991).

2. P. Rubin and S. McCampbell, "The Americans with Disabilities Act and Criminal Justice: Mental Disabilities and Corrections," *Research in Action* (Washington, DC: National Institute of Justice, 1995).

3. J. Morton, *An Administrative Overview of the Older Offender* (Washington, DC: National Institute of Corrections, 1992).

4. GRACE Project, *Incarceration of the Terminally Ill: Current Practices in the United States* (Alexandria, VA: Volunteers of America, 2001).

5. W. Gardner, J. Gracber, and S. Machkovitz, "Treatment of Offenders with Mental Retardation" in R. Wellstein, ed., *Treatment of Offenders with Mental Disorders* (New York, NY: The Guilford Press, 1998).

6. B. Anno et al., *Correctional Health Care: Addressing the Needs of Elderly, Chronically Ill, and Terminally Ill Inmates* (Washington, DC: National Institute of Corrections, 2004).

Chapter Resources

7. H. Hills, C. Siegfried, and A. Ickowitz, *Effective Prison Mental Health Services: Guidelines to Expand and Improve Treatment* (Washington, DC: National Institute of Corrections, 2004).

8. American Correctional Association, Declaration of Principles, available at http://www.aca.org/pastpresentfuture/principles.asp, accessed September 17, 2007.

Sex Offenders

34

Gilbert L. Ingram and Peter M. Carlson

Chapter Objectives

- Provide examples of different types of sex offenders.
- Explain why the identification of sex offenders is often a difficult task.
- Outline components of a viable sex offender treatment program.

Any thought of living near or working with a sex offender represents a nightmare to many in American society. Sex offenders have molested, raped, or otherwise victimized women, girls, men, or boys and are part of the dark underside of society that many do not comprehend or even want to acknowledge. Whenever a horrible sexually related crime is reported, the public outcry may be deafening. The usual response to sexual atrocities is to demand harsh justice in the form of a court sanction that involves significant prison time. In short, society typically loathes the sex offender more than other social outcasts who have been convicted of any one of a multitude of equally outrageous crimes.

The difficulties inherent in dealing with sex offenders are then passed on to the correctional setting. Because of fear of failure, lack of resources, or lack of commitment, some managers avoid the issue by not officially acknowledging sex offenders as a group requiring special attention. This tactic is shortsighted, unprofessional, and even unethical. It should never be an option.

■ Magnitude of the Problem

Sex offenders have attracted tremendous negative publicity during the past few years. Highly visible sex crimes have always generated an inordinate amount of public interest, but the recent intense media attention has created a public rage unprecedented in correctional history. Because sex offenders have no visible supporters, politicians feel free to clamp down hard, calling for longer sentences and lifelong public identification after release from prison.

Some people say the prevalence of sex offenses is growing, but valid data are not available. Many factors contribute to the lack of highly reliable data:

1. Embarrassment, fear, or self-blame on the part of victims deter many from ever reporting these acts.

2. Authorities, unless compelled to act or under unusual circumstances, frequently overlook sexual behavior that the public seems to condone. (Some sexual acts technically violate the law but are considered acceptable behavior when they take place between consenting adults in private settings.)

3. Even if arrested, a large number of sex offenders are not legally convicted after arrest or plead guilty to a lesser charge.

4. Only convicted sex offenders appear in sex crime statistics.

The collection of reliable data has been even more complicated in recent years. For example, the public spotlight on abuse by family members and a generally more supportive environment for reporting victimization has encouraged more people to step forward. Additionally, the definition of sexual offenses remains a continuing problem; behavior deemed unacceptable in one state may be acceptable in another. However, there are enough statistics available to conclude that the problem is fairly extensive. In 2005, there were 93,934 rapes reported in the United States, and 602,189 registered sex offenders.[1]

Data generally indicate that certain categories of sex offenders are more likely than other types of criminals to repeat their offenses (see **Table 34–1**). Recidivism

Table 34–1	**Characteristics of Sexual Offender Recidivists**

- Multiple victims
- Diverse victims
- Victims were strangers
- Juvenile sexual offenses
- Multiple paraphilias
- History of abuse and neglect
- Long-term separations from parents
- Negative relationships with their mothers
- Diagnosed antisocial personality disorder
- Unemployed
- Substance abuse problems
- Chaotic, anti-social lifestyles

Source: Center for Sex Offender Management, U.S. Department of Justice, *Recidivism of Sex Offenders*, available at http://www.csom.org/pubs/recidsexof.html, accessed July 25, 2007.

varies among different types of sex offenders, but is often related to specific characteristics of the offender and the offense. For example, federal studies confirm that rapists were 10.5 times more likely to be rearrested for rape than were other released prisoners.[2] A study of 54 rapists who were released before 1983 found that after four years, 28 percent had new sex offense convictions and 43 percent had a conviction for a violent offense. Studies of child molesters reveal relatively equal rates of reoffending. Extra-familial molesters were followed for an average time span of six years; during this period, 31 percent had a reconviction for a second sexual offense.[3] This is only a sampling of many generally accepted studies that confirm that deviant sex behavior is a large and continuing problem in this country. It is easy to appreciate the reluctance of some public administrators to take responsibility for managing a very unpopular group whose aberrant behavior is extremely difficult to change.

Even though successful treatment is possible, many sex offenders do repeat their inappropriate behavior after release from custody.[4] However, correctional administrators must keep in mind the greater public good that is sought. Many sex offenders may be deterred with proper intervention. Further, it is good management practice to involve all offenders in a meaningful activity to facilitate better control of the institution and to allow better use of public resources. As nearly all offenders will be returning to their communities, one very good reason to promote intervention for this group of offenders is that they will be someone's neighbor again in the future.

■ Basic Approach to Sex Offender Management

Each correctional situation is different, and there is no single formula for successful management. Consequently, many programs producing good results in one institution have failed completely when attempts were made to replicate them elsewhere. Differences in administrator personalities and levels of motivation as well as insufficient resources often make duplication impossible. However, certain basic management practices produce successful sex offender programs at local, state, and federal levels. In discussing these approaches, this chapter will focus on male offenders because they are the largest concern for correctional officials. Nonetheless, it must be kept in mind that instances of sex offenses by women have been increasing.

Institutions must focus on correctly identifying sex offenders as soon as possible after incarceration. For this to occur, an efficient classification and designation process must be in place. During classification, there is no need to try and uncover the underlying psychological reasons for offenders' inappropriate sexual behavior; this information will play an integral part later during treatment.

Immediately after classification, the offender should be separated from the general population as much as possible. At a minimum, a separate area for special treatment is necessary, and if possible, a separate housing area can minimize the many adjustment problems that sex offenders typically encounter. A special housing unit for sex offenders should be placed in an institution with a progressive, open-minded administration that can handle difficult cases easily. Inmate participation in other institution-wide activities outside the special unit seems to work well in this situation.

■ Classification

Early identification of sex offenders is not always an easy task. Sex criminals are usually reluctant to be candid about their activities, and they attempt to hide or alter facts in their favor. Official records are not necessarily the solution for classification purposes, because details of the commitment offense are frequently affected by lengthy legal maneuvers and sometimes not available to classification staff when the records are needed. Also, many offenders are incarcerated for apparent nonsexual activities such as breaking and entering or assault when their intent may have been the sex offense of rape.

Classification staff need be sensitive to these possibilities and convince inmates that a complete reporting of their activities is in their best interests. This is a formidable task but attainable if interviewers are trained to deal with sex offenders. Offense details frequently help staff identify this target group, and self-reports from those already motivated to seek help also are useful. Ideally, once inmates are made aware of the existence of a good sex offender treatment program in the correctional system, those who recognize their deviance as a problem will cooperate during classification.

Classification staff should record the information they collect on a standard checklist or inventory.[5] Many checklists provide a structured format to ensure that all relevant areas of inquiry have been assessed. If the identification of sexually deviant behavior remains in doubt after the initial interview process, additional attempts to obtain background information should be made, and a followup interview should be scheduled. If reluctant inmates learn more about the treatment program or have more time to talk, they may be less guarded in their responses.

Trained staff should use their knowledge about typical sex offender profiles. For instance, they should remember that almost all sex offenders have engaged in many minor transgressions before the current incarceration and that their current offense is part of an unhealthy cycle of behavior.[6] To gain cooperation during the interviews, staff should tell inmates that many people have been involved in all sorts of sex acts as they matured and that they should not be concerned, because the interviewer has heard just about everything. Staff should not condone such behavior but simply acknowledge its occurrence to elicit more information from the suspected sex offender.

Sex offenders will not necessarily have histories of physical aggression or attacks. In fact, most sex offenses do not involve forceful acts, and most sex offenders do not fit the image of the dangerous psychopath. Many but not all appear to be psychologically or emotionally impaired, neurotic, psychotic, or even brain-damaged. A trained professional will be able to make these kinds of determinations.

Intelligence level, age, and income level are not very helpful in identification at this early stage of classification. Even though many sex offenders are young and have somewhat low intelligence, the same can be said of inmates in general.

Similarly, most sex offenders started their inappropriate behavior at an early age and most come from poor family backgrounds, but the same can be said of other offenders.

■ Staff Issues

After classification has been completed and transfer to the treatment unit has been accomplished, the offender will be handled almost exclusively by staff who specialize in dealing with sex offenders. The most essential step in developing and running a useful program for sex offenders is having trained staff who have the right attitude toward these inmates. Staff should not be people who view sex offenders as horrible persons who deserve the worst punishment that the institution can create for them. On the other hand, staff should also not be people who consider sex offenders to be victims of poorly conceived social laws, innocent byproducts of dysfunctional families, or youngsters guilty only of normal youthful experimentation.

Staff members should be realistic and mature. They should be people who view the sex offense as illegal and inappropriate but believe the offender is capable of changing with proper motivation and assistance. People who are uncomfortable with the topic of sex or show too much interest in the area may not be right for this job.

It is best to hire professionally trained specialists. If this is not possible for all staff positions, additional personnel should be chosen carefully. Enough personnel should be available to give significant individual attention to each program participant. After staff are selected, they should be trained. They should learn about sex offenders in general, sexual behavior in all of its ramifications, and, of course, the particular treatment approach that will be followed. Staff should also be sensitized to the need for confidentiality and to the possible negative feelings that other staff who work in other areas of the institution may exhibit.

Staff must work well as a team. Total communication and cooperation are needed to monitor progress and to make important decisions about readiness for additional programs, privileges, and reentry to the community. Staff need to encourage positive behavior while also being able to react quickly, decisively, and appropriately to offender setbacks. Demonstrating acceptance but not approval of inmate transgressions is a very challenging balancing act, but it has been accomplished successfully in correctional environments.

Recruitment and financial limitations may make staff selection more difficult. The quality and quantity of professional treatment staff and resources available to administrators varies considerably in different locations. If professionals trained in sex offender treatment (usually psychologists and, less frequently, psychiatrists) cannot be hired, it may be best to defer implementation of the program. Attempting to run such a demanding program is difficult under the best of circumstances; if the right professionals are not present to train correctional staff or to provide specialized treatment, the program may fail. In most instances, a failing program is worse than no program and may cause significant legal liability.

That said, too often administrators have used a lack of human resources to justify not implementing programs when, in fact, they have not spent sufficient time or effort in seeking professional staff. The marketplace for suitable candidates is looking better as competition and cutbacks increase in the general medical area, and recent advances in treatment techniques for sex offenders produce more trained personnel interested in employment.

■ Evaluating and Admitting Sex Offenders

After arrival at the special treatment unit, all sex offenders identified through classification or by the courts as suitable candidates for treatment should be given a full explanation of the program requirements. An excellent example of a comprehensive handout for sex offenders is used by the Federal Bureau of Prisons.[7] Candidates are told that they will need at least enough time to complete all program requirements before release, and that they should be able to complete these within a reasonable period of time.

After offenders understand everything they need to know about the program, particularly their responsibilities in it, they are asked to volunteer for participation. This willing acceptance is necessary because sex offenders often become resistive during treatment; they have had a great deal of immediate gratification from their sexual behavior. Because the program demands intensive work from offenders, successful treatment will not occur if they do not wish to take part.

In determining offender suitability for the program, staff must remember that full participation is not possible if severe mental illness or basic deficiencies in areas such as reading and writing are present. These needs should be addressed before the offender is formally admitted to the treatment unit. Those offenders who are seeking admission for reasons other than treatment of sexual deviancy should not be accepted; inmates hoping for a diversion from routine prison life or seeking an earlier release opportunity should be screened out.

The evaluation phase of the program continues after the offender is officially admitted. Because every sex offender has a unique set of problems that needs to be addressed, it is vital to program success that the offender complete very extensive questionnaires pertaining to family background, education, social history, and sexual behavior. This information will help staff develop a workable plan for treatment and release into the community.

Extensive psychological testing is also necessary at this stage. These tests may vary according to staff preferences, expertise, and availability, but a basic assessment of personality, cognitive abilities, social attitudes, and sexual thoughts is necessary. If competent staff are present, and the cooperation of all concerned has been obtained (including, for example, top management's commitment to support and defend the use of a potentially controversial treatment tool), an assessment of the participant's sexual response to deviant and nondeviant themes is conducted. This assessment is accomplished by use of penile plethysmography initially and at regular intervals during treatment. The plethysmograph is a widely accepted instrument consisting of a small penile transducer (a circular gauge similar to a rubber band) that measures sexual arousal based on an offender's erectile response to certain sexual stimuli. The information is not only useful in determining the proper course of treatment but is frequently used during the actual treatment to condition appropriate sexual responses.

All of the information gained during the evaluation period is used by staff to formulate an overall treatment plan for each offender. This plan may be modified at any time as more information becomes available.

■ Treatment Program

Sex offender treatment programs throughout North America today use a combination of cognitive behavioral treatment and relapse prevention strategies. Programs typically include both individual and group therapy and focus on cognitive restructuring—helping the offender reconsider issues that surround sexual deviancy. Offenders are taught to be more aware of the victims of their offenses, are given empathy training, and learn about the sexual abuse cycle. Most programs include anger management training, relapse prevention, and how to improve social and interpersonal skill development. The programs emphasize changing deviant sexual arousal patterns.[8]

Teaching the offender to engage in meaningful social interactions, healthy recreation, and other self-improvement activities should be a high priority for staff. Like all inmates, sex offenders must be actively engaged in productive activities. While the basic premise of treatment—which must be accepted by participants— is that there is no cure for their problem, offenders should be encouraged that with their full participation and motivation, staff can teach them to control their deviant behavior.[9] Further, participants should accept that their deviant sexual behavior is totally inappropriate and unjustified. Holding sex offenders accountable for their behavior, past and present, is critical.

A comprehensive program for every sex offender includes specialized treatment activities in addition to the self-improvement and work activities assigned to all inmates. Naturally, treatment takes priority initially, occupying most of the offender's time during the intensive part of the sex offender program. However, treatment activities alone are insufficient for rehabilitation. Full programming is as essential to good institutional management as it is to good treatment.

Once a treatment plan is in place, the offender is assigned a staff member who serves as the lead therapist, providing individual counseling and coordination of the total treatment plan. Individual therapy is used to explore the dynamics of sex offenders' behavior, the difficulties they continue to have in relationships, and other issues raised in group therapies. As treatment progresses, discussions move to offenders' adjustment to the treatment program and finally to release planning. In addition to individual sessions at least once per week, various group therapy programs are essential.

Throughout the treatment program, offenders meet in group sessions aimed at resolving their longstanding difficulties and gaining understanding about their behavior. Regardless of the number of required therapy groups, one central group continues to focus on the basic deviant behavior, including a thorough discussion of inmates' offenses, victims, background, and present sexual thoughts and acts. The principal goal of this group is to examine how the offenders' behaviors, feelings, and thought processes led to their inappropriate sexual activity. The individual sessions complement this group work and add to this intensive self-examination. Many offenders must deal with extremely sensitive issues in private sessions before they can explore them in the group setting.

Several other key groups are significant components of a complete treatment effort. One such group deals with abuse and its ramifications in the offenders' lives. Most sex offenders were abused themselves as youngsters, frequently in a sexual way, but also physically or emotionally. In this group, offenders gain an understanding of how this abuse affected them and prepare for a later group designed to help them develop empathy for victims. Other groups might address anger management, social skills, sex education, and substance abuse, areas that cause difficulty for many sex offenders and are frequently factors in their sexually deviant behavior.

The final mandatory part of group treatment is relapse prevention training. Offenders have to accept the fact that the probability of offending again is very high if proactive steps are not taken. This group teaches offenders to recognize risk factors associated with deviant sexual behavior, to anticipate and modify risky situations, and to cope successfully with their postrelease environment.

■ Transition to the Community

A comprehensive treatment program must include a plan for reentry. Regardless of the offenders' level of program involvement, their success in completing the treatment goals or the development of a realistic relapse prevention plan, staff must attend to another important task: ensuring offenders' successful reentry.

Presumably, after successful program completion, offenders can anticipate and know how to cope with the challenges of life outside the institution. Knowing that they will need help from many others to maintain a nondeviant lifestyle, offenders must seek out socially approved support groups, positive recreational activities, and other forms of community assistance. However, staff responsibility goes beyond helping offenders with these preparations and should include informing the relevant local authorities when and where the offenders will be released. Until recently, ensuring that all contacts required by law had been made was relatively simple—notifying the courts and probation authorities usually satisfied all concerned. Today, however, staff also need to contact local law enforcement.

Additionally, staff must ensure that all sex offenders in their institutions are aware of required registration with the National Sex Offender Public Registry and the anticipated increased public attention that may result. This project, coordinated by the Department of Justice, is a cooperative effort between the state agencies hosting public sexual offender registries and the federal government and allows users to search for information about sex offenders.

■ Conclusion

Although successful techniques for treating sex offenders continue to be developed, dealing with sexual deviance will remain problematic for both staff and offenders. Increased public attention only exacerbates an already difficult task, in that no treatment program can promise that an offense will not reoccur. Fortunately, a realistic treatment effort can prepare the offenders to control their behavior and thereby better protect the community. Sex offenders who take responsibility for their behavior, discuss their past offenses openly, understand why their deviant acts were wrong, exhibit genuine remorse, and actively work in the treatment programs to acquire relapse prevention skills will undoubtedly pose less of a danger to the community than they did before their arrival at the institution.

DISCUSSION QUESTIONS

1. Do sex offenders require specialized treatment within correctional institutions?

2. Should individuals who have been convicted of sex offenses and have refused institutional treatment for these offenses be confined by the government after their sentence has been completed?

3. Should taxpayers pay for expensive sex offender rehabilitation programs?

4. Could you work with and provide supportive programming for sex offenders?

5. What unique problems do sex offenders pose in correctional settings?

ADDITIONAL RESOURCES

The Association for the Treatment of Sexual Abusers, http://www.atsa.com
Dru Sjodin National Sex Offender Public Registry, coordinated by the U.S. Department of Justice, http://www.nsopr.gov/

NOTES

1. Federal Bureau of Investigation, *Crime in the United States, 2005*, available at http://www.fbi.gov/ucr/05cius/, accessed September 17, 2007.

2. J. Chaiken, *Sex Offenses and Offenders: An Analysis of Data on Rape and Sexual Assault*, Bureau of Justice Statistics, available at http://www.ojp.usdoj.gov/bjs/pub/ascii/soo.txt, accessed September 17, 2007.

3. National Center for Missing and Exploited Children, *2006 Data*, available at http://www.missingkids.com/en_US/documents/sex-offender-map.pdf, accessed September 17, 2007; M. Rice, G. Harris, and V. Quinsey, "Sexual Recidivism Among Child Molesters Released From a Maximum Security Institution," *Journal of Consulting and Clinical Psychology* 59 (1991), pp. 381–386.

4. American Correctional Association, *A Directory of Programs That Work* (Landham, MD: American Correctional Association, 1996), pp. 124–136.

5. R. Borum, "Improving the Clinical Practice of Violence Risk Assessment," *American Psychologist* 3, no. 1 (1996), pp. 945–956.

6. L. Bays, R. Freeman-Longo, and D. Hildebran, *How Can I Stop? Breaking My Deviant Cycle* (Brandon, VT: Safer Society Press, 1990), pp. 19–48.

7. Federal Bureau of Prisons, *Program Participation Package* (Butner, NC: Federal Correctional Institution, 1996), pp. 1–36.

8. American Probation and Parole Association, the Center for Effective Public Management. *Myths and Facts About Sex Offenders*, August 2001, p. 7.

9. L. Bays and R. Freeman-Longo, *Why Did I Do It Again? Understanding My Cycle of Problem Behaviors* (Brandon, VT: Safer Society Press, 1989), p. 72.

Visitation

35

Reginald A. Wilkinson and Tessa Unwin

Chapter Objectives

- Explain the benefits and drawbacks of visitation in a prison setting.
- Identify the challenges presented to institution security by the inmate visiting program.
- Outline the role that visitations play in successful prison and jail administration.

The operation of a visitation program is integral to any correctional system. Hundreds of thousands of relatives and friends visit inmates each year. Experienced correctional managers know that visitation improves the prison environment, so all institutions should encourage visits from family and friends. Visits give inmates something to look forward to, an incentive to participate in rehabilitative programs, and a mechanism to help them cope with prison life. While visitations present challenges to security, an elaborate system of rules and regulations governs the process, and, overall, the benefits of visitations greatly outweigh the potential risks.

■ Benefits

Correctional institutions encourage visiting with family and friends for several reasons. The most important becomes evident after release. The prisoner who has maintained contact with supportive individuals has a safety net when he or she returns to the community. Family and friends provide a feeling of belonging to a group. They often help released offenders seek and find employment and conduct themselves in a positive, constructive manner after release. Newly released prisoners are more likely to see themselves (and be seen by others) with the stigma ascribed to former convicts if they do not maintain desirable social roles while incarcerated. These ex-convict roles are more likely to lead inmates back to criminal behavior.[1]

Visitation is also an incentive for good behavior, providing management with a powerful tool. Prisoners are fully aware that the visiting environment for general population inmates is significantly freer than for those in disciplinary status.

Offender reentry can only be enhanced by positive, nurturing experiences in the visiting room. The idea of family involvement is critical to successful community reintegration. Important, continuous relationships with the friends and family can play a positive role in inmates' rehabilitation. Many benefits can result if visitors maintain relationships following the release of the offender. If the inmate has children, the bonding experience can be continued in a healthy way, serving to lessen the shock and stigma of incarceration. It should be the duty of correctional administrators to help facilitate these visiting opportunities, even if special arrangements must be made.

■ Potential Risks

There are also drawbacks to allowing visits for a confined population, including contraband, illegal activities, and inmate tension.

Visitors are a primary pipeline for the smuggling of drugs and other contraband into a facility. Contraband is defined as anything not allowed into a particular facility and varies depending on the type and security level of that facility. Most systems divide contraband into two categories: major and minor. Major contraband consists of drugs and alcohol, tools, weapons, explosive ordnance, ammunition, currency, and the like, whereas minor contraband often consists of nuisance items such as excessive food. Inmates and their visitors have devised many ingenious ways to attempt to smuggle drugs into jails and prisons. Drugs have been found in many types of food as well as diaper linings, shoe heels, tubes of shampoo and toothpaste, felt-tip markers, stamps, greeting cards, and books. Illegal substances have been thrown over institution fences, taped to trash cans and toilets in the visiting room, sent in with packages of clothing, and left outside crew worksites. As such, significant numbers of alert staff are required, not only to supervise visits carefully but also to conduct background checks on visitors and to search packages or any items given to inmates.

Some prisoners use the visiting process to make contact with potential crime partners or gullible new friends whom they may later use to convey contraband into the prison or conduct illegal activities. In many instances, prisoners first make contact with outsiders through pen pal organizations, and then quickly exploit the friendship. Inmates also abuse the visiting privilege by convincing sympathetic visitors to bring them money or even participate in criminal activity requiring outside assistance. Certain visitors may be partners in crime who help the inmate to continue running illegal street enterprises during incarceration.

Some inmates schedule their visiting in order to get out of job assignments, while others become depressed because they have no visitors at all. A wise correctional manager knows that prison visiting is a sensitive and emotionally charged subject. Just as visiting provides a powerful incentive for good behavior, unfair treatment regarding visits, whether real or perceived, can create undesirable tensions. If a prisoner feels that his or her mother, spouse, or other family member has been unnecessarily hassled or in some way insulted, an outburst may result. Sensitivity training and professionalism are essential for staff involved in the visiting process.

It is incumbent on prison managers to be aware of the climate for visitors, both during the entry procedure and in the visiting room. A correctional officer who abuses his or her position by bullying and humiliating visitors creates a domino effect negatively affecting the visit, the inmates, staff, and prison security. Just as some employees excel in managing inmates or finding contraband, others have the skills and personality to work best with visitors. Corrections officers who know how to balance security and safety with dignity and empathy should be recognized as positive models for other officers.

■ Rules and Regulations

Prison systems have a myriad of rules, regulations, and procedures regarding visits. Rules vary widely according to tradition, security needs, and the availability of staff and visiting space. Informing staff, visitors, and inmates of the rules and regulations ensures a smooth operation; perceived or actual inconsistency and arbitrariness add unnecessary tension to the process. Many agencies include detailed visiting information in early correspondence mailed home by the inmate during the reception process. A courteous, informed, and professional staff can make the visiting experience positive for everyone. Additionally, every effort should be made to inform visitors of restrictions or delays caused by special circumstances (such as fog alerts, extra or extended counts, lockdowns, or disturbances) as soon as possible.

Searches

Searches are imposed to provide adequate safeguards against the introduction of contraband into correctional facilities. All searches should be conducted in a professional manner, without violating the legal rights of visitors and with respect for human dignity. Searches may include a body cavity search, strip search, pat-down, metal detector, or X-ray. In most correctional jurisdictions, strip and body cavity searches must be approved by the warden or designee and are performed only when there is a specific reason to suspect that the visitor has contraband. In most states, only a medical professional is permitted to conduct an intrusive body cavity search. Most corrections departments partner with local law enforcement in the interdiction of drugs or weapons into a prison and the arrest of people known to be carrying drugs. Some systems also utilize drug-detecting canines. High-tech drug detection devices such as ionizer systems are also being introduced into prison drug and explosive detection arsenals. More sophisticated technology is on the horizon.

Visit Terminations

Violation of prison rules may result in the termination of a visit. The visitor may also be suspended or removed from the approved visitation list. Violations include refusing to be searched, possessing contraband, attempting to convey contraband to an inmate, attempting to visit while intoxicated, presenting falsified identification, loaning identification to others, wearing inappropriate clothing, or engaging in prohibited physical contact, sex, or other behavior. If caught bringing illegal contraband into a prison, a visitor may be detained, arrested, and possibly prosecuted. Visitation obviously may be curtailed or terminated in an emergency.

■ Visitation List

Most prisoners develop their visiting list while still in the reception process. Lists generally include family, friends, attorneys, and clergy members. The list names the visitors, their address, phone number, and relationship to the inmate. After the individuals listed are screened and a background check is completed, some correctional systems actually interview visitors to determine their suitability for the prison setting. To avoid problems and conflict, prohibitions to visiting lists usually include known felons, former inmates, parolees and probationers, vendors, and prison volunteers. Some inmates are granted visits with individuals who are not on a visiting list for unusual situations such as to accommodate someone traveling a long distance or to address a family crisis.

General Population Visits

The number, frequency, and duration of visits are limited by space, personnel constraints, scheduling, and security considerations. Upon arriving at the prison, visitors are required to present photo identification and may need to be searched. Visitors are also informed of what constitutes contraband and the sanctions in place to punish those who attempt to convey contraband into the facility.

Conjugal Visits

The argument may be made that allowing conjugal visits between husbands and wives, further strengthens family bonds, helping to minimize the deleterious effects on the family of separation caused by incarceration. A stronger home and family relationship may also smooth the reentry of the offender into the community upon his or her release.

About half of America's prison inmates claim to be married, and six states (California, Connecticut, Mississippi, New Mexico, New York, and Washington) allow conjugal visitation.[2] In some jurisdictions, conjugal visits may be viewed as an unnecessary prisoner privilege and frowned upon by the general public and lawmakers. Jules Burstein compares this American conservatism to the more liberal attitudes in other countries and asserts that the acceptance of conjugal visits in other countries is generally attributable to two factors: a less puritanical and hypocritical attitude toward sex, and a greater emphasis on the family as a primary and vital social unit. Many foreign cultures view conjugal visits, along with home furloughs, as an individual's right.[3]

Visits with Children

Bringing families together is a laudable endeavor, yet one must wonder about the lasting effects on children of seeing a parent in prison. Robert R. Ross and

Elizabeth A. Fabiano have argued that the variety of prison visiting arrangements is a reflection of the complex social and moral issues involved in the question of whether children should be separated from their incarcerated mothers or exposed to prisons, including:

- Effects of such visits on the mother and child when the visit ends
- Feelings of other inmate mothers who do not have contact with their children
- Effects on the child of seeing the often-frightening physical structure of prisons
- Possible long-term effects on children who may live in a prison for short or long periods of time
- Effects of separating children, particularly infants, from their mothers while they are in prison[4]

Most correctional agencies, recognizing the value of keeping families together, try to make the visiting area conducive to child visitations by providing children's reading areas, toys, games, or outdoor picnic or playground areas. Partnerships with area churches, schools and charitable organizations can help maximize the resources available to enhance the visiting experience for children. A positive visit can help children see their incarcerated parent in a more normalized environment, allowing bonding and parenting to continue throughout the period of incarceration and beyond. A strong relationship with their children is a major incentive for many inmates to work toward a positive reentry to the community.

Innovative programs are springing up nationwide to counter the deleterious effects of prison visits on children. For example, the Girl Scouts of America has formed partnerships with some women's prisons to pilot scout troops behind bars. Incarcerated women and their daughters work together on projects, earning merit badges and learning how to be successful teammates. In some women's prisons, overnight and weekend visits are granted as rewards for successful completion of parenting programs. Other states—like Ohio and Nebraska—include unique incentive programs designed to increase an offender's awareness of responsibilities to the family with positive nurturing and interaction.

Legal Representation

The American Correctional Association requires that provisions be made to ensure attorney–client confidentiality, including special arrangements for such communications. These provisions include telephone communications, uncensored correspondence, and visits.[5] Whenever possible, separate visiting rooms are provided for inmates and their attorneys. This courtesy may also be extended to public officials and members of the media.

Concerns with High-Risk Inmates

In the case of death row, administrative segregation, disciplinary detention, and protective custody inmates, security concerns outweigh concerns about family closeness. High-risk prisoners or those in disciplinary housing are often restricted to non contact visits using screens, handcuffs, and leg irons at the discretion of the facility.

■ Role of the Community

Community agencies and businesses may be involved in the visiting process in various ways. Specialized bus companies offer regular charters to prisons from large cities, helping to alleviate the problems caused by long distances. Local businesses such as motels, restaurants, and gas stations benefit from visitors who patronize these establishments. In many communities, volunteer visitors offer friendship to prisoners providing guidance and support while the individual is incarcerated and assistance with finding employment and a place to live after release.[3]

■ Facility Design

The design of visiting areas should allow for adequate supervision and control. Visiting rooms should provide a comfortable visiting environment that is neat and clean, has adequate light and ventilation, and includes separate lavatory facilities for visitors and inmates. Today's visiting rooms must also be fully accessible for those with disabilities.

■ Conclusion

Correctional agencies, prisoners, visitors, and society in general can all benefit from an efficient, humane, and secure visiting program. Regular contact with visitors significantly enhances an inmate's quality of life and establishes a lifeline with the free community. Ties with family members, friends, and other loved ones are critical to inmates' successful return to the community, and visitations help inmates maintain these important relationships. Today's forward-thinking correctional agencies recognize that returning an offender to the community prepared to begin a productive new life is a hallmark of successful corrections. The seamless support of family and friends is crucial to that success.

DISCUSSION QUESTIONS

1. Why do prison administrators encourage visits to inmates by friends and family?
2. What are some of the drawbacks to allowing inmate visitation?
3. What controls can prison administrators put in place to minimize concerns around visitation?
4. In what ways can communities be involved in visitation?
5. Should young children be permitted in institutional visiting rooms? Why or why not?

ADDITIONAL RESOURCES

R. Tewksbury and M. DeMichele, "Going To Prison: A Prison Visitation Program," *The Prison Journal* 85 (2005), pp. 292–310.

NOTES

1. G. Dickinson and T. Seaman, "Communication Policy Changes from 1971–1991 in State Correctional Facilities for Adult Males in the United States," *Prison Journal* 74, no. 3 (1994), pp. 371–382.
2. C. Camp and G. Camp, *The Corrections Yearbook 2002* (Middletown, CT: Criminal Justice Institute, 2003), p. 149.
3. J. Burstein, *Conjugal Visits in Prison* (New York: Lexington Books, 1977), p. 24.
4. R. Ross and E. Fabiano, *Female Offenders: Correctional Afterthoughts* (New York: McFarland and Company, 1986), p. 58.
5. American Correctional Association, "Standard 3-4263," *Standards for Adult Correctional Institutions*, 3rd ed. (Lanham, MD: American Correctional Association, 1990).

Prison Work and Industry

36

Steve Schwalb, Robert C. Grieser, and J.C. Keeney

Chapter Objectives

- Describe the importance of written work policies and the benefits of inmate work assignments.
- Explain the role of administrators in work supervision and the benefits of prison industry programs.
- Outline problems associated with prison industry programs and the laws which have been enacted to address these concerns.

Prison administrators have long known the benefits of having inmates involved in meaningful work or program assignments, such as academic or vocational education. Institutions operate much better when most inmates have a detailed work assignment or a scheduled program to which they must report each day. There is much contentious debate about how prison industry programs should operate, yet such programs continue to thrive because everyone agrees that convicts should work while incarcerated. The benefits of prison industry programs may include:

- Reducing the debilitating effects of idleness and boredom
- Improving the safe management of prisons
- Teaching inmates valuable work skills
- Improving inmates' chances of success upon release

This chapter examines work policy and philosophy as well as the types of work inmates perform and potential concerns around work assignments. It also discusses the prison industry in general, dating back to the early history of corrections, its benefits and drawbacks, as well as the legal framework surrounding it and the public policy question it raises.

■ Work Policies

All prison and jail facilities with work programs must have a written philosophy in place clearly defining the departmental policies concerning inmate employment. Correctional agencies also should have a written departmental policy that delegates to various facilities the authority to develop institutional policies and procedures covering the areas of inmate maintenance assignments. The commissioner or director of the agency must clearly define the agency's philosophical position relating to inmate labor. It is also his or her responsibility to put this position in the form of a departmental policy or directive. These policies must concur with those of the legislative body to whom he or she reports.

In turn, the chief administrative officer of each facility must develop written institutional policies and procedures and make them available to all staff and to the incarcerated population. Such policies help establish the staff philosophy of inmate management, promote consistency of action by staff, and ensure that staff are operating within the scope of their responsibilities and the law. This written document must be based on the departmental philosophy statement and must be understood by all staff and inmates.[1]

A policy document that establishes an inmate work program should include:

- The types of maintenance jobs available
- The skill levels required for each position
- The pay ranges of the individual positions (if there is an applicable inmate pay system)
- Nonmonetary benefits associated with working, such as extra "good time"
- Any other items that merit defining

Privately owned and operated facilities must have a written corporate policy concerning inmate work assignments that allows the individual facilities to develop policies and procedures. When private facilities contract with public entities for inmates, this contract must spell out the expectations of both parties concerning inmate work assignments. Because privately operated facilities are usually contracted on a daily bed cost rate per inmate, it makes economic sense and is good correctional practice to use inmates in as many maintenance assignments as is feasible from a security standpoint.

Staff and inmates must be aware of the procedure by which inmates will be assigned to work details. In some jurisdictions, inmate assignments are under the purview of classification. In some facilities, there are assignment officers who take inmate applications and assign inmates to the various job openings. The system must be consistent, and both staff and inmates must understand how it works.

Inmate talents and skills should be utilized as much as possible. Presentence investigations and inmate history sections may give some indication of work experience and employment history in and outside institutions. In addition, there are a variety of vocational interest and ability tests given during the reception process at the beginning of the incarceration period. It is normally the case worker's or counselor's responsibility to assess an inmate's skill level, vocational training accomplishments, educational level, and work history and make a recommendation as to the type of institutional assignment that would be most appropriate.

■ Types of Work

The types of jobs available at a given facility will vary depending on the facility's security level. Obviously, it is easier to employ more inmates in lower custody level institutions than in maximum security facilities; maintenance assignments in higher security level facilities must be evaluated carefully.

Maintenance work assignments are among the most common jobs inmates perform. They not only help create a healthy atmosphere in the institutional setting but reduce operating costs. Common maintenance assignments include assistance with food service, cleaning, and custodial tasks.

One of the largest users of inmate labor in a facility is the food service department. Within this department there are a variety of positions, including porters, bakers, meat cutters, diet cooks, fry cooks, general cooks, servers, and janitors. There are many opportunities for inmates to start in food service in an entry-level, low-skill position and later advance to a more skilled position.

Inmates should be responsible for cleaning their own living areas. Staff should not have to clean up after inmates. It is degrading to the professional image of staff to have them sweep and mop floors that have been used by the incarcerated population. Staff should clean areas such as control rooms that are restricted from inmate access. Common use areas such as bathrooms, day rooms, and dining rooms should also be cleaned by inmates.

Supervisory staff must be alert and observe these activities because, on occasion, staff will allow overly ambitious inmates into restricted areas to clean, thus relieving employees of this responsibility. This activity can seriously jeopardize the security of the facility. Supervisors need to be observant and tour the facility on all shifts so that they are aware of staff activity. Detailed assignments must be given to inmate janitors so that they clearly understand their areas of responsibility. Staff must supervise housing unit janitors, and precautions need to be taken so that janitors are not able to access other inmates' personal property. In the event there is a problem of theft in the housing units, inmate janitors are instantly suspected. If the problem persists, inmate janitors may be the victims of violence.

Skilled inmates may practice their trades assisting electricians, painters, welders, and heating and air conditioning specialists. Plumbing services also may be provided by inmate maintenance plumbers working under the supervision of security staff. Inmates also may help paint to cover any graffiti or worn paint areas or maintain lawns, shrubs, and the grounds in general.

There are a variety of other areas that can benefit from inmate maintenance workers. The medical department as well as the administrative area of the facility will need inmate janitors. The motor pool can have inmates wash and service vehicles. The education department may use inmate clerks, library aides, teacher aides, janitors, and tutors.

■ History

The idea of having prisoners work dates back to the early history of corrections.[2] To reduce the debilitative effects of incarceration and to better prepare the inmate for employment upon release, inmates were allowed to work by themselves in their cells on shoemaking, tailoring, and other tasks. Thus administrators began to focus on the economics of prison industry—the goal being to generate revenue in excess of the cost of maintaining the prisoner. In the early 19th century, the state of New York, an early leader in many corrections reforms, developed the congregate system, whereby inmates worked together in prison factories under very rigid discipline. This group production technique allowed the manufacture of items that could not be made efficiently by an inmate in a cell. Items produced included carpets, clothing, barrels, and furniture. Products were sold on the open market to American customers or exported, and the proceeds were used to reduce prison operation costs.

As the 1800s progressed, some other states (particularly those in the Midwest and South, where agricultural production was prominent) developed prison labor programs in which inmates were leased to private businesses. The prisons benefited because the lease payments were used to reduce the cost of prison operations. The private businesses gained many long hours of virtually free labor. There were complaints, however, that inmates were abused by contractors. Despite these complaints, the prevalence of convict lease programs increased after the Civil War, as the southern states scrambled to partially replace slavery.

■ The Legal Framework

The open market sale of prisoner-made goods incited free labor and private business to unite to bring about legal restrictions on the sale of inmate products. They managed to eliminate certain prison industry operations. Several states enacted laws that restricted inmate work and training programs. At the federal level, several statutes were passed, including the Sumners-Ashurst Act. Passed in 1940, this law made it a federal crime to knowingly transport convict-made goods in interstate commerce for private use, regardless of what state law allowed.

In 1979, Congress passed the Justice Systems Improvement Act (commonly called the Percy Amendment, after its leading proponent), which permits waivers of the Sumners-Ashurst restrictions on the interstate sale of prison-made goods with several stipulations:

1. Inmates should be paid the prevailing wage with appropriate deductions for taxes, room and board, and court-ordered commitments such as restitution, child support, and alimony.
2. Local labor union officials should be consulted and must approve assignments.
3. Free labor must be determined to be unaffected by such programs.
4. Goods must be produced in an industry with no local unemployment.

This amendment also created the Prison Industry Enhancement certification program (commonly known as the PIE program) wherein the U.S. Justice Department certifies applicant programs for an exception from the Sumners-Ashurst restrictions.

Perhaps the most noteworthy development from the struggles among government and prison industry officials, private businesses, and labor leaders was the state use system. Initiated in New York, this system precludes the sale of prison-made products to the public but promotes their purchase by the state (that unit of government of which the prison industry is an integral part). Today, state use sales are by far the most common market for prison industry products across the country.

Benefits

There are many benefits to prison industry programs. First, they save taxpayers money. Most prison industry programs are at least partially self-sustaining, generating their income from the sale of goods and services. To the extent that these programs can be self-sustaining, a work program can be provided to the inmates without appropriated funds. Thus, the cost to the taxpayers of operating prisons is reduced. Further, some of the wages paid to inmate workers are applied to restitution, fines, child support, and alimony. Inmates also send some of their earnings home to their families. These revenues can reduce outlays for public assistance of various types. In this era of fiscal conservatism, reducing the taxpayer burden is an important attribute of prison industries.

Second, prison industries contribute to the safe management of prisons. By providing productive work and reducing inmate idleness, the presence of prison industry programs reduces the likelihood of disruptions and other violent inmate behavior, so prisons become safer places for staff and inmates. A well-managed prison is also a better neighbor to the host community than a poorly managed prison.

Third, prison industry experience improves inmate success upon release. Several research projects have tracked inmates upon release. One of the more comprehensive was conducted by the Federal Bureau of Prisons; over 7000 inmates with comparable characteristics were evaluated for as long as 12 years following release. The results indicated that inmates who worked in federal prison industries (FPI) while in custody were 20 percent more likely, upon release, to be crime-free, employed, and earning higher wages.[3]

When former inmates are employed and not engaged in criminal behavior, the obvious additional benefits include:

- A reduction in crime
- Improved public safety
- Greater contributions to the gross domestic product
- Increased tax revenues

In addition, prison industries may create jobs for law-abiding citizens, because they require raw materials, supplies, services, and equipment purchased from the private sector, thus creating jobs for various businesses. For example, supervision of inmate workers is provided by staff, whose salaries are normally paid from industrial revenues. Some of the monies paid to inmates are spent in prison commissaries, which stock items procured from local private sector vendors.

In summary, the number of civilian jobs created by prison industry revenues is substantial. This is an important consideration in evaluating the impact of prison industries on the private sector.

■ Concerns

A well-rounded work and program environment can assist greatly in maintaining a safe, secure, healthy institution, but there are several areas of potential concern:

Diversity. It is important that the chief executive officer of the facility have a written policy stating that all work assignments should be filled by inmates of all races or ethnicities, in proportion to the makeup of the inmate population. Traditionally, there have been certain work assignments that inmates of certain races or ethnicities prefer; this should not be allowed. Inmate gangs also attempt to control some work areas of an institution, so the intelligence staff must ensure that gang-affiliated prisoners are tracked closely.

Access to Information. Caution should be exercised to ensure that inmate clerks do not have unmonitored access to computers. Inmates may create all sorts of mischief on a computer system, including illegally accessing staff databases. If inmates are given access to computers, the machines should not be networked, offer access to sensitive information, or have a modem or another way to contact other networks.

Dependence. Inmates should not be assigned to the same job for an extended period of time. They may become possessive, and there is danger of staff becoming overly friendly or dependent upon an inmate who has been on the same assignment for years. Rotating job assignments helps to avoid these problems.

Authority. An inmate should never be put in a position of authority over other inmates. Staff should handle all direction, instruction, and supervision. If inmate supervisors are not well-controlled, prisoners may become abusive in the use of their power.

Manipulation. Putting policies in writing is critical in all aspects of prison and jail administration. Not having written policy and procedures covering work assignments can give the inmate population the perception that the system can be manipulated and that favoritism is the norm. All aspects of the process (e.g., work applications, waiting lists for specific assignments) need to be spelled out in detail.

Quality. The quality of an inmate work program is often ignored or given little attention by managers. This can be a fatal mistake due to the sensitive nature of many of the inmate work assignments. Food service, in particular, needs to be monitored on a daily basis. In the history of corrections in this country, there probably have been more incidents and disturbances relating to inconsistent or poor quality of food preparation or serving than any other single issue. The food service department should never be the dumping ground for malcontents or problematic inmates.

Overload. Most correctional facilities require all able-bodied inmates to have a work assignment. This requirement can easily overload inmate crews. If it only takes five inmates to do a job and ten are assigned, the extra inmates on the crew may make trouble. The facility would be better served working two five-person crews for four hours each rather than having ten inmates on a job doing little or nothing for eight hours.

Supervision. It is important that supervisors and senior managers of an institution remain sensitive to all aspects of the work assignment and selection process as well as the quality of work accomplished on a daily basis. Supervisors should tour the facility each shift to observe and assist staff in this endeavor. Effective chief executive officers are inside the facility regularly to observe and talk to staff and inmates.[4]

There are also arguments against prison industries that center on their adverse impact on the private sector. There are several concerns in this arena. First, antagonists of prison industry program argue that a provision commonly referred to as the mandatory source provides prison industries an unfair advantage. Because many prison industry programs are confined to state use sales, there are often statutes or regulations requiring that the government agencies buy first from the prison industry program. The private sector companies feel that this essentially locks them out of a percentage of the government market, which is unfair and has an adverse impact on their businesses. They advocate competition between prison industries and the private sector for all government business.

Second, critics argue that low inmate wages provide an unfair advantage to prison industries. These critics reject the assertion by prison industry officials that the constraints of prison work programs such as increased civilian supervision, tool control, pat searches, and unskilled workers substantially increase the total overhead costs. With the exception of the aforementioned PIE programs, inmates are paid only a small amount for their labor.

A third argument—that prison industries take too much work away from the community—points to a crucial debate, namely how to determine fairly the share of the market that prison industries should have. There is no magic formula, and the current practice varies widely. On the one hand, for instance, there is essentially no commercial production of license plates in the country. Virtually all are made in prison industries. On the other hand, there are numerous products purchased by the government that prison industries will never make, including computers, aircraft, ships, and weapons systems. However, on more common items such as office furniture, there needs to be a balance between the percentage of government contracts awarded to prison industries and to the private sector.

As the prison inmate populations rises, so will the need for more inmate labor, increasing the probability of elevated market share for prison industries. This elevation will come at a time when many private sector firms are facing increased competition, both domestically and from imports, and when many of them will rely more heavily on sales to the government.

■ Conflicting Mandates

Prison industries have a checkered history and a wide spectrum of opposition still exists. Therefore, it is no surprise that prison industries are deemed a success based on the extent to which they can be the most things to the greatest number of constituents. The FPI statute, for example, requires that FPI diversify production as much as possible to:

- Minimize impact on any one industry
- Employ as many inmates as practicable
- Perform work in a deliberately labor-intensive manner
- Be financially self-sustaining

- Sell only to the federal government
- Produce products that are comparable to those of the private sector in features, quality, and delivery and that do not exceed current market prices
- Produce no more than a reasonable share of the federal government purchases, so as to avoid an undue burden on private sector business and labor
- Teach inmates a marketable skill

Taken individually, almost everyone would agree with these mandates. Putting them together, however, creates what Warren Cikins, a criminal justice consultant and former senior staff member at the Brookings Institution, calls a "convergence of righteousness," describing the tension associated with simultaneously pursuing these competing demands in a balanced way.[5] Ultimately, the debate is not about whether prison industries should exist; rather, it is about the manner in which prisoners should work.

■ Public Policy Questions

There are many public policy questions associated with prison industries, and none of them have clear answers.

Should Prison Industry Programs Be Required to Be Self-Sustaining?

Those who answer "yes" would argue that this requirement encourages prison industries to be more efficient and businesslike. It also reduces the burden on taxpayers, because if the prison industry program were not self-sustaining, more appropriated funds would be required for an alternative program.

Those who answer "no" would suggest that the private sector businesses that compete with prison industries are bearing an unfair share of the burden. They would also suggest that if the manufacturing done by prison industries were turned over to the private sector, the additional taxes collected from the increased sales would offset much if not all of the increased appropriated funds required to fund replacement programs.

Should Prison Industries Be Labor-Intensive?

Idleness reduction is a critical contribution of prison industry programs. The less labor-intensive programs are, the more idleness there will be. On the other hand, as the economy becomes less and less labor-intensive, the extent to which inmates are being provided market-based skills becomes questionable, yet this also argues that the effects of prison labor on private sector jobs wanes with each passing year. Those who support labor intensity in prison industries argue that inmates are being taught a work ethic, which is one of the most important skills that employers desire in new hires.

Should Prison Industries Have a Mandatory Source?

Those in favor of retaining this preference suggest that when it was agreed to limit prison industry sales only to the government, there was consent between private industry and labor that a certain amount of the government's business would be reserved for prison industries. Opponents argue that having a mandatory source keeps the government from getting the most "bang for its buck" through full and open competition.

Should Inmates Be Paid Minimum Wage?

There is no support for inmates being paid minimum or higher wages without deductions for room and board, taxes, and so on. If an employer had to pay minimum wage, the number of inmate jobs would reduce dramatically because the labor costs would be far too high to be competitive. This could affect the total number of inmates employed and increase idleness. An argument can also be made that when the government provides room and board, education, and medical care at no cost to the inmate, lower tax-free inmate wages can be paid and the net income for the inmate is virtually the same as minimum wage with deductions.

Should Prison Industries Be Permitted to Sell Their Products to the Private Sector?

Advocates of this approach (which would essentially constitute a repeal of the Sumners-Ashurst Act) argue that the United States imports products from foreign countries that pay their workers wages comparable to prison industry wages in this country. Some economists would argue that any value added in the economy, regardless of the source, is advantageous. Opponents, however, envision a return to the abuses of the past. Trade experts also question whether statutes and foreign trade treaties should disallow the sale of prison-made goods in the domestic economy.

To What Extent Should the Private Sector Be Allowed to Operate Prison Industries?

Some advocate granting greater access to inmate labor by private industry. Others suggest that private companies should operate the current industrial programs run by the government, injecting the business principles at which the private sector is arguably more adept.

■ Conclusion

The provision and maintenance of a sense of order in the correctional environment are critical to institution management. Staff must provide daily routines for prisoners that facilitate a normal, calm, and stable atmosphere. Work assignments contribute a great deal to this quality of life. Inmates should be assigned to meaningful work assignments to enhance the operations of a correctional institution. Maintenance work assignments for inmates make an institution more cost effective and contribute to other important aspects of corrections including assisting inmates in the process of reentry and teaching a positive work ethic.

The Reentry Policy Council has recommended that training and job assignments in correctional facilities provide meaningful work experience that will assist offenders in preparing for employment opportunities in the free community.

Training and work should be conceived and developed with the job opportunities that will be available for inmates in their home communities in mind. All institutional programs and activities should help inmates prepare for success in the future as they return home.[6]

Prison administrators should understand the critical contributions that prison industry programs make to the safe operation of correctional facilities. Legislators should appreciate that prison industries may reduce costs and make prisons more manageable. The private sector is legitimately concerned that prison industry growth will come at an expense, and there are no easy choices among the various options and competing interests. Successful industrial programs are the byproduct of good communication, quality production programs, and skill in juggling these competing interests.

DISCUSSION QUESTIONS

1. What act allows for the interstate sale of prison-made goods? What are the major provisions of this act?
2. What arguments are presented against prison industry programs?
3. Should legal restrictions exist for the sale of prison made goods?
4. Should prison-made goods be sold in the open market?
5. What role does the private sector played in prison industries?

ADDITIONAL RESOURCES

J. DiIulio, *No Escape: The Future of American Corrections* (New York: Basic Books, 1991).

Federal Prison Industries, http://www.unicor.gov

National Correctional Industries Association, http://www.nationalcia.org

M. Reynolds, *Factories Behind Bars* (Dallas, TX: National Center for Policy Analysis, 1996).

NOTES

1. M. Fleisher, *Warehousing Violence* (Newbury Park, CA: Sage Publications, 1989), p. 85.

2. J. Roberts, *Work, Education and Public Safety: A Brief History of Federal Prison Industries* (Washington, DC: Federal Prison Industries, 1996).

3. W. Saylor and G. Gaes, "Post-Release Employment Preparation Study Summary" (Washington, DC: Federal Bureau of Prisons, 1991, updated June, 1996), pp. 4–8.

4. W. Cikins, Remarks before the Subcommittee on Crime, House Judiciary Committee, May 19, 1994.

5. *Ibid.*

6. Re-Entry Policy Council, "Charting the Safe and Successful Return of Prisoners to the Community. Council of State Governments," (New York: Council of State Governments, January 2005), available at http://www.reentrypolicy.org/reentry/THE_REPORT.aspx, accessed June 2, 2006.

Drug Treatment

37

James A. Inciardi, James E. Rivers, and Duane C. McBride

Chapter Objectives

- Explain why detoxification is a necessary component of effective drug treatment.
- Outline how methadone maintenance assists patients.
- Describe the benefits of residential-based drug treatment programs.

There are several major modalities of substance abuse and addiction treatment available for prison and jail inmates:

- Chemical detoxification
- Methadone maintenance
- Drug-free outpatient treatment
- Self-help groups
- Residential therapeutic communities

Each modality has its own view of substance abuse and addiction, and each affects the patient in different ways.

Chemical Detoxification

Designed for persons dependent on narcotic drugs, chemical detoxification programs typically involve inpatient settings and last between one and three weeks. The rationale for using detoxification as a treatment approach rests on two basic principles:

1. *The concept of addiction.* Addiction is viewed as drug craving accompanied by physical dependence that motivates continued usage. Addiction

results in a tolerance to the drug's effects and a syndrome of identifiable physical and psychological symptoms when the drug is withdrawn abruptly.

2. *The concept of withdrawal.* Negative aspects of the abstinence syndrome discourage many addicts from attempting withdrawal. This makes addicts more likely to continue using drugs.

Given these principles, the aim of chemical detoxification is the elimination of physiological dependence through a medically supervised procedure.

Methadone, a synthetic narcotic that produces many of the same effects as morphine or heroin, is the drug of choice for detoxification from heroin addiction. Generally, a starting dose of the drug is reduced gradually in small increments until the body adjusts to the drug-free state. While many detoxification programs address only the addict's physical dependence, some provide individual or group counseling in an attempt to address the problems associated with drug abuse, while a few refer clients to other longer-term treatments. For drug-involved offenders in prisons and jails, the mechanism of detoxification varies by the client's major drug of addiction. For opiate users, methadone or clonidine is preferred; for cocaine users, desipramine has been used to ameliorate withdrawal symptoms.[1] Almost all narcotic addicts and many cocaine users have been in a chemical detoxification program at least once. Studies document, however, that in the absence of supportive psychotherapeutic services and community followup care, virtually all addicts relapse.

In all detoxification programs, success depends upon following established protocols for drug administration and withdrawal. Research shows increasing rates of program completion, yet many clinicians feel that mere detoxification from a substance is not drug abuse treatment because it does not help people stay off drugs.[2]

From this perspective, for detoxification to be successful, it must be the initial step in a comprehensive treatment process.[3] Thus, detoxification is a temporary regimen that gives addicts the opportunity to reduce their drug intake; for many, this means that the criminal activity associated with their drug taking and drug seeking is interrupted. Finally, given the association between injection drug use and the human immunodeficiency virus (HIV) and acquired immune deficiency syndrome (AIDS), detoxification also provides counseling to reduce AIDS-related risk behaviors.

■ Methadone Maintenance

Methadone maintenance uses methadone as a substitute drug that prevents symptoms of withdrawal from morphine or heroin. More important, however, methadone is orally effective, making intravenous use unnecessary. In addition, it is longer acting than heroin, with one oral dose lasting up to 24 hours. These properties have made methadone useful in the management of chronic addiction.[4] During the first phase of methadone treatment, the patient is detoxified from heroin on dosages of methadone sufficient to prevent withdrawal without either euphoria or sedation. During the maintenance phase, the patient is stabilized on a dose of methadone high enough to eliminate the craving for heroin. Although this process would appear to substitute one narcotic for another, the rationale behind methadone maintenance is to stabilize the patient on a less debilitating drug and make counseling and other treatment services available.

Studies have demonstrated that methadone maintenance patients have favorable outcomes in a number of areas. However, they also indicate that few patients remain drug-free after treatment. More specifically, a number of investigations have found that individuals on methadone maintenance continued to use high levels of such nonopiate drugs as cocaine and marijuana.[5] On the other hand, much of the research has concluded that those on methadone maintenance have been more likely to reduce their criminal activity, become employed, and generally improve in psychosocial functioning.[6] Well-designed programs tend to be integrated with other forms of treatment and social services.[7]

As such, methadone maintenance is effective for blocking heroin dependency. However, methadone is also a primary drug of abuse among some narcotic addicts, resulting in a small street market for the drug. Most illegal methadone is diverted from legitimate maintenance programs by methadone patients. Hence, illegal supplies of the drug are typically available only where such programs exist.

The role of methadone in prison settings is complex, and only a few such programs are available for incarcerated populations. The difficulties seem twofold. First, there is a major security concern. Prison officials are uncomfortable with the general distribution of methadone to potentially large numbers of their inmates. Moreover, many feel that treatment should be "drug-free" and that methadone simply continues drug dependence.

One of the few methadone maintenance programs in a jail is the Key Extended Entry Program (KEEP) at New York City's Rikers Island correctional facility. KEEP's client population includes those awaiting trial as well as sentenced prisoners, with some 3000 receiving treatment each year. The program meets all federal guidelines. An evaluation of KEEP found that those in the program were more likely to continue treatment after leaving jail than those in other types of drug treatment.[8]

Perhaps the major argument for methadone maintenance in a jail setting is that, unlike stays in state and federal prisons, jail stays tend to be short. In KEEP, for example, the treatment period is only 45 days. As such, a strong case can be made for offering methadone maintenance as a means for continuing or initiating treatment for those returning to the street relatively soon.

■ Drug-Free Outpatient Treatment

Drug-free outpatient treatment encompasses a variety of nonresidential programs that do not employ methadone or other pharmacotherapeutic agents. Most have a mental health perspective. Primary services include individual and group therapy, and some programs offer family therapy and relapse prevention support. An increasing number of drug-free outpatient treatment programs are including case management services as an adjunct to counseling. The basic case management approach is to assist clients in obtaining needed services in a timely and coordinated manner. The key components of the approach are assessing, planning, linking, monitoring, and advocating for clients within the existing nexus of treatment and social services.

Evaluating the effectiveness of drug-free outpatient treatment is difficult because programs vary widely—from drop-in centers to highly structured arrangements that offer counseling or psychotherapy. A number of studies have found that outpatient treatment has been moderately successful in reducing daily drug use and criminal activity. However, the approach appears to be inappropriate for the most troubled and the antisocial users.

The number of rigorously designed studies of corrections-based outpatient programs is quite small.[9] One of the few examples involves a relatively well-funded and designed program known as "Passages"—a 12-week nonresidential program for women incarcerated in the Wisconsin correctional system.[10] Although the treatment staff and correctional administrators agreed that the program improved clients' self-esteem and their ability to deal with important issues, evidence of subsequent reduced drug use and criminal activity was not reported.

■ Self-Help Groups

Self-help groups, also known as twelve-step programs, are composed of individuals who meet regularly to stabilize and facilitate their recovery from substance abuse. The best known is Alcoholics Anonymous (AA), in which sobriety is based on fellowship and adhering to the twelve steps of recovery, which include belief in a greater power or being, prayer or meditation, and admission of wrongdoing. The steps move group members from a statement of powerlessness over drugs and alcohol to a resolution that members will carry the message of help to others and will practice the AA principles in all affairs. In addition to AA, other popular self-help groups are Narcotics Anonymous, Cocaine Anonymous, and Overeaters Anonymous. All these organizations operate as stand-alone fellowship programs but are also used as adjuncts to other modalities. Although few evaluation studies of self-help groups have been carried out, the weight of clinical and observational data suggest that they are crucial to recovery.

Research has failed to demonstrate that anonymous fellowship meetings by themselves are effective with heavy drug users.[11] Further, there are few known evaluations of prison-based self-help programs, for a variety of reasons:

- Prison administrators and treatment professionals tend to prefer other types of programs.
- The model contains variables that are extremely difficult to operationalize and measure.
- Members and leaders often view scientific studies of their groups as intrusive threats to anonymity and therapeutic processes.
- Evaluation research funding is more often available for innovative programming than for such well-established services.[12]

Nevertheless, self-help programs are widespread in correctional settings. There is a widely shared belief that they work. The meetings are organized and run by volunteers at no cost to the prison authorities, and the meetings appear to help inmates make the transition from correctional to community-based settings.[13]

■ Residential Therapeutic Communities

The therapeutic community (TC), is a total treatment environment in which the primary clinical staff are typically former substance abusers—"recovering addicts"—who themselves were rehabilitated in therapeutic communities. The treatment perspective of the TC is that drug abuse is a disorder of the whole person—that the problem is the person and not the drug, and that addiction is a symptom and not the essence of the disorder. In the TC's view of recovery, the primary goal is to change the negative patterns of behavior, thinking, and feeling that predispose a person to drug use. As such, the overall goal is a responsible, drug-free lifestyle. Recovery through the TC process depends on positive and negative pressures to change. This pressure is brought about through a self-help process in which relationships of mutual responsibility to every resident in the program are built. In addition to individual and group counseling, the TC process has a system of explicit rewards that reinforce the value of earned achievement. As such, privileges are earned. In addition, TCs have their own rules and regulations that guide the behavior of residents and the management of their facilities. Their purposes are to maintain the safety and health of the community and to train and teach residents through the use of discipline. There are numerous TC rules and regulations, the most conspicuous of which are total prohibitions against violence, theft, and drug use. Violation of these cardinal rules typically results in immediate expulsion from a TC. TCs have been in existence for decades, and their successes have been well documented.

Hillsborough County Sheriff's Office Substance Abuse Treatment Program

The Hillsborough County jail program in Tampa, Florida, was established in 1988 to address the short-term treatment needs of pretrial jail inmates.[14] The program provides services for 60 inmates—48 males in a direct supervision pod, and 12 females housed in a unit with women who are not in treatment. In a treatment milieu emphasizing recovery, cooperation, and interdependence, treatment is provided in groups of 8 to 12 inmates using a cognitive-behavioral, skills-based approach that focuses on relapse prevention. The goals of the program are to encourage long-term abstinence and to involve participants in ongoing treatment services after release from jail. Aftercare is accomplished through linkages with the local programs.

The evaluation of the program examined 535 admissions from June 1988 through January 1991, and 422 untreated "controls" who requested treatment but were not admitted either because of a lack of space or because they were released from jail prior to treatment entry. At two months after release from jail, 16 percent in the treatment group and 33 percent in the control group had been rearrested; at six months, 46 percent in the treatment group and 58 percent in the control group had been rearrested. During the year after release, those in the treatment group had a mean elapsed duration of 221 days prior to rearrest, as compared with 180 days for those in the control group.

Stay 'N Out TC

The Stay 'N Out TC in New York's Arthur Kill Correctional Facility was established in 1974 and follows the traditional TC model.[15] Treatment occurs in prison, and no aftercare services are provided. A follow-up study compared a group of several clients to a no-treatment control group on the following major variables: arrest and parole outcome.[16] For both men and women, the lowest proportion arrested were those in the TC group.

Cornerstone

Cornerstone, founded in 1976 on the grounds of the Oregon State Hospital, is a prerelease TC program with a six-month aftercare program. Clients are referred from the state's three prisons. In a three-year follow-up study, 144 Cornerstone residents who graduated between 1976 and 1979 were compared with three other groups—inmates who dropped out of Cornerstone within 30 days of entry; all Oregon parolees with a history of alcohol or drug abuse who were released during 1974; and a similar population released in Michigan at the same time.[17] The Cornerstone dropouts had the highest rates of recidivism, with 74 percent returning to prison within three years after release. The Michigan group had between 45 and 50 percent who returned, and the Oregon parolees, 37 percent. The Cornerstone graduates had the lowest rate of recidivism, with 29 percent returning to prison within three years after release.

KEY and CREST Outreach Center

Delaware's KEY/CREST program is a three-stage continuum of treatment that begins in the institution and extends to community-based aftercare. Treatment begins in the KEY, a prison-based therapeutic community for male inmates established in 1988 and located at the Multi-Purpose Criminal Justice Facility in Wilmington, Delaware. During the closing months of 1990, CREST Outreach Center was established, also in Wilmington, under a five-year National Institute on Drug Abuse (NIDA) treatment demonstrations grant as the nation's first work release TC. CREST is a transitional facility with a six-month residential program that continues the basic TC treatment approach combined with work release. After six months at CREST, clients proceed to community-based aftercare.[18] Follow-up studies have demonstrated this treatment continuum to be highly effective in that those who have received the full continuum of treatment are three times more likely to be drug-free and almost twice as likely to be arrest-free 18 months after release than the no-treatment controls.[19] This suggests that an integrated continuum of treatment may be a highly promising approach.

Federal Bureau of Prisons' Residential Drug Abuse Treatment Program

The Federal Bureau of Prisons' Residential Drug Abuse Treatment Program attempts to identify, confront, and alter the attitudes, values, and thought patterns that led to criminal behavior and drug or alcohol abuse. The program consists of three stages. First, there is a unit-based treatment within the confines of a prison where prisoners live together and undergo therapy (generally for 9 to 12 months). Second, following completion of the residential portion, inmates continue treatment for up to 12 months while in the general population of the prison through monthly group meetings with the drug abuse program staff. Third, if transferred to community-based facilities prior to release from custody, inmates who completed the residential portion (and the institutional transition portion, if time allows) participate in regularly scheduled group, individual, and family counseling sessions.

The Federal Bureau of Prisons (BOP) conducted a rigorous analysis of their Residential Drug Abuse Treatment Program using a comparison group of inmates who were matched with the inmates who participated in the program on a variety of important background characteristics. The BOP concluded that the program is highly effective. Inmates who participated in the program were significantly less likely to recidivate and significantly less likely to relapse to drug use, for as long as three years after release from prison. The large sample size (1800 offenders), the research methodology, and the multi-site sample make these outcomes particularly noteworthy.[20]

Although TCs and residential treatment programs are the most visible drug abuse treatment programs in prison settings, there are numerous other types, many of which are grounded in individual and group counseling and twelve-step approaches. However, there is limited information about these programs in the substance abuse literature.

■ Conclusion

A legacy of the "war on drugs" has been a criminal justice system that is drug-driven in many respects. In the legislative sector, new laws have been created to deter drug use and increase penalties for drug-related crime. In the police sector, drug enforcement initiatives have been expanded, which, in turn, has increased the number of arrests for drug-related crimes. In the judicial sector, the increased flow of drug cases has resulted in overcrowded dockets and courtrooms and the creation of new drug courts, special dispositional alternatives for drug offenders, and higher conviction and incarceration rates. In the correctional sector, there has been the further crowding of already overpopulated jails and penitentiaries.

In response to this situation, criminal justice systems throughout the United States have been structuring and implementing treatment programs at every level—for arrestees (those released before trial), those on probation, jail and prison inmates, parolees, and those individuals under other forms of post-release surveillance. Although some highly visible programs have been highlighted in the literature, little is known about what is being accomplished in most jurisdictions.

Although it is generally agreed that criminal justice systems throughout the United States are overwhelmed with drug users, little is known about the actual health services and treatment needs of drug-involved offenders. It is important that treatment needs assessments be conducted to determine what services and how many treatment slots are needed by criminal justice clients at all levels. In addition, only minimal attention has focused on the issue of co-morbidity, which might apply to those inmates with dual diagnoses. Additional research is also needed to determine the number and type of treatment services available to probation departments, diversion programs, and other criminal justice entities and how effective these services are, alone or in combination.

DISCUSSION QUESTIONS

1. How is drug addiction defined?
2. Are drug programs effective when operating inside prisons or jails?
3. Should taxpayers fund drug treatment programs?
4. What are they key components of a successful correctional drug treatment program?
5. Why is methadone considered problematic in the prison environment?

ADDITIONAL RESOURCES

M. Anglin, M. Prendergast, and D. Farabee, "The Effectiveness of Coerced Treatment for Drug Abusing Offenders," (Washington, DC: paper presented at the Office of National Drug Control Policy's Conference of Scholars and Policy Makers, 1998), available at http://www.ncjrs.gov/ondcppubs/treat/consensus/anglin.pdf.

J. Inciardi, S. Martin, and C. Butzin, "Five Year Outcomes of Therapeutic Community Treatment of Drug-Involved Offenders After Release from Prison," *Crime & Delinquency* 50, no. 1 (2004), pp. 88–107.

B. Sims, *Substance Abuse Treatment with Correctional Clients: Practical Implications for Institutional and Community Settings* (Binghamton, NY: Haworth Press, Inc., 2005).

B. Wallace, *Making Mandated Addiction Treatment Work* (Blue Ridge Summit, PA: Jason Aronson Publishers, 2005).

NOTES

1. D. Gerstein and H. Harwood, eds., *Treating Drug Problems*, vol. 1 (Washington, DC: National Academy Press, 1990); R. Mattick and W. Hall, "Are Detoxification Programmes Effective?" *The Lancet* 347 (1996), pp. 97–100.

2. Mattick and Hall, "Are Detoxification Programmes Effective?"

3. Gerstein and Harwood, eds., *Treating Drug Problems*, pp. 174–176; D. McBride and C. VanderWaal, "An Evaluation of a Day Reporting Center for Pre-Trial Drug-Using Offenders," *Journal of Drug Issues* 27 (1997), pp. 377–397; S. Magura et al., "The Effectiveness of In-Jail Methadone Maintenance," *Journal of Drug Issues* 93 (1993), pp. 75–99.

4. V. Dole and M. Nyswander, "A Medical Treatment for Diacetylmorphine (Heroin) Addiction: A Clinical Trial with Methadone Hydrochloride," *Journal of the American Medical Association* 193 (1965), pp. 80–84.

5. C. Chambers, W. Taylor, and A. Moffett, "The Incidence of Cocaine Abuse among Methadone Maintenance Patients," *International Journal of the Addictions* 7 (1972), pp. 427–441.

6. J. Ball and A. Ross, *The Effectiveness of Methadone Maintenance* (New York: Springer-Verlag, 1991), p. 202.

7. Ball and Ross, *The Effectiveness of Methadone Maintenance*, p. 162.

8. Magura et al., "The Effectiveness of In-Jail Methadone Maintenance," pp. 75–99.

9. R. Peters, "Drug Treatment in Jails and Detention Settings," in *Drug Treatment and Criminal Justice,* ed. J. Inciardi (Newbury Park, CA: Sage Publications, 1993), pp. 44–80; J. Wellish, D. Anglin, and M. Pendergast, "Treatment Strategies for Drug Abusing Women Offenders," in *Drug Treatment and Criminal Justice*, ed. J. Inciardi (Newbury Park, CA: Sage Publications, 1993), pp. 5–29.

10. G. Falkin et al., *Treating Prisoners for Drug Abuse: An Implementation Study of Six Prison Programs* (New York: Narcotic and Drug Research, Inc., 1991).

11. Falkin et al., *Treating Prisoners for Drug Abuse: An Implementation Study of Six Prison Programs*; Wellish et al., "Treatment Strategies for Drug-Abusing Women Offenders," pp. 6–9.

12. B. Brown, "Program Models," in *Drug Abuse Treatment in Prisons and Jails*, C. Leukefeld and F. Tims (eds), NIDA Research Monograph, no. 118 (Rockville, MD: National Institute on Drug Abuse, 1992), pp. 31–37.

13. Peters, "Drug Treatment in Jails and Detention Settings," p. 65; Brown, "Program Models," p. 35.

14. Peters, "Drug Treatment in Jails and Detention Settings," p. 55.

15. H. Wexler and R. Williams, "The Stay 'N Out Therapeutic Community: Prison Treatment for Substance Abusers," *Journal of Psychoactive Drugs* 18 (1996), pp. 221–229.

16. H. Wexler et al., "Outcome Evaluation of a Prison Therapeutic Community for Substance Abuse Treatment," *Criminal Justice and Behavior* 17 (1990), pp. 71–92.

17. G. Field, "The Cornerstone Program: A Client Outcome Study," *Federal Probation*, June 1985, pp. 50–55.

18. R. Hooper, D. Lockwood, and J. Inciardi, "Treatment Techniques in Corrections-Based Therapeutic Communities," *The Prison Journal* 73 (1993), pp. 290–306; J. Inciardi, "Prison Therapeutic Communities," in *Examining the Justice Process*, J. Inciardi, ed. (Fort Worth, TX: Harcourt Brace, 1996), pp. 397–409.

19. J. Inciardi et al., "An Effective Model of Prison-Based Treatment for Drug-Involved Offenders," *Journal of Drug Issues* 27 (1997), pp. 261–278.

20. B. Pelissier et al., "TRIAD Drug Treatment Evaluation Project," *Federal Probation* 65 no. 3, December (2001).

Chapter Resources

Prisoner Access to the Courts

Kenneth C. Haas

38

Chapter Objectives

- Describe the history of the constitutional right of access to the courts as it pertains to prisoners and the role of the U.S. Supreme Court in prison law and in right-of-access law.
- Outline the pivotal role that the lower federal and state courts play in interpreting and applying the decisions of the Supreme Court.
- Explain the challenges prison and jail administrators face in trying to balance the rights of inmates with the need for institutional safety and security.

Since 1941, the U.S. Supreme Court has announced many important decisions concerning the extent to which prisoners retain a constitutionally protected right of access to the courts. The lower federal and state courts have issued many hundreds of decisions applying the principles articulated by the Supreme Court to cases brought by state and federal prisoners and by inmates in local jails. Prison and jail administrators have been left with the daunting task of complying with these decisions while maintaining the proper balance between inmate rights and the need for institutional safety and security. The purpose of this chapter is to provide an overview of the leading cases and developments in this area of law.

■ Constitutional Right of Access to the Courts

The right of access to the courts may be considered to be the most important of all inmate rights, because it is the right upon which all other rights turn. Without it, most prisoners would be unable to appeal their convictions or sentences in a meaningful way. Furthermore, they would find it virtually impossible to file lawsuits challenging prison policies or conditions that violate their statutory or constitutional rights. These include the Eighth Amendment right to be protected against cruel and unusual punishment, the First Amendment rights to freedom of speech and religion, and the Fifth and Fourteenth Amendment rights to equal protection under the law and to due process protections in prison disciplinary proceedings.

Prior to the 1960s, American courts generally followed a policy of declining jurisdiction in nearly all suits brought by prisoners. Known as the "hands-off doctrine," this policy reflected the traditional view of the prisoner as a "slave of the state" without enforceable rights.[1] As a practical matter, the judiciary's reluctance to become involved in the internal operations of prisons made it extraordinarily difficult for prisoners to seek judicial relief from mistreatment or harsh conditions of confinement.[2]

Generally, courts based refusals to review inmate petitions on one or more of the following rationales:

1. Prison management is exclusively an executive branch function and judicial intervention would violate the separation of powers doctrine.

2. Judicial intervention may subvert prison discipline.

3. Judges lack expertise in penology.

4. Opening the courthouse doors to prisoners would lead to a flood of inmate litigation.

5. Considerations of federalism and comity should compel federal courts to refrain from considering claims brought by state prisoners.[3]

The last of these rationales remained surprisingly strong until the Supreme Court took an active interest in prison litigation in several key cases. *Monroe v. Pape* held that Section 1983 of the Civil Rights Act of 1871 gives federal courts original jurisdiction over claims alleging violations of federal constitutional or statutory rights by state or local officials.[4] *Cooper v. Pate* affirmed that state prisoners could bypass the state courts and bring Section 1983 suits against state correctional officials in federal courts.[5] Until these holdings, the vast majority of state-prisoner lawsuits were brought under the federal habeas corpus statute, which, unlike Section 1983, permits no monetary damages and requires petitioners to go through the time-consuming process of exhausting state judicial remedies.[6]

A strict version of the hands-off doctrine prevailed among most courts until the early 1960s. Nevertheless, in 1941—while the hands-off doctrine remained strong—the Supreme Court for the first time invalidated a prison policy on the ground that it interfered with an inmate's right to send petitions to the courts. In the case of *Ex Parte Hull*, the Court struck down a Michigan prison regulation that required inmates to submit all their legal petitions to prison officials for approval.[7] Whenever prison authorities felt that inmate petitions were frivolous, inaccurate, or poorly written, they would refuse to mail them to the courts. The Supreme Court held that this procedure amounted to an impermissible denial of the right of access to the courts. The justices told prison officials, "Whether a petition for a writ of habeas corpus addressed to a federal court is properly drawn and what allegations it must contain are questions for that court alone to determine."[8]

The *Hull* decision did not precisely identify the source of the right to mail a habeas corpus petition to a court without screening and approval by prison officials. However, in subsequent decisions, the Court found several constitutional sources for protecting the right of access to the courts for prisoners and others who wished to seek judicial resolution of their claims. The following year, in *Cochran v. Kansas*, the Court declared that if state prison officials refused to forward an inmate's appeal of his or her conviction, they violated the Fourteenth Amendment right to equal protection of the law.[9] More frequently, the Court has held that the right of access to the courts is implicit in the due process clauses of the Fifth and Fourteenth Amendments.[10] The Court sometimes has found the right of access to the courts to be included in the First Amendment right to petition the government for redress of grievances.[11]

In the years following *Hull* and *Cochran*, however, most courts remained reluctant to interfere with prison policies restricting inmate access to the courts. Many courts approved such prison practices as refusing to allow prisoners to purchase or receive law books, allowing the confiscation of an inmate's legal documents found in another inmate's cell, refusing to permit a prisoner to type his or her own legal papers, and censoring or withholding legal correspondence between prisoners and attorneys. Moreover, even when prison regulations were more accommodating to the right of access, other factors such as ignorance, illiteracy, and poverty kept prisoners from filing their complaints.[12]

Today's prisoners also are likely to find that legal barriers and personal handicaps can make court access difficult. The right to a state-supplied attorney does not extend to inmate actions attacking prison conditions or to discretionary appeals of a criminal conviction.[13] Most prisoners lack the money to hire attorneys, and only in the rarest cases would an attorney take an inmate's case without a fee or on a contingent fee basis. Consequently, most cases brought by prisoners originate from either the petitioning prisoner or from a jailhouse lawyer (or "writ writer")—a prisoner who claims to have expertise in law and prepares legal documents for fellow inmates. Thus, it is not surprising that the first major right-of-access case after *Hull* involved the limitations that prison officials could place on jailhouse lawyers.

■ Legitimizing Jailhouse Lawyers

In *Johnson v. Avery* (1969), the Supreme Court invalidated prison regulations that prohibited jailhouse lawyers from helping other prisoners with their legal problems.[14] The Court acknowledged that jailhouse lawyers may burden the courts with frivolous complaints and undermine prison discipline by establishing their own power structures and taking unfair advantage of gullible prisoners. These concerns, however, were outweighed by the importance of ensuring that prisoners have reasonable access to the courts. Because most prisoners possess neither the funds to hire attorneys nor the educational background to write their own appeals, their only recourse in most cases, reasoned the justices, was to seek the help of a fellow inmate. Accordingly, the Court concluded that prison officials could no longer enforce no-assistance rules unless the prison itself provided some type of legal services program that was reasonably effective in helping prisoners pursue their legal claims.[15]

The *Johnson* decision, more than any other case, paved the way for more effective efforts by prisoners seeking access to the courts. Following this decision, it became increasingly difficult to escape the logic that if inmates have the right to the assistance of another inmate in the preparation of legal documents, they cannot be absolutely restrained from acquiring the requisite legal materials and due process protection needed to assist in the preparation of petitions or to acquire an attorney or some other type of competent assistance to help them seek an appropriate and speedy judicial remedy.

However, the *Johnson* decision lacked precision. It provided prison officials with only the basic parameters of the right of access. Since 1969, the Supreme Court has resolved several important issues left unsettled by this case. In 1971, the Court affirmed a lower court ruling that required prison officials to provide inmates with an adequate law library—one that contained enough books and materials to ensure that literate prisoners could do meaningful research in support of their petitions.[16] Three years later, the Court struck down a California prison policy that barred law students and legal paraprofessionals from working with prisoners.[17] Also in 1974, the justices invalidated a Nebraska regulation stating that prisoners could seek legal assistance only from a single inmate legal advisor. This advisor had to be appointed by the warden and was permitted to provide assistance in preparing only habeas corpus petitions and not civil rights complaints brought under Section 1983.[18]

■ Requirement of Legal Assistance

Arguably the most important of the Supreme Court's post-*Johnson* decisions was the 1977 case of *Bounds v. Smith*.[19] In this case, the Court held that even when prison policies allow jailhouse lawyers to operate, prison officials nevertheless must provide prisoners with either an adequate law library or adequate legal assistance from persons trained in the law. Like the *Johnson* opinion, the *Bounds* opinion was far from specific in explaining what it would take to provide adequate legal services and materials for prisoners. The *Bounds* majority noted that:

> While adequate law libraries are one constitutionally acceptable method to assure meaningful access to the courts, our decision here . . . does not foreclose alternative means to achieve that goal. . . . Among the alternatives are the training of inmates as paralegal assistants to work under lawyers' supervision, the use of paraprofessionals and law students, either as volunteers or in formal clinical programs, the organization or volunteer attorneys through bar associations or other groups, the hiring of lawyers on a part-time consultant basis, and the use of full-time staff attorneys, working either in new prison legal assistance organizations or as part of public defender or legal services offices.[20]

Not surprisingly, there is still a great deal of confusion and continuing litigation about what prison officials must do to guarantee inmates meaningful access to the courts. Questions involving the adequacy of particular prison law libraries or legal services programs must be answered by the state and federal courts on a case-by-case basis. Most courts have held that *Bounds* is satisfied when states provide inmates with adequate law libraries and access to materials with some quasi-professional help.

The case of *Hooks v. Wainwright* provides an excellent example of the difficulties lower courts (and prison officials) face when determining whether a particular prison legal service program meets the reasonable legal needs of the inmate population and thus satisfies the requirements of *Bounds*.[21] In *Hooks* a federal district court in Florida held that the state's plan to provide prisoners with law libraries staffed by inmate law clerks and librarians was insufficient to guarantee prisoners access to the courts. The court ordered the Florida Department of Corrections to provide some form of attorney assistance as part of its legal services plan. However, this decision was reversed by the Eleventh Circuit Court of Appeals, which held that the lower court had interpreted *Bounds* too broadly and that attorneys were not required. In other words, a combination of law libraries and inmate law clerks will meet the *Bounds* mandate. The Ninth Circuit also approved the use of inmate law clerks rather than lawyers, but it added that the clerks must have received at least some sort of legal training.[22]

Most courts also agree that prison officials do not always have to provide access to an adequate law library to satisfy the *Bounds* requirements. For example, in *Carper v. Deland,* the U.S. Court of Appeals for the Tenth Circuit found that by eliminating prison law libraries and instead employing two private attorneys to help prisoners prepare and file post-conviction appeals and civil rights actions challenging conditions of confinement, the Utah State Prison had done all that was necessary to comply with *Bounds*.[23] In *White v. Kautzky* (2003), a federal district court in Iowa stressed that prisoners' *Bounds* rights were not violated when the Iowa Department of Corrections failed to keep prison law libraries up to date and instead elected to use attorneys to provide legal assistance to inmates.[24] Nevertheless, the court found that inmate Duane White successfully established a genuine issue of material fact—whether the poor performance of these attorneys stymied his effort to bring a legitimate claim to the courts.[25] An up to date prison law library may be essential to ensuring meaningful access to the courts, but prisons in a growing number of states have discontinued their law libraries and switched to a system relying upon attorneys or trained legal assistants to aid inmates in drafting habeas corpus petitions and civil rights claims.[26]

Prison and jail administrators should proceed carefully when considering the elimination of law libraries or any other methods of providing inmates with legal assistance that seem to be working well and have been upheld by the courts. Many judges will not hesitate to reject legal assistance programs that fail to meet the reasonable legal-assistance needs of a particular inmate population. For example, in *Morrow v. Harwell*, a Fifth Circuit panel held that using a weekly bookmobile service accompanied by minimal assistance from two law students in lieu of a law library was not adequate to ensure meaningful access to the courts for the 160 inmates in a county jail in Texas.[27] Similarly, a federal court in Illinois declared that a prison law library that was staffed by two inmate clerks and two inmate trainees, none of whom had received adequate legal training, was insufficient to meet the legal needs of a prison population of over 1200 inmates.[28]

Since *Bounds*, many courts have been sympathetic to claims that prison and jail officials must provide indigent inmates with free paper and pens to use in preparing legal petitions and free stamps to mail documents to courts.[29] Most courts have not yet gone so far as to hold that officials are constitutionally obligated to permit inmates to possess or use typewriters, word processors, or computers to prepare legal documents.[30] However, some courts have held that the denial of typewriters and other supplies should be taken into account in determining the extent to which prisoners are entitled to access to a photocopying machine or to other resources.[31] Although the courts are divided on the issue, some courts have held that if inmates are given access to a photocopier, indigent inmates are entitled to a reasonable number of free photocopies or photocopies at reduced prices.[32]

Implementing the right of access to the courts raises especially difficult problems—both for prison officials and for courts—when the prisoner has been placed in disciplinary segregation or some other type of administrative segregation designed for particularly disruptive or dangerous inmates. It would be quite costly to establish a full-fledged prison law library in every special housing unit in a prison, and releasing the prisoner from restrictive confinement for a visit to the prison's main law library may pose significant security and logistical problems. On the other hand, a prisoner placed in restrictive confinement does not lose his or her right of access to the courts.

In an important 1993 decision, a Third Circuit panel approved a Delaware prison policy that did not permit inmates in a special maximum security unit to go to the prison's main law library.[33] Instead, the maximum security inmates were provided with three types of legal resources and services:

1. A very small satellite law library in the maximum security unit
2. A paging system through which inmates could request photocopies of materials contained in the main law library
3. Varying degrees of legal assistance provided by paralegals who had been trained by an attorney

Although a federal district court found that this combination of services fell below the requirements established in *Bounds*, the appellate court declared that prison officials had done all that was necessary to comply.

The parameters of *Bounds* remain unclear, but at a minimum, it appears that most courts will require either an adequate law library or assistance from persons who have some demonstrable understanding of the legal process.[34] When determining whether the inmates of a prison have the assistance necessary to prepare court documents with reasonable adequacy, courts generally will examine such factors as:

- The number of prisoners entitled to legal assistance
- The types of claims these inmates are entitled to bring
- The number of persons rendering assistance
- The training and qualifications of those who render assistance

At least one clear principle has emerged in the years following *Bounds*. It is unconstitutional for prison or jail officials to retaliate against a prisoner for bringing or pursuing litigation. It is well-established that retaliating against an inmate for filing a lawsuit or an administrative grievance against correctional employees violates the right of access to the courts.[35] To prevail on a claim of retaliation, an inmate must establish that he or she "engaged in conduct protected by the Constitution or by statute, the defendant took an adverse action against [him or her], and this adverse action was taken (at least in part) because of the protected conduct."[36] The adverse action need not be constitutionally prohibited in and of itself. It could range from a relatively minor change in an inmate's job assignment to a major transfer from the general population to administrative segregation or to another prison with more onerous conditions of confinement.[37] Some courts have ruled that once the inmate proves that his or her constitutionally protected conduct was a motivating factor for the adverse action, the burden falls on prison officials to prove by a preponderance of the evidence that they would have taken the same action even if the inmate had not engaged in the protected activity.[38] Other courts, however, place the burden on the inmate to prove that the adverse action would not have occurred but for the retaliatory motive.[39]

■ Imposing the Actual Injury Requirement

Although *Bounds* remains substantially intact, a significant 1996 decision, *Lewis v. Casey,* has narrowed the scope of the prisoner's right to court access in a way that gives prison officials more breathing room in establishing a constitutionally permissible system of providing legal assistance to prisoners by imposing the actual injury requirement.[40] In a 5-to-4 vote, the *Lewis* decision held that an inmate alleging a violation of *Bounds* must show that shortcomings in the prison's library or assistant program caused him or her "actual injury"—that these shortcomings stymied or interfered with his efforts to pursue a legitimate legal claim.

Writing for the majority, Justice Scalia argued that *Bounds* did not create "an abstract, freestanding right to a law library or legal assistance."[41] As Justice Scalia saw it, permitting an inmate to bring a right-of-access claim simply by demonstrating that his prison's library or legal services program is "sub-par in some theoretical sense" would be just as pointless and wasteful of scarce resources as allowing a healthy inmate to bring an Eighth Amendment medical care claim because of the alleged inadequacies of the prison infirmary.[42] Therefore, a prisoner asserting a *Bounds* violation must show not only that the prison library or legal assistance program is deficient but that the deficiencies obstructed his or her efforts to file a legal claim. Scalia wrote:

> [An inmate] might show, for example, that a complaint he prepared was dismissed for failure to satisfy some technical requirement which, because of deficiencies in the prison's legal assistance facilities, he could not have known. Or that he had suffered arguably actionable harm that he wished to bring before the courts, but was so stymied by inadequacies of the law library that he was unable even to file a complaint.[43]

Justice Scalia added that an inmate could not satisfy the actual-injury requirement by pointing to "just any type of frustrated legal claim."[44] Prisoners, he asserted, are entitled to adequate legal assistance only when they bring a nonfrivolous claim that either challenges the conditions of their confinement or attacks their sentences, directly or collaterally. The impairment of any other kind of legal action, Justice Scalia stressed, "is simply one of the incidental (and perfectly constitutional) consequences of conviction and incarceration."[45] In Justice Scalia's words, "*Bounds* does not guarantee inmates the wherewithal to transform themselves into litigating engines capable of filing everything from shareholder derivative actions to slip-and-fall claims."[46]

This part of Justice Scalia's *Lewis* opinion indicates that prison and jail officials have no constitutional obligation to assist inmates in general civil matters. Prison and jail officials, however, would be wise to tread carefully here. Justice Scalia's comments arguably amount to *dicta*—remarks and observations in a judicial opinion about issues that were not argued by the contending parties, were not at issue or under consideration in the case, and therefore do not amount to an actual holding of the Court. In *Lewis*, the Court did not actually hold that inmates have no right of access to the courts in matters such as divorce proceedings, parental rights actions, or workers' compensation claims involving preincarceration injuries. Indeed, one legal commentator recently pointed out that such an interpretation of *Lewis* is inconsistent with the realities of inmates' actual needs of court access:

> [A]n inmate who has been denied any visitation with her children or who was threatened with termination of her parental rights would have a compelling need to gain access to court to contest these matters. Although not directly related to the inmate's living conditions, the rights at stake are arguably of critical importance.[47]

It would make little sense and it would invite unnecessary litigation to deny inmates the legal assistance or library materials they need to deal with legal problems that preceded their imprisonment.

Another controversial part of Justice Scalia's *Lewis* opinion is the assertion that *Bounds* does not require the state to enable inmates to "litigate effectively" once in court.[48] This comment has drawn sharp criticism from legal scholars, but it too is best understood as *dicta*, not an actual holding of the Court.[49] Prison and jail administrators, accordingly, should not read *Lewis* as limiting the duration of the requirement for legal assistance to the preparation of initial pleadings, thus permitting them to ignore an inmate's need to submit or respond to motions, pleadings, or discovery requests.

A close examination of the *Lewis* majority opinion suggests that Justice Scalia wanted to emphasize that *Bounds* was not meant to transform inmates into miniature law firms with access to the same kinds of extraordinary legal resources that real law firms have.[50] He also wanted to stress that if a court found that only a few inmates in a particular prison could show that they could not pursue nonfrivolous litigation because of inadequate legal assistance, the court should remedy only the particular inadequacies and should not issue a systemwide injunction mandating major improvements in every aspect of the prison's law library or legal assistance program.[51]

It is important to consider that Justice Scalia's majority opinion emphasized that it is ultimately the responsibility of courts "to provide relief to claimants, in individual or class actions, who have suffered, or will imminently suffer, actual harm."[52] Obviously, a court cannot provide relief based solely on a prisoner's initial complaint. A prisoner who brings a nonfrivolous claim to court will be required to defend the claim, respond to motions to dismiss or for summary judgment, and to move the case toward judgment and ultimately trial. It makes no sense to interpret *Lewis* as giving prison administrators the power to hinder or thwart nonfrivolous suits as soon as they are filed.

Although the *Lewis* decision imposed a new hurdle for prisoners seeking improvements in prison legal services programs, it did not rescind other well-established concomitants of the right of access to the courts. As noted earlier, it remains illegal for prison staff to retaliate against prisoners for exercising their right of access to the courts. Similarly, *Lewis* does not overrule lower court decisions holding that out-of-state prisoners must be provided with basic legal materials such as statutes and case law from the sending state or with adequate legal assistance relevant to the law of the sending state.[53]

It is especially important to note that *Lewis* did not abrogate the longstanding consensus that prison officials, without good cause, cannot read or interfere with outgoing or incoming inmate-attorney mail or with other legitimate legal mail. In 1974, the Supreme Court, citing the right of access to the courts, upheld the constitutionality of a Nebraska policy that prohibited prison staff from reading mail from an attorney to an inmate, but permitted them to open and inspect legal mail for contraband in the presence of the inmate.[54] Since then, most lower courts have agreed that inmates have a constitutional right not to have their inmate-attorney mail read and that inmates must be given the opportunity to be present when such mail is opened and inspected for contraband.[55] *Lewis v. Casey* has not undermined this consensus.[56] In fact, in an important 2003 case, the U.S. Court of Appeals for the Sixth Circuit held that both inmate-attorney mail and mail from a court constitute legal mail that cannot be opened outside the presence of the inmate.[57]

It would be a mistake, however, to underestimate the impact of *Lewis v. Casey* on prison litigation. Since *Lewis*, federal and state courts have dismissed hundreds of inmate right-of-access claims for failure to meet the actual-injury requirement.[58] There can be little doubt that *Lewis* has reduced both the number of frivolous lawsuits filed by prisoners and the costs of responding to inmate litigation. However, it is also likely that nonfrivolous and meritorious inmate claims have gone unheard as a result of the actual-injury requirement and the difficulties of demonstrating a need for systemwide relief when legal assistance programs are under judicial scrutiny.

■ Prison Litigation Reform Act of 1995

Another development in the mid-1990s was intended to lead to a reduction in prisoner lawsuits. On April 26, 1996, President Clinton signed into law the Prison Litigation Reform Act of 1995 (PLRA).[59] The PLRA was enacted in an effort to respond to what Senator Robert Dole called an "alarming explosion" in the number of cases filed by state and federal prisoners.[60] The law requires, among other things, that before a prisoner can file a civil rights action in a federal court, he or she must exhaust all available administrative remedies, including the prison's inmate grievance system.

The act also introduced a provision commonly known as the "three-strikes" provision. This provision prohibits a prisoner from filing an *in forma pauperis* petition (a petition to permit an indigent litigant to proceed with a case when the court determines that he or she lacks the funds to pay the full filing fee) if the prisoner has filed three or more federal petitions that were dismissed as frivolous or malicious, or if he or she has failed to state a claim for which relief can be granted. Additionally, the act bars prisoners from filing suits seeking damages for mental or emotional injuries suffered while in custody unless they can prove that they also suffered some kind of physical injury.

The PLRA's exhaustion requirement has been widely criticized by legal scholars on the grounds that exhaustion is an unusually complex area of law, that the language of the PLRA is particularly vague, and that the resulting procedural hurdles prevent inmates with nonfrivolous and meritorious claims from bringing those claims to a federal court.[61] The PLRA provides, in part, that prisoners must exhaust "such administrative remedies as are available" before filing a lawsuit challenging prison conditions.[62]

In *Booth v. Churner* (2001), the Supreme Court considered whether this provision applied to prisoners seeking monetary damages for allegedly unconstitutional conditions of confinement even when the only available prison grievance procedures do not permit monetary relief.[63] By a unanimous vote, the Court answered the *Booth* question in the affirmative, reasoning that even if the prisoner is seeking only monetary relief, the prison administrative grievance process might provide some other relief or positive response to the grievance. Thus, before a prisoner can bring suit for monetary damages over allegedly unconstitutional prison conditions, he or she must first complete the prison grievance process. In the aftermath of *Booth*, many questions remain unsettled in this area, and lower courts are dealing with a flood of litigation concerning PLRA-related exhaustion issues.[64]

The three-strikes provision of the PLRA also has generated a good deal of controversy and litigation. Some legal analysts believe that this section of the PLRA is unconstitutional because it has the effect of foreclosing meritorious claims brought by indigent inmates and thus violates the fundamental right of access to the courts.[65] This argument clearly deserves consideration. Under the three-strikes provision, indigent inmates who previously filed three nonmeritorious claims are denied *in forma pauperis* status for future lawsuits, even if their new suits are meritorious. A court can grant an exception to this rule only when the prisoner can prove that he or she is in "imminent danger of serious physical injury," a standard that is vague and extraordinarily difficult to meet.[66]

The Supreme Court has not yet ruled on the question of whether the three-strikes provision violates the right of access to the courts or is unconstitutional in any other way. However, the lower federal courts have dealt with a very important constitutional issue—whether the provision violates the equal protection rights of indigent prisoners in that it prohibits them from filing nonfrivolous and meritorious claims after a third strike while permitting nonprisoners and nonindigent prisoners who also have previously filed three or more nonmeritorious claims to bring new claims. Although many legal scholars have argued that the three strikes policy violates the right of equal access to the courts, most of the federal courts that have addressed this argument have rejected it.[67]

The PLRA's physical injury requirement also has been condemned by many legal scholars but left untouched by the Supreme Court. The physical injury rule stipulates that "No federal civil action may be brought by a prisoner confined in a jail, prison, or other correctional facility, for mental or emotional injury suffered while in custody without a prior showing of physical injury."[68] Legal scholars have denounced the physical injury requirement as having no impact in reducing frivolous or meritless lawsuits, but having an odious impact in preventing prisoners from bringing compelling and clearly meritorious claims to the courts.[69] A leading scholar in corrections law recently declared that the physical injury rule defines inmates as "corporeal persons [with] little more than body and appetite to the exclusion of dignitary interests."[70]

In the absence of guidance from the Supreme Court, the lower courts have struggled with the question of what constitutes a "mental or emotional" injury and whether the physical injury requirement applies to all kinds of constitutional complaints.[71] Some courts have reasoned that the requirement does not apply to certain constitutional claims, such as due process, equal protection, and First Amendment claims. These courts have taken the position that such claims are brought primarily to vindicate the constitutional right in question and thus should not be treated only as claims for mental or emotional injury.[72] Other courts have disagreed, holding that the physical injury requirement does not permit the recovery of compensatory damages for constitutional violations unless the inmate can show that he or she sustained a physical injury.[73] However, these courts generally agree that even in the absence of a physical injury, an inmate who can prove that he or she suffered an emotional or mental injury can obtain declaratory and injunctive relief and is eligible to recover punitive damages (damages awarded to a prevailing plaintiff that are not meant as compensation for an injury or loss and are instead intended to punish and make an example of the defendant).[74]

■ The First Amendment and the Right of Access to the Courts

In *Shaw v. Murphy* (2001), the Supreme Court decided a case that raised the question of whether prisoners have a special right of access to the courts, grounded in the First Amendment, that protects inmate-to-inmate legal correspondence.[75] The complaint was originally brought by Kevin Murphy, a Montana prisoner who served as a law clerk providing legal assistance to fellow prisoners. When Pat Tracy, a prisoner who was housed in the prison's maximum-security unit, was charged with assaulting a guard, Murphy sent him a letter of advice as to how to defend himself in the upcoming prison disciplinary hearing. The letter, which was inspected and read in accordance with prison policies governing inmate-to-inmate correspondence, was found to contain accusations that the guard had repeatedly made homosexual advances towards inmates and had a personal agenda to harass inmates. On the basis of the letter, Murphy was charged with violating prison rules against insolence and interfering with prison due process hearings.[76]

After being found guilty of both charges and receiving a suspended sentence of ten days in segregation and demerits that could affect his custody level and privileges, Murphy filed a suit alleging that his discipline violated, among other things, what he described as his First Amendment right to provide legal assistance to other inmates. Writing for a unanimous Court, Justice Thomas asserted that the question to be resolved was whether anything in the First Amendment or the Court's past precedents on the rights of prisoners guarantees inmates enhanced First Amendment protection simply because legal advice is in correspondence that otherwise violates prison rules.[77] He answered this question in the negative, stressing that there was no basis for giving inmate-to-inmate correspondence that includes legal advice any more First Amendment protection than correspondence that does not include legal advice.[78]

Justice Thomas relied heavily upon the Supreme Court's 1987 decision in *Turner v. Safley*.[79] In *Turner*, the Court made an effort to reconcile contradictory past precedents and formulate a reasonably clear standard of review for cases in which prison officials contend that a regulation or policy that has the effect of giving inmates less constitutional protection than free people enjoy is nevertheless constitutional because of the special needs of institutional safety and security.[80] The *Turner* Court rejected the argument that such rules or policies should be found constitutionally permissible only if prison officials can prove that the rules in question have an important or substantial relationship to safety and security goals.[81] Instead, the Court found that the proper standard to be applied is whether the disputed rules are reasonable and rational: "[W]hen a prison regulation impinges on inmates' constitutional rights, the regulation is valid if it is reasonably related to legitimate penological interests."[82]

The *Turner* majority explained that its "reasonableness" test, rather than a more demanding standard of review, was necessary to ensure that prison officials, not the courts, make the day-to-day difficult decisions that have to be made to operate safe and effective prisons.[83] In *Turner*, the Court used the reasonableness test to review the constitutionality of certain prison regulations in Missouri, one of which barred inmates from corresponding with inmates in other prisons without special approval, unless the correspondence concerned legal matters or was with members of the inmate's family. Finding that the regulation served legitimate security goals and that it would be unreasonable to expect prison officials to read every inmate-to-inmate letter in search of evidence of criminal or gang-related activities, the *Turner* Court held that the regulation did not violate prisoners' First Amendment rights.[84]

In *Shaw v. Murphy*, Justice Thomas emphasized that the Court was not holding that prisoners have no First Amendment rights or no right of access to the courts.[85] The relevant question was whether *Turner* allowed prisoners an increase in constitutional protection whenever their correspondence includes legal advice, and the Court concluded that it did not.[86] Justice Thomas also noted that Tracy had been housed in a maximum-security unit that was off-limits to Murphy and other lower-security prisoners and that prison officials had assigned another inmate law clerk to help Tracy with his legal troubles—thus, Tracy had not been deprived of legal advice.[87] He also noted that the Court has never held that prisoners have a constitutional right to provide legal advice to other prisoners.[88]

In one respect, *Shaw v. Murphy* amounts to something of a setback for prisoners and others who would like the Court to take a more expansive view of the right of access to the courts. The Court's refusal to give special protection to inmate correspondence that includes legal advice arguably gives prison officials greater latitude in limiting the kinds of legal assistance that jailhouse lawyers can dispense to other prisoners.

The negative impact of *Shaw*, however, should not be exaggerated. In *Johnson v. Avery*, the Court acknowledged that jailhouse lawyers "are sometimes a menace to prison discipline," and the majority made it clear that prison administrators could place reasonable time, place, and manner restrictions on jailhouse lawyers and other inmates who offer legal advice or assistance to their fellow inmates.[89] The Supreme Court has never held that illiterate or otherwise disadvantaged inmates have a right to be assisted by a particular jailhouse lawyer or by any other inmate they might choose. The right of access to the courts, as defined by *Johnson v. Avery* and *Bounds v. Smith*, requires only that correctional officials ensure that prisoners have a reasonably adequate opportunity to bring nonfrivolous legal claims to the courts. The Supreme Court has never prescribed any one method for doing this, and, as previously explained, has never held that every constitutionally sound system for providing legal services must include assistance from fellow inmates. *Shaw v. Murphy* did not overrule any of the Court's prior decisions in this area of law and does not change the legal landscape significantly.

In fact, the *Shaw* decision ultimately may place prisoners' right of access to the courts in a more secure and protected constitutional position than it has enjoyed in the past. Critics of the Supreme Court's right-of-access jurisprudence always have been quick to argue that the constitutional source and the precise contours of the right of access to the courts are obscure and unsettled. In *Bounds v. Smith*, Chief Justice Burger and Justice Rehnquist wrote dissenting opinions complaining that "the fundamental constitutional right of access to the courts which the Court announces today is created virtually out of whole cloth with little or no reference to the Constitution from which it is supposed to be derived."[90] Similarly, in *Lewis v. Casey*, Justice Thomas authored a long concurring opinion that challenged the very existence of a constitutional right of access to the courts. He pointed out that over the years, the Court had discovered the source of the "supposed right" in a variety of constitutional provisions, usually the due process or equal protection clauses of the Fourteenth Amendment.[91] "It goes without saying," he declared, "that we ordinarily require more exactitude when evaluating asserted constitutional rights."[92]

In *Lewis*, Justice Thomas raised serious doubts as to the continuing vitality of the constitutional right of access to the courts, and his arguments must be taken seriously. An implied constitutional right—a right that is not specifically mentioned in the Constitution but is found by the Supreme Court to exist in a broad constitutional provision such as the due process clause or the equal protection clause—is inherently more vulnerable to shrinkage or even extinguishment than is a right that is clearly and undeniably spelled out in the Constitution. This is why those who support the Court's controversial *Roe v. Wade* decision to decriminalize most anti-abortion laws are concerned that a future Court may very well overrule *Roe*.[93] The holding is fragile because it is based on the right to privacy—a right that appears nowhere in the Constitution but was found by the *Roe* majority to be implicit in the due process clauses of the Fifth and Fourteenth Amendments.[94]

Until *Shaw v. Murphy*, the right of access to the courts, like the right to privacy, was in considerable peril. However, by granting review of Kevin Murphy's First Amendment claim and by deciding *Shaw v. Murphy* as a case that legitimately raised cognizable questions about the scope of prisoners' First Amendment rights to exchange legal correspondence, the Court arguably has given the right of access to the courts a more secure constitutional foundation than it previously had attained. Moreover, the majority opinion conceded that prisoners have a First Amendment right to exchange legal correspondence, but not a special right to do so—only one that, like all other prisoners' constitutional rights, can be limited by the *Turner* reasonableness test.[95]

Ironically, Justice Thomas' majority opinion in *Shaw v. Murphy* does not identify the particular provision of the First Amendment that provides protection to inmates' legal activities. However, in the aftermath of *Shaw*, the lower courts increasingly have located the prisoner's right of access to the courts in the First Amendment clause that protects the right to petition the government for redress of grievances.[96] (Many courts simply cite the First Amendment generally as the source of the right of access to the courts;[97] some cite both the First Amendment and other constitutional provisions;[98] others cite only the due process clauses;[99] and a few simply refer to the right without specifying a constitutional source.[100])

Robert Tsai has proposed a broad "speech-centered" theory of court access that conceptualizes the right of access to the courts, above all, as a vindication of both the prisoner's and the free person's right to question the legitimacy of government actions.[101] Tsai argues that by treating people's efforts to protect their constitutional rights as the equivalent of anti-government expression, the courts can "breathe new life into the right of access."[102] Tsai's speech-based theory of the right of access goes beyond the right to petition the government for redress of grievances and stresses that the central purpose of the First Amendment is not only to protect dissident speech, but to facilitate "critique and reformation of the state."[103]

It is doubtful that Justice Thomas had any intention to reinvigorate the constitutional right of access to the courts in a manner consistent with Tsai's expansive vision. Nevertheless, the Supreme Court's willingness to analyze Kevin Murphy's right-of-access claim as a First Amendment question constitutes a step in the direction advocated by Tsai. If the Supreme Court follows *Shaw v. Murphy* with a holding that clearly and formally locates the right of access to the courts in the First Amendment, the right of access will be more constitutionally secure than it has been in the past. Moreover, such a holding hardly seems unwarranted. When prisoners or free people bring civil rights claims to the courts, they indisputably are doing nothing more—and nothing less—than petitioning the government for the redress of grievances.

■ Conclusion

In *Ex Parte Hull* (1941), the U.S. Supreme Court recognized that prisoners have a constitutional right of access to the courts. However, the Court has been unclear and inconsistent in pinpointing the constitutional source of the right of access and in delineating its scope. From 1969 to 1977, the Court's decisions had the effect of significantly enlarging the parameters of prisoners' rights in this area of law. The Court's 1977 holding in *Bounds v. Smith*, requiring prison officials to provide inmates with adequate law libraries or other kinds of assistance in bringing nonfrivolous claims to the courts, was an especially important victory for prisoners. Since *Bounds*, the Court has refused to expand the contours of the right of access.

Two developments in the 1990s—the Court's decision in *Lewis v. Casey* (1996) and the enactment of the Prison Litigation Reform Act—unmistakably signaled the end of an era in which prison officials' obligations to provide legal assistance to prisoners were expanded. *Lewis* and the PLRA have resulted in conflicting decisions among lower courts, but the pendulum clearly has swung in the direction of making it more difficult for prisoners to bring their claims to the courts. Some legal commentators are in favor of this trend, pointing out that the number of frivolous or malicious lawsuits filed by inmates has been reduced. On the other hand, it has been argued that many meritorious inmate claims undoubtedly have been thwarted as well.

In *Shaw v. Murphy* (2001), the Supreme Court held that prisoners do not have an unfettered right to engage in legal correspondence with other prisoners. But in doing so, the Court recognized that the First Amendment provides some protection for such correspondence. If the Court eventually makes it clear that the right of access to the courts stems from the First Amendment and, in particular, falls under the First Amendment right to petition the government for redress of grievances, the right of access almost certainly will be safer from judicial or legislative impairment than ever before. In the meantime, it is important to remember that the line of decisions from *Ex Parte Hull* to *Bounds v. Smith* remains substantially intact. As a result, American prisoners continue to retain a constitutionally protected right of access to the courts and prison and jail administrators must comply with these rulings.

DISCUSSION QUESTIONS

1. Why is the question of the constitutional source of the right of access to the courts such an important question?

2. What is the significance of *Johnson v. Avery* (1969) and *Bounds v. Smith* (1977) in expanding the legal rights of prisoners?

3. Do *Lewis v. Casey* (1996) and the Prison Litigation Reform Act of 1995 represent major setbacks for prisoners' rights?

4. Does *Shaw v. Murphy* (2001) signal the willingness of the Supreme Court to locate the right of court access in the First Amendment?

5. Why should prison and jail officials be familiar with contemporary corrections law?

ADDITIONAL RESOURCES

ACLU National Prison Project, http://www.aclu.org/prison/index.html

L.S. Branham, *The Law and Policy of Sentencing and Corrections,* 7th ed. (St. Paul, Minnesota: West, 2005).

Prison Legal News, http://www.prisonlegalnews.org

NOTES

1. *Ruffin v. Commonwealth*, 62 Va. 790, 796 (1871).

2. K. Haas, "Judicial Politics and Correctional Reform: An Analysis of the Decline of the 'Hands-Off' Doctrine," *Detroit College of Law Review* (1977), p. 795.

3. *Ibid.*

4. *Monroe v. Pape*, 365 U.S. 167 (1961).

5. *Cooper v. Pate*, 378 U.S. 546 (1964).

6. K. Haas and G. Alpert, "American Prisoners and the Right of Access to the Courts," in *The American Prison*, L. Goodstein and D. MacKenzie, eds. (New York: Plenum Publishing, 1989), pp. 65–76.

7. *Ex Parte Hull*, 312 U.S. 546 (1941).

8. *Ibid*, 549.

9. *Cochran v. Kansas*, 316 U.S. 255 (1942).

10. *Procunier v. Martinez*, 416 U.S. 396, 419 (1974).

11. *California Motor Transport Co. v. Trucking Unlimited*, 404 U.S. 508, 510 (1972).

12. D. Edmonston, "The Expansion of a Prisoner's Right of Access to the Courts," *Capital University Law Review* 1 (1972), p. 192.

13. *Ross v. Moffitt*, 417 U.S. 600 (1974).

14. *Johnson v. Avery*, 393 U.S. 483 (1969).

15. *Ibid*, 488–490.

16. *Younger v. Gilmore*, 404 U.S. 15 (1971).

17. *Procunier v. Martinez*, 416 U.S. 396, 419-422 (1974).

18. *Wolff v. McDonnell*, 418 U.S. 539, 577-580 (1974).

19. *Bounds v. Smith*, 430 U.S. 817 (1977).

20. *Ibid*, 828.

21. *Hooks v. Wainwright*, 536 F.Supp. 1330 (M.D. Fla. 1982) *rev'd* 775 F. 2d 1433 (11th Cir. 1985).

22. *Lindquist v. Idaho Board of Corrections*, 776 F.2d 851 (9th Cir. 1985).

23. *Carper v. Deland*, 54 F.3d 613 (1995).

24. *White v. Kautzky*, 269 F.Supp. 2d 1054 (2003).

25. *Ibid*, 1060–1063.

26. J. Schouten, "Not So Meaningful Anymore: Why a Law Library is Required to Make a Prisoner's Access to the Courts Meaningful," *William and Mary Law Review*, 45 (2004), p. 1195.

27. *Morrow v. Harwell*, 768 F.2d 619 (5th Cir. 1985).

28. *Walters v. Thompson*, 615 F.Supp. 330 (N.D. Ill. 1985).

29. *Wade v. Kane*, 448 F.Supp. 678 (E.D. Pa. 1978), *aff'd*, 591 F.2d 1338 (3d Cir. 1979).

30. *Roberts v. Cohn*, 63 F.Supp. 2d 921 (N.D. Ind. 1999), *aff'd*, 215 F.3d 1330 (7th Cir. 2000).

31. *Harrell v. Keohane*, 621 F.2d 1059 (10th Cir. 1980).

32. *Giles v. Tate*, 907 F.Supp. 1135 (S.D. Ohio 1995).

33. *Abdul-Akbar v. Watson*, 4 F.3d 195 (1993).

34. C.E. Smith, "Examining the Boundaries of *Bounds*: Prison Law Libraries and Access to the Courts," *Howard Law Journal*, 9 (1987), p. 27.

35. *Gomez v. Vernon*, 255 F.3d 1118 (9th Cir. 2001); *Allah v. Seiverling*, 229 F.3d 220 (3d Cir.2000) (en banc); *DeWalt v. Carter*, 224 F.3d 607 (7th Cir. 2000).

36. *Thaddeus-X v. Blatter*, 175 F.3d 378, 386–387 (6th Cir. 1999).

37. *Rauser v. Horn*, 241 F.3d 330 (2001).

38. *Babcock v. White*, 102 F.3d 267 (7th Cir. 1996); *Graham v. Henderson*, 89 F.3d 75 (2d Cir. 1996).

39. *Peterson v. Shanks*, 149 F.3d 1140 (10th Cir. 1998); *Woods v. Smith*, 60 F.3d 1161 (5th Cir. 1995).

40. *Lewis v. Casey*, 518 U.S. 343 (1996).

41. *Ibid*, 351.

42. *Ibid*.

43. *Ibid*.

44. *Ibid*, 354.

45. *Ibid*, 355.

46. *Ibid*.

47. J. Gerken, "Does *Lewis v. Casey* Spell the End to Court-Ordered Improvement of Prison Law Libraries," *Law Library Journal*, 95 (2003), pp. 491, 498.

48. *Lewis v. Casey*, 518 U.S. 343, 354 (1996).

49. D. Steinberger, "*Lewis v. Casey*: Tightening the Boundaries of Prisoner Access to the Courts," *Pace Law Review*, 18 (1998), pp. 377, 413–417; J. Gerken, "Does *Lewis v. Casey* Spell the End to Court-Ordered Improvement of Prison Law Libraries," pp. 491, 497–502.

50. *Lewis v. Casey*, 343, 351–355.

51. *Ibid*, 356–363.

52. *Ibid*, 343.

53. *Petrick v. Maynard*, 11 F.3d 991 (10th Cir. 1993); *Story v. Morgan*, 786 F.Supp. 523(W.D. Pa. 1992).

54. *Wolff v. McDonnell*, 418 U.S. 539, 577 (1974).

55. *Bieregu v. Reno*, 59 F.3d 1445, 1450–1457 (3d Cir. 1995) (analysis and list of cases); *Royse v. Superior Court*, 779 F. 2d 573, 575 (9th Cir. 1986); *Jensen v. Klecker*, 648 F.2d 1179, 1182 (8th Cir. 1981).

56. *Davis v. Goord*, 320 F.3d 346, 351 (2d Cir 2003); *Cody v. Weber*, 256 F.3d 764, 769 (8th Cir. 2001); *Kalka v. Megathlin*, 10 F.Supp.2d 1117, 1123 (D. Ariz. 1998) (list of cases).

57. *Sallier v. Brooks*, 343 F.3d 868, 873–878 (6th Cir. 2003).

58. *Cody v. Weber*, 256 F.3d 764, 770 (8th Cir. 2001); *McBride v. Deer*, 240 F.3d 1287, 1290 (10th Cir. 2001); *Akins v. United States*, 204 F.3d 1086, 1090 (11th Cir. 2000); K. Westwood, "Meaningful Access to the Courts' and Law Libraries: Where Are We Now?" *Law Library Journal*, 90 (1998), p. 193.

59. Pub. L. No. 104–134, 110 Stat. 1321–66 (1996) codified as amended in U.S.C. sections of 18, 28, and 42.

60. 141 CONG. REC. 14,570 (1995).

61. A. Slutsky, "Totally Exhausted: Why a Strict Interpretation of 42 U.S.C. section 1997e(a) Unduly Burdens Courts and Prisoners," *Fordham Law Review*, 73 (2005), p. 2289; C. Chen, "The Prison Litigation Reform Act of 1995: Doing Away with More Than Just Crunchy Peanut Butter" *St. John's Law Review*, 78 (2004), p. 203.

62. Pub. L. No. 104–134 110 Stat. 1321–66 (1996), 42 U.S.C. section 1997e (a).

63. *Booth v. Churner*, 532 U.S. 731 (2001).

64. *Thomas v. Woolum*, 337 F.3d 720 (6th Cir. 2003); *Kozohorsky v. Harmon*, 332 F.3d 1141 (8th Cir. 2003); *Ortiz v. McBride*, 323 F.3d 191 (2d Cir. 2003).

65. J. Lukens, "The Prison Litigation Reform Act: Three Strikes and You're out of Court—It May Be Effective, But Is It Constitutional?" *Temple Law Review*, 70 (1997), p. 475.

66. Pub. L. No. 104–134, 110 Stat. 1321–66 (1996), 28 U.S.C. section 1915(g); B. Costello, Jr., "'Imminent Danger' Within 28 U.S.C. section 1915(g) of the Prison Litigation Reform Act: Are Congress and Courts Being Realistic?" *Journal of Legislation*, 29 (2002), p. 1.

67. R. Jeffrey, "Restricting Prisoners' Equal Access to the Federal Courts: The Three Strikes Provision of the Prison Litigation Reform Act and Substantive Equal Protection," *Buffalo Law Review*, 49 (2001), p. 1099; J. Franklin, "Three Strikes and You're Out of Constitutional Rights? The Prison Litigation Reform Act's Three Strikes' Provision and Its Effects on Indigents," *University of Colorado Law Review*, 71 (2000), p. 191; *Higgins v. Carpenter*, 258 F.3d 797 (8th Cir. 2001); *Abdul-Akbar v. McKelvie*, 239 F.3d 307 (3d Cir. 2000); *Rodriquez v. Cook*, 169 F.3d 1176 (9th Cir. 1999).

68. Pub. L. No. 104–134, 110 Stat. 1321–66 (1996), 42 U.S.C. section 1997(e).

69. J. Winslow, "The Prison Litigation Reform Act's Physical Injury Requirement Bars Meritorious Lawsuits: Was It Meant to?" *UCLA law Review*, 49 (2002), p. 1655; J. Robertson, "Psychological Injury and the Prison Litigation Reform Act: A 'Not Exactly' Equal Protection Analysis," *Harvard Journal on Legislation*, 37 (2000), p. 105.

70. J. Robertson, "A Saving Construction: How to Read the Physical Injury Rule of the Prison Litigation Reform Act, "*Southern Illinois University Law Review*, 267 (2001), pp. 1, 13.

71. *Perkins v. Kansas Department of Corrections*, 165 F.3d 803 (10th Cir. 1999).

72. *Cannell v. Lightner*, 143 F.3d 1210 (9th Cir. 1998).

73. *Royal v. Kautzky*, 375 F.3d 720, 723 (8th Cir. 2004).

74. *Ibid.*

75. *Shaw v. Murphy*, 532 U.S. 223 (2001).

76. *Ibid*, 225–227.

77. *Turner v. Safley* 482 U.S. 78 (1987).

78. *Ibid*, 230.

79. *Ibid*, 482 U.S. 78 (1987).

80. *Ibid*, 84–86.

81. *Ibid*, 86–89.

82. *Ibid*, 89.

83. *Ibid*, 89–91.

84. *Ibid*, 91–93.

85. *Shaw v. Murphy*, 532 U.S. 223, 228 (2001).

86. *Ibid*, 228–230.

87. *Ibid*, 225, n.1.

88. *Ibid*, 231, n.3.

Chapter Resources

89. *Johnson v. Avery*, 393 U.S. 483, 488, 490 (1969).

90. *Bounds v. Smith*, 430 U.S. 817, 840 (1977) (Rehnquist, J., dissenting).

91. *Lewis v. Casey*, 518 U.S. 343, 366–367 (1996) (Thomas, J., concurring).

92. *Ibid*, 367.

93. *Roe v. Wade*, 410 U.S. 113 (1973).

94. *Ibid*, 152–153.

95. *Shaw v. Murphy*, 532 U.S. 223, 228–231 (2001).

96. *King v. Zamiara*, No. 04-1366, 150 F. Appx. 485, 494 (6th Cir. 2005); *Romansky v. Stickman*, No. 04-3036, 2005 U.S. App. LEXIS 20083 (3d Cir. Sept. 19, 2005); *Chappell v. McCargar*, 152 Fed. Appx. 571 (9th Cir. 2005).

97. *Asad v. Crosby*, No. 04-13825,158 Fed. Appx. 166 (11th Cir. 2005); *Siggers-El v. Barlow*, 412 F.3d 693, 699 (6th Cir. 2005).

98. *Bourdon v. Loughren*, 386 F.3d 88, 92–93 (2d Cir. 2004).

99. *Gomez v. Vernon*, 255 F.3d 1118, 1122 (9th Cir. 2001).

100. *Breshears v. Brown*, No. 04-41749, 150 Fed. Appx. 323 (5th Cir. 2005).

101. R. Tsai, "Conceptualizing Constitutional Litigation As Anti-Government Expression: A Speech-Centered Theory of Court Access," *American University Law Review*, 51 (2002), p. 835.

102. *Ibid*, p. 838.

103. *Ibid*.

Compliance with the Constitution

39

Judith Simon Garrett and Christine D. Salmon

Chapter Objectives

- Compare and contrast the constitutional rights that are most significant to corrections law.
- Describe the effects of the Prison Litigation Reform Act of 1995.
- Identify common legal issues that arise in correctional institution settings and how the courts have responded to these issues.

The old adage that a lawyer who represents himself has a fool for a client is even more true for a non-lawyer. Prison wardens and jail administrators should not assume that their expertise can substitute for the legal training and knowledge of a lawyer. The prison or jail should employ an attorney to be available to consult with the warden or superintendent on an ongoing basis. If this is not possible, an attorney employed by the state, city, or county should be readily available for consultation. It is always preferable to discuss potential problems with an attorney before they arise rather than after a formal complaint has been filed in court. It is also helpful for the institution to have a formal grievance procedure in place. This allows the administration, with advice from an attorney, to attempt to resolve issues, thereby limiting the number of inmate lawsuits. In addition, the facts compiled through the grievance process will facilitate the more expeditious dismissal or resolution of a suit.

This chapter provides a general overview of the types of major legal issues that arise in the corrections setting and how these issues have been resolved by the courts. It would be impossible to address every legal issue in this area, and thus specific legal questions should be referred to an attorney.

■ Constitutional Rights

Inmates constitutional rights are not extinguished upon admission to prison. Indeed, they maintain many rights while in prison and are constrained from exercising other rights only to the extent that to do so would interfere with the safe and orderly operations of the prison. The rights most important to inmates are enumerated in the Constitution. There are also important rights created by federal and state statutes.

Recently, Congress has enacted legislation that limits prisoners' rights in certain areas. In addition, the judiciary has taken a more hands-off attitude toward corrections, similar to one it had prior to the 1960s. For example, the Supreme Court gave the Federal Bureau of Prisons (BOP) considerable discretion in its interpretation of a federal statute granting permission to reduce an inmate's sentence based on participation in a drug treatment program.[1] The Court showed deference to the BOP by finding the BOP's policy of categorically denying eligibility to violent offenders as a permissible construction of the statute in line with congressional intent.[2]

First Amendment

The First Amendment protects freedom of expression. Historically, inmates have been concerned primarily with their right to receive publications, their right to correspond with friends and family, and their right to practice their religion.[3] The general rule on prisoners' First Amendment rights was articulated by the Supreme Court in *Turner v. Safley* and *O'Lone v. Estate of Shabazz,* in which the Court ruled that an inmate's constitutional rights may be restricted if reasonably related to legitimate penological interests.[4] This standard has proven to be quite easy for prison administrators to meet, as courts have defined a host of legitimate penological interests:

- Safety and security
- Rehabilitation
- Budgetary constraints
- Prison order
- Equal opportunity

The courts have ruled repeatedly that the judgment of prison administrators should be given substantial deference, as they are best suited to determine security and safety needs in particular. For example, in *Beard v. Banks,* the Supreme Court applied the framework established in *Turner* to uphold a Pennsylvania Department of Corrections (DOC) policy that prohibits inmates in the most restrictive housing unit from receiving photographs or any publications that are not religious or legal in nature.[5] The Court showed deference to the decision of prison authorities, holding that the DOC's policy is sufficiently related to providing increased incentives for better prison behavior and is not a violation of the inmates' First Amendment rights.

Accordingly, inmates may not be denied all means of communicating with friends, family members, and others in the community. The prison may not suspend or eliminate phone calls, correspondence, and visits simultaneously and indefinitely, except in some very unusual and dangerous situations. However, the courts have permitted prisons to limit inmates' avenues of communication to correspondence and phone calls while prohibiting visiting, or to permit visiting but not phone calls. Moreover, restrictions frequently limit the people with whom inmates may correspond or visit. For example, many states prohibit inmates from corresponding with one another and in few instances are former inmates permitted to return to visit those remaining in prison.

An inmate's right to practice his or her religion was strengthened considerably through the Religious Freedom Restoration Act (RFRA), which was intended to restore the level of religious freedom that preceded the Supreme Court's decision in *Employment Division v. Smith*.[6] Specifically, this act prohibited the government from "substantially burdening" an individual's religious exercise unless such burden was necessary to further a compelling governmental interest and the burden was the least restrictive alternative to further that interest. The RFRA gave rise to a substantial number of lawsuits and resulted in significant changes in the types of inmate religious activities permitted behind prison walls. However, in *Boerne v. Flores*, the Court ruled that RFRA was unconstitutional with respect to application of the law to state and local governments.[7] Presumably, the standard established in *O'Lone v. Estate of Shabazz* will again become the test for determining the constitutionality of prison rules and policies that allegedly interfere with inmates' religious practices in state and local correctional facilities.

Fourth Amendment

The Fourth Amendment prohibits unreasonable searches and seizures. In the community, this constitutional provision is often relied upon to challenge searches of people, their homes, or their cars, particularly in the absence of a search warrant. In prisons, the Fourth Amendment has limited applicability, due to the Supreme Court's repeated rulings that prisoners have a limited expectation of privacy in their cells or in their person.[8] As a general rule, prison administrators are on sound legal ground in routinely searching inmate cells for contraband and requiring inmates to submit to shakedowns for the same purpose. However, such searches would not be proper if conducted for another purpose, such as to harass the inmate, or if exercised in a discriminatory fashion, such as to search only Hispanic inmates.

Prisoners of both genders have used the Fourth Amendment to challenge searches of their person by correctional officers of the opposite gender, but with little success. Specifically, these searches are permissible so long as they are conducted in a professional manner with no sexual connotations or innuendoes.[9] The courts have urged administrators to make some effort to protect inmates' privacy as much as possible, such as permitting only limited viewing of inmates in the shower or while dressing. Courts generally prohibit cross-gender strip searches except under exigent circumstances.[10] It is worth noting that female inmates in Washington State prevailed in objecting to a cross-gender search based on an Eighth Amendment claim of cruel and unusual punishment. An overwhelming number of these inmates had extensive histories of sexual and physical abuse. Thus, they successfully argued that the searches by the male officers caused them substantial emotional trauma.[11]

Recently, inmates have used the Fourth Amendment rights to challenge the constitutionality of federal and state DNA indexing statutes that require the collection of DNA samples from individuals convicted of specific felonies. While the Supreme Court has yet to rule on this issue, state and lower federal courts have held that while the extraction of a DNA sample is an invasion of an inmates' constitutionally protected right to bodily integrity, the search and seizure is permissible because of the states' compelling interest in maintaining a DNA database.[12]

Fifth Amendment

The Fifth Amendment, in conjunction with the Fourteenth Amendment, prohibits the state from depriving any person of life, liberty, or property without affording that person due process of law. Due process is often raised in the context of inmate discipline, inmate classification, and inmate transfer decisions. In 1995, the Supreme Court substantially altered the standard for determining the types of situations for which inmates must be granted due process. In *Sandin v. Conner,* the Court ruled that "discipline in segregated confinement did not present the type of atypical, significant deprivation in which a state might conceivably create a liberty interest," and thus the inmate was not entitled to due process.[13] After *Sandin,* in order to determine whether a liberty interest has been created, thus giving rise to a right to due process, one must ask whether the restraints on the prisoner "impose atypical and significant hardship on the inmate in relation to ordinary incidents of prison life."[14] If they do not, prison administrators may take action without providing due process.

Generally, correctional employees are protected from suits seeking personal liability so long as they act reasonably and within the scope of their employment. Those who behave recklessly or take actions outside the scope of their employment could be held personally liable for violating inmates' statutory or constitutional rights. However, the Supreme Court recently ruled the corrections officers in private firms are not entitled to the qualified immunity enjoyed by state employees in cases where inmates claim violations of their civil rights under Section 1983. In *Richardson v. McKnight*, the Court ruled that Section 1983 was intended to deter "state actors" from depriving persons of their rights, but that private corrections employees are not "state actors" and thus the deterrence argument does not apply.[15]

Eighth Amendment

The Eighth Amendment protects inmates from suffering cruel and unusual punishment. Prisoners rely upon this amendment to challenge conditions of confinement that involve their basic human needs such as medical care, housing, food, clothing, safety, and exercise. The Supreme Court has held that in order for conditions of confinement to violate the Eighth Amendment, they must present an unreasonable risk of harm to the inmates, and the prison administrators must be "deliberately indifferent to these dangerous conditions."[16] The courts find that most cases filed under the Eighth Amendment do not rise to the level of a constitutional violation.

With respect to medical care, the Eighth Amendment prohibits medical care that is so inadequate as to show "deliberate indifference to serious medical need."[17] While the federal government, many states, and some localities enjoy at least some immunity from civil actions seeking monetary damages, and thus cannot be sued under a theory of medical malpractice, many prisons employ medical providers on a contract basis, and these providers could be liable in cases of negligence or other wrongdoing.[18]

Prisoners also rely on the Eighth Amendment to challenge instances where prison officials use force against them. In *Hudson v. McMillan,* the Supreme Court ruled that inmates need not suffer severe physical abuse in order to sustain an Eighth Amendment claim, so long as the force involved the "wanton and unnecessary infliction of pain" or force was used "maliciously or sadistically for the very purpose of causing pain."[19]

Lastly, inmates use the Eighth Amendment in cases regarding the death penalty. While the constitutionality of a death sentence remains firmly in place since the Supreme Court's 1976 decisions upholding the capital punishment statutes in Georgia, Texas, and Florida, the method of the execution is still open to Eighth Amendment challenges.[20] In *Hill v. McDonough* (2006), death row inmate Clarence E. Hill challenged that the method of lethal injection could cause unnecessary pain and was therefore a violation of the Eighth Amendment.[21] Although the Supreme Court did not rule on the merits of Hill's Eighth Amendment claim, the Court unanimously held that the inmate could bring such a claim under section 1983. While the ruling does not immediately affect Eighth Amendment jurisprudence, in that it has not declared a specific method of execution unconstitutional, the decision is significant as it opens the door to future litigation. Hill's claim was remanded to the 11th Circuit for an examination of the merits of his claim.

Section 1983

Specifically, section 1983 of the Civil Rights Act of 1964 creates individual liability for state officials who interfere with prisoners' rights guaranteed by the Constitution or by federal statute. Inmates have used Section 1983 to initiate some of the most significant cases in corrections law.[22] In addition to these federal provisions, some rights created by state statutes may also be significant.

Prison Litigation Reform Act of 1995

The Prison Litigation Reform Act of 1995 (PLRA) contains provisions that:
- Limit inmates' ability to file lawsuits
- Require inmates to exhaust all available administrative remedies prior to filing suits
- Require the payment of full filing fees in some instances
- Impose harsh sanctions (including the loss of good time credits for filing frivolous and malicious lawsuits)
- Require that any damages awarded to inmates be used to satisfy pending restitution orders[23]

Additionally, the PLRA limits the scope of prospective relief that courts may order such that only changes necessary to correct the violation of law are permitted. Even then, courts must ensure that the relief is narrowly drawn, and the courts must consider any potential adverse impact on public safety and prison operations that could result from the relief ordered. Courts are also substantially limited in their ability to grant prisoners early release. The PLRA provides the government with new means to modify or terminate relief that was previously ordered in prison condition cases.

The PLRA and similar sorts of legislation enacted in several states are not intended to inhibit prisoners from challenging prison conditions that fail to afford them basic rights and privileges protected by federal laws and the Constitution. However, the Supreme Court has specified that the PLRA's exhaustion requirement means that an inmate must exhaust all administrative remedies prior to pursuing a formal complaint.[24] The Court has rejected inmates' arguments that such a requirement would be overly harsh for prisoners and noted that the proper exhaustion requirement gives prisoners the incentive to make use of the internal grievance procedure, thereby providing prison administrators with a means to fix their mistakes.

■ Grievance Process

A formal grievance process can be extremely helpful in avoiding lengthy and complicated legal proceedings. This process provides an opportunity for prison administrators to create a clear record of the facts in a particular case for the court and to address a worthy complaint appropriately before litigation begins. For a formal grievance process to be viewed by the courts as credible, it must document the facts behind the grievance and provide an explanation for the action taken.

■ Conclusion

The constitutional and statutory provisions that apply to corrections are both numerous and complex. However, administrators can avoid most serious problems by keeping a few fundamental principles in mind. First, prisoners are human beings that deserve to be treated with respect and dignity. Second, although inmates are deprived of their liberty, they remain citizens of this country and are entitled to most of the rights to which other citizens are entitled. Third, inmates should be granted free access to the courts. Prison administrators should let the courts judge whether inmates' claims have merit, not try to prevent inmates from filing.

DISCUSSION QUESTIONS

1. What restrictions should be permitted on an inmate's First Amendment rights?

2. How does the advent of DNA technology and the promulgation of statutes requiring the submission of a DNA sample from convicted felons fit in with Fourth Amendment jurisprudence?

3. How can corrections staff facilitate the administrative remedies process while allowing inmates to move forward with valid claims?

4. How did the PLRA affect inmates' ability to file lawsuits?

5. What sort of grievances fall under inmates' Eighth Amendment rights?

NOTES

1. *Lopez v. Davis*, 531 U.S. 230 (2001); 18 U.S.C. § 3621(e)(2)(B).

2. *Ibid*, 531 U.S. 241.

3. *Thornburg v. Abbott*, 490 U.S. 401 (1989); *Procunier v. Martinez*, 416 U.S. 396 (1974).

4. *Turner v. Safley*, 482 U.S. 78 (1987); *O'Lone v. Estate of Shabazz*, 482 U.S. 342 (1987).

5. *Beard v. Banks*, 542 U.S. 406 (2004), 126 S.

6. Religious Freedom Restoration Act (RFRA), 42 U.S.C. 2000bb (1993); *Employment Division v. Smith*, 494 U.S. 872 (1990).

7. *City of Boerne v. Flores*, 117 Sup. Ct. 2157 (1997).

8. *Hudson v. Palmer*, 468 U.S. 517 (1984); *Block v. Rutherford*, 468 U.S. 576 (1984); *Bell v. Wolfish*, 441 U.S. 520 (1979).

9. *Bagley v. Watson*, 579 F.Supp. 1099, 1103 (D. Or. 1983); *Timm v. Gunter*, 917 F.2d 1093, 1100 (8th Cir. 1990).

10. J.D. Ingram, "Prison Guards and Inmates of Opposite Genders: Equal Employment Opportunity versus Right of Privacy," *Duke Journal of Gender Law and Policy* 3 (2000).

11. *Jordan v. Gardner*, 986 F.2d 1521 (9th Cir. 1993).

12. *Nicholas v. Goord*, 430 F.3d 652 (2d Cir 2005); *Padget v. Donald*, 401 F.3d 1273 (2005).

13. *Sandin v. Conner*, 515 U.S. 472 (1995).

14. *Ibid*.

15. *Richardson v. McKnight*, 117 Sup. Ct. 2100 (1997).

16. *Wilson v. Seiter*, 111 Sup. Ct. 2321 (1991); *Helling v. McKinney*, 113 Sup. Ct. 2475 (1993).

17. *Estelle v. Gamble*, 429 U.S. 97 (1976).

18. Federal Tort Claims Act, 18 U.S.C., Section 2672.

19. *Hudson v. McMillan,* 112 Sup. Ct. 995 (1992).

20. *Gregg v. Georgia,* 428 U.S. 153 (1976); *Jurek v. Texas,* 428 U.S. 262 (1976); *Proffitt v. Florida,* 428 U.S. 242 (1976).

21. *Hill v. McDonough,* 126 S. Ct. 2096 (2006).

22. Civil Rights Act of 1964, P.L. 88-353, 78 Stat. 241 (1964).

23. P.L. 134, 104th Cong., 2d Session (April 1996).

24. *Woodford v. Ngo,* 548 U.S. Ct. —, 2006 WL 1698937 (U.S.).

Chapter Resources

Rehabilitation

40

James Austin

Chapter Objectives

- Identify the major factors that affect criminal behavior.
- Explain the role of correctional treatment programs in reducing crime.
- Describe common challenges faced by ex-offenders returning to the community.

There is much debate regarding the potential of correctional treatment interventions to reduce crime in general and among offenders who are under the control of the criminal justice system, in particular. But correctional treatment interventions do not operate in a vacuum. Personal, social and economic factors also affect criminal behavior. In addition, the political climate influences whether correctional administrators support rehabilitation and treatment programs, regardless of their efficacy. Given all of these influencing factors, it can be difficult to determine whether rehabilitation programs alone reduce crime.

Factors Relating to Criminal Careers

Criminologists have conducted considerable research that shows that criminal careers for both adults and juveniles do not follow predictable patterns. Many youths who are delinquent during their adolescent years cease their criminal activities by adulthood. Most adult offenders have not had juvenile crime careers (or at least not extensive ones) and are unlikely to continue their criminal behaviors indefinitely. Many factors other than treatment influence patterns of criminal behavior.

The causal factors affecting criminal activity can be grouped into two categories:

1. Structural factors—demographic factors (such as gender and age) that are largely static and cannot be modified

2. Situational factors—societal influences (such as employment, marriage, and societal and economic considerations) that are more dynamic and flexible

Structural Factors

The vast majority of crimes are committed by young men between the ages of 15 and 24.[1] The hereafter rate of offending declines dramatically, so that by age 30, many offenders have effectively "aged out" of crime and are no longer considered high risk. There are, of course, exceptions to this generalization, but these exceptions tend to be adults with extensive juvenile and adult criminal histories who are unable or unwilling to pursue a normative lifestyle.

The fact that most active criminals are young is especially relevant to the significant number of adult prisoners, probationers, and parolees. Many offenders who are under correctional supervision are well above their peak years of criminal behavior and can be expected to significantly reduce their recidivism rates due solely to maturation. With today's longer prison terms, increasing numbers of inmates will be less likely to recidivate due to their age. Several studies have shown that only a small proportion of the crimes committed each year can be attributed to released prisoners or parolees. These data underscore that correctional treatment and punishment initiatives will have minimal, if any, affect on crime rates.

Situational Factors

The ability of a former offender to maintain stable employment (coupled with the aging process) will significantly reduce that person's criminal tendencies. One important study in California found that providing even very modest economic assistance to released prisoners greatly reduced the rates of recidivism.[2]

Another major factor that reduces offenders' probability of continuing their criminal behavior is maintaining a stable and supportive marriage.[3] A stable marriage helps ex-offenders maintain jobs and places to live as well as reduce drug and alcohol abuse.[4]

Criminologists also have examined the societal and economic influences on crime rates. Some factors produce social stress that in turn affects crime rates. States with high rates of violent crimes, mental illness, and suicide tended to have high rates of the following social and economic factors:

- Business failures
- Unemployment claims
- Workers on strike
- Personal bankruptcies
- Mortgage foreclosures
- Divorces
- Abortions
- Illegitimate births
- Infant deaths

- Fetal deaths
- Disaster assistance
- State residency of less than five years
- New houses authorized
- New welfare cases
- High school dropouts

It is also noteworthy that studies did not find an association between incarceration rates and crime rates. In fact, there is a well-established negative correlation (i.e., states with high incarceration rates tend to have high crime rates). Some experts believe that high rates of incarceration may contribute to high levels of social stress and thus increase crime rates.

The Role of Incarceration

Numerous studies have examined the relative effects of incarceration on crime rates in general and on individual offenders. In their pioneering study, Sampson and Laub followed 880 juveniles from adolescence through adulthood and found that neither the number of incarcerations nor the length of incarceration had a direct impact on a person's criminal career.[5] They went on to note that incarcerations actually have a deleterious effect on recidivism, as they severely disrupt efforts to maintain relationships with loved ones and secure stable employment.

A review of numerous studies of the many early release programs that are now operating throughout the country found that neither moderate increases nor moderate decreases (three to six months) in prison terms have an impact on either crime rates in general or an individual offender's rate of reoffending. Again, age is a dominant factor; older inmates (age 35 and above) have by far the lowest rates of recidivism.

In addition, studies have shown little evidence that adjustments in the use of incarceration have an independent impact on crime rates. For example, states with the highest crime rates have the highest incarceration rates, and states with the lowest incarceration rates have the lowest crime rates. Historical fluctuations in crime rates and incarceration rates reveal no clear relationship between incarceration rates and crime rates. However, the tripling of not only the prison population but also the probation, jail, and parole populations over the past 15 years has had some impact on crime rates. At the same time, other developments such as an improving economy, reductions in unemployment and drug use, increases in deportation of illegal aliens, tighter gun control measures, and a growing number of prevention programs sponsored by private sector organizations have substantially reduced crime rates.

The Role of Correctional Treatment

For years criminologists have debated whether correctional treatment helps juvenile and adult offenders. The debate began with a 1974 publication by Robert Martinson that left the unfortunate impression that "nothing works."[6] Martinson's publication was based on a review of existing evaluations of prison treatment programs by himself and his colleagues—one of the first meta-analyses that attempted to summarize the findings of numerous experimental and quasi-experimental studies of rehabilitation programs. This pioneering work has been followed by several other major meta-analyses that reach a different conclusion: that under certain conditions, some treatment interventions can have a significant impact on recidivism rates. In other words, many treatment programs fail, but a sizeable number succeed.[7]

Many have disagreed with these later findings; they argue that the meta-analyses are suspect and overstate the merits of rehabilitation. In particular, the studies cited by these meta-analyses tend to have small sample sizes (under 250 cases for experimental and control groups). Also, in many of the studies, the differences between the recidivism rates of control and experimental subjects were minimal (5 to 10 percent). Furthermore, the recommended conditions necessary for treatment to succeed are difficult to define and replicate in other sites.

Well-designed and -administered correctional treatment programs are the exception rather than the rule. Program integrity is often weak, which may explain the absence of strong treatment effects for many treatment programs. Correctional agencies are often ill-equipped to design and implement effective treatment programs. Most agencies do not themselves believe that effective treatment is possible or that it is part of their mission.[8] A survey showed that most prison wardens believe that only 25 percent of their inmates are amenable to treatment. The wardens also stated that involving inmates in rehabilitation programs was not a high priority for their organization. However, they do view such programs as having an important place in a prison setting.[9]

The Federal Bureau of Prisons (BOP) has concluded, based on thorough research and analysis, that work experience and vocational training programs in federal correctional institutions have significant effects on offenders' ability to successfully reintegrate into the community following release from prison. Specifically, it found that prison programs can have a positive effect on postrelease employment and arrest in the short run, and on recommitment in the long run. The research was based on data from more than 7000 offenders who had been released from prison for 8 to 12 years. Inmates who had worked in federal prison industries or participated in a vocational training program (the study group) were significantly more likely to be employed than inmates who had done neither (the comparison group). The study group inmates were also significantly less likely than the comparison group to recidivate. Inmates who worked in prison industries were 24 percent less likely to recidivate than comparable inmates who had not worked in industries, and inmates who participated in vocational training or apprenticeship training were 33 percent less likely to recidivate than inmates who had not.[10]

Based upon another study that also employed a very rigorous methodology, the BOP has concluded that residential drug abuse treatment has a positive effect on inmates' propensity to recidivate. A 1998 study involving 1800 offenders who had participated in intensive substance abuse therapy for either 9 or 12 months revealed that offenders who completed the program were 15 percent less likely than inmates who did not complete the program to be rearrested during the first three years after reentry into the community. Similarly, graduates of the program were less likely to use drugs following release. The large sample size (1800 offenders), rigorous research design, and multisite sample make these findings particularly noteworthy.[11]

▪ Conclusion

Among the various forms of rehabilitation programs, interventions (or treatment programs) that help equip an individual to secure meaningful employment in today's increasingly competitive economy will be the most successful. Programs that simply offer drug treatment or cognitive learning-based interventions that do not enhance the offenders' ability to perform basic tasks essential for any form of employment are unlikely to reduce recidivism. Further, the private and public sector must recognize the need to provide employment opportunities for this segment of the population.

Based on research to date, the following conclusions can be made regarding the impact of treatment and punishment on crime rates and individual offenders:

- The vast majority of crimes are not committed by persons released from prison. Consequently, prison-based treatment programs and punishment will have little impact on crime rates in general. There is also evidence from one study that crimes committed by probationers do not contribute significantly to a jurisdiction's crime rate.[12]
- Under certain circumstances, treating offenders can have positive results. These positive results are strongest for programs that provide for long-term aftercare and increase the offender's ability to secure employment (part- or full-time).
- Under certain circumstances, punishing offenders can have positive results.
- Under certain circumstances, treating or punishing offenders can have negative results.
- Change (positive and negative) also can occur and often does occur based on other factors that have nothing to do with treatment (e.g., maturation, random events, etc.).
- The vast majority of correctional treatment programs have not been evaluated.
- Most correctional treatment programs are not well-administered, target the wrong clientele, and are too small to have any impact on crime rates or public safety.
- It is extremely rare to find a well-administered treatment program that has been evaluated properly and has demonstrated dramatic treatment effects.
- Factors that will undoubtedly reduce the likelihood of maintaining a criminal lifestyle are age, no juvenile crime career, no history of violence, no evidence of drug use or abuse, the ability to secure employment, and the ability to maintain a meaningful marriage or relationship.
- Treatment and punishment will have only moderate effects on crime rates.

Finally, one must take into account the current political climate. Although considerable debate exists among criminologists and correctional administrators regarding the merits of rehabilitation, there is little if any support for such programs among leading politicians. Rather than supporting the funding of more or different treatment programs, the current climate seems to encourage truth-in-sentencing, three strikes laws, boot camps, chain gangs, lowering of the age for waiving juveniles into adult court, and austere prisons without programs and recreation.

Given this atmosphere, it will take more powerful evidence that treatment reduces rates of recidivism and increases public safety to persuade those who favor a more punitive approach. However, advocates of more and stronger forms of punishment have no conclusive scientific findings to justify their policies either. For this reason alone, the use of rehabilitation and treatment programs that help ensure a more humane and less costly correctional system is warranted and should be expanded. Most important, correctional agencies should become far more accountable (both fiscally and administratively) for such programs and engage in more studies of well-administered programs that better prepare offenders to secure and maintain meaningful employment.

DISCUSSION QUESTIONS

1. What are some strategies that communities might consider to reduce crime?
2. To what extent should correctional treatment programs be judged by the recidivism rate of its graduates?
3. What are the strongest predictors of criminal behavior?
4. What are some of the structural factors that affect criminality?
5. What are some of the situational factors that affect criminality?

ADDITIONAL RESOURCES

National Council on Crime and Delinquency, http://www.nccd-crc.org
National Institute of Justice, http://www.ojp.usdoj.gov/nij
Urban Institute, http://www.urban.org

NOTES

1. T. Hirschi and M. Gottfredson, "Age and the Explanation of Crime," *American Journal of Sociology* 89 (1987), pp. 552–584; Federal Bureau of Investigation, *Age-Specific Arrest Rates and Race-Specific Rates for Selected Offenses* (Washington, DC: U.S. Department of Justice, 1990); T. Flannagan and K. Maguire, eds., *Sourcebook of Criminal Justice Statistics 1989* (Washington, DC: U.S. Government Printing Office, 1990).

2. J. Braithwaite, *Crime, Shame, and Reintegration* (Cambridge, England: Cambridge University Press, 1989); R. Crutchfield, "Labor Stratification and Violent Crime," *Social Forces* 68 (1989), pp. 489–512; R. Sampson and J. Laub, *Crime in the Making: Pathways and Turning Points through Life* (Cambridge, MA: Harvard University Press, 1993); N. Shover, *Aging Criminals* (Beverly Hills, CA: Sage Publications, 1985).

3. Sampson and Laub, *Crime in the Making: Pathways and Turning Points through Life.*

4. T. Gibbens, "Borstal Boys after 25 Years," *British Journal of Criminology* 24 (1987), pp. 49–62; B. Knight, S. Osborn, and D. West., "Early Marriage and Criminal Tendency in Males," *British Journal of Criminology* 17 (1977), pp. 348–360; A. Rand, "Transitional Life Events and Desistance from Delinquency and Crime," in *From Boy to Man: From Delinquency to Crime*, M. Wolfgang, T. Thornberry, and R. Figlio, eds. (Chicago: University of Chicago Press, 1987), pp. 134–162.

5. Sampson and Laub, *Crime in the Making: Pathways and Turning Points through Life.*

6. R. Martinson, "What Works? Questions and Answers about Prison Reform," *The Public Interest* 35 (1974), pp. 22–54.

Chapter Resources

7. D. Andrews et al., "Does Correctional Treatment Work? A Clinically Relevant and Psychologically Informed Meta-Analysis," *Criminology* 28 (1990), pp. 369–404; W. Davidson, et al., *Interventions with Juvenile Delinquents: A Meta-Analysis of Treatment Efficacy* (Washington, DC: National Institute of Juvenile Justice and Delinquency Prevention, 1984); C. Garrett, "Effects of Residential Treatment on Adjudicated Delinquents: A Metanalysis," *Journal of Research in Crime and Delinquency* 22 (1985), pp. 287–308; D. Gendreau and P. Ross, "Revivification of Rehabilitation: Evidence from the 1980s," *Justice Quarterly* 4 (1987), pp. 349–407; R. Gottschalk, et al., "Community-Based Interventions," in *Handbook of Juvenile Delinquency*, H. Quay, ed. (New York: Wiley and Sons, 1987); Lipsey, "The Efficacy of Intervention for Juvenile Delinquency" (Reno, NV: paper presented at the meeting of the American Society of Criminology, 1989); T. Palmer, *The Re-Emergence of Correctional Intervention* (Newbury Park, CA: Sage Publications, 1992).

8. J. Austin, "Using Early Release To Relieve Prison Crowding: A Dilemma in Public Policy," *Crime & Delinquency* 32 (1990), pp. 404–502.

9. F. T. Cullen et al., "The Correctional Orientation of Prison Wardens: Is the Rehabilitative Ideal Supported?" *Criminology* 31 (1993), pp. 69–92.

10. W. Saylor and G. Gaes, "Training Inmates through Industrial Work Participation and Vocational and Apprenticeship Instruction," *Corrections Management Quarterly* 1, no. 2 (1997), pp. 32–43.

11. *TRIAD Drug Treatment Evaluation Project: Six Month Interim Report* (Washington, DC: Federal Bureau of Prisons Office of Research and Evaluation, 1998).

12. M. Geerken and H. Hayes, "Probation and Parole: Public Risk and the Future of Incarceration Alternatives," *Criminology* 31 (1993), pp. 549–564.

Reentry

41

Peter M. Carlson and Lior Gideon

Chapter Objectives

- Understand the concept of reentry and its goals.
- Explore the paradox of corrections as an agent of change.
- Outline possible impediments to reentry and treatment programs.

The Bureau of Justice Statistics (BJS) estimates that federal and state prisons and local jails currently house 2.2 million prisoners.[1] Taking into consideration that 93 percent of all inmates will eventually return home, it is estimated that nearly 700,000 inmates will be released back to their home communities in 2007, at a rate of more than 1700 inmates each day.[2] Unfortunately, soon after these offenders are released, many of them are rearrested and reincarcerated. Studies show that nearly 80 percent of released prisoners are back behind bars within 10 years of release. As such, reentry has become the new buzzword in correctional reform and an important concern (along with rehabilitation and reintegration) for policymakers, criminologists, and the entire field of corrections.[3]

The offender's future, the impact on his or her family, the cost of returning the individual to prison, and the impact on already overcrowded penal facilities are serious outcomes in which all parties have a stake. The success or failure of an inmate's transition back to his or her home community has many significant effects on public safety and health as well. Reentry is obviously most critical to the offender; his or her future turns on the ability to succeed in the free community.

■ Responsibility

Whose responsibility is it when a released offender returns to jail or prison? In many cases, it is very easy to place the blame for failure directly on the offender. After all, according to the classical school of criminology, each individual makes choices in life, and the violator of society's laws has chosen this course of behavior. Many will agree that it is not possible to change another person's attitude or moral standards; an individual either chooses to follow the law or chooses to violate it. The principle of free will asserts that an individual contemplates a variety of actions and then selects the most desirable one by weighing the benefits of committing a criminal act against the negative outcomes of getting caught.

However, according to the positivist school of criminology, other circumstances may interfere with an individual's ability to choose the right course in life. Genetic mental deficiencies, social or economic circumstances, drug or alcohol addiction, or one's choice of associates may cause an individual to override his or her good judgment and create a pathway to crime.

Research shows that many offenders were unemployed prior to arrest, are functionally illiterate, or have some form of mental health problems. Similarly, studies have demonstrated that most inmates are poorly educated, lack vocational skills, struggle with drugs and alcohol abuse, and suffer from some form of mental illness.[4] Despite these issues, offenders are still released back into society, trying to succeed without a system to assist them in dealing with the problems that are often associated with future criminality.

Correctional systems must make every effort to prepare a convicted offender for a successful return to the community. In truth, the process of helping to prepare an inmate for his or her eventual release is a key element in the criminal justice system in the United States. The National Commission on Safety and Abuse in America's Prisons conducted a comprehensive review of correctional facilities and concluded that a reinvestment in programming for prisoners to reduce recidivism is an essential reform for correctional programs and operations nationwide.[5]

In their current state, many prisons and jails are not adequate places for rehabilitation. The current model of prison operations is based solely on incapacitation and not on inmate rehabilitation or preparation for a successful release.[6] However, scholars have argued that even with rehabilitation goals in mind, prisons still fail to effectively deter offenders. In fact, the corrections system resembles a "revolving door of justice."[7] Similarly, as prisons become overcrowded, many correctional officials are forced to release offenders early, thus creating a backfire effect.[8] Policy in the United States, which commits billions of dollars to lock up offenders for increasingly longer periods of time, has proven to be ineffective. Some critics propose that these dollars would be better spent on programs designed to improve the success rate of prisoner reentry. Furthermore, the high recidivism rates should suggest that the existing model for corrections is not solving the crime problem, and if anything, it is only perpetuating the crime crisis. This paradox should be acknowledged, as it lies in the heart of any reintegration and reentry discussion.

◼ Inmate Reentry Programs

Reentry is not a single event, but is a process—a series of interrelated events that culminates in the physical release of the individual to the community. Joan Petersilia defines reentry as the collective impact of all institutional programs and activities that work together to help prepare offenders for a successful return to their home communities where they can live within the law. Petersilia believes that an effective reentry process relies on inmate participation in a continuum of activities during confinement, an effective release process, and the accountability/supervision of the offender once he or she is back in the free community.[9]

The reentry process should begin when a newly convicted prisoner first starts serving a sentence of confinement at a prison facility and has several subsequent stages. The first step, inmate classification or assessment, includes classification of the inmate and staff determination, in partnership with the inmate. It determines what programs may help prepare that inmate for his or her return to the community and outlines the inmate's participation in all planned activities. Essentially, one can argue that an inmate reentry program consists of every program offered in the correctional system in which the inmate participates. Programs such as teaching an individual to read and write, drug abuse treatment, counseling, work assignments, and anger management groups are all good examples of institutional activities and can all be classified as helping prepare for reentry.

Inmate Programming

Several aspects of inmate programming can assist in the process of reentry:
- ◼ Educational and vocational training
- ◼ Drug treatment programs
- ◼ Therapeutic communities
- ◼ Faith-based/religious programs

Studies have concluded that participants who had correction-based education, vocation, and work programs recidivate at a lower rate than non-participants and that such programs may help provide opportunities for a successful path after release.[10] Recent research also suggests that prisoners are more likely to re-offend if they are unemployed, use drugs or abuse alcohol, or have extensive criminal histories.[11] Therapeutic communities have proven to be especially effective in a correctional institution setting due to their characteristic as a total treatment environment that isolates participants from the rest of the prison population, segregating them from drugs, violence, and other aspects of prison life commonly associated with ineffective rehabilitation efforts.[12] Alternatively, faith-based programming may offer inmates a way to cope with the harsh prison environment and ward off negative emotions, though studies examining the effect of religious programming are just developing.

Prerelease Activities

Prerelease programs should focus on the issues the individual will have to deal with in the future such as finding a residence, looking for employment, dealing with expectations of parole officers, etc. These programs are generally conducted in the last 180 days prior to an inmate's release. Effective programs often use practitioners who will work with inmates upon their release, including parole officers, halfway house supervisors, and employers.

Transition

There are many facets in a good transitional program. Inmates being released should have a prearranged residence and community assistance in terms of seeking employment. Mental health inmates should be released with a 30-day supply of any medication that has been part of their treatment regimen, and should have prearranged appointments with a community mental health program.

■ Motivation

The challenge for correctional staff is to motivate the inmate to prepare for his or her future, even when reentry is years away. At least 95 percent of all state prisoners are released from prison at some point, and this fact alone should be enough to justify the development and continuation of rehabilitative programs within the correctional environment.[13]

One might think that inmates generally have poor attitudes and are not motivated to prepare for or make a successful reentry to their home areas. However, research by the Urban Institute found that prior to release, most offenders expressed a strong desire to change and held positive attitudes, especially feelings of high self-esteem and control over life.[14] This evaluation confirms what correctional practitioners already know—the vast majority of incarcerated inmates, if given the opportunity, will engage in self-help programs and work hard at improving their personal situations. Even the most difficult and challenging offenders are often interested in making productive use of their time while confined and find hope and excitement in solid institutional self-help programs.

Researchers have concluded that excellence in institutional leadership results in high-quality correctional operations; the quality of inmate care and security are in direct proportion to the quality of a facility's organizational and management practices.[15] These same factors have a huge influence on the ability of staff to positively influence inmates. Quality inmate program opportunities create a better institutional environment and lead to inmates who are statistically more likely to succeed when they return to their homes. Inmate attitudes are improved by participation in programming, and the outcomes upon reentry are significant.

■ Methods of Release

Parole

The concept of parole began in the 1700s as a means of releasing military prisoners of war and first appeared in the United States in the 1800s. New York first authorized this reduction in time based on meritorious conduct and hard work. Over time, other states adopted this practice of releasing inmates back to the community under the threat that these same inmates would be returned to confinement if again caught in criminal acts.

As the correctional model shifted towards an emphasis on rehabilitation, indeterminate sentencing increased. According to this model, offenders were sentenced to a term of unspecified length, such as four to eight years, with parole release after four years if the individual demonstrated good behavior, completed institution programs, and demonstrated a sincere desire to conduct himself or herself according to the law. However, as society lost faith in the capacity of rehabilitation and the effectiveness of subsequent parole release, state and federal laws shifted toward greater determinacy in sentencing. As such, parole has been discontinued in federal jurisdictions for all offenses that have occurred since November 1984.

Expiration of Sentence

Offenders who complete their entire term of confinement must, by law, be released to their home community or to the jurisdiction of the local area where they were originally sentenced. Inmates who have completed their entire term of confinement are typically released without the requirement of community supervision.

Halfway House

Regardless of whether an inmate is released by parole or by completion of his or her sentence, most correctional jurisdictions will attempt to place the offender in a halfway house for the last three to six months of the term of confinement. These supervised residential institutions allow the offender to decompress from the routine of a prison in a more gradual manner and give the inmate a place to reside as he or she transitions to freedom. Community halfway houses also assist inmates in finding employment and help them with other mundane but necessary tasks such as getting a driver's license. Some offer special assistance to deal with special situations such as drug problems or mental illnesss.

■ Aggravating Factors

Many inmates leaving correctional facilities have complex social and medical problems. Some have lengthy histories of mental illness, and a number have serious and potentially debilitating medical concerns. Others report histories of drug and/or alcohol abuse, while many are functionally illiterate or have learning disabilities and have no plans or capability for meaningful employment. These factors are ample evidence that correctional personnel have a major responsibility to offer rehabilitative and retraining programs to inmates in their care. These programs must be made available to inmates early in their confinement and be of sufficient number and quality to offer all inmates the opportunity to improve their futures.

It is important that correctional authorities target all offenders with program options, especially the more difficult and high-risk group of inmates. Research has demonstrated the efficacy of intervening with the neediest offenders early in their incarceration.[16] Policies and programs are necessary to ensure that the future of our public's safety is served by intervening with these offenders as early as possible.

◼ Conclusion

Former U.S. Supreme Court Chief Justice Warren Burger once said, "We must accept the reality that to confine offenders behind walls without trying to change them is an expensive folly with short-term benefits—winning the battles while losing the war. It is wrong. It is expensive. It is stupid."[17]

Society has yet to realize that merely locking up offenders is a partial solution to the problem of crime. Serving time in prison is not adequate to rehabilitate offenders, and may even interfere with reintegration and successful reentry. According to Ortmann, the prisonization process experienced by inmates during their incarceration prevents them from being rehabilitated while incarcerated.[18] Incapacitation is limited and the results of it have a strong effect on both individual offenders and society. In fact, Elliot Currie argues that the tendency for incarceration to make some criminals worse is one of the best-established findings in criminology.[19] Yet there is strong evidence in support of prison-based treatment and rehabilitation.[20] Prison-based drug treatment programs, therapeutic communities, educational and vocational training, and faith-based programs all show promising results in terms of their effect on rehabilitation, reintegration and successful reentry.

As such, policymakers should invest more resources in treating and rehabilitating offenders while they are incarcerated. This investment will help ease the process of reentry and provide released offenders with true opportunities to rehabilitate and reintegrate back into the normative non-criminal community as functioning members. Inside correctional facilities, reentry planning must commence early and be a guiding principle during confinement.

Today's challenging inmate has a very high chance of becoming tomorrow's failure as offenders legally exit the secure perimeters of correctional institutions. The task of correctional professionals is to use every possible means to turn the statistics in favor of each individual inmate's success and end the revolving door phenomenon. The safety and well-being of all of our communities depend upon it.

DISCUSSION QUESTIONS

1. How would you respond to a citizen who believes rehabilitative programs are a total waste of tax dollars?

2. What should serve as the primary goal of confinement: punishment or rehabilitation?

3. When should planning for an inmate's release begin?

4. What impediments may be faced by released offenders in their effort to reintegrate back into society?

5. Do you agree with this statement by former Supreme Court Chief Justice Warren Burger that, "We must accept the reality that to confine offenders behind walls without trying to change them is an expensive folly with short-term benefits—winning the battles while losing the war." Why or why not?

ADDITIONAL RESOURCES

American Correctional Association, *Reentry Today: Programs, Problems, and Solutions* (Alexandria, VA: American Correctional Association, 2006).

Prisoner Reentry Institute at John Jay College of Criminal Justice, available at http://www.jjay.cuny.edu/centersinstitutes/pri/pri.asp

Rehabilitation and Reentry, The National Institute of Justice, available at http://www.ojp.usdoj.gov/reentry/publications/inmate.html

R. Seiter, "Inmate Reentry: What Works and What to Do About It," *Corrections Compendium* 29, no. 1, January/February 2004.

NOTES

1. Bureau of Justice Statistics, "Reentry Trends in the United States" available at http://www.ojp.usdoj.gov/bjs/reentry/reentry.htm, accessed May 1, 2007.

2. J. Petersilia, *When Prisoners Come Home: Parole and Prisoner Reentry* (New York: Oxford University Press, 2003); B. Krisberg and S. Marchionna, "Attitudes of U.S. Voters toward Prisoner Rehabilitation and Reentry Policies" available at http://www.nccd-crc.org/nccd/pubs/2006april_focus_zogby.pdf, accessed on May 9, 2007; J. Travis, *But They All Come Back: Facing the Challenges of Prisoner Reentry* (Washington DC: The Urban Institute Press, 2005).

3. J. Austin, "Prisoner Reentry: Current Trends, Practices, and Issues," *Crime & Delinquency,* 47 no. 3, (2001): 314–334.

4. J. Petersilia, *When Prisoners Come Home: Parole and Prisoner Reentry;* J. Travis, *But They All Come Back: Facing the Challenges of Prisoner Reentry.*

5. Commission on Safety and Abuse in America's Prisons, "Confronting Confinement" (Washington, DC: Vera Institute of Justice, 2006).

6. R. Seiter, "Inmates Reentry: What Works and What to Do About It" in *Reentry Today: Programs, Problems, and Solutions* (Alexandria, VA: American Correctional Association, 2006), pp. 77–90.

7. J. Petersilia, *When Prisoners Come Home: Parole and Prisoner Reentry;* R. Freeman, "Can We Close the Revolving Door?: Recidivism vs. Employment of Ex-Offenders in the United States," *Urban Institute Reentry Roundtable*, May, 2003; J. Travis, *But They All Come Back: Facing the Challenges of Prisoner Reentry.*

8. S. Walker, *Sense and Non-Sense about Crime and Drugs: A Policy Guide* (Washington DC: The Urban Institute Press, 2006).

9. J. Petersilia, *When Prisoners Come Home: Parole and Prisoner Reentry.*

10. E. Tischler, "Making a Place of Change," *Corrections Today* 61, no. 4, p. 74; C. Davis, *Education: A Beacon of Hope for the Incarcerated. Enhancing Public Safety* (Washington, DC: U.S. Department of Justice Office of Community Oriented Policing Services, December 2000) available at http://www .urban.org/UploadedPDF/411061_COPS_reentry_monograph.pdf, accessed May 9, 2007.

11. N. La Vigne, A. Solomon, and K. Beckman, *Prisoner Reentry and Community Policing: Strategies for Enhancing Public Safety* (Washington, DC: U.S. Department of Justice, Office of Community Oriented Policing Services, 2006) available at http://www.urban.org/UploadedPDF/411061_COPS_ reentry_monograph.pdf, accessed May 9, 2007; J. Petersilia, *Prisoner Reentry: Public Safety and Reintegration* (New York: Oxford University Press, 2001); J. Petersilia, *When Prisoners Come Home: Parole and Prisoner Reentry;* J. Travis, *But They All Come Back: Facing the Challenges of Prisoner Reentry.*

12. M. Prendergast and H. Wexler, "Correctional Substance Abuse Treatment Programs in California: A Historical Perspective," *The Prison Journal* 84, no. 1 (2004), pp. 8–35.

13. T. Hughes and D.J. Wilson, *Reentry Trends in the United States: Inmates returning to the community after serving time in prison.* U.S. Department of Justice, Bureau of Justice Statistics, available at http:// www.ojp.usdoj .gov/bjs/reentry/reentry.htm, accessed June 3, 2006.

14. C. Visher, N. LaVigne, and J. Travis *Returning Home: Understanding the Challenges of Prisoner Reentry. Maryland Pilot Study: Findings from Baltimore* (Washington DC: The Urban Institute Press, 2004).

15. P. Carlson, "Something to Lose: A Balanced and Reality-Based Rationale for Institutional Programming," in *Crime and Employment: Critical Issues In Crime Reduction For Corrections*, J. Krienert and M. Fleisher, eds. (Walnut Creek, CA: Altamira Press, 2004).

16. C. Lowencamp and E. Latessa, "Understanding The Risk Principle," in *Topics In Community Corrections* (Washington, DC: U.S. Department of Justice, National Institute of Justice, 2004).

Chapter Resources

17. Quoted in J. Petersilia, *When Prisoners Come Home: Parole and Prisoner Reentry*, p. 93.

18. R. Ortmann, "The Effectiveness of Social Therapy in Prison: A Randomized Experiment," *Crime & Delinquency,* 23, (2000), pp. 591–601.

19. E. Currie, "Rehabilitation Can Work," in *Exploring Corrections: A Book of Readings*, T. Gray, ed. (Boston, MA: Allyn & Bacon, 2002).

20. D. Andrews, "Does Correctional Treatment Work? A Clinically Relevant and Psychologically Informed Meta-Analysis," *Criminology* 28, no. 3, (1990), pp. 369–404; F. Cullen, "Attribution, Salience, and Attitudes Toward Criminal Sanctioning," *Criminal Justice and Behavior* 12, no. 3 (1985), pp. 305–331; P. Gendreau and B. Ross, "Revivification of Rehabilitation: Evidence from the 80s," *Justice Quarterly* 4, no. 3 (1987), pp. 349–408; P. Greenwood and F. Zimring, *One More Chance: The Pursuit of Promising Intervention Strategies for Chronic Juvenile Offenders* (Santa Monica, CA: U.S. Office of Juvenile Justice and Delinquency Prevention, 1985); M. Hamm and J. Schrink, "The Conditions of Effective Implementation: A Guide to Accomplishing Rehabilitative Objectives in Corrections," *Criminal Justice and Behavior* 16, no. 2 (1989), pp. 166–182; J. Inciardi, S. Martin, and C. Butzin, "An Effective Model of Prison- Based Treatment for Drug-Involved Offenders," *Journal of Drug Issues* 27, no. 2 (1997), pp. 261–278; D. Lipton, G. Franklin, and H. Wexler, "Correctional Drug Abuse Treatment in the United States: An Overview," in *Drug Abuse Treatment in Prison and Jails*, F. Tims, and C. Leukefeld, eds. (Rockville, MD: National Institute on Drug Abuse, 1992), pp. 8–29.

Chapter Resources

Emergency Preparedness

V

YOU ARE THE ADMINISTRATOR
Rioting in New Castle

Riots broke out in Indiana's privately run New Castle Correctional Facility on April 24, 2007. For two hours, about 500 inmates ran wild, burning mattresses, smashing windows, and throwing furniture. Before response teams arrived on the scene, several guards and maintenance workers barricaded themselves in a room to avoid dozens of raging inmates wrapped in masks and wielding sticks and knives. Two staff and seven inmates suffered minor injuries, including cuts, scrapes, and tear gas exposure.

According to Indiana Department of Corrections Commissioner J. David Donahue, the rioting began after several newly transferred inmates took off their shirts in the prison's recreation area in protest. Later, a spokeswoman for the prison management company said that inmates had been complaining about the lack of recreation and other inmate programming.

After the incident, the state analyzed the events and recommended pressing charges against 26 inmates, most of whom were recently transferred from another facility in Arizona. Their report concluded that the transfer happened too quickly as an effort to combat overcrowding in the privately managed Arizona prison. The Department of Corrections insisted that in the future more time be allowed to hire and train staff to prepare for large transfers.

- *What does this event show about causes of institutional unrest?*
- *What steps could have been taken to prevent this riot?*
- *What use of force is appropriate in circumstances like this?*

Sources: Charles Wilson, "Prison Riot Charges May Come in July" *Associated Press*, June 26, 2007, available at http://www.indystar.com/apps/pbcs.dll/article?AID=/20070626/LOCAL/706260430& template, accessed July 26, 2007; "Indiana Halts Transfer of more Ariz. Inmates," *Associated Press*, April 25, 2007, available at http://www.msnbc.msn.com/id/18294136/, accessed July 26, 2007.

Causes of Institutional Unrest

42

John J. Armstrong

Chapter Objectives

- Describe the basic mission of a prison under a confinement model.
- Give examples of the varied sources of unrest in correctional facilities.
- Explain the relationship of prison management practices to prison unrest.

Prison riots resulting from unrest have occurred in the United States since before the Revolutionary War. If correctional professionals fail to recognize, understand, and control the causes of unrest, unrest will remain an unfortuate hallmark of correctional administration.

■ Cause of Unrest

The sources of unrest are as varied as the correctional facilities that have experienced violence. Researchers suggest several causes, including:

- Crowding
- Insufficient funding
- Gang activity
- Racial and cultural conflict
- Changes in policy
- Poor management practices
- Insufficient staff training
- Inadequate facility security
- Poor living conditions
- External events

Despite the complexity and diversity of disturbances, the American Correctional Association has narrowed the causes of prison unrest into three general categories: inmates, conditions of confinement, and correctional management. Unrest often involves a failure in more than one category.

Inmates

Felons incarcerated as a consequence of a lack of self-control present special challenges. These men and women often are angry and antisocial and seek immediate gratification when faced with a problem. Inmate-created unrest does not always spring from a sudden or specific cause. An unhealthy facility climate frequently sparks unrest. Effective staff can recognize subtle verbal and nonverbal changes in inmate behavior. If they do not, investigation and control measures may be limited to reactive, after-disturbance efforts, rather than proactive ones.

The presence and influence of gangs also contribute to inmate-created unrest. Gang activity in a prison focuses on dominance and turf control, often along racial or geographic lines. Gang rivalries spawn a climate of tension, violence, and coercion. New gang members are recruited by force. Nonaligned inmates often arm themselves for protection against gang activities. Staff can control this threat to the well-being of a facility by installing mechanisms to gather intelligence and to identify gang members and leaders. These mechanisms are reviewed regularly to check that they are performing well and incorporate due process protections. Staff recognize gang-affiliated inmates by their tattoos, association with other known gang members, personal property, involvement in group activity of a negative nature (assaults, etc.), and by their own admission.

Unrest also results when predators are not separated from particularly vulnerable inmates, such as those who are very young or very old, are weak, are retarded or have disabilities. An effective inmate classification system, in addition to staff that promptly recognize such problems and respond appropriately, undermines this type of threat.

Conditions of Confinement

Some unrest may appear inmate-created but instead is rooted in the conditions of confinement. These roots of unrest are related to sanitation, food, idleness, inadequate facility maintenance, cell space, access to medical or mental health care, work, school or addiction programs, or innumerable other service delivery areas.

Crowding has been the most explosive condition in recent correctional history. Unprecedented prison population growth has plagued jurisdictions throughout the country at a time when public sector funding has continued to decline as a social priority. Population growth affects everyone—staff, inmates, and the public. Overcrowding reduces the quality of life in a facility, burdens its physical plant, and increases staff stress. These environmental, physical, and social pressures serve as a seedbed for destructive perceptions that lead to a sense of hopelessness in inmates and staff alike.

Correctional Management

Correctional managers promulgate policies and procedures based on established correctional standards and continually review them against set performance measures. Effective management practices lead to staff success, high employee retention rates, and little or no unrest.

Facility and departmental policy and procedure manuals and post orders supply staff with the organization's mission, values, and expectations; the basis of authority; and an outline of each position's responsibilities. Consistent policies that govern inmate conduct and accountability are the foundation of facility security, order, and safety. In a secure facility, staff and inmates perceive consistency, fairness, and justice as uniform standards applied in the absence of discrimination.

To be effective, disciplinary, grievance, and classification appeal processes must be easily understood by inmates and supported by staff. Fair and uniform appeal process outcomes are important to avoid unrest. By investigating and reviewing incidents, an administrator can evaluate circumstances leading to unrest. Additionally, institutions should have a clear mission statement, the foundation for all policy development and decision making. It serves as the benchmark for all operations, as the prime reference and resource for agency and facility administrative directives. When staff apply policies, procedures, and support structures to climate-related problems, they balance their actions against the agency standards and values established in the mission statement.

■ Prison Management

Almost every major prison riot in recent history has involved a security breach. To maintain facility security, staff must be able to rigidly control all operational elements. A comprehensive program of security audits ensures compliance with established standards and the requirements stated in the operations manual. Security auditing often will reveal vulnerabilities and offer correctional management an ongoing method by which to document, correct, and consistently improve operations and security. The consistent application of the principles of auditing—specifically, evaluation, correction, and follow-up—reduce the potential of the security breach as a precursor to a critical incident.

A unit's operations manual is the blueprint that outlines several key areas of management:

- Essential plans, systems, and post orders
- Proper reporting procedures
- Practices for issuing and controlling keys, tools, and weapons
- Techniques for conducting routine and random searches for contraband
- Methods for conducting urinalyses, inspections, tours, and visiting
- Systems for maintaining inmate documentation.

Managers also carefully establish the framework in which they communicate with each other, with their subordinates, and with the news media. A clear chain of command with clean lines of authority reduces inmate unrest resulting from mismanaged information or indecisiveness. Established channels of communication to staff and inmates eliminate confusion, misinformation, and destructive rumors.

For example, changes in operating policies disrupt highly structured prison routines and the expectations of inmates and staff. The prudent distribution of information, accompanied by explanation when necessary, lightens the impact of the message. Nevertheless, change—whether in policy, privileges, or living conditions—always heightens tension in a prison. Consequently, timing becomes a paramount concern when implementing change.

Accessible and responsive administrators and staff lead inmates to feel free to communicate and to know that they will receive appropriate and timely feedback. Reasonable questions and concerns are met with reasonable responses. Staff effectively counter externally driven causes of inmate unrest (such as issues reported in the media) with open and honest communication. Managers also counter externally driven causes by explaining to reporters, legislators, and governmental officials the impact of their messages on corrections. For example, reports of "get tough on crime" proposals often are interpreted by staff and inmates as fact, even before legislation has been introduced.

Effective managers also recognize that agency and facility stability results from recruiting quality applicants and maintaining high-quality pre-service and in-service training programs. The challenges facing modern corrections also require training in cultural sensitivity. For example, a common problematic social dynamic found in many American prisons is that staff from predominantly suburban or rural areas supervise urban inmates. (Inevitably, most inmates are from major metropolitan areas, and most prisons are located in isolated areas.) It is difficult to develop open communication between staff and inmates from such diverse backgrounds. Staff must be taught to view all offenders the same way; fairness is the key. Instances of perceived disrespect or injustice involving staff-driven or inmate-driven cultural or racial stereotypes create conditions resembling a tinderbox waiting for a spark.

■ Actively Gauging the Climate

Staff in direct contact with offenders are a facility's eyes and ears, its primary resource for identifying unrest indicators. Their evaluation of the inmate population is dynamic and continual. Facility security depends on how effectively these professionals deal with offenders and how the professionals use their training and interpersonal skills to evaluate the institution's climate.

Among all correctional professionals, correctional officers have the most daily interaction with the largest number of inmates. When the health, safety, and welfare of inmates are preserved, when their lifestyles of criminality are altered, chances are, responsibility lies not with the psychiatrists or social workers, but with skilled and concerned correctional officers. Dedicated and vigilant correctional professionals are the linchpin in the effort to achieve institutional safety and security.

Facility tours can act as another barometer of facility climate. All employees—including medical and program staff and administrators—should conduct area inspections and evaluate trends in incident and disciplinary reports, offender grievances, confiscated contraband, recreational grouping, and commissary activity.

Effective management depends on rumor control, and rumor control depends on rumor identification. Responsive line staff monitor rumors. Disregarding rumors places a prison climate at risk. Most rumors start with a grain of truth; otherwise, they would not be transmitted. Some, however, are propelled by personal issues. After filtering the information they gather informally, staff must be empowered to investigate and defuse a rumor or a consequent condition. Such empowerment should be tempered by the observance of communication priorities and the chain of command.

Management by Walking Around

Managers frequently walk through a facility, not just to be seen, but also to observe, listen, respond, and evaluate. Meeting the correctional mission requires management by walking around: paying attention, interacting with staff and inmates, identifying and addressing problems.

An unannounced or random walk often helps an administrator check on the eight dimensions of criminal justice performance measures for prisons: security, order, activity, conditions, safety, care, justice, and management.[1] All management staff are encouraged to do this. If prison wardens want correctional staff to be their eyes and ears, they must set the example by daily visits throughout the facility. Staff tend to emulate the leadership style and organizational values and expectations that are demonstrated regularly by senior management officials.

The agency commissioner and the facility warden hold the most visible positions in corrections and determine whether a department is successful. They ensure public protection, staff safety, and the maintenance of a secure, safe, and humane facility in a climate promoting high standards of professionalism, respect, integrity, dignity, and excellence.

Interpreting the Indicators and Achieving Performance

Indicators of unrest and prison performance measures appear dissimilar (see **Table 42–1**) yet one set of indicators cannot be assessed without considering the others. Assessment is crucial to the effective and active management of a prison climate; all such data are linked and, together, provide a valid overview of the institutional environment.

Conclusion

Prevention is an everyday event in a prison—on every shift for every employee. An agency attuned to detecting problems can identify a problem in its early stages and resolve it before it becomes a crisis. Forestalling unrest requires active monitoring of the prison environment, detection of unrest predictors, investigation of problems, and quick and appropriate resolution.

Table 42–1	Predictors of Riots and Disturbances

Separation of inmates by racial or ethnic groups

Increase in purchases of food items at inmate canteens

Transfer requests

Staff requests for sick leave

Inmates gathering with point people facing away from the group

Increase in disciplinary cases

Increase in voluntarily lockups

Inmate–employee confrontations

Direct and indirect inmate intimidation of officers

Threats against officers

Inmate sick calls

Inmate violence against other inmates

Increase in number of weapons found in shakedowns

Harsh stares from inmates

Drop in attendance at movies or other popular functions

Unusual and/or subdued actions by inmate groups

Appearance of inflammatory and rebellious materials

Warnings to "friendly" officers to take sick leave or vacation

Employee demands for safety

Staff resignations

Letters and/or phone calls from concerned inmate families demanding protection for inmates

Unusual number of telephone inquiries about facility conditions

Outside agitation by lawyers or activists

Increase in complaints and grievances

Source: Reprinted with permission from *Preventing and Managing Riots and Disturbances*, p. 131, American Correctional Association, Alexandria, VA.

DISCUSSION QUESTIONS

1. Are riots an inevitable consequence of the prison environment?
2. What are the most effective strategies to reduce prison unrest and the violent disturbances that may result from such unrest?
3. How should a prison administrator develop an awareness of inmate tension?
4. Why is prevention of institutional unrest considered a key component of good correctional leadership?
5. Are public expenditures better focused upon crisis response planning, disturbance suppression training and riot equipment, or prevention centered activities?

ADDITIONAL RESOURCES

K. Wright, *Effective Prison Leadership* (Birmingham, New York: William Neil Publishing, 1994).

NOTES

1. C. Logan, "Criminal Justice Performance Measures for Prisons," in *Performance Measures for the Criminal Justice System*, J. DiIulio, ed. (Washington, DC: U.S. Bureau of Justice Statistics, 1993).

Emergency Management

E.A. Stepp

Chapter Objectives

- Describe the role a prison administrator plays in emergency preparedness.
- Explain the stages of response: planning, active management, and aftermath.
- Differentiate between emergency response and emergency preparedness.

Prison and jail riots or disturbances continue to be the nightmare of institution administrators. While events necessitating full-scale emergency response are rare, they are ever-present in prison managers' minds. Any number of small, isolated, and seemingly unimportant events that occur frequently can mushroom into full-blown emergencies. Emergencies can occur in isolated portions of a facility or, when not immediately contained, engulf the entire prison.

Disturbances range from passive demonstrations by inmates (i.e., food or work strikes) to violent acts against property, staff, or other inmates. Whether an event is planned by a few individuals or orchestrated by a large group, it can escalate quickly into a worst-case scenario, such as a riot, hostage situation, or escape.

How then do prison administrators operate safe and secure institutions? How can staff move from routine daily operations to the highest level of emergency response? After the events of September 11, 2001, how can prison administrators prepare for traditional responses to prison emergencies and ensure that emergency plans incorporate new response standards established by the federal government? To answer these questions, one must take a look at the entire process of prison emergency preparedness and the new coordinated, all-hazards approach to emergency management.

■ Planning

The planning stage of emergency response includes development of established, specific emergency response plans, training, and assignment of resources. Before anything happens, management must ensure that staff are prepared to recognize and respond to a multitude of different emergencies. Each type of situation may require an individualized, specific response plan. If the first stage of emergency response is emphasized, a system of emergency preparedness can be established. There is a huge difference between the *emergency preparedness* (preparation, planning, training, and budgeting) and *emergency response* (active management, intervention, containment, and resolution). The goal of any administrator must be to establish an emergency preparedness program to ensure that actual emergency response is effective.

The first step in preparing a response to emergency situations is to develop site-specific plans for each type of emergency. An overall emergency plan for an institution contains several contingency plans for specific emergencies. These include plans to resolve riots, escapes, hostage situations, weather emergencies, food or work strikes, fires, and situations that could necessitate evacuation of inmates and staff from the institution. Planning for these and other types of emergencies will require establishing a contingency planning team consisting of different staff experts from various departments within the institution.

Internal Cooperative Contingency Planning

Planning for emergency response requires a "big picture" approach that examines all available resources. All contingency plans must define the responsibilities of staff from various departments. This is true when planning for even small situations. Plans must be developed through cooperative means, using the collective experience of staff from all departments within the institution. In cases where emergency plans are written by security staff, with little or no input from other departments, unrealistic or dangerous assumptions may be made.

All emergency plans should cover certain general elements clearly:
- *Communication of the initial alarm.* Once an emergency situation is identified, all staff must know how to alert management and others of the situation.
- *Securing the scene and initial containment of incident.* Most emergencies begin small and spread quickly if initial containment actions are not effective. It is critical to the final resolution of any incident to contain the incident to the smallest area possible by locking the affected area or building.
- *Command structure.* All plans should specify responsibilities for immediate emergency response command. Everyone must know who the initial on-scene commander is prior to senior management staff arrival.
- *Notification and call back procedures.* Emergency situations must be communicated swiftly and clearly to the facility chief executive officer. A

standard process of timely and urgent recall of staff to the institution should be in place.

- *Command center location and operation.* The location of the command center must be established clearly. Appropriate equipment and instructions for activating a command center must be in place in a secure location. Because predicting the location of an incident is impossible, the command center should be located in an area inaccessible to inmates. In most cases, this requires establishing a command center outside the secure portion of the institution. Emergency plans, communication equipment, and enough space for several staff to work are essential when planning for a command center.

- *Preparation of emergency response teams.* Before correctional administrators can begin to develop strategies to resolve emergencies within their facilities, it is critical that decisions be made regarding resources that will be not only needed, but dedicated to the emergency response.

Emergency Response Teams

Different levels of response are required, contingent upon the type of emergency faced. For each level of response, different teams with incident-specific training are required. Regardless of team membership and levels of specialty expertise, common qualities apply to all. All teams must be composed of volunteers, receive specialized training in specific skills and be highly trained in agency policy regarding use of force. All staff selected for membership on any team should be required to pass physical, academic, and psychological screening.

Traditional Disturbance Control Teams

The first level of response to disturbances is the traditional disturbance control team (DCT). DCTs are trained in riot control formation and the use of defensive equipment such as batons, stun guns, and chemical agents. DCTs are trained to control and contain both large and small groups of inmates involved in disturbances. Application of "less lethal" technologies usually falls to DCTs. Extensive training is necessary and should include minimum proficiency standards. A certification process must be included that requires a knowledge of agency policy and emergency plans as well as proficiency in the use of related equipment.

Armed DCTs

While traditional DCTs should be viewed as the primary emergency response team for incidents requiring containment and control, a higher level of response may be necessary to deal with a difficult scenario that has escalated to one that requires the use of deadly force. This type of incident requires managers to respond with a specially trained team; the use of an armed DCT should be limited to situations where staff or inmate lives are in imminent danger. Training and certification standards increase as the manager of the incident follows an escalating use-of-force continuum. Use of armed DCTs implies that management has accepted that lethal force may be used.

Tactical Response Teams

Tactical response teams are the most highly trained and skilled emergency response staff. These teams, known as correctional emergency response teams, or special operations response teams, are similar to traditional special weapons and tactics teams found in most police departments. Tactical teams required advanced levels of training in barricade breaching, hostage rescue tactics, and precision marksmanship with pistols, rifles, and assault weapons.

External Cooperative Contingency Planning

Most emergency situations in correctional institutions require assistance from outside law enforcement sources. Since September 11, 2001, there has been a renewed emphasis on cooperative contingency planning. This requires effective cooperation with entities outside of correctional agencies or departments. Historically, correctional facilities have developed emergency plans with an emphasis on self-directed and self-sustained response. Agreements with local, state, and federal agencies were often ill-defined or nonexistent. Failure to establish relationships with outside law enforcement agencies will contribute to an already chaotic situation and prolong incident resolution.

Managers must ensure that institution emergency plans include mutually agreed-upon cooperative contingency plans with outside agencies. Plans may be as simple as assistance with site access and traffic control, or as complicated as meshing institution tactical teams with outside tactical resources and hostage rescue attempts. Regardless of the level of assistance, preexisting memoranda of understanding establishing limits of assistance and command and control will reduce confusion and delay. Managers must be able to implement plans and take actions to resolve the situation and not debate questions of philosophy regarding expertise, resources, or how to accomplish a given goal.

Cooperative contingency plans should be signed by the outside agency representative and the institution warden. Once this is accomplished, role definition is achieved. Institution managers must also realize that coordination with local or state emergency action centers is important. Intelligence briefings regarding both domestic and international terrorism threats can be provided so long as mutual assistance agreements exist. Outside groups such as terrorist organizations may believe it newsworthy to target government facilities, including prisons. Prison administrators can no longer take an "inside out" approach to emergency planning. Planning must also include an "outside in" review and plan.

Training and Mock Exercises

A realistic training program must include all key staff. Because not all staff will be available to respond during an emergency, managers must ensure that cross training of staff occurs. Staff will be much more prepared and effective when they can perform multiple tasks and assume a variety of roles during an emergency.

For any emergency plan to succeed, all staff must be familiar with their responsibilities and management expectations. The only way to determine whether a plan is effective is to conduct regular tests and analyze results. Management must evaluate emergency plans, devise training scenarios, and analyze staff performance. Major exercises involving all staff and outside agencies should be conducted at least annually. This type of mock exercise requires planning, use of resources, and coordination with each cooperative contingency plan outside of the agency. In most cases, however, emergency plans can be tested with little or no disruption of normal institution operations.

Small, internal training exercises that test specific parts of plans also should occur regularly. These tests include communication devices, staff recalls, command center set-up, and area containment. Having staff who are trained properly in their roles and in the requirements to implement emergency plans is the goal, as it limits the amount of time required to contain and ultimately resolve a crisis. The beginning of any emergency is the most critical time. There is often an early window of opportunity to resolve, or at a minimum, contain the incident. Managers should demand effective, well-rehearsed emergency plans to ensure that confusion and delay do not become the reality of the response.

One of the most important aspects of emergency preparedness is knowing the capabilities and limitations of various emergency response teams. Practice scenarios where each team participates with other groups, including outside agencies, will provide this critical information to crisis managers.

Testing emergencies plans should include the following steps:

1. Identify those staff who are to respond and those to be used as role players.
2. Ensure that staff are cross-trained.
3. Ensure that staff know how to assume their various roles.
4. Assign monitors to evaluate and criticize the exercise.
5. Provide a method of terminating the exercise should a real incident occur.
6. Establish a code word or signal to alert staff should an actual emergency occur.
7. Conduct a debriefing with all staff participants.

The debriefing is a critical component. This evaluation of the exercise and should provide useful feedback from participants. Gathering information from training participants allows management to assess the organization of emergency teams, the adequacy of communication, and the overall use of resources. Based on this feedback, administration can revise plans to achieve better results in actual emergencies.

Meeting Legal Requirements

A final step in the training portion of preparation is ensuring all prearranged agreements with other agencies have been reviewed by the respective legal representatives. Short notice implementation of these agreements will be required. Prior to implementation, everyone in the institution should know what staff from other agencies legitimately can and cannot do during a crisis.

Tactical Options

Tactical planning and preparation for crisis resolution is an important element of overall emergency response training. Standard tactical plans will not be incident specific but will be based on certain known factors. These include agency philosophy and goals and tactical capabilities of teams. Flexibility of response is necessary. Standard, preexisting tactical plans based on agency management philosophy can then be tailored to specific situations as intelligence and information from the crisis develop.

A crisis manager can use any of the following three options to resolve major incidents in a correctional setting:

1. Negotiation—involves the least amount of force
2. DCTs—involves non-lethal technologies and options such as chemical agents and distraction devises
3. Use of deadly force—involves intervention by an armed team

An effective emergency management structure relies on the use of all resolution options, since options only work properly when used together. Negotiations cannot succeed without tactical options. Tactical options cannot succeed without information and intelligence obtained by negotiators. All components of the emergency management structure must cooperate and train together to ensure success. Success for one is success for the entire emergency management structure.

It is nearly impossible to plan a specific response to all the different crisis situations that could arise in a correctional setting. However, an effective emergency preparation program should include plans for accessing all areas of the institution during an emergency. Preplanned breach points and means of entry provide the needed flexibility to develop an incident-specific tactical plan during active management of the emergency situation. Preestablished breaching plans should include the following:

- *Identification of a staging area.* There should be primary and secondary staging areas for teams that are not visible to inmates, media, or the public.
- *Plan of approach.* Primary and secondary approaches should exist to areas that provide concealment.
- *Preparation for entry.* Again, primary and secondary points of entry into all buildings should be established in advance of a crisis.
- *Required physical hardware and equipment necessary to effect entry.* Every building inside the secure perimeter of the institution should have blueprints that identify doors, windows, hatches, tunnels, and all locking devices. The type of construction and strength of walls, doors, etc., must be known in advance.
- *Identification of the method-of-entry options for opening or removing by force.* Each entry point must be planned in advance. All possible entry methods should be documented, including key rings, cutting tools, torches, saws, and explosives.
- *Assignment of primary entry teams.* Tactical teams must work together and practice the techniques before attempting forced entry in an actual situation. Incident-specific plans can then be developed based on known variables in the preexisting breach and forced entry plans.

All staff members must understand the institution's policy on use of force. Managers plan the resolution phase during active management of a crisis in accordance with a use of force continuum. A use of force (particularly deadly force) policy is based on the reasonable response from staff to actions by the inmate. A sequence of escalating steps lead to the top of a continuum. At each level the policy should define clearly what response from staff is acceptable in response to actions initiated by an inmate or group. Response by staff must meet the threat level presented. The response and the policy are governed by law. Active management of a crisis in a correctional setting does not include rules of engagement—a military term that permits a more flexible strategy of response. The response to each emergency is always governed by preexisting use of force policy that remains consistent, and legally and morally acceptable, regardless of the nature of the crisis or incident.

■ Prediction

Prediction is the phase of emergency preparedness that precedes the implementation of a well-planned emergency response. This phase will develop the methods to identify the possibility of an emergency, the probable type of incident, and the best response. The prediction phase requires daily communication of risk factors identified within the prison population. Emergencies can often be predicted if disturbance factors can be identified and evaluated properly.

Prison administrators should develop a risk analysis mechanism designed to predict the degree of possibility for an emergency. This risk analysis should include intelligence information, an assessment of inmate grievances and complaints, and a review of any possible common ground issues (factors that affect the entire inmate population) discovered. These issues may include:

- Medical care
- Food service and preparation
- Disciplinary programs
- Inmate work assignments
- Safety
- Sanitation
- Grievance procedures

■ Prevention

Prevention is the phase of emergency preparedness that enables the prison administrator and the staff to maintain or restore safe, humane, and professional conditions of confinement of the inmate population. This phase, like prediction, requires constant monitoring and communication among staff. Prevention includes appropriate programs for inmates and effective safety, security, and sanitation programs. The most important factor in prevention is consistent enforcement of rules, policies, and directives. In almost every prison in the United States, establishing a daily routine and communicating expectations must be a priority. Communication of any changes to the inmate routine reduces the risk factors and serves as a means of prevention. The bottom line is that prevention requires complete reporting of disturbance risk factors and responding with actions that mitigate these factors.

New and Additional Emergency Preparedness Elements

With the likelihood of intense scrutiny both during and after any prison disturbance, utilization of all available outside resources will be required. Additionally, managing a disturbance or emergency within a prison requires planning and execution of strategies that are familiar to other law enforcement and emergency response agencies.

Continuity of Operations Plan

A continuity of operations plan (COOP) is designed to ensure continuing facility operations of personnel and technical infrastructure. Terms of the plan should be tailored to meet the requirements of the National Incident Management System (discussed below). The COOP is a concept that requires evolution of traditional response into protection of electronic and communication, water, power, and sewer systems, along with designations of key personnel who would assume direction of various operations in worst-case scenarios. COOP plans are broken down into three main topical sections. Each section may contain as many as 12 subsections of specific required activities and coordination. These plans are designed to cover a wide range of potential incidents.

An example taken from the Florida Division of Emergency Management includes the following areas:

1. Contact information (state, county, and agency and facility name)—All primary response personnel and alternates should be listed with primary and secondary contact information.

2. Policy and administration—Objectives, responsibilities, planning assumptions, plan execution, and post-incident review and remedial actions should be documented.

3. Essential elements criteria—This includes procedures; clearly defined mission; essential functions; delegations of authority; orders of succession; alternate facilities; interoperable communications; vital records and data bases; logistics and administration; security, physical, and operational coordination of personnel; and testing, training, and implementation of emergency plans.[1]

National Incident Management System

The National Incident Management System (NIMS) was established by the Secretary of Homeland Security to standardize emergency responses from differing jurisdictions and disciplines. The theory behind NIMS is to enable responders to work together more quickly and effectively when responding to natural disasters and other large scale emergencies. NIMS focuses on a unified approach to incident management, command structures, and emphasizes mutual aid agreements among agencies.

While it may be difficult to view a local prison disturbance in the same light as a disturbance that occurs on a national scale, the concept of NIMS does translate to local jurisdictional mutual aid agreements as detailed earlier. The national plan incorporates best practices (lessons learned) from incident managers at the federal, state, and local levels as well as in the private sector. All emergency preparedness planning and training should incorporate the elements of the NIMS into local plans. Just as the prison administrator wants to eliminate common ground issues from the inmate population, law enforcement, through NIMS, attempts to lay a common ground framework for all potential emergency responders.

National Response Plan

The National Response Plan (NRP), developed by the Department of Homeland Security, is a comprehensive approach to emergency management that goes hand in hand with COOP and NIMS planning. It establishes protocols for state and local jurisdictions and the private sector to coordinate responses with the federal government. The goals of the National Response Plan are:

1. To save lives and protect the health and safety of the public, responders, and recovery workers
2. To ensure security of the homeland
3. To prevent an imminent incident, including acts of terrorism, from occurring
4. To protect and restore critical infrastructure and key resources
5. To conduct law enforcement investigations to resolve the incident, apprehend the perpetrators, and collect and preserve evidence for prosecution and/or attribution
6. To protect property and mitigate damages and impacts to individuals, communities, and the environment
7. To facilitate recovery of individuals, families, businesses, governments, and the environment[2]

The NRP establishes goals from an emergency response point of view and provides insight to prison administrators in preparing local plans.

■ Dealing with the Aftermath

After a crisis situation is contained and resolved, the prison administrator's toughest challenge may just be beginning. In all major crises in a correctional setting, returning the institution to a normal, pre-event status is a unique challenge for administrators. Administrators must have a plan to deal with inmates and their needs. Just as important, they will need a plan to deal with staff and their needs. Feelings of guilt, disbelief, and failure are common among staff following a major prison incident. Support functions for both staff and the inmate population will be critical in returning the facility to normal operations. Staff from both the chaplain's department and the psychology department should play an important role in dealing with and resolving deep-seated feelings among staff and inmates. Support and counseling for everyone involved must be available and be a part of the institution's emergency preparedness planning and response.

■ Conclusion

Emergency preparedness plans require utilization of all available resources to ensure an effective and practical system of response that can also be implemented in concert with outside resources. The use of NIMS, COOP, and a review of the NRP will result in a better and more realistic emergency plan. However, planning for emergencies in a prison setting still requires the use of traditional first responders.

DISCUSSION QUESTIONS

1. Do you support use of deadly force for resolution of disturbances?
2. Does the cost of preparation and training affect active emergency management and crisis resolution?
3. Have prisons and jails been affected by domestic and international terrorism?
4. What are the differences in emergency planning in today's world compared to the planning done prior to September 11, 2001?
5. What role do Homeland Security and the National Response Plan play in local planning?

NOTES

1. County Coordination Checklist—Continuity of Operations, Division of Emergency Management, Department of Community Affairs, "County Coordination Checklist for Agency COOP Plans in accordance with Chapter No. 2002-43: Relating to Disaster Preparedness—Amends 252.365," Florida Division of Emergency Management (2004), available at http://www.floridadisaster.org/documents/COOP/CountyCoordination Checklist.pdf, accessed September 23, 2007.

2. Department of Homeland Security, National Response Plan, available at http://www.dhs.gov/xprepresp/committees/editorial_0566.shtm, accessed September 23, 2007.

Hostage Situations

44

Gothriel "Fred" LaFleur, Louis Stender, and Jim Lyons

Chapter Objectives

- Describe the goals of hostage negotiation and outline characteristics of successful hostage negotiators.
- Identify what characteristics drive the use of negotiation.
- Explain the importance of intelligence gathering in a hostage situation.

The safety, security, and orderly operation of a correctional facility can be severely interrupted by inmate disturbances. Such events are particularly threatening and stressful when they involve the taking of hostages. The potential for major disturbances, including hostage situations, exists at any time in any facility, regardless of whether the facility houses maximum or minimum custody inmates. Proactive strategies—effective administrative attention to institution operations and staff awareness of the overall climate, including any signs of inmate unrest—can reduce the potential for major crises significantly. However, even a small event can result in a prison riot and a hostage situation. When faced with such a situation, administrators must be prepared to respond. To do so, they must have a coordinated plan to address the incident. The way in which prison administrators prepare for and respond to this type of crisis can make the difference between a small-scale disturbance and a full-fledged riot.

■ Responding to Hostage Situations

There are four ways, outlined by Thomas Strentz, to respond to hostage incidents:

1. **Assault of the Location: A Preemptive Strike**

   This method can produce a rapid conclusion but it poses significant risk to the hostages, tactical team, and captor. Hostages are most likely to be harmed during the initial phase of a disturbance. Certain hostage takers may leave administrators no option other than a straightforward tactical response if homicidal or threatening actions have occurred or are expected.

2. **Selected Sniper Fire: Shooting the Perpetrator**

   If hostages have been harmed or are in serious danger, sniper fire could end the situation quickly. Use of a sniper may be considered if members of the tactical team are in imminent danger, but hostage takers have been known to switch clothes with hostages, and the incorrect person could be shot. The size and complexity of the disturbance are factors to consider; riotous situations of large magnitude may not be resolved easily by selective lethal force. Sniper fire is generally not considered an initial option.

3. **Use of Chemical Agents: Flushing the Hostage Taker into the Open**

   Chemical agents may force a hostage taker out of the area, allowing for hostage escape or rescue, but they may not reliably produce the desired effect and could be harmful to the hostages' health. Chemicals usually have been used for diversion or as a supporting element, introduced at the same time that a tactical assault is initiated.

4. **Containment and Negotiation: Talk and Endless Patience**

   The containment and negotiation process has proven to save lives.[1] The success rate of negotiations is very high when communication has been isolated and the captor speaks only with the negotiator; if the captor is allowed to speak with friends, family, members of the media, and others, the situation has the potential to regress quickly.[2]

■ Purpose and Theory of Hostage Negotiation

The basic purposes for negotiating during a hostage incident are to preserve lives and regain control of the correctional facility and the inmates. First and foremost, there is a need to negotiate because of the value of human life. Psychologists Michael McMains and Wayman Mullins have prioritized negotiators' objectives in correctional facilities, in order of priority:

1. Attempt to save the lives of hostages, citizens, prison staff, and hostage takers
2. Regain control of the prison environment
3. Prevent escape
4. Minimize casualties
5. Apprehend the hostage takers
6. Recover property[3]

Though negotiation is the preferred method to resolve a hostage situation, negotiation may not be a viable alternative in some cases. The Federal Bureau of Investigation (FBI) has pinpointed characteristics that an incident must have to be negotiable:

- The hostage taker must want to live.
- Authorities must threaten to use force.
- The hostage taker must have clear demands.
- The negotiator must be viewed as one who can hurt or help the captor.
- There must be time to negotiate and a reliable channel of communication between captor and negotiator.
- The location and communication of the incident must be contained.
- All communication must be channeled to and through the negotiator.
- In multiple-captor scenarios, the negotiator must be able to deal with the captor who is the decision maker.[4]

If all of these characteristics are not present, negotiation becomes a less viable option, and hostages are in more danger.

While it might seem counterintuitive, one of the main benefits of the negotiation process is that it takes time. The passage of time helps decrease stress levels, increase rationality, allow development of rapport, increase trust, clarify communication, exhaust the hostage taker, and increase the probability that hostages will be released unharmed. With more time, important intelligence can be gathered, tactical teams can prepare, and command personnel can organize; captors' expectations will drop; and hostages will have more chances to escape. Additionally, with time, the Stockholm syndrome may develop. This psychological condition may develop in hostage situations in which people taken hostage and their captors begin to identify with each other. This helps decrease the likelihood that hostages will be harmed because hostages will be cared for by the captors. If the syndrome is not present, hostages may be in great danger, and negotiating efforts will focus on keeping hostages alive.[5]

■ Structure and Role of Response Teams

At the onset of any hostage situation, the tactical team should establish a command post in a quiet area between the inner and outer perimeters. Large numbers of people will be involved in the incident, so an effective process of communication is imperative.

- The *on-scene commander* (OSC) will direct all activities from this site. An OSC should be situated in the command post with a few advisors in the roles of second in command, negotiation team leader, and special weapons and tactics (SWAT) team leader (see **Figure 44–1**).
- The *second-in-command* supervises interaction with the media, legal representatives, public officials, or other entities. This person organizes all incoming information to present to the commander.
- The *negotiation team leader* briefs the commander on the negotiation process, intelligence gathered, and the hostage takers' current mental status, including input obtained from specialists such as a psychological consultant. Additional members of the negotiation team may include a supervisor, primary and secondary negotiators, intelligence gatherers, and mental health consultants.
- The *SWAT team leader* informs the commander about the position and readiness of his or her team members and the feasibility and likelihood of success for various assault options.

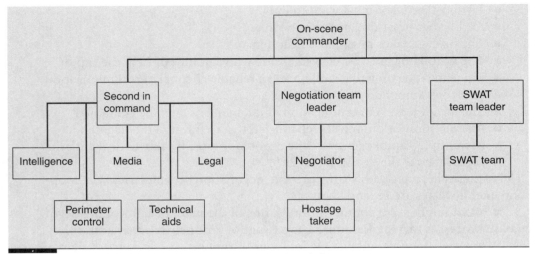

Figure 44–1 Recommended Lines of Communication—Hostage Negotiation
Source: Reprinted from G. Fuselier, Lines of Communication FBI Law Enforcement Bulletin, March 1961, Federal Bureau of Investigation.

It is not possible to have a negotiating posture without tactical support and containment. Tactical and negotiation teams must share information and intelligence.[6] Both teams gather intelligence (e.g., hostages harmed, medical needs, weapons involved, mental stability of hostage taker) invaluable to the other team, and strategies are altered based on this information. Actions taken without the negotiator's knowledge will make it difficult for the negotiator to provide a cover story to the hostage takers and maintain their trust. Should negotiations prove unsuccessful, the negotiator can confirm the hostage takers' location or otherwise provide distraction at the time of any planned assault.

Negotiation Teams

A peaceful resolution for everyone involved is the goal for the negotiator, regardless of the situation. Responsibilities of the negotiation team include the following:

- Gathering intelligence
- Formulating tactics to defuse an incident
- Communicating with hostage takers
- Recording intelligence
- Tracking negotiations
- Coordinating with the commander and tactical team

Team members should be non-managers and non-uniform staff. They should have a minimum of one year of correctional experience and be carefully selected and trained in negotiation.[7] The FBI requires negotiators to complete the FBI's Basic Hostage/Crisis Negotiations course, a 40-hour training program that is available to correctional and law enforcement personnel. Additionally, they should possess the following characteristics:

- Emotional maturity
- Good listening and interviewing skills
- Ability to make logical arguments and be persuasive

- "Street smarts"
- Experience
- Ability to communicate well with many different types of people
- Ability to take responsibility for the negotiation
- Understanding of the key principles of negotiation
- Ability to cope with uncertainty
- Flexibility
- Commitment to the negotiation approach
- Ability to plan and support an assault[8]

Negotiation team members should become familiar with typical inmate hostage takers and be trained to communicate effectively with each type. Hostage takers are generally categorized into three main groups:

- Mentally unstable (e.g., suicidal, paranoid, antisocial)
- Criminals (e.g., robbers, murderers, rapists, gang members, those whose escape attempts failed)
- Crusaders (e.g., white supremacists, militant Muslim groups, terrorists, religious fanatics)

Victims' Assistance Team

Prison hostage situations may become sieges in which the prison's armed force surrounds a hostage situation and isolates it for an extended period of time. Because hostages are often employees of the facility, it is advisable to establish (prior to the incident) a victims' assistance team that would communicate with hostages' families and provide an area for them to gather. The victims' assistance team should provide emotional support to the families, disseminate information and intelligence on the hostage situation, and shield family members from the news media.[9] This team may be used to provide emotional debriefing to the released hostages.

■ Negotiation Strategies

The goals of hostage negotiation are to:

- Open lines of communication
- Reduce stress and tension
- Build rapport
- Obtain intelligence
- Stall for time
- Allow hostage takers to express emotion and ventilate
- Establish a problem-solving atmosphere

In addition, hostage negotiations must hold the following principles to be inviolate:

- No hostage should be exchanged for release or reduction in sentence for any inmate
- No weapons should be supplied to hostage takers
- No intoxicating substances should be exchanged for the release of hostages
- No hostage should be exchanged for a different hostage

Prison negotiators need to be concerned about minimizing property damage. Prison hostage takers often take over and destroy the very settings needed to maintain the inmate population. Before beginning any negotiation, the negotiators should ensure that hostage takers are contained. Tactical teams should, to the extent possible, minimize the area occupied by the hostage takers.

In a riot, it is impossible to negotiate. An assault may be necessary if the situation does not stabilize. It may also be preferable to assault before leaders begin to emerge. The courts have ruled that using force to quell a riot is not unlawful unless the assault team acts maliciously and sadistically with intent to cause harm.[10]

■ Return to Normal Operations

Administrators need to plan and systematically implement steps to return the facility to normal operations at the conclusion of any disturbance. Especially following hostage situations, tension among inmates and staff may be high. Resolution of an incident may produce a tendency to believe the worst is over and to reduce attention to valid concerns. The following issues should be addressed after any disturbance:

- Short-term responsibilities—searching for contraband, securing inmates, assessing damages, counting inmates, providing medical care to hostages and inmates, and collecting evidence for future prosecutions.
- Medium-term efforts—providing continued support and counseling to employees in coping with their experiences, repairing damage to facilities, working toward normalization of institutional operations, and undertaking a thorough investigation of the causes of the crisis, and an evaluation of how it was handled. Actions such as publicly recognizing the sacrifice made by hostages and expressing appreciation for exemplary action by staff may help reintegrate the corrections community.
- Long-term solutions—developing policy reflecting what was learned from the disturbance. Institutions may discover better ways to forecast problems, improve the flow of information, fix previously unrecognized problems, improve relationships with other agencies, boost morale, and meet challenges.

■ The Importance of Training

It does little good to have an institution disturbance control plan unless staff and supervisors have been trained to activate it. Administrators should give training a high priority. For a coordinated response to be successful, people in each component need to understand clearly the functions of people in other components. To enhance cooperation, negotiators and personnel from tactical teams should train together regularly. Command and supervisory personnel must be trained in the nature of a negotiable incident so they can make informed decisions.[11]

To protect themselves from liability, administrators should develop policies and procedures that document required training, as well as records that show who was trained during what period of time. Institutions should be able to prove in court, if necessary, that officers received training. All staff members should receive training in disturbance control and hostage incidents, including realistic simulations of incidents. Many states have centralized training academies that should coordinate with institutions to meet their training needs. The way corrections administrators prepare for and handle a disturbance can make the difference between a peaceful resolution and a true disaster.

■ Conclusion

Correctional administrators may face the difficult task of dealing with inmate disturbances. Many disturbances can be avoided or minimized by taking appropriate proactive measures to prevent them from escalating to violence, riots, or hostage situations. Staff should monitor the institutional climate continually looking for signs of unrest to reduce the possibility of inmate disturbances. In addition, all correctional facilities should have a coordinated emergency response plan that outlines the roles and responsibilities of staff as well as individuals from coordinating agencies. This plan must cover the use of tactical responses and negotiation strategies. Administrators who have developed comprehensive disturbance control plans and trained staff on these plans can deal more successfully with a disturbance or hostage situation.

DISCUSSION QUESTIONS

1. How do prison conditions contribute to inmate disturbances?
2. How would a negotiator determine if the Stockholm syndrome is in effect? What problems does the Stockholm syndrome present in terms of resolving the crisis?
3. Do you think prison riots are an effective way for inmates to force change in their prison conditions?
4. Should more restrictive inmate confinement strategies be employed as a preemptive response or as a way to reduce potential riots and hostage situations?
5. How would you balance improving inmate conditions as a way to decrease disturbances with public and political demand for punishment and accountability?

ADDITIONAL RESOURCES

S. Romano, "Achieving Successful Negotiation in a Correctional Setting," *Corrections Today* (April 2003), pp. 114–118.

T. Fagan, "Negotiating by the Numbers," *Corrections Today* (October 2000), pp. 132–136.

T. Fagan, *Negotiating Correctional Incidents: A Practical Guide* (Lanham, MD: American Correctional Association, 2003).

NOTES

1. H. Schlossberg, "Psychiatric Principles of Negotiations," *The U.S. Negotiator* (Spring 1996).
2. T. Strentz, "Law Enforcement Policies and Ego Defense of the Hostage," *FBI Law Enforcement Bulletin* 48, no. 4 (1979), pp. 1–12.
3. M. McMains and W. Mullins, *Crisis Negotiations: Managing Critical Incidents and Hostage Situations in Law Enforcement and Corrections* (Cincinnati, OH: Anderson Publishing, 1996), p. 290.
4. G. Noesner, "Negotiation Concepts for Commanders," *FBI Law Enforcement Bulletin* 68, no. 1 (1999).
5. McMains and Mullins, *Crisis Negotiations: Managing Critical Incidents and Hostage Situations in Law Enforcement and Corrections*, p. 286.
6. R. Loudin and G. Leak, "Command and Control during Crisis Incidents," *The U.S. Negotiator* (Spring 1996).
7. *Standards for Adult Correctional Institutions*, 2nd ed., Standard #2-4094 (College Park, MD: American Correctional Association, 1981).

8. G. Fuselier, "What Every Negotiator Would Like His Chief To Know," *FBI Law Enforcement Bulletin* 55, no. 3 (March 1986), pp. 12–15.

9. P. Miller et al., "Lesson Learned: The Oakdale and Atlanta Riots—Interviews with BOP Wardens Johnson and Petrovsky," *Corrections Today* 50 (1988), pp. 16–18.

10. *Whitley v. Albers*, 475 U.S. 312 (1986).

11. B. Wind, "A Guide to Crisis Negotiations," *FBI Law Enforcement Bulletin* (October 1995), pp. 1–7.

Chapter Resources

Use of Force

45

Marie L. Griffin and John R. Hepburn

Chapter Objectives

- Describe circumstances in which an officer may use force in a correctional setting.
- Discuss critical elements of an agency's use of force policy.
- Identify relevant constitutional amendments and how they have been applied to cases involving the use of force.

Correctional institutions are extremely coercive organizations in which all activities are carried out in an environment of uncertainty. In both jails and prisons, where staff are usually unarmed and always outnumbered by the population of resistant prisoners, the ability of staff to control the prisoners is a matter of major importance.

For the most part, staff rely on their *legitimate power*; that is, prisoners accept that staff have the authority to give reasonable instructions related to inmates' daily activities.[1] Of all the types of power, legitimate power ensures compliance by the largest number of prisoners, over the widest scope of prisoner activities, and over the greatest amount of time and effort involved in those activities. In contrast, *coercive power* is most effective when it is always available but seldom used. In prisons and jails, coercive power is an ever-present resource that can be mobilized to provide the force necessary to support legitimate power.

Lethal force is a rarely used type of coercive power that represents the extreme end of the continuum of force. Lethal weapons are issued routinely only to those who guard the perimeter of an institution. Few, if any, officers working within the population are armed with lethal weapons unless they are responding to an internal disturbance. Nonlethal force is used much more often than lethal force; officers routinely rely on direct physical contact with prisoners to maintain control and security. For the most part, this use of force involves only some form of hands-on contact with the prisoners. However, such nonlethal weapons as stun devices and chemical sprays are becoming more prevalent within jails and prisons.[2]

■ Incidence of the Use of Force

It is generally agreed that deadly force is rarely used and that nonlethal force is frequently used, but there are few studies that provide data about how often the different types of force are used. Each institution keeps its own records of the use of force incidents that occur, but totaling these incidents across institutions or making comparisons between institutions is difficult, if not impossible. Institution policies have different definitions of force, different requirements about when incidents involving force should be reported, and different specifications about the type and completeness of such reports.

In 1993, the American Correctional Association (ACA) conducted a national survey of the use of force in 325 prison facilities, representing 49 state correctional systems and the Federal Bureau of Prisons. This survey showed that in the 12 preceding months, the number of incidents ranged from 0 (in 17 facilities) to more than 200 (in 8 facilities, with one facility with 652 incidents). Most facilities fell between these extremes, however, reporting between 7 and 90 incidents. Overall, facilities reported a mean of 70 and median of 34 incidents.[3]

In prisons, the incidence of the use of force is greater within larger facilities and within maximum security facilities. In contrast, force may be used more frequently in jails, especially in the intake units that receive arrestees who are intoxicated, angry, or frightened. A study of use of force incidents in the Maricopa County, Arizona jail system found 2995 reported use of force incidents (used or threatened) over a two-year period.[4] Over half of these incidents (1808) occurred in the intake facility, while the remaining 1187 incidents occurred in the remaining six jail facilities.

Lethal force is most likely to be used against escaping inmates and to control group disturbances, whereas nonlethal force is most likely to be used when officers become involved in inmate-on-inmate fights or when an inmate refuses to comply with lawful orders.[5] Most incidents of nonlethal force are spontaneous, use only hands-on force, occur in housing units, and involve only one inmate. Although lethal force is designed to have deadly consequences, nonlethal force rarely results in injuries to either officer or inmate. Most inmate injuries and nearly all officer injuries from nonlethal force are minor abrasions or scrapes.

■ Use and Effectiveness of Nonlethal Weapons

Studies by the ACA and the Institute for Law and Justice reach similar findings with regard to the availability of nonlethal weapons in correctional facilities. These reports indicate that:

- Nonlethal weapons are present in most facilities, although prisons are more likely than jails to have such weapons.
- Chemical irritants and batons (or some type of impact weapon) were available for use in nearly all the prisons studied, but in only about half of the jails.
- Less-than-lethal projectile guns were available in nearly half the prisons and less than 20 percent of the jails, and a stun device was available in approximately one third of both the prisons and the jails.[6]

Not all officers are routinely equipped with these weapons, however. Instead, these weapons are more likely to be stored in a central arsenal or distributed only to certain, perhaps supervisory, staff. As a result, many jails and prisons report that the weapons were not used during the preceding year. Even in those facilities in which these weapons had been used at least once during the past year, only the chemical irritants were used an average of 10 or more times.

When surveyed about the effectiveness of nonlethal weapons, prison and jail administrators considered all the following nonlethal options to be effective, rated in order of effectiveness: less-than-lethal projectile gun, chemical irritants (such as oleoresin capsicum—pepper spray), stun devices, batons, and other nonlethal weapons.[7] After overcoming initial resistance by jail personnel, stun devices and pepper spray quickly became integral tools for officers' responses to altercations with inmates. One study noted that the mere display or threat of a nonlethal weapon often was enough to control inmates and terminate an altercation. As such, pepper spray use was found to be infrequent and more likely to be used when only one officer was involved with more than one inmate, especially when controlling inmate-to-inmate altercations. Stun devices and pepper spray were likely to totally incapacitate noncompliant inmates, and their level of effectiveness was consistently high regardless of gender or size of inmate or the degree of resistance encountered.[8] Research has demonstrated that nonlethal weapons are effective in gaining control over noncompliant inmates except when used against inmates with a mental or substance impairment.

■ Officer Attitudes Toward the Use of Force

Unlike the policing literature, few studies examine correctional officers' attitudes toward the use of force.[9] Studies have found that detention officers with increased levels of punitive and custodial orientation, greater satisfaction with the quality of supervision, or more concerns about role conflict, fear of victimization, and personal authority reported an increased readiness to use force.[10] It is important to note, however, that readiness to use force on the part of correctional officers is neither necessarily punitive nor indicative of the potential for misuse of force. A competent corrections officer is required to maintain a certain level of readiness to use force. Both correctional training and policy emphasize that officers should be prepared to use legitimate force to gain the compliance of or maintain control over inmates.

■ Civil Liability of Lethal Force

Different standards are used to judge the civil liability of the use of lethal force by police and correctional officers. For police, the appropriate use of lethal force is judged in terms of the Fourth Amendment's prohibitions against unreasonable seizure. The standard is defined by the Supreme Court's decision in *Tennessee v. Garner* (1985), which specifies that deadly force is only appropriate to seize a fleeing suspect when, under the totality of the circumstances, "the officer has probable cause to believe that the suspect poses a significant threat of death or serious physical injury to the officer or others."[11]

In contrast, civil liability for correctional officers is defined in terms of the Eighth Amendment, which provides that "Excessive bail shall not be required, nor excessive fines imposed nor cruel and unusual punishments inflicted." Although the cruel and unusual punishment clause is an explicit "intention to limit the power of those entrusted with the criminal law function of government," it protects against only those actions that are "repugnant to the conscience of mankind."[12]

As a result, correctional officers have more latitude than police officers in the use of force. First, the courts begin with the (often false) assumption that all escaping prisoners are dangerous, so evidence of a threat is not required to justify the use of lethal force. Second, warning shots and shooting to maim cannot be justified by current standards that govern police, but can be justified, and may even be preferred by the courts, in correctional settings. Warning shots may be more safely used when prisons are separated from the neighboring populace by open fields. If the situation permits less-than-deadly force, then deadly force should not be used—disabling force is always preferred to deadly force.[13]

Finally, the courts have established that only the unnecessary and wanton infliction of pain constitutes cruel and unusual punishment as forbidden by the Eighth Amendment.[14] In *Whitley v. Albers*, a case involving an officer who shot an inmate during a disturbance, the Supreme Court concluded that "whether the particular measure undertaken inflicted unnecessary and wanton pain and suffering ultimately turns on whether force was applied in a good faith effort to maintain or restore discipline or maliciously and sadistically for the very purpose of causing harm."[15]

Civil Liability of Nonlethal Force

Correctional officers may use force lawfully within a correctional institution in defense of self or others or to enforce prisoner rules and regulations, prevent a crime from occurring, or prevent escape. In all cases, the degree of force used must be shown to have been reasonable under the totality of the circumstances known at that time. A successful claim of excessive force must demonstrate that, either intentionally or through gross negligence or recklessness, officers used excessive force that inflicted bodily harm under circumstances when officers knew (or should have known) that it was an unnecessary and wanton infliction of pain.[16] Force applied in a good faith effort to maintain discipline is not excessive; a successful plaintiff must prove that the force was applied maliciously and sadistically to cause harm.

Litigation typically occurs under one of four conditions:[17]

1. Those cases that assert that an officer reacted improperly to an inmate, either by overreacting to a resistant inmate or by using or threatening to use force against a nonresistant inmate; the use, or continued use, of a nonlethal weapon is examined in terms of the officer's reasonable belief that such force was necessary.

2. Negligent use of nonlethal weapons contrary to the manufacturer's recommendations; this implies carelessness, not wanton harm, and is judged against the standards of use expected of any trained officer.

3. Failure to provide timely medical aid to an inmate who has been injured. In *Estelle v. Gamble*, the Court ruled that there must be a deliberate indifference to the inmate's medical needs and not simply an inadvertent failure to provide medical care.[18] Indifference implies culpability whenever a reasonable person would have known that medical aid was needed. Further, the deliberate indifference must result in substantial harm to the inmate.[19]

4. Legal liability for any misuse of a nonlethal weapon may extend to supervisors or administrators. Supervisors may be liable whenever it can be established that they (1) failed to intervene at the scene to prevent an officer's excessive use of force, (2) assigned an officer to a duty or issued to the officer a weapon that he or she was not trained to use, or (3) failed to investigate or discipline an officer who was known to have misused nonlethal weapons. Administrators who fail to provide training or to establish clear policies regarding the proper and appropriate uses of lethal and nonlethal weapons also may be liable.[20]

■ Policies and Training

Both the National Sheriffs Association and the National Association of Chiefs of Police have recommended a few basic actions that agencies can take to reduce their legal liability for excessive force claims. Miller offers a similar, but more detailed, set of recommendations.[21] Further, model policies have been drafted by the ACA and the Commission on Accreditation of Law Enforcement Agencies. At a minimum, each agency should have a written use of force policy, use of force training, and systematic institutional review of all incidents in which force is used.

Use of Force Policy

One recommendation is to create, and then to periodically review, written policies regarding the appropriate application of force. The use of force policy should clarify the levels of force that are appropriate for various situations that arise, always calling for the use of the minimum force needed. In many cases, the levels of force are portrayed as a continuum, with higher levels of force being used in response to a higher threat or level of resistance from inmates. These policies also should include reporting requirements, review procedures, and a statement of disciplinary actions that may result from the excessive use of force. Finally, the policy must include a clear statement pertaining to medical aid and medical review following the use of force.

If nonlethal weapons are available, the policy should indicate which weapons are authorized for use, which officers are authorized to carry or use such weapons, and when and how officers will be trained to use such weapons. The policy also should:

- Include a statement that locates the weapons on the continuum of force (e.g., is a chemical spray less severe than, more severe than, or the same severity as the use of hands-on contact with an inmate?)
- Give specific directions or limitations for their use (e.g., stun devices may not be applied to the head or the genitals and batons are not to be applied to the head or the neck)
- Define any situations in which, or persons (e.g., pregnant women, older persons, or persons who are mentally ill) against whom, nonlethal weapons are not to be used

The use of lethal weapons also must be included in any use of force policy. These policies are likely to be tailored to the individual needs of each correctional facility. The Federal Bureau of Prisons, for instance, stipulates that deadly force may be used to prevent escapes from secure facilities or to stop an inmate whose actions present imminent danger to others. Moreover, firearms generally are not to be used in a minimum security institution; they may be used to prevent an escape from a minimum security institution only when specifically authorized by a warden.[22]

In addition, deadly force policies should clarify expectations with regard to such related matters as the use of warning shots and shooting to maim, as both practices are preferred by some jurisdictions and prohibited by others. Indeed, Nebraska, Texas, and some other states have policies that explicitly state that disciplinary actions may be taken against officers who fail to use deadly force to prevent an escape.

Use of Force Training

Agencies should provide all officers with entry-level and in-service training in the use of force. Training should include a review of all institutional policies pertaining to the use of force as well as some general guidelines about the officer's civil liability when using force. Training should include technical information and physical training regarding the proper use for each authorized force technique or weapon, whether handholds, physical restraints, or such nonlethal weapons as stun devices and chemical irritants. Trainees also should receive basic information about rendering temporary medical assistance to those who may be harmed by the use of force, especially by chemical irritants. Finally, the training should acquaint officers with a large number and variety of situations through role-playing activities that require them to exercise discretion along the entire use of force continuum. It is important to train for deescalation of force and to emphasize that verbal communication skills are the preferred means of inmate control.[23]

Systematic Institutional Review

Agencies should develop and maintain a process by which all instances of the use of force are systematically recorded and reviewed.[24] All incidents of any physical contact with an inmate, regardless of whether they result in an injury, should be recorded and reviewed, as should all incidents in which an inmate is threatened with a lethal or nonlethal weapon. Systematic and formal review will help to identify problem areas and problem inmates within the institution and officers who may need further training or formal sanction. Unacceptable behaviors by officers or supervisors must be addressed and, if appropriate, sanctioned according to a set of written disciplinary guidelines.

■ Utility and Costs

Use of force policies, practices, and training vary widely across correctional institutions.[25] There is no single standard or model that fits all institutions. For that matter, there is no consensus about what is defined as use of force or what is an appropriate response to a given situation. Systematic data on use of force incidents are almost nonexistent. The state of current knowledge about the use of force in corrections is best illustrated by the fact that these simple questions remain unanswered:

1. How often, and in what contexts, is lethal (or nonlethal) force used annually?
2. What types of inmates and officers are involved, and what situational factors are present, in use of force incidents?
3. How many inmates and staff are injured annually as a result of lethal and nonlethal force incidents, and what is the extent of their injuries?

Despite the fact that the use of force is a common and expected occurrence in correctional settings, little is known about its utility and its costs.

The question of utility revolves around the issue of what type or level of force is most useful in maintaining safety and control. Today, the focus is on the utility of nonlethal weapons in corrections. The Science and Technology Division of the National Institute of Justice supports many efforts to examine the usefulness of chemical irritants, electronic stun devices, intense pulsating lights, sticky foam, capture nets, projectile launchers, stun grenades, and other less-than-lethal weapons in prisons and jails. The limited data available suggest that nonlethal weapons can be used effectively in corrections, but more research is needed to discern which weapons are effective and which weapons are ineffective (or worse, counterproductive) for specific situations or settings.[26] The following questions still need to be addressed:

1. Should such nonlethal weapons as chemical irritants and handheld stun devices be issued to all officers?
2. Will the presence or threat of a nonlethal weapon calm or exacerbate the situation?

3. Will the use of a nonlethal weapon be more effective than conventional hands-on force in gaining control over the inmate and in reducing injuries to the inmate and officer involved in the incident?

The use of force involves many potential costs. One is the direct financial cost associated with injuries that occur to inmates or officers in the application of force. For inmates, these direct costs comprise the medical treatment received; for officers, these costs include medical treatment, workers' compensation claims, and lost work days. Indirect financial costs to the institution accrue in the form of the time and dollars spent to process inmate grievances, to respond to civil lawsuits alleging excessive force, and to address governmental (e.g., U.S. Department of Justice) and nongovernmental (e.g., Amnesty International) investigations into the misuse of force.

The use of force also has emotional costs for those who use force, especially deadly force. Law enforcement agencies routinely require some level of counseling following shooting incidents. How (if at all) does the use of force, especially deadly force, by correctional officers lead to feelings of guilt, remorse, or despair that affect an officer's quality of work and interpersonal relationships with family and friends? Unfortunately, these questions remain largely unanswered.

■ Conclusion

Available data on the incidence of the use of lethal and nonlethal force in correctional institutions suggest a wide disparity within and among federal, state, and local jurisdictions. It is generally agreed, however, that lethal force is restricted to use against escaping inmates and in response to group disturbances, while nonlethal force generally is applied to individual inmates as a means of control. Nonlethal weapons most commonly used in prisons and jails are chemical irritants and batons. Less-than-lethal projectile guns and stun devices are available in about one third of all correctional institutions and, together with chemical irritants, are generally considered to be most effective in controlling inmate behavior.

Issues of civil liability revolve around both lethal and nonlethal force. Although given more latitude than the police in resorting to lethal force, correctional officers are held to the cruel and unusual punishment prohibition contained in the Eighth Amendment to the U.S. Constitution as interpreted by several federal courts. Appropriate nonlethal force must generally meet the tests of reasonableness, non-negligence, timely medical aid for injuries resulting from the use of force, provisions for adequate training, and the existence of clear policies regarding the use of force.

To avoid liability for claims of excessive force, agencies should have a written policy on the use of lethal and nonlethal force. Such a policy should establish a continuum of force available to correctional officers and associate the levels of force with the levels of inmate resistance or threat. Training in the use of hands-on techniques for control of inmates, as well as lethal and nonlethal weapons, should be mandatory and updated regularly. A process of supervisory review is also helpful.

Issues surrounding the utility and costs of the use of force by correctional officers remain largely unresolved. The level of force and type of weapon best suited to a particular incident remain the subject of continuing debate. Injuries to both inmates and correctional officers and the economic impact of medical treatment, processing inmate grievances, and civil lawsuits should be seen as real costs likely to be incurred when even reasonable, nonexcessive force is used.

DISCUSSION QUESTIONS

1. What costs are associated with the misuse of force by correctional officers?

2. How does the court assess allegations of misuse of force by a correctional officer?

3. What are the relative advantages and disadvantages of alternative, non-lethal weapons in a correctional setting?

4. In what instances may correctional officers use force lawfully within a correctional institution?

5. How might officer attitudes toward the work environment influence their readiness to use force against an inmate?

ADDITIONAL RESOURCES

R. Johnson, *Hard Time: Understanding and Reforming the Prison*, 3rd ed. (Belmont, CA: Wadsworth, 2002).

J. Marquart, "Prison Guards and the Use of Physical Coercion as a Mechanism of Prisoner Control," *Criminology* 24 (1986), pp. 174–188.

P. Zimbardo, *The Stanford Prison Experiment: A Simulation Study of the Psychology of Imprisonment Conducted at Stanford University*, available at http://www.prisonexp.org

NOTES

1. J. Hepburn, "The Exercise of Power in Coercive Organizations: A Study of Prison Guards," *Criminology* 23, no. 1 (1985), pp. 145–164.

2. C. Hemmens, J. Maahs, and T. Pratt, "Use of Force in American Jails: A Survey of Current Policies." *American Jails*, January/February (2000), pp. 47–52; P. Henry, J. Senese, and G. Ingley, "Use of Force in America's Prisons: An Overview of Current Research," *Corrections Today* 56 (1994), pp. 108–114.

3. P. Henry, J. Sense, and G. Ingley, "Use of Force in America's Prisons: An Overview of Current Research," *Corrections Today* 56, no. 4 (1994), pp. 108–114.

4. J. Hepburn, M. Griffin, and M. Petrocelli, *Safety and Control in a County Jail: Nonlethal Weapons and the Use of Force.* A report submitted to Maricopa County Sheriff's Office, National Sheriffs' Association, and the National Institute of Justice (September 1997).

5. J. Senese, *Summary Report: Institutional Use of Force Reports.* A paper presented at the American Correctional Association Open Symposium on Use of Force (Orlando, FL, 1994), pp. 6–7.

6. Institute for Law and Justice, *Less Than Lethal Force Technologies in Law Enforcement and Correctional Agencies* (National Institute of Justice, Alexandria, VA: National Institute of Justice,1993), pp. 3–12.

7. J. Onnen, *Oleoresin Capsicum: Executive Brief* (Alexandria, VA: International Association of Police Chiefs, 1993).

8. Hepburn et al., *Safety and Control in a County Jail: Nonlethal Weapons and the Use of Force.*

9. M. Griffin, *The Use of Force by Detention Officers* (New York: LFB Scholarly Publishing, 2001).

10. M. Griffin, "The Influence of Organizational Climate on Detention Officers' Readiness to Use Force in a County Jail," *Criminal Justice Review* 24, no. 1 (1999), pp. 1–26; M. Griffin, "The Influence of Professional Orientation on Detention Officers' Attitudes Toward the Use of Force," *Criminal Justice and Behavior* 29, no. 3 (2002), pp. 250–277.

11. *Tennessee v. Garner*, 471 U.S. 1 (1985).

12. *Estelle v. Gamble*, 429 U.S. 97, 98 (1976).

13. J. Walker, "Police and Correctional Use of Force: Legal and Policy Standards and Implications," *Crime and Delinquency* 42 (1996), pp. 144–156.

14. *Wilson v. Seiter*, 501 U.S. 294 (1991).

15. *Whitley v. Albers*, 475 U.S. 312 (1986).

16. *McRorie v. Shimoda*, 795 F.2d 780, 9th Cir. (1986).

17. Institute for Law and Justice, *Less Than Lethal Force Technologies in Law Enforcement and Correctional Agencies.*

18. *Estelle v. Gamble,* 429 U.S. 97, 98 (1976).

19. *May v. Enomoto*, 633 F.2d 164 (CA9 1980).

20. D. Daane and J. Hendricks, "Liability for Failure To Adequately Train," *Police Chief* 58, no. 11 (1991), pp. 26–29.

21. N. Miller, "Less-Than-Lethal Force Weaponry: Law Enforcement and Correctional Agency Civil Law Liability for the Use of Excessive Force," *Creighton Law Review* 28, no. 3 (1995), pp. 733–794.

22. Federal Bureau of Prisons, *Program Statement 558.12 Firearms and Badges*, (Washington, DC: Federal Bureau of Prisons, 1996), p. 6.

23. J. Nicoletti, "Training for De-Escalation of Force," *Police Chief* 57, no. 7 (1990), pp. 37–39.

24. D. Lyons, "Preventive Measures Cut Physical Force Suits," *Corrections Today* 52 (1990), pp. 216–224.

25. Hemmens et al., "Use of Force in American Jails: A Survey of Current Policies."

26. Institute for Law and Justice, *Less Than Lethal Force Technologies in Law Enforcement and Correctional Agencies*; Hepburn et al., *Safety and Control in a County Jail: Nonlethal Weapons and the Use of Force.*

Creating the Future

VI

YOU ARE THE ADMINISTRATOR
College Students Assaulted

Consider how you would sentence the offender in this fictionalized account:

Two freshmen at a local university, Jim Carroll and Whit Caulkins, were jogging near the campus when three homeless men approached them and demanded money "for dinner." Jim responded first by saying truthfully that they did not have any money with them. One of the homeless men immediately became angry, and shouted: "You little rich boys better cough up some bucks, or both of you are going to be hurt!"

Whit's temper flared, and he stepped toward the homeless man, threatening: "Get lost you old piece of garbage or we'll teach you some manners." Jim tried to pull Whit back, but as he did, one of the men pulled a knife and stabbed Whit in the chest. The men then ran off, and Whit crumpled to the ground. Two other students witnessed the assault from a distance and ran to the nearest emergency phone to call for help. Law enforcement and medical providers reacted quickly, apprehending the homeless men and transporting Whit to the hospital emergency room. He had a serious injury, but not a fatal one.

The offenders were transferred to the city detention facility. The one charged with the knife assault, Charles Spellman, had been diagnosed with schizophrenia and had been in and out of mental institutions for the past 25 years. He had many previous arrests for vagrancy and petty assault, but few convictions due to his history of mental instability. The night of the attack he had been out drinking with the two other offenders before they confronted Jim and Whit.

- *What types of sanctions are appropriate for Mr. Spellman, given the overwhelming cost of confinement and increasingly longer sentences?*
- *What purpose should guide the sentencing decision (e.g., rehabilitation, incapacitation, punishment, or general deterrence)?*
- *How has the treatment of cases like this and the handling of inmates like Mr. Spellman changed in recent years? How will it change in the future?*

The Future of Sentencing

46

Julius Debro

Chapter Objectives

- Explain how sentencing in criminal cases has changed in recent decades.
- Differentiate between determinate and indeterminate sentencing and identify the benefits and drawbacks of each.
- Discuss the purposes of sentencing guidelines.

There has been a massive increase in U.S. jail and prison populations in recent years. By 2005, the total incarceration rate (prison and jails) reached 738 per 100,000, making America's incarceration rate the highest in the world.[1] One of the major reasons for the increased prison population is that sentencing policy has changed both at the state and federal level. The public views crime as one of the nation's most important issues, and legislators are demanding stiffer penalties for criminal offenders.

■ Sentencing Disparity

Historically, decisions about types of punishment were made by the legislature, with judges attempting to follow the will of the legislature but often making decisions based on their review of the law. Judges' decisions quite often depended upon information received from the prosecutor or the probation department. There was never any consistency in sentencing or sentencing legislation throughout the country. Judges were never given clear goals, and they had broad powers to impose sentences.[2]

Quite often, sentences imposed by judges were harsh, discriminatory, and inconsistent.[3] Persons committing the same offense but appearing before a different

503

judge would receive vastly different sentences. Judges' sentencing practices had three major goals: to deter others from committing crimes, to incapacitate individuals who were considered dangerous to the community, and to rehabilitate offenders. However, these goals were often incompatible with public needs. Offenders were not deterred by long sentences, and those persons who had been incapacitated for long periods of time ended up being released much earlier than the public expected.

Indeterminate Sentencing

By the 1960s, every state in the nation had an indeterminate sentencing system.[4] Indeterminate sentencing was a form of sentencing in which the judge sentenced a person to a range of years. The range could be from one year to life, which essentially meant that the defendant had to serve a minimum of one year but could serve his or her entire life in prison. As long as a judge sentenced within a statutory range, there were no rules to govern the amount of time a judge could impose.

An administrative committee called a parole board generally determined the amount of time a person served in prison. The parole board had its own set of guidelines for decision making, and a person could serve many years in prison without any knowledge of when he or she would be released. Over the years, there were complaints about the indeterminate sentence that lead to an investigation by the U.S. Senate Judiciary Committee. The Senate Judiciary Committee noted that the indeterminate sentencing resulted in an outdated and unworkable model of rehabilitation, stating that

> Recent studies suggest that this approach has failed . . . the rehabilitation model is not an appropriate basis for sentencing decisions. We know too little about human behavior to be able to rehabilitate individuals on a routine basis or even to determine accurately whether or when a particular person has been rehabilitated.[5]

This ruling was a major setback for indeterminate sentencing and rehabilitation. A final study released in 1975 concluded that, "Nothing works . . . . We haven't the faintest clue about how to rehabilitate offenders and reduce recidivism."[6] In response, legislators soon introduced sentencing guidelines at both state and federal levels.

Sentencing Guidelines

The major purpose of sentencing guidelines is to eliminate disparity and move away from rehabilitation as the goal for punishment. The guidelines have also been used in some cases to limit prison population growth by tailoring sentences to prison capacity.

In a move towards presumably-appropriate ranges of sentences within which judges must sentence offenders, ranges have been established by sentencing commissions. For the most part, sentences are based mainly on the severity of the crime and the defendant's prior history. Judges can depart from these guidelines, but they must state their reasons for the departures. When a judge does depart, there can be an appeal.

The federal government has been using sentencing guidelines since 1987. Sentencing guidelines also have been adopted in many states. Some states enacted mandatory minimum terms that may override the guidelines, but most did not. Some states also eliminated parole, as recommended by the American Bar Association policy group. Other states moved to descriptive guidelines designed to help judges follow existing sentencing norms more consistently.[7]

Problems with Sentencing Guidelines

According to Judge Jose Cabranes of the United States Court of Appeals for the Second Circuit, "The sentencing guidelines system is a failure—a dismal failure, a fact well known and fully understood by virtually everyone who is associated with the federal justice system."[8] One of the major problems with sentencing has been the problem of racial disparity. According to the Sentencing Project, a national advocacy organization that promotes reforms in sentencing law and practice,

> More than 60% of the people in prison are now racial and ethnic minorities. For Black males in their twenties, 1 in every 8 is in prison or jail on any given day. These trends have been intensified by the disproportionate impact of the "war on drugs," in which three-fourths of all persons in prison for drug offenses are people of color.[9]

Determinate sentencing structures can exacerbate racial disparities by imposing sentences that are disproportionate to the crime for which the offender was convicted. For example, while the *Journal of the American Medical Association* states that there is no medical difference between powder and crack cocaine, there are significant racial disparities in both usage and sentencing guidelines for these substances.[10] Perhaps as a result of similar disparities, more than three quarters of crack cocaine defendants in 2005 were African American.[11]

Attempts to establish or refute discrimination in criminal justice processes—especially in the courts—have been fraught with methodological difficulties. Sample sizes are generally too small, legal and nonlegal variables are difficult to control, and the problem of generalizing from studies of one court to another makes the research vulnerable to criticism. Different findings will always exist in different courts; thus any charges of discrimination must be attributed to the exercise of discretion at some point from apprehension to sentencing.

In summary, while federal guidelines were intended to make sentencing more predictable and less discretionary, the system has essentially transferred substantial discretion from judges to prosecutors.

■ Restorative Justice Model

American judicial systems are now moving toward restorative justice, a new concept that shifts the focus away from the offender and puts it on the victim and the community. In this model, crime is viewed as an act against people in the community and not just against the state, so justice involves the criminal making amends with the community. In addition to incarceration, sentencing in this model includes plans for restitution and mediation programs that involve the victim, the community, and the offender.

■ Structured Sentencing

The U.S. Supreme Court has made several recent rulings that have significantly influenced the use of federal and state sentencing guidelines in sentencing decisions.

First, in *Apprendi v. New Jersey* (2000), the Court questioned the fact-finding authority of judges in making sentencing decisions.[12] In this case, Charles Apprendi plea bargained with the state prosecutor and had pleaded guilty to unlawful possession of a firearm—a crime that carried a prison term of 5 to 10 years in New Jersey. However, the sentencing judge noted that in the offense, Apprendi had fired several shots into the home of an African American family in an effort to convince them to move from his neighborhood. The sentencing judge deemed this action a hate crime and utilized sentencing enhancement procedures to add to Apprendi's sentence of confinement.

The U.S. Supreme Court overturned the judge's decision, ruling that a judge may not impose an enhanced sentence based on facts that were not included in the plea bargain nor a finding of the jury process. The Court determined that the lower court had erred in applying sentencing enhancements based on aggravating factors that were not findings of fact from a jury or admitted by the defendant and that this was not permitted under the due process requirements of the Fifth Amendment and the jury guarantees of the Sixth Amendment.

In a second notable case, *Blakely v. Washington* (2004), the U.S. Supreme Court again ruled sentencing enhancement.[13] In this case, Blakely, an estranged husband, kidnapped his wife and transported her in a wooden box in the back of his pickup truck from her home in Washington to Montana. Blakely ordered their teenage son to follow in a separate car, threatening to shoot his mother if he did not comply. While traveling, the teenager managed to notify the police, and Blakely was arrested. He subsequently pleaded guilty to kidnapping, which carried a maximum sentence of 53 months under state sentencing guidelines.

However, the judge departed from these guidelines and sentenced Blakely to 90 months, based a belief that Blakely's offense was deliberately cruel. Upon appeal, the U.S. Supreme Court found that the facts in support of Blakely's enhanced sentence were neither admitted by the defendant nor found by a jury and were therefore a violation of the Sixth Amendment right to trial by jury.

Finally, the U.S. Supreme Court ruled in *U.S. v. Booker* (2005) and *U.S. v. Fanfan* (2005) that the United States Sentencing Guidelines were unconstitutional as applied and that the problem could be remedied only by making the guidelines voluntary for federal judges.[14] In the *U.S. v. Booker* case, a federal judge had again added time onto the offender's sentence based on "additional facts by a preponderance of the evidence." The U.S. Supreme Court ordered the judge to either sentence within the guidelines or to hold a separate sentencing hearing before a jury. In the *Fanfan* case, the federal sentencing judge was mindful the *Blakely* decision, and concluded that he could not impose an expanded sentence. The government appealed this decision, but again, the Court sided with the judge, concluding that the Sixth Amendment, as construed in *Blakely*, does apply to federal sentencing guidelines.

Conclusion

In the future, the American system of justice may gradually return more sentencing discretion to judges. Judges now have access to the complete historical record of the offender and are in a better position to make sentencing decisions as compared to the legislators or sentencing commissioners who establish sentences without regard to the individual characteristics of an offender. In fact, between 2004 and 2006, 22 states enacted some form of sentencing reform.[15] The United States Sentencing Commission also has recommend changes to federal sentencing guidelines. One such change would reduce the disparity between sentences for powder and crack cocaine.[16]

Elected representatives must be willing to review the effectiveness of the legislated sentencing guidelines and seek adjustment where necessary. It is extremely important that sentencing guidelines at the federal and state levels be reevaluated routinely. Legislative bodies need this feedback, and the credibility of the entire American judicial system is at stake.

DISCUSSION QUESTIONS

1. What are the primary differences between determinate and indeterminate sentencing?
2. What are sentencing guidelines and why were they created?
3. How can the problems with sentencing guidelines be resolved?
4. Why were determinant sentencing structures enacted?
5. What are some of the problems attributed to determinate sentencing?

ADDITIONAL RESOURCES

Families Against Mandatory Minimums, http://www.famm.org
The Sentencing Project, http://www.sentencingproject.org
United States Sentencing Commission, http://www.ussc.gov

NOTES

1. U.S. Department of Justice, Bureau of Justice Statistics, *Prison and Jail Inmates at Midyear 2005* (May 2006).
2. I. Nagel, "Structuring Sentencing Discretion: The New Federal Sentencing Guidelines," *Journal of Criminal Law & Criminology* 80 (1990), pp. 883, 913–939.
3. L. Katz, *The Justice Imperative: An Introduction to Criminal Justice* (Cincinnati, OH: Anderson Publishing Company, 1980), p. 32.
4. Nagel, "Structuring Sentencing Discretion: The New Federal Sentencing Guidelines."
5. D. Lipton, R. Martinson, and L. Wilks, *The Effectiveness of Correctional Treatment* (New York: Praeger Publishing, 1975).
6. The National Commission on Reform of Federal Criminal Law (Brown Commission), *Hearings before the Subcommittee on Criminal Laws and Procedures of the Senate Judiciary Committee,* 92d Congress, 1st session, (1971).
7. R. Fraser, "Prison Population Growing under Minnesota Guidelines," *Overcrowded Times* 4, no. 1 (1993).
8. J. Cabranes, "Sentencing Guidelines: A Dismal Failure, *New York Law Journal* 227 (1992), p. 2.
9. The Sentencing Project, "Racial Disparity" available at http://www.sentencingproject.org/IssueAreaHome.aspx?IssueID=3, accessed September 11, 2007.
10. D. Hatsukami and M. Fischman, "Crack Cocaine and Cocaine Hydrochloride: Are the Differences Myth or Reality?" *Journal of American Medical Association* 276 (1996), pp. 1580–1588.

11. United States Sentencing Commission, *2005 Sourcebook of Federal Sentencing Statistics* (Washington, DC: U.S. Sentencing Commission, 2006), Table 34.

12. *Apprendi v. New Jersey*, 530 U.S. 466 (2000).

13. *Blakely v. Washington*, 542 U.S. 296 (2004).

14. *U.S. v. Booker*, 543 U.S. 220 (2005); *U.S. v. Fanfan*, 25 S.Ct. 738 (2005).

15. R. King, *Changing Direction: State Sentencing Reforms 2004–2006* (Washington, DC: Sentencing Project, 2007).

16. United States Sentencing Commission, Report to Congress: *Cocaine and Federal Sentencing Policy* (Washington, DC: U.S. Sentencing Commission, 2007).

Growth of the Private Sector

Douglas C. McDonald

Chapter Objectives

- Describe the factors that gave rise to the growth of the private prison industry in the 1980s.
- Explore research findings about the cost and quality of private prison operations.
- Discuss the complexities of the relationship around corrections between the government and the private sector.

Whether to turn the management and operation of entire prisons and jails over to private firms has become a hotly contested issue. Contracts for more narrowly focused services such as healthcare or food services have existed for some time and have raised few objections. However, in the mid-1980s, a vocal private correctional industry emerged and offered to take over entire correctional facilities and, indeed, entire state systems. Since then, the private correctional industry has grown rapidly, and although many still see the issue as government *versus* the private sector, a more apt description of current affairs in many jurisdictions is government *and* the private sector. Private facilities are now an established part of the correctional landscape, and the further growth of the industry is likely to be strong.

Private Corrections—Repeating History

Private imprisonment is not a new invention; privately operated jails were commonplace in England until the 19th century.[1] In the United States, the government took sole responsibility for prisons and jails until wardens began leasing out convicts for work and housing assignments with private businesspeople in the early 19th century.[2] This practice was largely swept away by reform movements in the wake of scandals, but privately operated facilities continued to survive in low-security and community-based facilities and juvenile correctional systems.[3]

In the mid-1980s, a newly formed private firm offered to take over the entire state prison system in Tennessee and to run it more efficiently—an offer that was considered but ultimately declined. With slightly less notice, a number of small firms began to contract with local governments for private management of jails and with states for low security and some medium security prisons.[4] In many quarters, however, opposition to privatization remains powerful, as does the sense of battlement. For example, a national organization of public correctional employees, the Corrections and Criminal Justice Reform Task Force, declared at its third annual meeting in 1997 that privatization was the "number one threat to our profession in the nation."[5]

The emergence of the private imprisonment industry was the result of several factors:

- Demand for prison and jail beds had been growing, largely because of tougher sentencing laws and the war on drugs.
- State and local governments were slapped with expenditure caps.
- Public debt ceilings were being reached.
- Voters were declining to approve increases in public debt for prison and jail construction.[6]

Faced with conflicting demands caused by these phenomena, many public administrators welcomed the solution offered by private entrepreneurs. Private firms would build the needed facilities using their own capital and then charge the government a price that would recoup both the capital investment and ongoing operating costs. Governments could pay for these services using funds appropriated for operations, thereby avoiding the need to gain voters' approval of increased public debt.

In the mid-1980s it looked as if the private sector was threatening to take over huge parts of the public correctional industry. This did not happen, but growth of the private correctional industry has still been strong. Between 1986 and 1996, the number of beds in private facilities in operation or under construction increased in this country at an average annual rate of 45 percent. Few of these beds were empty. The occupancy rate of all private adult facilities averaged 96 percent during 1996. In 1987, there were about 3000 prisoners in private facilities. By 1996, the number had soared to more than 85,000. During the 12 months ending December 31, 1996, the number of prisoners increased 30 percent. By the end of that year, there were 132 prisons and jails in operation, with another 39 under construction or in the planning stages. Seventeen state governments and the District of Columbia have at least one contract with a private facility within their borders, and some states (California, Florida, and Texas) contract for large numbers of facilities. Twelve other states also contract with facilities located in other states; these "exported" prisoners accounted for about 13 percent of all those under custody in private facilities at the close of 1996.[7]

As a result of these trends, growth in the private correctional industry's revenues has been explosive: from about $650 million in 1996 to over $1 billion expected in 1997.[8] By mid-1997, Wall Street and individual investors were impressed with these growth statistics and with the apparently bright prospects for future growth (private facilities have only about 3 percent of the market share of prisoners in the United States). Stock prices of the four publicly traded firms had consequently been bid up very high, providing these companies with substantial amounts of cash to finance further expansion.

The industry has long been dominated by a few big players. By the end of 1996, the Corrections Corporation of America, based in Nashville, held 49 percent of all prisoners in private adult facilities. The Wackenhut Corrections Corporation held 27 percent.[9] Today, the industry appears to be experiencing still further consolidation, as well as some diversification. Smaller firms are being acquired by larger ones, and some are developing new capabilities—such as drug treatment services—to augment their core competencies. Firms that have been focusing on the adult correctional market are also moving into juvenile corrections.

Whereas the Corrections Corporation of America offered in 1986 to pay Tennessee $250 million for a 99-year lease on the state's prisons, most private facilities have either been newly constructed or are government-owned and operated under a service contract. Unlike in the United Kingdom, government entities in the United States have not generally divested themselves of public properties. In 1997, however, exactly that happened. The District of Columbia sold its correctional treatment facility to the Corrections Corporation of America for $59 million, and the firm will operate it for the district under a 20-year leaseback arrangement. Proposals to do the same were being floated again in Tennessee in 1997, and in Florida as well. By 2005, 7 percent of all inmates were housed in private facilities, and among the federal inmate population, nearly 15 percent were in private prisons.[10]

The National Performance Review also has stimulated greater attention to improving government performance. The Review specifically supports privatization as one means of improving effectiveness and efficiency. Although the federal government cannot dictate programmatic preferences of state and local governments, this broad movement to improve government operations and support privatization of services is no doubt very influential. Indeed, surveys of state and local governments have found increased use of contracting for a broad range of social services.[11]

The effects of this increasing reliance upon private imprisonment have not been studied extensively. Most research attention has focused on whether private facilities are less costly than public ones and, to a lesser extent, whether their services are better or worse. Little systematic research has addressed other questions such as whether privatization has furthered government objectives other than cost containment, or how the experience of relying upon both government and privately operated facilities has changed or not changed correctional administrators' approaches to managing imprisonment services. Nor has there been systematic study of what some call a new "correctional-industrial" complex, in which a well-financed private correctional industry lobbies for criminal sentencing legislation that expands the supply of prisoners and, by extension, the potential for greater profits. Concerns about such self-interested distortion of penal policy-making have been voiced for years, but no studies have sought to determine the extent to which lawmakers are actually swayed by private industry lobbying.[12] The political pressures to pass "get-tough" legislation are already powerful in this country, even without any obviously self-interested lobbying by private correctional firms.

■ Cost Comparisons: Public versus Private

Studies comparing the cost of private and public facilities have not reported consistent findings. Several reports find private facilities to be less costly. These include studies of Illinois work release centers and Logan and McGriff's study of a 350-bed minimum security work camp in Tennessee.[13] (The latter study estimated that contracting was 3 to 8 percent less costly than the public alternative.) In 1989, the Urban Institute conducted a study which compared a private minimum security facility in Kentucky, a similar facility operated directly by the state, and two secure treatment facilities for serious juvenile offenders in Massachusetts—one public and the other private. It concluded that government-run facilities were 20 to 28 percent more expensive than private ones.[14] Similarly, a study by the state of Texas estimated that the cost of contracting for four different privately operated prisons was about 15 percent lower than what direct government operation would have cost.[15] A study by Archambeault and Dies of two privately operated facilities in Louisiana and one government-operated one concluded that the privately operated facilities were cheaper by 12 to 14 percent over a 5-year period.[16]

Still other studies found small or insignificant differences in costs. For example, an early study of the Eckerd Foundation's operation of the Florida School for Boys at Okeechobee found no cost savings attributable to private management compared with the cost of another publicly operated training school.[17] In his study of public and privately operated custodial facilities for juveniles in the United States, Donahue calculated the average cost per resident to be $22,600 in the public facilities and to be $22,845 in the private ones—an insignificant difference.[18] A statewide Tennessee study compared costs of a private multicustody facility and two state-run medium security prisons and also found an insignificant difference.[19]

At least one study found private facilities to be more costly than their public counterparts. In a 1985 study, the Pennsylvania General Assembly's Legislative Budget and Finance Committee examined cost data provided by the U.S. Immigration and Naturalization Service (INS) for government-run centers and privately run centers for detaining illegal aliens. The study concluded that the average daily cost per inmate was 17 percent higher in the private facilities.[20] However, another study of INS detention centers estimated that private centers were 7 to 19 percent less costly to the government.[21] This was confirmed in yet another study, which found the private facilities to be substantially less costly than INS-operated ones on average.[22]

Fewer studies have attempted to compare the quality of services delivered. In the Urban Institute's study, the authors reported that "by and large, both staff and inmates gave better ratings to the services and programs at the privately-operated facilities; escape rates were lower, there were fewer disturbances by inmates; and in general, staff and offenders felt more comfortable at the privately operated ones."[23] Another study by Logan of public and privately operated facilities in New Mexico reported equivocal findings.[24] The Tennessee study found no difference in the level of performance among the privately operated and public facilities studied.[25]

One reason the findings regarding costs are inconsistent is that determining these costs is surprisingly difficult. Analysts have not always recognized the extent of this difficulty, which in part explains the variable findings in the research literature. Accounting practices followed in the public and private sectors differ in significant ways, which frustrates direct comparisons of costs.[26] In the private sector, accounting methods have been designed to value all inputs used to produce a good or a service, and most costs are thereby captured. In contrast, public sector accounting systems were designed not to identify costs but to monitor expenditures to ensure that funds are used for their intended purposes. The focus of accounting is therefore on the agency rather than the service being delivered. In many jurisdictions, a number of different agencies and government accounts provide funds or other resources used for correctional institutions; counting only those expenditures by correctional agencies produces an undercount of the true cost of publicly delivered imprisonment. For example, retirement fund contributions for employees in some places are paid not by the correctional agency's funds but from an overhead government account.[27] Other departments may provide medical or psychiatric care, utilities, transportation, or educational services. In nearly all governments, separate accounts are not kept for capital expenditures (as opposed to ongoing operating expenditures), and determining the cost of the capital assets "consumed" during a particular period of service is nearly impossible.

On the private side, other obstacles exist to identify the costs of contracting. The price charged to governments may not cover all costs, as firms may elect to experience shortfalls in hopes of winning more work in the longer run or may subsidize operations in one place with earnings obtained elsewhere. Still other costs of contracting, not always counted, include the government's expenditures to procure the contracts and to monitor their operations. A fair calculation would include all of the government's costs of contracting, not just payments to contractors.

An additional methodological challenge faced by researchers is that comparable public and private facilities may not exist in the same jurisdiction for study. Consequently, researchers have had to estimate the costs of hypothetical public facilities or have compared costs from other facilities that are not precisely equivalent—which raises doubts about the validity of the inferences drawn.

With respect to costs and savings, students of the evaluation literature have not drawn consistent conclusions. For example, a study by the U.S. General Accounting Office examined five evaluations and reported that:

> We could not conclude from these studies that privatization of correctional facilities will not save money. However, these studies do not offer substantial evidence that savings have occurred. . . . These studies offer little generalizable guidance for other jurisdictions about what to expect regarding comparative operational costs and quality of service if they were to move toward privatizing correctional facilities.[28]

Proponents of privatization, in contrast, read the research literature as showing "still more evidence that operating cost savings in the general range of 10 to 20 percent are typical" and that "these cost savings are often matched with performance improvements (e.g., few disturbances, few escapes, increased prisoner involvement in work programs, and more programs aimed at reducing recidivism)."[29]

There does not seem to be a universally prevalent cost advantage to public or private entities. Rather, because of various constraints (labor availability, restrictions on employee salary levels, regulatory requirements, government procurement procedures, etc.), private firms may be able to exploit opportunities in specific niches and find ways to deliver services at lower costs. In other places, such opportunities may not exist, and the publicly operated facilities may be less costly at a given level of service. Some governments are also more sophisticated than others in their contracting practices, which may result in lower relative costs and higher performance by the private firms.

In 2005, Abt Associates completed a research report on behalf of the National Institute of Justice which examined the private sector's operation of the federally owned Taft correctional institution (private operation of this facility had been mandated by Congress in 1997). The study compared both the operational cost and quality to similar facilities operated by the Federal Bureau of Prisons (BOP).

Abt concluded that the net cost to the government of contracting for the operation of the facility was less than what the government would have spent to operate the facility itself. Abt also noted that the private sector's performance under the contract was consistent with what was expected by the government, noting that the BOP renewed the 3-year base contract each subsequent year.[30]

■ Legal and Moral Issues

Critics have argued that privately operated correctional facilities are of questionable constitutionality, or are improper, regardless of whether they are more cost-efficient. For example, in 1989, the American Bar Association (ABA) House of Delegates passed a resolution urging jurisdictions to proceed "with extreme caution in considering possible authorization of contracts with private corporations or other private entities for the operation of prisons or jails."[31] The accompanying report declared that:

> There can be no doubt that an attempt to delegate total operational responsibility for a prison or jail would raise grave questions of constitutionality under both the federal constitution and the consti-

tutions of the fifty states. The more sweeping the delegation, the more doubtful would be its constitutionality.[32]

For 50 years, the courts have allowed the federal government to delegate broad powers to private actors, and thus at the federal level, "private exercise of federally delegated power is no longer a federal constitutional issue." [33] Nor has delegation by state and local governments been seen as a federal constitutional issue since the 1920s. Not surprisingly, no federal court has found private imprisonment to be unconstitutional, despite the ABA's warning. Nor are there bans in most state constitutions against private delegation of correctional authorities. Nonetheless, judges in state courts have ruled inconsistently on issues regarding private delegation of state powers. To clarify this, legislatures in several states have passed laws authorizing delegation of correctional authority to private individuals or firms.[34] At least one state legislature (Washington) has passed laws explicitly banning privatization of formerly public functions.

Some critics argue that imprisonment is a core function of government, something that is intrinsically governmental in nature and should not be delegated to private actors.[35] However, the definition of what constitutes an intrinsically governmental function is being changed. Many policymakers find it entirely appropriate to delegate the administration of this function, while maintaining at the same time that the government has the responsibility for ensuring its provision. Accordingly, the courts have ruled that private imprisonment on behalf of government agencies constitutes "state action," and that governments retain the ultimate responsibility for what goes on in them.[36] Private facilities must comply with the same standards and laws that apply to public ones; they must conform to law and established standards.[37]

Still other objections to privatization have been voiced, arguing that the legitimacy of governmental authority is weakened in inmates' eyes by having private (and especially for-profit) corporations administer imprisonment.[38] Whether this actually occurs, however, is an empirical question that has not been studied. Moreover, it is reasonable to suspect that inmates' perceptions of legitimacy have more to do with whether the actions of the correctional workers conform to law and norms of fairness than whether they are public or private employees.

In short, in most states, the question of whether to privatize turns not on matters of constitutionality or statutory law but on matters of policy. And whether one thinks it proper to delegate imprisonment authority to private actors depends ultimately upon an individual's fundamental values and principles, as well as a consideration of the direct material interests of engaged in delivering correctional services. Given that there is no clear national consensus for or against private delegation of imprisonment services, policy battles in legislative chambers are likely to continue in coming years.

■ A Sensitive Relationship

The performance of a contractor depends to some extent upon the relationship between the contractor and the government and upon the government's management of the contractor. In at least some jurisdictions, contracting out facility operations has been handled poorly; the government's specification of the services to be delivered has been poorly defined, silent on objectives, and long on procedures to be followed. Even when governments turn to contracting in hopes of reducing costs, they sometimes fail to establish an initial benchmark—the cost of direct government operation. Evaluation of bidders' proposals has sometimes emphasized cost over more general value, with the result that "low ballers" have been chosen against the better interests of the government. Monitoring has not always been adequate, despite the federal courts' clear insistence that governments cannot evade responsibility and liability for correctional services by delegating them to private actors. Contracting for both facilities and operations also gives the winning firm an edge on future competitions (because it will then own a facility while others will have to build one) and may reduce competition in the marketplace.

Therefore, rather than focusing on whether existing private facilities are less or more cost effective than public ones, public managers should ask the following important question: How can government agencies obtain the results they seek to achieve? In contracting, governments have a tool for accomplishing any number of strategic objectives, which may include lower costs or improved correctional services. For example, a government seeking to lower correctional costs while maintaining service quality could establish a cost above which offers would not be entertained, explicit performance standards could be specified in the contract, monitoring systems could be designed to measure compliance with these standards, and actual costs could be monitored to ensure that targets have not been overrun. Governments can choose to terminate a contract if performance is not satisfactory.

Real life is more complicated and constrained than this simplified model suggests. For example, a commonplace observation is that government and businesses differ fundamentally in their purposes: Whereas private firms can be relatively single-minded in their pursuit of revenues or profit, government agencies and programs often strive to achieve multiple goals. This multiplicity of purposes stems, in part, from the genesis of programs in politics. That is, public programs are designed and enacted following a process by which different interests are accommodated. But this does not mean that government programs must operate with conflicting (or worse, unstated) missions. These missions can be clarified, and priorities can be established where multiple purposes exist.

■ Conclusion

How these developments in privatization will play out will probably depend as much upon politics as upon the merits or faults of private correctional facilities. Because prisons and jails claim such a large share of governments' budgets (especially at the state level), pressures for cost efficiency will probably continue. The threat of privatization may encourage public managers to find alternative ways of delivering correctional services, and organized public employees may succeed in staving off calls for their positions to be contracted out to private prison and jail operations. However, given the broad interest in relying upon private firms to deliver public services, calls for privatization are likely to continue.

DISCUSSION QUESTIONS

1. What factors gave rise to the growth of the private prison industry in the 1980s?
2. What legal and moral issues surround the topic of prison privatization?
3. Why is the contractual relationship between the government and private prison operators sensitive?
4. What have research findings contributed to the discussion about the cost and quality of private prison operations?
5. Should privatization be encouraged or discouraged in the future?

NOTES

1. A. Crew, *London Prisons of Today and Yesterday* (London: I. Nicholson & Watson, 1933), p. 50; M. Feeley, "The Privatization of Prisons in Historical Perspective," in *Privatization and Its Alternatives*, W. Gormley (ed.) (Madison, WI: University of Wisconsin Press, 1991), p. 397; W. Holdsworth, *A History of English Law*, vol. 4, 3rd ed. (London: Cambridge University Press, 1922–1924).

2. B. McKelvey, *American Prisons: A History of Good Intentions* (Montclair, NJ: Patterson Smith, 1977).

3. D. McDonald, "Private Penal Institutions," in *Crime and Justice: A Review of Research*, M. Tonry, ed. (Chicago: University of Chicago Press, 1992).

4. A. Press, "The Good, the Bad, and the Ugly: Private Prisons in the 1980s," in *Private Prisons and the Public Interest*, D. McDonald, ed. (New Brunswick, NJ: Rutgers University Press, 1992).

5. Corrections and Criminal Justice Reform Task Force, *Report on the 3rd Round Table Conference*, May 1997.

6. D. McDonald, "Introduction," in *Private Prisons and the Public Interest*, D. McDonald, ed. (New Brunswick, NJ: Rutgers University Press, 1990).

7. C. Thomas et al., *Private Adult Correctional Facility Census*, 10th ed. (Gainesville, FL: Center for Studies in Criminology and Law, 1997).

8. N. Xiong, "Private Prisons: A Question of Savings," *New York Times*, July 13, 1997.

9. Thomas et al., *Private Adult Correctional Facility Census*.

10. P. Harrison and A. Beck, *Bureau of Justice Statistics Bulletin: Prisoners in 2005* (Washington, DC: U.S. Department of Justice, Office of Justice Programs, 2006).

11. Council of State Governments, *State Trends and Forecasts: Privatization*, vol. 2, no. 2 (Lexington, KY: 1993); R. Miranda and K. Andersen, "Alternative Service Delivery in Local Government, 1982–1992," *Municipal Year Book 1994:* (Washington, DC: International City/County Management Association 1994).

12. K. Schoen, "Private Prison Operators," *New York Times*, March 28, 1985.

13. K. Chi, "Private Contractor Work Release Centers: The Illinois Experience," in *Innovations* (Lexington, KY: Council of State Governments, 1982); C. Logan and B. McGriff, "Comparing Costs of Public and Private Prisons: A Case Study," *NIJ Reports* no. 216 (1989).

14. Urban Institute, *Comparison of Privately and Publicly Operated Corrections Facilities in Kentucky and Massachusetts* (Washington, DC: U.S. Department of Justice, National Institute of Justice, 1989).

15. State of Texas, *Recommendations to the Governor of Texas and Members of the Seventy-Second Legislature* (Austin, TX: Sunset Advisory Commission, 1991).

16. W. Archambeault and D. Dies, Jr., "Cost Effectiveness Comparisons of Private versus Public Prisons in Louisiana: A Comprehensive Analysis of Allen, Avoyelles, and Winn Correctional Centers" (Baton Rouge, LA: Louisiana State University, School of Social Work, 1996).

17. A. Brown et al., *Private Sector Operation of a Correctional Institution: A Study of the Jack and Ruth Eckcerd Youth Development Center, Okeechobee, Florida* (Washington, DC: U.S. Department of Justice, National Institute of Corrections, 1985).

18. J. Donahue, *The Privatization Decision* (New York: Basic Books, 1990).

19. Tennessee Legislative Fiscal Review Committee, *Cost Comparison of Correctional Centers* (Nashville, TN: 1995).

20. Joint State Government Commission, *Report of the Private Prison Task Force* (Harrisburg, PA: General Assembly of the Commonwealth of Pennsylvania, 1987).

21. D. McDonald, "The Costs of Operating Public and Private Correctional Facilities," in *Private Prisons and the Public Interest*, D. McDonald, ed. (New Brunswick, NJ: Rutgers University Press, 1990).

22. D. McDonald, *Contracting for Private Detention Services: The Costs of Private and Government Detention Facilities* (Cambridge, MA: Abt Associates, Inc., forthcoming).

23. Urban Institute, *Comparison of Privately and Publicly Operated Corrections Facilities in Kentucky and Massachusetts*.

24. C. Logan, *Well-Kept: Comparing the Quality of Confinement in a Public and a Private Prison* (Washington, DC: U.S. Department of Justice, National Institute of Justice, 1991).

25. Tennessee Legislative Fiscal Review Committee, *Cost Comparison of Correctional Centers*.

26. McDonald, "Private Penal Institutions"; D. McDonald, *The Cost of Corrections: In Search of the Bottom Line* (Washington, DC: National Institute of Corrections, 1989).

27. D. McDonald, *The Price of Punishment: Public Spending for Corrections in New York* (Boulder, CO: Westview Press, 1980).

28. U.S. General Accounting Office, *Private and Public Prisons: Studies Comparing Operational Costs and/or Quality of Service* (Washington, DC: 1996).

29. Thomas et al., *Private Adult Correctional Facility Census.*

30. D. McDonald, "Contracting for Imprisonment in the Federal Prison System: Cost and Performance of the Privately Operated Taft Correctional Institution," NCJ 211990 (Washington, DC: National Institute of Justice/NCJRS, 2005). See also "Evaluation of the Taft Demonstration Project: Performance of a Private Sector Prison and the BOP," Federal Bureau of Prisons, unpublished report 2005; P. Harrison and A. Beck, Bureau of Justice Statistics Bulletin: *Prisoners in 2005* (Washington, DC: U.S. Department of Justice, Office of Justice Programs, 2006).

31. American Bar Association, "Report to the House of Delegates" (Chicago, 1989).

32. American Bar Association, "Report to the House of Delegates."

33. *Carter v. Carter Coal Company*, 298 U.S. 238 (1936); D. Lawrence, "Private Exercise of Governmental Power," *Indiana Law Journal* 61 (1986).

34. *Private Sector Involvement in Financing and Managing Correctional Facilities* (Washington, DC: National Criminal Justice Association, 1987).

35. I. Robbins, *The Legal Dimensions of Private Incarceration* (Washington, DC: American Bar Association, 1988).

36. *Medina v. O'Neill*, 569 F.Supp. 1028 (S.D. Texas, 1984); *Ancata v. Prison Health Services, Inc.*, 769 F.2d 700, 702 (11th Cir., 1985).

37. C. Logan, *Private Prisons: Pro and Con* (New York: Oxford University Press, 1990); McDonald, "When Government Fails: Going Private as a Last Resort."

38. J. DiIulio, Jr., "The Duty To Govern: A Critical Perspective on the Private Management of Prisons and Jails," in *Private Prisons and the Public Interest*, D. McDonald (ed.) (New Brunswick, NJ: Rutgers University Press, 1990); I. Robbins, *The Legal Dimensions of Private Incarceration* (Washington, DC: American Bar Association, 1988).

Chapter Resources

Corrections in the 21st Century

48

Martin F. Horn

> ## Chapter Objectives
>
> - Describe the role and relevance of corrections to society as a whole.
> - Identify the challenges faced by corrections systems today and in the future.
> - Explain the importance of sound correctional leadership.

The field of corrections, and imprisonment in particular, is large. Over 2 million Americans (one of every 136) are in prison or jail in the United States.[1] This includes one of eight black men between the ages of 25 and 29 and one in 27 Hispanic males in that same age group.[2] Black men have a 32 percent chance of serving time in prison at some point in their lives—a greater chance than that of going to college. Local, state, and federal spending on corrections exceeds $60 billion annually, and corrections employs about 750,000 people nationally.[3] Recent estimates suggest state and federal prisons will grow by more than 192,000 persons in the next five years, and the cost of incarceration will balloon by an additional $27.5 billion.[4]

So many people enter and leave prisons and jails that their experiences of confinement and the administration of the facilities in which they are incarcerated are often ignored, at grave risk to the quality of civic life. Increasingly, the public is demanding more transparency about how prisons and jails are managed. There is growing concern that the growth and operation of correctional systems are antithetical to democratic ideals. Following the atrocities at Abu Ghraib prison, the *New York Times* editorial board wrote, "The sickening pictures of American troops humiliating Iraqi prisoners have led inevitably to questions about the standards of treatment in the corrections system at home, which has grown tenfold over the last 30 years."[5]

It would be nice to live in a world where there were no prisons or jails, a world of perfect justice and harmony. But despite our best efforts, crime persists, and persons who would do harm to others have to be separated from the rest of society. It is important to remember that in its day, the prison was the great social reform of its time. Prior to that reform, penal institutions dealt with crime through corporal punishment, banishment, and social degradation. If the United States continued to respond to crime this way today, it would need fewer prisons. Since that is not the direction most Americans wish to go, correctional administrators and policymakers are faced with the challenge of making the prison and the jail places that serve shared democratic values and enhance public safety.

Leadership

The challenge of corrections administration is the challenge of leadership. According to John DiIulio, Jr., "prison management may be the single most important determinant of the quality of prison life."[6] Leadership determines what happens inside a prison. An institution as self-contained as a prison or jail, left to its own devices, can drift into a netherworld of danger and vice. Strong administrators and professional staff enable prisons and jails to be places where offenders are kept safely and are ultimately returned to their homes—no worse off, and hopefully better, than the day they arrived.

U.S. prisons and jails do an extraordinarily good job at performing their most basic missions. A recent study by the Bureau of Justice Statistics (BJS) found that the mortality rate among state prison inmates was almost 20 percent lower than for similar groups in the general population, with African Americans in prison experiencing a mortality rate 57 percent lower than their peers on the outside.[7] In a similar vein, an article in the *New England Journal of Medicine* recently found that persons are more likely to die when they leave prison than while they are in prison.[8] Such studies indicate that America's prisons are safe and getting safer.

Legitimacy

Prisons and jails are dependent on the legitimacy conferred on their leadership by the inmates. Legitimacy refers to the consensus among the inmates that the administration is fair and has the best interests of the inmates in mind. Inmates, by and large, are adult and sentient; it is a mistake to infantilize them or treat them as incapable of understanding. They understand regimentation and rules, as long as those rules are evenly applied and reasonably related to their welfare. That is why effective managers speak about being firm, fair, and consistent. It is also why they recognize the importance of providing quality food and medical care to inmates. Correctional administrators do these things not only because they are the right thing to do and are constitutionally required, they do it to show inmates that they care about them as people.

Anne Owens, Her Majesty's Inspector of Prisons, has identified four key aspects of legitimacy:

1. Safety—All prisoners, even the most vulnerable, are held safely.
2. Respect—Prisoners are treated with respect for their human dignity.
3. Purposeful activity—Prisoners are able, and expected, to engage in activity that is likely to benefit them.
4. Resettlement—Prisoners are prepared for release into the community, and helped to reduce the likelihood of reoffending.[9]

◾ Safety

Human beings are social by nature, interested in belonging to a group. People join groups for many reasons, safety not the least among them. Accordingly, one way correctional managers earn legitimacy is by keeping inmates safe. If administrators do not keep them safe, inmates will find alternative means, often in the form of arming themselves or joining a group, such as a gang, to gain protection.

Corrections administrators are keenly aware of the need to protect vulnerable inmates. In a diverse country such as the United States, members of any minority are potentially vulnerable in a prison setting. It is the responsibility of the conscientious prison administrator to be attentive to the risks associated with racial and ethnic biases. Changing definitions of gender and sexual orientation pose new challenges to accommodate the identity of individuals while still acknowledging and dealing with the prejudices and hostility they may face in a prison environment. In the future, protecting vulnerable inmates from harm without compromising their right to move freely in prison population will continue to pose problems for administrators.

The prison must achieve a balance of the rights of the few and the needs of the many. Unfortunately, a single predatory inmate can make all the other inmates feel unsafe. When inmates feel unsafe, management loses its legitimacy, and inmates will do what they have to in order to feel safe again. Thus, the institution may have to isolate and contain predators in order that the majority of inmates have a safe environment.

The Mentally Ill

In the 1960s the United States embarked upon a remarkable social transformation with respect to its treatment of the mentally ill. The invention of psychotropic medications and the gradual loss of confidence in the traditional "talking therapies" combined with the high cost of running large state hospitals led to a national movement to deinstitutionalize mentally ill persons. Unfortunately, in the rush to deinstitutionalize, there was inadequate time to build the community safety net necessary to support these individuals. Pursuant to the law of unintended consequences, the burden of treating persons with mental illness today falls heavily on prisons and jails.

The prevalence of mental illness within correctional institutions is high, with a recent BJS study finding mental health problems in 56 percent of state prisoners and 64 percent of local jail inmates.[10] This is in part due to the transinstitutionalization that occurred in the last century. Transinstitutionalization is defined as:

> *The movement of mentally ill from publicly funded mental health hospitals to nursing homes and correctional institutions. The increase in mental health illnesses in prisons not only burdens the prison health care system, but it further compromises the mental health status of prisoners with mental health diseases. Mental health care providers are also in short supply within correctional institutions, despite the fact that courts have mandated the treatment of mentally ill offenders. Additionally, correctional officers, who are in charge of security issues, often lack an understanding about appropriate management of mental health illnesses. . . .*[11]

Bernard E. Harcourt, Professor of Law and Criminology at the University of Chicago recently observed, "It should be clear why there is such a large proportion of mentally ill persons in our prisons: individuals who used to be tracked for mental health treatment are now getting a one-way ticket to jail."[12] When one looks at the combined institutionalization rates for prisons and jails as well as mental health and retardation facilities over the last half of the 20th century, it is obvious the overall rate has not changed. What has changed is where these persons are held.

The experience of imprisonment for a mentally ill person can be an especially hard one. Many small jails in rural counties do not have access to psychiatric hospitals or physicians. Persons confined in those jails are condemned to a cruel and isolating existence made all the more difficult by the inability of the jailer to meet their needs. Untreated, the mentally ill inmate becomes a strain on the very fabric of the institution. Other inmates are disturbed by his or her symptoms, and staff are frustrated and exasperated by their inability to make the mentally ill individual and the other inmates comfortable. In this troubling mix, the seeds of bad outcomes are sown, and violence and suicide can ensue.

It is a sad commentary on 21st century America if jails are the best healthcare option offered to the mentally ill. Prison and jail administrators need to be in the forefront of advocacy for diversion of the mentally ill from jail and prison to the kinds of caregiving settings where they can find appropriate treatment.

Sexual Violence and Harassment

The deprivation of contact with members of the opposite sex and the prohibition on all sexual activity in confinement create another dangerous dynamic where integrity is easily lost. Allowing sex among inmates is wrong for many reasons, not least that it creates destabilizing jealousies and competition among inmates. Because sex is a basic human need, it is highly sought and valued. Unfortunately, many members of today's corrections staff do not understand the simple concept that inmates are incapable of consent to sexual activity with a staff member. A prison or jail is, by its very nature, inherently coercive. It is naïve and uninformed to believe that inmates can consent to sexual contact as an act of free will. Staff need to understand this concept, as it is another area where prison and jail administrators have been silent for too long.

More importantly, leadership must set an example for appropriate human relations in the institutional setting. Sexual harassment and sexual relationships among managers and subordinates send a mixed message to staff and are often at the root of sexual liaisons between staff and inmates. In today's work environment where cross-gender supervision is more common, the need to address this issue is acute.

The enactment of the Prison Rape Elimination Act reflects a growing concern around this issue.[13] Correctional administrators have zero tolerance for sexual abuse of inmates. However, they believe that some of the estimates of prevalence and frequency are exaggerated. If inmates were experiencing widespread rape, they would rebel. The anger and terror engendered by widespread sexual abuse would lead to more violence, suicide, homicide, and riots. The respect for inmates—respect for them as individuals and for their personal safety—is the most fundamental and critical correlate of legitimacy. Inmates would soon withdraw their grant of legitimacy from any correctional organization where such violations of their personal safety were widespread and condoned.

Nonetheless, because of society's homophobia and the shame and embarrassment that victims experience, there is no doubt that the incidence of inmate sexual encounters and sexual assault is underreported. Creating a prison or jail environment where these behaviors are not tolerated and where reporting is safe and easy for victims continues to be a challenge. As a first step, staff members must integrate the abhorrence of this behavior into their culture, combined with aggressive prosecution of violators.

Drugs and Alcohol

The overwhelming majority of prison and jail inmates are addicted to alcohol and other drugs.[14] Often, drug and alcohol use is related to mental illness. Sometimes, the addiction is an attempt at self-medication. The need to keep inmates sober while confined remains an ongoing challenge to corrections agencies.

Recent federal law mandated the creation of random drug testing protocols in prisons receiving federal residential substance abuse treatment funds.[15] Addiction is a primary problem for most prison and jail inmates, and if they continue to get high upon release, they will most certainly fail. If they are getting high in jail and prison, the likelihood that they will continue to get high upon release is near 100 percent. Moreover, if inmates are getting high while confined, it suggests larger problems within a correctional facility.

The presence of drugs in a correctional facility means someone, usually an inmate or gang of inmates, is in control of their sale and distribution. It means that there will be fights and extortion and that the control of the facility will be in the hands of inmates, not staff. More disconcertingly, it may indicate that staff is aware of the presence of drugs and, at best, is turning a blind eye, or worse, is actually participating in the offense. Inmates know which staff members participate in or allow illegal activities to occur. The permission or participation by corrections officials of the sale and use of illegal drugs within a facility eliminates any pretense of inmate rehabilitation.

Corrections officials must take the strongest possible steps to eradicate drugs from a prison or jail environment. This requires strategic thinking, creative approaches, and most importantly, the creation of a value system among staff at all levels that drug use will not be tolerated. Only by creating a top-to-bottom culture that does not permit drugs to be introduced and used in the correctional setting can one begin to talk about "corrections."

No amount of searching inmates, visitors, staff, and packages can substitute for the dedication of staff that understand the reasons drug use must be eradicated in confinement. Staff need to understand that their safety and that of their fellow officers, as well as their mission and reputation, depend upon the elimination of drugs from the facility. Too often drugs are an institution's "dirty little secret" and are not discussed. If organizations are to eliminate drugs, senior leadership must first acknowledge that a problem exists and openly discuss it with staff.

■ Rehabilitation

The debate continues about whether prisons and, to a lesser extent, jails, have a contribution to make to rehabilitation. Latessa and others argue that properly constructed efforts specifically directed to the most high-risk offenders and designed to eradicate "criminogenic" behaviors can affect outcomes.[16] Latessa and his colleagues may be correct, but current research is limited in quantity and in scope. What does seem clear is that in order for inmates to succeed upon release, several issues need to be addressed, namely, sobriety, education, work, and housing.

An individual who does not stay sober upon release will fail. That is one reason drug-free prisons and jails must be the norm. It is also why institutions must help the individual stay sober in the days immediately following release. Accordingly, reentry work programs must focus on helping inmates to understand their addiction and connecting them with organizations that can help them stay sober upon release. Prisons and jails must be places where sobriety is taught and addictions are treated. Post-release supervision agencies have to understand that addiction is a recurring disease of the brain and that recovery does not proceed in a straight line.[17] Post-release supervision strategies should be designed with this understanding.

In the 21st century if you cannot read, you cannot work. It is that simple. The vast majority of inmates admitted to prisons and jails continue to lack a high school diploma and many still read below the eighth grade level. If there is one positive thing prisons can do, it is to teach people how to read. Before an inmate can participate in optional prison or jail activities, he or she should first be taught how to read. That said, it would be unwise to predicate an inmate's release upon attainment of an educational milestone. How much time an individual serves in confinement should be based on the crimes for which he or she was incarcerated. Nonetheless, prison and jail administrators have tools at their disposal that can provide inmates with an incentive to obtain an education, and these tools should be utilized. Investments in educational resources for prisons and jails are sound investments of public monies.

If prison and jail administrations are serious about reentry, they have to be serious about helping inmates find work upon release. More importantly, probation and parole agencies need to invest in finding work for the persons subject to their supervision. The seriousness with which a probation or parole agency approaches its work can be determined quickly by the time, effort, and money it invests in finding work for offenders. Simply training probation or parole agents how to connect their clients to work or agencies that help them find work can often pay big dividends in offender employment outcomes. Prisons can assist with this challenge by offering prison work assignments and vocational opportunities. To work effectively, whether one is a neurosurgeon or a short order cook, requires pride in one's work, the ability to work with others, reliability, and acceptance of supervisory criticism. These basic skills can be taught in prison and jails; expensive vocational training is not the only way to help inmates enter the world of work.

Additionally, the prospect of finding adequate, affordable housing is especially relevant today in many large urban areas where this concern can pose a significant barrier to reentry. This, too, should be a concern of correctional staff and leadership.

■ The Correctional Workforce

Today, recruitment and retention of correctional staff is a growing challenge.[18] In an increasingly diverse society, staff should be diverse as well. Forging a unified team from a diverse workforce can be a daunting challenge for managers. Much is written about workers' desire for benefits, autonomy, and creativity. Nonetheless, the needs of today's workforce are, at heart, no different from that of yesterday. To recruit and retain qualified and motivated corrections workers, institutions need to pay them well. There is no substitute for a good living wage, good benefits, and the opportunity for career advancement in a safe workplace. If any of those components is missing, no amount of rewards will compensate.

Unions representing corrections workers can be an asset. In those jurisdictions where responsible collective bargaining has resulted in good salary arrangements, good medical benefits and a good pension, recruitment and retention are likely to be easier and more successful than in jurisdictions where those things are absent. Unions can focus on corrections policy or on working conditions. Any undue attention to the former can undermine the latter. The smart corrections administrator will not view a union as a menace or an enemy. Rather, labor relations can help achieve shared values and goals like a safe workplace, good training, and advancement opportunities. An environment of trust between administration and labor relations can set an example for staff and inmates about how individuals can relate to each other and settle their differences in civil society.

Integrity is at the heart of effective leadership and a professional workforce. Drug trafficking and sexual contacts are examples of the ways in which integrity can be compromised. The top administrator of any corrections agency must put integrity first and foremost among his or her list of priorities. Integrity is easily lost and takes many forms. The simple failure to pay attention to what is going on is its simplest form. As mentioned previously, left to their own devices, total confinement institutions can devolve into places where bad things happen. Thus, the first duty of the administrator is to be aware.

Finally, integrity is lost when managers allow inmates to run the facility. Often understaffing is blamed, but poor training and lack of supervision are often contributing factors. The investment in a trained workforce—too often the first thing sacrificed in tight budget times—is the best insurance against a loss of integrity. A well-trained, well-motivated workforce that understands its mission is less likely to be corrupted than an ill-trained, unmotivated one.

■ Conclusion

The emergence of private corrections organizations and the growth of that industry in the last quarter of the 20th century present both a challenge and an opportunity to corrections today. Clearly, the private corrections organization is here to stay. These corporations have demonstrated what most professionals in corrections already knew—that running prisons and jails is mostly good common sense, good will, and good management. The distribution of poorly operated prisons and jails is evenly divided among public and private institutions.

The private operators do present a challenge to the monopoly held by corrections agencies and unions over the operation of correctional facilities. They thereby give the public a choice. Competition is a good thing for the field. It motivates corrections administrators to do better, to think more creatively, and to focus on outcomes. It also provides a welcome bargaining chip in the collective bargaining arena. In many situations where it is impractical or unnecessary for a jurisdiction to create a facility for the long term, the private sector can meet surge capacity needs or bridge a gap. Where public financing limitations prevent the creation of sufficient new capacity, private equity can fill the financial void.

Nonetheless, the management of corrections always has been and always will be the responsibility of government. The government must aggressively manage the use of private corrections. Private operators must be subject to the same oversight and accountability mechanisms as public institutions, and the oversight agency must demand high levels of service from both.

Correctional administrators must find ways to make the operation of America's prisons and jails sufficiently transparent to satisfy the citizenry. The public must be confident that the deprivation of liberty corrections institutions exercise reflects the values of constitutional democracy. The emergence of the correctional standards movement certainly began the process. However, too few correctional agencies, particularly local jails, are accredited. Critics argue that self-regulation by the profession leads to standards and an accreditation process that cannot satisfy the demand for transparency. The administration of accreditation must itself be legitimate and transparent. There are models to learn from in this area, including the system in place for higher education.

Increasingly there are calls for oversight by public or private bodies possessing the "golden key," that is, the authority to enter a prison or jail unannounced any time. There is a federal requirement that each state have such a body to oversee mental health facilities receiving federal funds.[19] Some jurisdictions already have public bodies that perform this function in prisons and jails.[20] It is critical that, however the function is performed, it be accountable to elected officials and to the public. Care of the inmate is an important concern, but not the only consideration in the administration of correctional facilities. The public also holds correctional administrators accountable for the protection of the public and staff and for running an efficient, cost-effective organization.

Confinement of those accused of crimes and those serving sentences will continue to present challenges in a free society. The manner in which society address these challenges will defines its values, as it has throughout history. In the early years of the 19th century, scholars and philosophers from throughout the world visited the United States to witness the revolution in punishment that began at the Walnut Street Jail in Philadelphia. Today, America's correctional leadership is being questioned. The struggle to regain legitimacy lies ahead, but American traditions create a sound basis on which to proceed; values-based leadership and integrity will take us the rest of the way.

DISCUSSION QUESTIONS

1. What are some of the greatest difficulties facing corrections departments today? How are these challenges likely to change in the future?

2. What key objectives does society expect corrections systems to address? Can corrections systems meet these objectives? Why or why not?

3. Why is leadership important in corrections? What are some of the key decisions that leaders have to make?

4. Why is the presence of drugs in an institution particularly problematic? What steps can staff take to prevent this problem?

5. What major issues must inmate programming address in order to improve the likelihood that releasing offenders will not recidivate?

NOTES

1. P. Harrison and A. Beck, *Prisoners in 2005* (Washington, DC: U.S. Department of Justice, Bureau of Justice Statistics, 2006).

2. P. Harrison and A. Beck, *Prison and Jail Inmates at Midyear 2005* (Washington, DC: U.S. Department of Justice, Bureau of Justice Statistics, 2006).

3. U.S. Department of Justice, Office of Justice Programs, "Direct Expenditures by Criminal Justice Function, 1982–2005" available at http://www.ojp.usdoj.gov/bjs/glance/tables/exptyptab.htm, accessed September 23, 2007.

4. Pew Charitable Trusts, *Public Safety, Public Spending* (Washington, DC: Public Safety Performance: A Project of the Pew Charitable Trusts, February 2007).

5. D. Jehl, "The Struggle for Iraq: Prisoners, Some Iraquis Held Outside Control of Top General," *New York Times*, May 5, 2004, A1.

6. J. DiIulio, Jr., *Governing Prisons* (New York: The Free Press, 1987), p. 255.

7. C. Mumola, *Medical Causes of Death in State Prisons, 2001–2004* (Washington, DC: U.S. Department of Justice, Bureau of Justice Statistics, 2006).

8. I. Binswanger et al., "Release from Prison—A High Risk of Death for Former Inmates," *New England Journal of Medicine* 356, no. 2, (2007), pp. 157–165.

9. HM Inspectorate of Prisons, *Expectations* (London: HM Inspectorate of Prisons, 2004), p. 5.

10. D. James and L. Glaze, *Mental Health Problems of Prison and Jail Inmates* (Washington, DC: U.S. Department of Justice, Bureau of Justice Statistics, 2006).

11. "Understanding Prison Health Care: Mental Health" revised June 13, 2002, available at http://movementbuilding.org/prisonhealth/mental.html, accessed July 26, 2007.

12. B. Harcourt, "The Mentally Ill, Behind Bars," *New York Times*, January 15, 2007.

13. Prison Rape Elimination Act, 42 U.S.C. § 15601 et. Seq.

14. J. Karberg and D. James, *Substance Dependence, Abuse, and Treatment of Jail Inmates, 2002* (Washington, DC: U.S. Department of Justice, Bureau of Justice Statistics, July 2005); C.J. Mumola and J. Karberg, *Drug Use and Dependence, Federal and State Prisoners 2004* (Washington, DC: U.S. Department of Justice, Bureau of Justice Statistics, October 2006).

15. Residential Substance Abuse Treatment for State Prisoners, 42 U.S.C. § 3796 ff.

16. E. Latessa, "What Works in Correctional Intervention," *Southern Illinois University Law Review* 23 (1999).

17. N. Volkow, "Drugs, Brains, and Behavior—The Science of Addiction" (Bethesda, MD: U.S. Department of Health and Human Services, National Institute on Drug Abuse, 2007) available at http://www.drugabuse.gov/scienceofaddiction/sciofaddiction.pdf, accessed October 2, 2007.

18. Workforce Associates, Inc., *A 21st Century Workforce for America's Correction Profession* (Lanham, MD: American Correctional Association, 2004).

19. New York State 42 U.S.C. §10805, MHL §45.03(a).

20. Constitution of the State of New York, Art. XII, §5; N.K. Teeters, "The Pennsylvania Prison Society: A Century and a Half of Penal Reform," *Journal of Criminal Law and Criminology* 28, no. 3 (Sept–Oct 1937), pp. 374–379.

Glossary

Access to care: The opportunity for an individual to meet with medical personnel regarding any illness or medical issue.

Accountability: Responsibility and awareness of inmate movement and activities.

Accreditation: The process of review and certification by the American Correctional Association (ACA), the National Commission on Correctional Health Care (NCCHC), and the Joint Commission for the Accreditation of Healthcare Organizations (Joint Commission), which promulgate standards for accreditation and rate correctional and health care facilities against these standards.

Acquired immune deficiency syndrome (AIDS): An infectious disease that renders the body's system of immunity virtually unable to destroy invading disease or other illness.

Administrative remedy: A process by which a confined offender may appeal a decision of a staff member or file a grievance about a disciplinary action or any institutional policy.

Administrative segregation: A confinement status in which the inmate is separated from the institution's general population to (1) provide additional security for an inmate who may be a security risk, (2) house a prisoner who may be under investigation for a violation of facility regulations, or (3) separate an inmate who may be awaiting classification or transfer.

Admission and orientation (A&O): The initial phase of confinement in which a prisoner is given an overview of institutional operations, rules, program availability, and other general data. Some correctional agencies provide A&O in reception centers and then transfer inmates to another facility; others have a separate A&O housing unit within each facility and then place the prisoner in a general housing unit upon completion of the program.

Adult basic education: An educational program designed to assist students by, teaching fundamental reading, writing, and mathematics skills.

Affirmative action program: A management program designed to ensure the fair recruitment and promotion of women and minorities throughout an organization.

American Correctional Association (ACA): An organization of professional correctional staff who are employed in adult and juvenile prisons and jails as well as probation and parole staff.

Americans with Disabilities Act (ADA): A law passed by Congress in 1990 that establishes specific requirements of accessibility and employability for those with qualifying disabilities.

Appeal: A process by which an individual may request a higher-level review of a decision; this can describe a legal examination of a court decision or a request for reassessment of an administrative decision in the institutional setting.

Appellant: The individual who requests an appeal.

Architectural Barriers Act: A federal law passed in 1968 that requires all federal buildings and facilities that receive federal funding to be accessible by anyone with physical disabilities.

Ashurst–Sumners Act: A 1935 federal law that prohibits the interstate transportation of prison-made goods.

Auburn system: An early approach or philosophy of confinement in the United States that began at the New York State Prison in Auburn, New York; this penal system emphasized congregate work, silence, and punitive discipline.

Authoritarian leadership model: An organizational management structure in which power and legitimate authority are consolidated in one person or a few members of an agency's hierarchy.

Bail: A financial guarantee pledged by an individual or his or her family that the person arrested and in custody will, if temporarily released, return for a scheduled court appointment. If the subject does not appear as pledged, the bail is forfeited.

Benchmarking: A process that enables comparison of key data between or within correctional institutions (or parts of institutions) or within a single institution over time.

Bill of Rights: The first 10 amendments to the U.S. Constitution.

Bivens case: In the prison context, litigation in federal court that alleges a civil rights violation by federal correctional officials. This class of litigation was named after a case decided by the Supreme Court and permits a legal challenge that is equivalent to lawsuits authorized under U.S. Code, Title 42, Section 1983 brought by state inmates for alleged violations of their civil rights by state prison employees.

Body cavity searches: Security checks or searches of body orifices such as the mouth, vagina, and anus.

Bond: A written document that assures that an individual will be present at a future criminal proceeding. Money or other property may be required by a bond agent to insure the defendant's appearance. If the defendant does not appear as scheduled, the bond is forfeited.

Boot camps: An alternative correctional program designed for first-time offenders and intended to be used in lieu of prison confinement. This relatively short-term program is structured like the military's boot camps. It emphasizes extensive discipline and intensive program involvement.

Boundary spanning: A management concept that emphasizes one's ability to work productively across traditional bureaucratic lines of an organization and to develop ties with other organizations, both governmental and private, that affect the agency's work environment and mission.

Bridewell: A term for jails (or gaols) in Great Britain in the 16th century. They were intended to detain people until they could pay debts or had a trial.

Budget development: The detailed process of creating a budget proposal for a future period of time, which lists what an agency or program will expect to spend in a future budget time frame.

Budget execution: The process of spending funds during a fiscal year.

Budget oversight: The supervision or accountability of those responsible for authorizing the expenditure of funds.

Bureaucratic management model: An organizational management structure that is formal, hierarchical, and steeped in required process in which system of supervision and communication that flows from top to bottom.

Calculated use of force: The physical response of staff to a controlled situation in which the inmate does not represent an imminent threat to him- or herself or others (e.g., an inmate in a locked cell threatening to destroy property).

Callout: An appointment schedule published daily in a correctional institution that staff members use to arrange activity for inmates. (e.g., meeting with a staff physician).

Capital expenses: Money spent over a predetermined amount for major items in the institution such as equipment, furniture, and machinery.

Capital improvement expenses: Money spent for new construction of prison or jail facilities or for the maintenance of existing facilities.

Capital offense: An offense punishable by the death penalty.

Capital punishment: A criminal sanction of the death penalty.

Case management: The oversight of a prisoner's classification, programming, and release preparation. This involves the planning, coordinating, and tracking of inmates with various services based on assessed need. Inmates generally contribute to this planning and programs utilized may involve such activities as education, counseling, health care, and work or vocational training.

Cell blocks: Inmate housing units in the prison environment. A block is generally one housing unit with many individual or group cells.

Census: An informal accounting of inmates in the correctional setting. At an unscheduled time, the institution is "locked down" (all doors are locked and staff check all prisoners in the area and write incident reports for all who are found in an area where they should not be.

Centralized management: The focus and concentration of power and decision making with one person in the correctional setting; this model is often associated with the bureaucratic theory of organization.

Chain of command: The hierarchical structure of management control in an organization.

Chief executive officer (CEO): The head of an organization; in the correctional context, the CEO is called a warden or superintendent.

Civil Rights Act: Federal legislation passed in 1964 that prohibited personnel or operational decisions based on race, color, sex, religion, or national origin. Affected areas include employment, public accommodations, and the provision of state or local government services.

Class action suits: Lawsuits that allow a person or a group of individuals to bring litigation against an individual or organization on behalf of all those similarly situated.

Classical school of criminology: A philosophy that explains aberrant or illegal behavior as a rational choice made by an individual who is capable of making a logical decision; this philosophy assumes all individuals will make decisions that serve their best interest.

Classification: The process (within a correctional facility) of sorting or classifying inmates so that they go to the appropriately secure prison, with the necessary custodial supervision; this process also involves the approval of prisoner work assignments, program planning, and release planning. Reclassification occurs at scheduled intervals throughout the offender's confinement as he or she makes progress.

Classification committee: A group of institutional staff members who make up a central committee charged with the responsibility for inmate classification.

Closed-circuit television (CCTV): Video cameras that transmit a signal to a specific, limited set of monitors.

Code of silence: The requirement that an individual withhold what is believed to be vital or important information; this code may be voluntary or involuntary, and often describes the informal requirement that an individual not provide information about the activities of his or her colleagues and associates.

Communicable disease: A disease that can be transmitted from one person to another; examples of these diseases are tuberculosis, hepatitis, and sexually transmitted diseases.

Community relations board: A committee of representatives of the local city closest to a correctional facility that volunteer to serve as an advisory team for the institutional management staff and a liaison with the community.

Compassionate release: The early release of a prisoner based on extraordinary and new factors; this type of release may be considered in some jurisdictions if an inmate develops a terminal illness while confined.

Compelling government interest: A legal term that specifies a significant government requirement that is critical to the mission of the agency involved. In the correctional context, this term generally refers to the security and good order of the facility.

Compliance audit: The assessment of institutional compliance with policy and established procedures; requires positive and negative reinforcement to ensure that rules are being followed.

Concurrent sentence: A sentence that is imposed by the court and served at the same time as another sentence.

Conditional release: Release from a correctional facility in which the sentencing authority maintains a form of supervision over the releasee (e.g., parole). If the specified conditions of release are violated, the individual is returned to the institution to complete the term that otherwise would have been served in the community.

Conditions of release: The requirements of the correctional system that supervise a prisoner's conditional release; typical conditions include prohibitions against violations of the law and restrictions that affect travel, association, and work.

Congregate system: The philosophy of prison operation that developed at the New York State Prison at Auburn and involved housing prisoners in isolation but permitting them to work together during the day in silence.

Congregate work: The practice of allowing prisoners to work together.

Conjugal visitation: Visiting privileges for inmates in some jurisdictions that allow for total privacy for the inmate and his or her visitor(s) to allow for emotional and sexual intimacy without staff supervision.

Consecutive sentence: A sentence that is imposed by the court to be served after another sentence is completed.

Consent decree: A judicial order that ratifies an agreement made between two or more parties in a legal dispute. This process is commonly used to resolve litigation between inmates, the correctional jurisdiction, and other interested parties.

Contact visiting: Visiting privileges for inmates that permit casual contact, such as a greeting kiss or hand holding, is permitted under the supervision of staff members.

Contempt of court: A legal finding that an individual or an agency has purposefully hindered a court after a judge has ruled in a case; a court can sanction an individual or a correctional jurisdiction with fines or imprisonment.

Contraband: An unauthorized item in a prison or jail; contraband can be defined as dangerous (for example, a weapon) or as nuisance (for example, too many articles of clothing).

Convict: A prisoner or one convicted of violating the law and sentenced to a period of confinement.

Corporal punishment: Any form of physical punishment.

Correctional emergency response team (CERT): See special operations response team.

Count: An institutional procedure in a prison or jail that involves staff physically counting the inmates. Counts are scheduled throughout the day but can also be called spontaneously by a senior administrator; all routine activity stops during an institution count until the control center indicates the count is correct and all inmates are appropriately accounted for.

Crisis intervention: The provision of trained personnel to respond swiftly to any personal or institutional emergency; generally is used to describe a mental health crisis and requires the timely response by a trained individual to deal with an acute and time-critical situation such as a suicidal prisoner.

Cross-gender supervision: The practice of allowing staff of the opposite gender to supervise inmates.

Cruel and unusual punishment: A term to denote institutional practice or conditions that are considered to be significantly harsh and that fall below constitutional standards.

Day fines: A court-ordered monetary fine based on the amount of earnings the individual makes in one day.

Decentralized management: The dispersal and sharing of power and decision-making within the correctional setting; this involves the delegation of authority from the warden to personnel at lower levels of the organization.

Delegation: The sharing of power that involves authorizing decision making to occur at other levels of management within the correctional environment.

Deliberate indifference: A legal term used in prison litigation to identify official conduct that ignores a threat to an inmate's safety or well-being.

Deportation: The process of removing a foreign national from the country.

Detainee: An individual confined in pretrial, unconvicted status.

Detainer: A legal hold or warrant placed against an individual already confined that signifies that another jurisdiction seeks to take the prisoner into its custody once the person is released from his or her current detention.

Determinate sentence: A judicial sanction for a specific amount of time (for example, a sentence to confinement for a period of five years).

Deterrence: A philosophy or goal of a judicial sentence; the imposition of a sanction of confinement is intended to specifically prevent the law violator from reoffending. Knowledge that an individual has been sanctioned for the offense may generally deter others from committing the offense.

Diagnostic and Statistical Manual IV (DSM-IV): A system of classification for those who suffer from mental illness published and updated by the American Psychiatric Association.

Discharge: A release from correctional supervision; the term can refer to a release from parole supervision or from confinement.

Disciplinary report: A formal, written, investigated report that documents a prisoner's infraction of an institutional regulation. If found to have merit, the report is forwarded to a disciplinary committee for a formal hearing. The committee or disciplinary hearing officer may impose an appropriate sanction if the report is found credible.

Disciplinary segregation: The inmate housing in a correctional institution that is used for punishment; placement in this status is the result of a formal disciplinary hearing and is for a specific period of time. While in this special housing unit, an inmate receives minimal privileges.

Discrimination: Bias or preference often based on race, religion, creed, color, national origination, age, gender, or disability.

Disturbance control team (DCT): A riot control team of highly trained staff members who are fully equipped to respond to institutional disturbances.

Due process: A legal term that refers to an individual's constitutional right to certain procedures and processes.

Electric fences: Perimeter fences of correctional institutions that use electricity to deter inmates from escape. Some electrified fences may be set to stun while others may be set to electrocute inmates attempting to escape; these fences are used in lieu of staffed perimeter towers.

Emergency plans: Written documents that specify how staff should respond to emergencies in the institutional setting. Contingency planning is generally required for crises of all types, including riots, hostages, food or work boycotts, natural disasters, and external assaults on the correctional facility.

Equal employment opportunity (EEO): A management program designed to ensure that fair and unbiased treatment in work opportunities is extended to all employees regardless of their race, gender, age, or ethnicity.

Excessive force: Force used above and beyond that which is necessary to control a situation.

Execution: The process of carrying out an order of the court; also the implementation of the death penalty.

Executive clemency: An official pardon by the governor of a state or the President of the United States.

Executive management team: The senior management staff of a correctional institution that is generally composed of the warden and associate wardens responsible for specific operational divisions of the facility.

Federal bureau of prisons (BOP): A component of the U.S. Department of Justice, this federal agency is responsible for the confinement of those offenders sentenced for violation of federal laws and the operation of all federal correctional facilities across the United States.

Federal prison industries (FPI): A government-owned corporation that sells goods made by prisoners in factories located within correctional facilities; the trade name for this corporation is UNICOR.

Felony: A criminal offense that is punishable by the death penalty or confinement of a year or more in a state prison.

Fine: A financial penalty imposed by a court on a convicted individual.

Formulary: The list of pharmaceutical medications available for prescription within the institutional medical operation.

Free will: The concept that each individual is able to determine right from wrong, and when a person makes the choice to violate the law it is done as a rational, cognitive act.

Furlough: An authorized release from prison or jail for a short duration. Furloughs can be escorted or unescorted. They may be approved for emergency, educational, social, or prerelease purposes.

Furman v. Georgia: This 1972 legal decision by the U.S. Supreme Court objected to the lack of consistency in the decision to seek the death penalty for criminal acts.

Gang locking system: Mechanical or electrical locking systems that enable a staff member to remotely open or close all of the cell doors, individually or collectively, on one range of cells.

Gang: A group of individuals who have a formal or informal organization intended for the purpose of criminal activity or other acts of intimidation and violence in support of their goals.

Gaol: Early form of a jail in England.

Gender bias: Discrimination against an individual because of his or her gender.

General equivalency diploma (GED): A high school diploma earned by the successful completion of a battery of tests that demonstrate knowledge of reading, writing, and mathematical concepts.

Geriatric: Descriptive term associated with the elderly or the issues of old age.

Good time award: A statutory award that reduces the length of time an inmate must serve on a sentence of confinement; this incentive is given to prisoners if their behavior is meritorious.

Government by proxy: A term that refers to the growing partnerships between government and private enterprise (e.g., a privately-run prison). Many public sector programs are operated in part or completely by nonprofit and for-profit organizations. This term is often used to describe public-private partnerships that occur when specific functions of government have been privatized to a private sector, for-profit company.

Government Performance and Results Act (GPRA): A congressional act passed in 1993 intended to improve government accountability and performance. The GPRA was to focus on the actual results of government activity and services of agencies within the federal government. This reform emphasized "downsizing" and "reinventing" federal government, devolution of federal activities to states, and the privatization of many federal government activities.

Grievance: A formal grievance; in the labor-management context, a dispute between an employee labor organization and the employer.

Habeas corpus: A legal process that allows a prisoner to seek relief from a court for allegations of illegal conditions of confinement.

Habitual offenders: Inmates sentenced to harsher sanctions because of their prior record. These repeat offenders are subject to greater penalties in many jurisdictions because of their repetitive involvement in specific crimes identified by state or federal legislation.

Halfway house: Community corrections facilities intended to permit prisoners to spend the last few months of confinement in a transitional program located near their homes; this process is designed to allow an inmate to "decompress" from the rigid and controlled environment of a correctional facility yet have restrictions and supervision during this critical period of preparing for release.

Hands-off doctrine: The attitude of the courts prior to the 1960s in response to the filing of inmate litigation regarding the alleged conditions of confinement within correctional institutions. The prevailing philosophy was that the judicial system had neither the institutional management expertise nor the authority to define the constitutional rights of prisoners.

Hawes-Cooper Act: Federal legislation created in 1922 that was designed to restrict the transportation of prison-made goods from one state to another.

High-mast lighting: The lighting system for prisons and jails for the outside of buildings; these bright lights are generally elevated over 100 feet above the facility's compound and perimeter fence or wall.

Hospice care: Care that provides support to individuals in the final phase of a terminal illness. This care is intended to provide comfort as opposed to a cure.

Hostage negotiation team: A highly trained team of staff who are called upon to resolve hostage situations in the correctional environment.

House arrest: A sanction or status of confinement that restricts an individual to his or her home or residence and is regulated by telephone checks or electronic monitoring; the individual may be permitted to depart the specified area for a specific purpose such as work.

Hulks: Abandoned ships used for the confinement of prisoners in England during the 1700s.

Immediate use of force: The instant physical force applied by staff when a prisoner acts out with little or no advance warning.

In-custody death: The death of an inmate while he or she is confined in the correctional environment.

Inpatient services: Programs and services, typically medical or psychiatric-related, offered to individuals who have been admitted to a hospital environment.

Incapacitation: A philosophy or goal of correctional confinement that seeks to prevent crime by placing the offender in a position where he or she cannot violate the law.

Incarceration: Confinement in a correctional facility.

Incentive award: An honor presented to personnel to express appreciation for exceptional performance and to motivate the continuation of such behavior.

Incident report: A written record of inmate misbehavior. When an inmate violates institutional rules, the behavior is documented with a formal write-up and referred for investigation and a formal disciplinary hearing.

Incompetent: A legal term used for individuals who are unable to comprehend the nature of judicial proceedings against them and who lack the capacity to assist an attorney with their defense.

Indeterminate sentencing: A judicial sanction for an unspecified amount of time.

Index crimes: The Federal Bureau of Investigation's annual Uniform Crime Report that provides statistical measurement of specific criminal activity for homicide and non-negligent manslaughter, rape, robbery, aggravated assault, burglary, larceny, car theft, and arson.

Informant: An individual who provides information against another person.

Informed consent for treatment: A patient's authorization to receive medical or mental health care; all treatment, as well as the absence of treatment, offers the possibility of benefit and risk; patients must be given a full explanation of the ramifications of care and grant their approval.

Injunction: A judicial order that requires the defendants in litigation to cease an identified activity that is directly connected to the court case in question.

Inmate code: A term used to describe the solidarity among confined prisoners; these unwritten rules reflect the values of inmates at a specific institution.

Inmate disruptive group: See security threat groups.

Inmate grievance system: The formal institutional inmate complaint system allowing a prisoner to file a written request for review and redress of a concern. Most such remedy systems have a two- or three-step system that permits grievances to be reviewed by the warden and appeals of this decision to be made to a headquarters office.

Insane: A legal term defining an individual's mental state. Insanity generally is used to identify a person who does not have a firm grasp on reality and cannot be held responsible for his or her criminal activity at the time it took place.

Inservice training: Agency training for employees that is intended to enhance staff knowledge, skill, and job performance.

Inside cell: A physical design of a prison or jail housing unit in which cells are built back to back along a center plumbing chase of the cell house; cells are not aligned on an outside wall of the building.

Institutional culture: Organizational beliefs that establish shared values about staff, inmates, and interaction between the two groups within a prison or jail environment.

Intelligence: Pertinent information overtly and covertly gathered that provides the basis for inmate management decisions.

Intensive probation: A specific form of conditional freedom granted as a sanction by the court in lieu of a sentence to confinement. The convicted offender is supervised by a probation officer and is subject to specific restrictions; violating these rules can result in a probation violation and the invocation of a sentence of incarceration. Intensive probation utilizes probation officers with smaller caseloads and the ability to provide better accountability for the probationer.

Intermediate sanctions: An alternative punishment considered less severe than a sentence to prison or jail confinement but more severe than probation (e.g., short-term placement in a halfway house or home confinement).

Intermittent sentence: A sanction that is interrupted by intervals of liberty (e.g., a sentence to weekend confinement in the local jail).

Internal controls: This term refers to an agency's internal processes and procedures to assure integrity in a program or system.

Interstate compact: A legal contract between state jurisdictions and the federal government that enables each participating jurisdiction to transfer confined inmates or paroled inmates from one area to another. While the state with original custody maintains legal authority for custody, the state with the subject provides supervision.

Jail: A detention facility, generally operated by county or municipal government, where both pretrial and sentenced prisoners are confined.

Just deserts: A phrase that refers to the philosophy of retribution as a goal for the criminal justice system (e.g., offenders receive their proper punishment when they are given an appropriate sanction).

Justice model: A philosophy for justice administration that reflects the intent to punish the transgressions of an offender appropriately. This model emphasizes the societal desire to punish one who has been found guilty.

Labor-management relations: The formal relationship between the organized labor force of an institution, if unionized, and management. If the line (nonsupervisory) staff have organized and formed a union, the process of collective bargaining is regulated by federal law, and formal rules of negotiation are required for specific work-related areas.

Law Enforcement Assistance Administration (LEAA): A federal agency established by the Omnibus Crime and Safe Streets Act of 1968. This agency provided funding to federal, state, and local authorities for technical assistance and the development of new programs in corrections. It also offered scholarships for students interested in the field of correctional administration. LEAA was dismantled by Congress in 1982.

Lease system: Often found in the South during the nineteenth century, this program allowed prison administrators to lease inmates to outside businesses or to agricultural work.

Leg irons: Metal restraints that connect and restrict an individual's ankles. These restraints are usually a chain similar to handcuffs.

Less than lethal weapons: Weapons that are intended to assist staff in controlling prisoners who are violent or riotous (e.g., batons, mace and other chemical aerosols, flash-bang stun grenades, and electrical stun devices).

Lex talionis: A term that means an "eye for an eye" and refers to retribution for a crime that has been inflicted upon a victim and limits the punishment to a sanction equal to the damage originally done.

Liberty interest: A legal reference to an institutional process that affects a prisoner and is of such importance that it can potentially alter an individual's release date or freedom. If a staff decision (e.g., the forfeiture of a parole date due to inmate misconduct) can affect an inmate's release, then the subject must be protected with specific procedural requirements (e.g., written advance notice of a disciplinary hearing and the right of an inmate to call witnesses).

Line staff: The nonsupervisory correctional personnel who directly provide supervision and support services for the inmate population.

Linear indirect configuration: A prison architectural design that is rectangular in shape and does not afford direct visual supervision of all areas of the housing unit or cellblock.

Literacy program: The educational program designed to teach remedial reading and writing skills to those who cannot meet basic testing standards.

Living will: A legal document that expresses an individual's decision on the use of artificial life support systems if he or she is incapacitated or unable to make such decisions.

Lockdown: The process of securing all prisoners in their cells for an extended period of time usually during emergencies and times of extreme inmate tension.

Lockstep: A term describing the controlled, single-file marching of prisoners in which prisoners are required to move in unison with one arm extended forward and resting on the shoulder of the next convict and talking is not permitted.

Lockup: A temporary holding facility used by police and sheriff departments to detain arrestees pending transfer to a more permanent facility or release.

Malfeasance: Wrongdoing by a government or public official that is illegal or otherwise a violation of the public trust.

Malingering: Faking symptoms of an illness.

Malpractice: A legal term generally applied to an unacceptable response by a medical provider to a patient; this is often defined as a failure to meet the appropriate standard of care.

Management by walking around: A philosophy and practice of organizational management that requires involvement by senior administrators in daily "walking and talking" throughout the correctional facility so that they know what is going on in the facility.

Mandatory minimum sentence: A court sanction in which the minimum sentence is specified by law and require that specific criminal acts are punished by a minimum amount of time in prison.

Mandatory release: A form of prison or jail release after a prisoner has served two thirds of his or her term. The prisoner completes the rest of the sentence under the supervision of a parole officer. Violations of regulations or new arrests can cause the status to be revoked and the individual returned to confinement to serve the remainder of the sentence.

Mandatory sentences: A judicial sentence required by statute to specify the amount of time to be served; the intent of lawmakers in establishing such sentencing was to limit the court's discretion in sentencing.

Manslaughter: A lesser degree of homicide; the killing of another intentionally but after being provoked—used to describe an unintentional killing that is caused by inexcusable negligence.

Mark system: A correctional innovation developed by Alexander Maconochie at the British penal island known as Norfolk Island in Australia. This system was an early version of a behavioral modification program: inmates earned "marks" by their behavior and eventually were able to earn their release by an early form of parole known as a "ticket of leave."

Medical model: The philosophy of prison management associated with the goals of rehabilitation, which presumes staff can identify the factors that led a prisoner to a life of crime and offer institutional programs to repair these personal deficiencies.

Meet and confer: A statutory requirement for representatives of a public agency and representatives of employee organizations to meet for a reasonable period of time and in good faith to attempt to resolve differences and reach agreement on issues related to working conditions..

Mental illness: Refers to one of many mental health conditions characterized by inadequate mental functioning, strange behavior, poor emotional response to others, and other behavior that indicates an inability to adapt to one's environment.

Mental retardation: A condition characterized by reduced mental function or slow learning of motor or language skills.

Mentoring: A formal or informal process of developing staff for future assignments and to help them adjust to new roles and responsibilities in the prison or jail environment.

Methadone maintenance: Rehabilitation technique using a synthetic prescription narcotic to detoxify individuals addicted to illegal substances.

Misconduct codes: A list of disciplinary offenses that are subject to punishment in a prison or jail.

Misdemeanor: An offense considered less serious than a felony and punishable by sanctions up to the imposition of jail confinement for relatively minimal periods of time, usually less than one year.

Misfeasance: An act of misusing or abusing one's power and authority.

Mobile patrols: Armed institutional perimeter security vehicles.

Modular building: A system of construction in which the buildings are factory made and transported to the building site for assembly and merging with other elements of the structure.

Motion detector: A device that will sound an alarm if it senses motion.

National Institute of Corrections (NIC): A division of the Federal Bureau of Prisons that provides financial and technical assistance to local, state, and federal correctional agencies and is responsible for the operation of the National Academy of Corrections.

Needs assessment: An evaluation that ascertains the requirements of an individual or a group of individuals. In the context of inmate classification, a survey of an inmate's personal needs permits valid program planning for the individual.

Nonfeasance: Failure to perform a task that has been assigned or is required by virtue of one's position.

Offender: An individual convicted of a criminal offense.

Organized crime: A term that is used to reference the traditional Italian mafia crime families; also refers to any sophisticated criminal activity of a gang involved in illegal acts.

Outside cells: A physical design of prison or jail housing in which cells are placed along the inside of an external wall of a cell house building and have barred windows or polycarbonate windows that are not breakable.

Outpatient services: Programs and services, typically medical or psychiatric-related, offered to individuals that have not been admitted to a hospital environment.

Paramilitary organization: A term used to describe an organization or agency management style if it is hierarchical and similar to the military's; most correctional agencies are organized in this manner.

Parole: A conditional release from confinement prior to the expiration of a sentence. The parolee must abide by specific rules or be returned to an institution to complete service of the sentence.

Parolee: An individual who has been conditionally released from a correctional institution and is subject to parole supervision.

Parole guidelines: Scoring matrix that establishes general guidelines that determine when an offender is eligible for parole consideration; the points accumulated in scoring are generally related to the offense characteristics as well as the personal history of the applicant.

Participative leadership model: An organizational management structure that is democratic by nature; all personnel are expected to join in decision making.

Penal system: A correctional system.

Penitence: Regret for one's misdeeds; encouraging a prisoner to feel remorse and to have a subsequent desire for reformation.

Penitentiary: See Prison.

Penological interest: A legitimate expectation of a correctional institution (e.g., security) often used within legal decisions to reference tasks that an institution should be expected to perform.

Pennsylvania system: An early approach or philosophy of confinement in the United States that was developed by the Quakers in Philadelphia, which established separate, isolated housing for each prisoner, required silence and penitence, and permitted each inmate to read only the Bible.

Pepper spray: A nonlethal chemical agent (oleoresin capsicum) used to control inmates who are acting out.

Per capita cost: The cost per day of institutional operations; can be expressed as a daily cost for specific items such as food, medical, or maintenance expenses or for the general cost of confinement for each prisoner.

Performance accountability: A system for government programs that measures outcomes or results. See also Benchmarking.

Perimeter security: The fence line or wall that surrounds a correctional facility. It is usually enhanced by razor wire, electronic detection systems, and armed staff.

Personal recognizance: A court release, as on bail, predicated on an individual's personal promise to return to court at an appointed time and does not require a financial pledge to guarantee his or her presence.

Personality disorder: A psychological diagnosis that describes an individual who has developed a pattern of behavior that greatly impairs his or her ability to relate to and care for others (e.g., antisocial personality disorder); such an individual functions poorly in society (as demonstrated by law violations), is impulsive, and is often aggressive toward others.

Plea: A defendant's formal response in court to criminal charges.

Podular housing units: Jail or prison housing units designed for direct supervision of prisoners; staff are not isolated behind security glass or bars but supervise by moving among the inmates.

Positivist school of criminology: The theory explaining criminal behavior as caused by internal and external factors outside of the individual's control (e.g., mental retardation that impairs an individual's ability to make a rational decision to avoid criminal behavior).

Postconviction relief: A judicial determination supportive of a prisoner's petition for release or a reduction of sentence; inmates may seek this court action in appeal of their conviction or sentence as well as challenge an action of the correctional agency and staff.

Presentence investigation report: A report prepared by a probation or parole officer at the request of the court. This document provides a relatively brief summary of the crime and the personal history of the convicted offender and is intended to provide background for a judge after a defendant is found guilty and before the individual is sentenced.

Preservice training: Introductory training for new correctional agency employees intended to establish the agency philosophy of operation as well as teach specific techniques.

Pretrial detainees: Individuals detained by court authority while waiting for trial or other court proceedings who are usually held in jail facilities and not mixed with sentenced offenders.

Prison: A long-term institution of confinement for those offenders that have been sentenced to more than one year of detention.

Prison Litigation Reform Act (PLRA): A 1995 federal law that, among other things, limits the ability of inmates to file lawsuits by requiring the payment of full filing fees and requires inmates to exhaust all institution administrative remedies before filing.

Prison Rape Elimination Act (PREA): A 2003 federal law regarding the sexual assault of prisoners, which requires all prison administrators to gather annual statistics about institutional sexual assault and provides grant funding to states to deal with the issue.

Prisoner: An inmate in official custody of a prison or jail facility or under the direct supervision of a correctional or law enforcement official.

Prisoner movement: A reform crusade that occurred during the 1960s within and outside U.S. correctional institutions. Many inmates and citizens within society pressed for more rights for the confined. As a result, there were many changes in routine institution operations and the overall philosophy of institutional management.

Prisoner rights groups: Citizens outside the correctional institutions who pressure prison and jail managers to liberalize rules and regulations within the facilities; these groups also support legal challenges that would increase recognition of the constitutional rights of those confined.

Prisonization: The process of acculturation in which a prisoner takes on the beliefs and customs of other inmates in the institution.

Private corrections: See Privatization.

Privatization: The performance of government functions by private, for-profit organizations.

Probation: A form of conditional freedom granted as a sanction by the court in lieu of a sentence to confinement in which convicted offender is supervised by a probation officer and is subject to specific restrictions; violating these rules can result in a probation violation and the invocation of a sentence of incarceration.

Pro se: The legal process of acting as one's own attorney.

Probation officer: An employee of a probation agency whose primary responsibility is the supervision of a caseload of probationers.

Prosecutor: An attorney in the employ of the government and assigned to initiate and pursue criminal proceedings against individuals charged with violations of criminal law.

Protective custody (PC): An administrative detention status in a jail or prison that requires the separation of an inmate from the general population and is assigned if an inmate is threatened by or scared of other prisoners.

Psychopath: A psychiatric term that refers to the psychopathic personality; psychopaths are dangerous, emotionally unstable individuals with no moral sense of right and wrong.

Psychosis: A mental disorder in which the patient's thought process is extremely disordered.

Psychotropic medication: A type of prescription medication for the mentally ill.

Public defender: An attorney who is employed by a government agency and who is responsible for the representation of criminal defendants who are unable to pay for private counsel.

Qualified immunity: A legal term defining the exemption of government employees from lawsuit damages if the plaintiff cannot prove the employee intended to violate the plaintiff's constitutional rights. Wardens and senior administrators are often exempted from lawsuits based on their nonpersonal involvement in the alleged wrong being litigated.

Radial plan: An institutional architectural plan that places housing units arranged around a circular central area.

Rap sheet: An individual's arrest record.

Rated capacity: The number of inmates that the correctional institution is designed to accommodate based upon staffing levels, the square footage of housing areas, and the size of the support and program space within the facility.

Reasonable accommodation: A term that references the requirement of federal law for employers to make existing facilities accessible and usable by individuals with disabilities.

Reasonable force: The amount of force deemed necessary to control an inmate in a specific situation.

Reasonable suspicion: A set of circumstances or facts that establishes a reasonable person's belief that criminal activity may be taking place.

Receiving and discharge (R&D): The section of a jail or prison that is responsible for the reception of new or transferred prisoners and the release of those offenders who have satisfied the legal obligation of their sentence to confinement.

Reception and diagnostic center: Intake facilities for the assessment, classification, and orientation of new prisoners before inmates are transferred to a prison or jail for which they are appropriately classified.

Recidivism: The repeating of criminal behavior that results in new arrests, convictions, or the return to confinement.

Reentry: The transitional process of returning a prisoner to his or her home community, generally through a halfway house or other community program designed to support the individual during the transition from the structured life of confinement to the freedom of the community.

Reformation: A philosophy of justice administration based on the belief that an offender can be reformed by incarceration or other rehabilitation activities.

Reformatory: A correctional institution designed for younger, first-time offenders who are considered able to be rehabilitated and trained.

Rehabilitation: The process of restoring an offender to a legal and lawabiding lifestyle.

Release: The legal and authorized process of departure from confinement.

Release planning: The ongoing process of preparing a convicted and confined offender for his or her return to the community by offering programs that facilitate the development of skills and attitudes that create the ability and desire to live within the law.

Religious accommodations: Exceptions or new programs developed by institutional authorities that allow an inmate to abide by specific requirements of his or her faith.

Religious Freedom Restoration Act (RFRA): A federal law that established certain minimum standards for the expression of one's religious beliefs; it was subsequently ruled unconstitutional by the Supreme Court.

Residential treatment program: A halfway house or other community residential center designed to assist with the reintegration of a prisoner to society; can also serve as an alternative sanction to prison and function as a halfway house.

Restitution: A judicial sanction that requires a convicted offender to pay the victim or the local community as recompense for the illegal act committed; can be in the form of money or mandated time of service to others.

Restorative justice: A new concept of defining the function and philosophy of the criminal justice system that focuses less on the offender and more on the victim and concentrates on assisting those who have been seriously affected by crime.

Rethermalization: The reheating of an item. This term often refers to the heating of chilled or frozen precooked food.

Retribution: A correctional philosophy that emphasizes the punishment of someone convicted of a crime.

Revocation: The act of taking something previously given; usually refers to the cancellation of parole or mandatory release status.

Road gangs: Prison or jail inmates assigned to work on outside crews that maintain roads, parks, or other government property. Some crews are made up of minimum security trustees, while others are made up of higher security prisoners supervised by armed correctional staff.

Road prisons: Correctional institutions, generally located in the South, whose prisoners were used for agricultural work, road maintenance, and other assignments outside the prison grounds.

Ruiz v. Estelle: This 1990 Texas class-action law suit contested the overcrowding, healthcare, work conditions, security, and disciplinary procedures of the Texas state prison system. This case reversed the "hands-off" doctrine that federal courts had previously applied to inmate complaints about state prison conditions.

Sally ports: Entrances to correctional institutions have two gates that are never opened at the same time; this security feature prevents a prisoner from escaping through a gate when it is open.

Scalar principle: A principle of management that describes the vertical organization of command within an organization which outlines the difference of responsibilities between various levels within a correctional organization (e.g., the difference between serving as a lieutenant and as a correctional officer).

Seclusion: Separate administrative housing for inmates who are mentally ill or have demonstrated an intent to harm or actually harmed themselves or others or are so mentally ill that they cannot function on their own.

Section 1983: A federal statute (U.S. Code, Title 42, Section 1983) that authorizes litigation in federal court for claims of violations of an individual's civil rights by state agents.

Security threat group (STG): Prison or street gang that has banded together in the correctional environment and represent a threat to the welfare, safety, and good order of the institution, staff, or inmates.

Segregated confinement: Separate security cells in the special housing unit (SHU) that are intended as a sanction for the punishment of inmates who have been found guilty of violation of institution rules.

Self-surrender: The status of a sentenced offender who has been granted permission of the court to turn him- or herself in to a designated correctional institution on a specific time and date for service of sentence.

Sentence: The sanction or punishment for a crime that has been imposed by a court.

Sentencing disparity: The difference in sentences given by the courts for similar crimes.

Sentencing guidelines: A set of guidelines that establishes the sentencing options for a court. The matrix of guidelines establishes the range of options based on the characteristics of the offense and the personal history of the convicted offender.

Separate system: The philosophy of prison operation that developed in Pennsylvania at the Walnut Street Jail and Eastern Penitentiary, which required convicts to be housed apart from others in individual cells. Inmates were forbidden to communicate with each other, occasionally were given piecework to perform in their cells, and were expected to reflect on their misdeeds (be penitent).

Shakedown: The search of an inmate or a place in a prison or jail conducted frequently and spontaneously to assist in the control of contraband.

Shock confinement: See Boot Camps.

Sick call: An institutional process by which inmates who are ill visit the medical clinic to seek health care.

Silent system: A philosophy of prison management that required inmates to remain silent and not communicate with each other; intended to prevent prisoners from corrupting each other, punish them for their past misdeeds, and enhance staff control of the population.

Simsbury mine: The earliest prison developed in the United States in the Simsbury mine in Connecticut that was closed because the conditions were so wretched that there were several disturbances.

Sovereign immunity: The doctrine that states that the state or government cannot commit a legal wrong and is immune from civil law suit or criminal prosecution.

Span of control: The public administration concept that defines the effective limitation of how many staff can be supervised directly by one person.

Special diets: Specific food requirements to accommodate medical needs or religious beliefs.

Special housing unit (SHU): The most secure inmate housing unit in a correctional facility; often used to describe disciplinary housing.

Special masters: Representatives appointed by the court to oversee correctional institutions that are or have been involved in conditions of confinement litigation who gather information and monitor operations within the prison or jail.

Special needs offenders: Inmates with unusual requirements or unique disabilities; often used to identify the special institutional support necessary for inmates with physical disabilities, those who are mentally or medically ill, and older inmates.

Special operations response team (SORT): Emergency tactical team trained for highly volatile prison or jail crisis situations.

Special weapons assault team (SWAT): The tactical team used by community law enforcement for unusual crisis situations.

Staff liability: A term that refers to the vulnerability of staff members to inmate law suits.

Standard of care: A term often used to describe the normal expectation of an individual with a specific license or certification in a medical field; the duty to use the care and skill ordinarily exercised by medical professionals practicing under similar circumstances

Standing mainline: A management technique that insures senior staff members of a correctional facility are available to the inmate general population on a daily basis; staff members attend the lunch or evening meal (mainline) and inmates may seek answers to any questions or issues that may be pertinent to them.

Strategic planning: The process of preparing for the future by anticipating the issues and problems that will be faced several years in the future; effective strategic planning is accomplished by involving staff in the planning process at all levels of the organization.

Strip search: A full search of a prisoner or institution visitor that requires all clothing to be removed and all areas of the body searched.

Subpoena: A court order that mandates the appearance of an individual in a specific court at an appointed time and ensures that witnesses are in court for judicial proceedings.

Summary judgment: A ruling by the court based on the presentation of facts without the need for a trial.

Superintendent: See Warden.

Supermaximum: The most secure and highly controlled prison environment.

Therapeutic community (TC): A highly structured, self-governed residential program (e.g., for drug treatment) designed around a treatment program geared to clients with similar problems and intended to shape individual behavior to be socially acceptable.

Three strikes laws: These enhanced sentencing sanctions (often life sentences) have been enacted by legislators for offenders who have been convicted of felony criminal acts for the third time; often used for those who have committed criminal acts.

Ticket of leave: An early form of parole or conditional release.

Time served: A court decision to sentence a convicted individual to the time already spent in confinement; also refers to the amount of time an inmate has been confined on his or her sentence.

Tort: A civil wrong; an injury or breach of duty to the person or property of another. Whereas crimes are offenses against society, torts are civil injuries to private persons and may be litigated in civil court proceedings.

Total institutions: A term describing an institution in which large numbers of individuals live, work, and recreate together in a highly supervised and scheduled manner in a confined environment; refers to correctional facilities, mental institutions, military institutions, and many hospitals.

Totality of conditions: The sum or aggregate of institutional conditions that, when considered as a whole, may violate the prohibition against cruel and unusual punishment as specified in the Eighth Amendment of the Constitution.

Transfer: The movement of a prisoner from one institution to another.

Transitional/intermediate services: Programs and activities that help facilitate an individual's transfer or shift from an inpatient to an outpatient status.

Transportation: A British criminal sanction in which offenders were punished by banishment from England to the New World and later Australia.

Trusty: A minimum security inmate who is assigned to work outside the secure perimeter of a correctional institution, or one who is given more responsibilities within the facility.

Truth in sentencing: A phrase that is associated with determinate sentencing and the fact that an inmate will actually serve the amount of time to which he or she was sentenced by the court.

Turnkey: A work assignment that involves opening and closing a security gate within a correctional institution.

Unfair labor practice: An action, in the labor-management context, that is an alleged abridgement of either party's rights.

Uniform Crime Report (UCR): An indexed crime analysis statistical report published by the Federal Bureau of Investigation annually on homicide and non-negligent manslaughter, rape, robbery, aggravated assault, burglary, larceny, car theft, and arson.

Unit management: A decentralized form of inmate management in which a team of staff from various disciplines are assigned to one housing unit to work exclusively with inmates. The staff offices are located in the unit, and this team is responsible for the case management and oversight of the unit's inmates.

Unity of command: A principle of management that specifies that each employee of an organization should have only one supervisor.

Universal precautions: The usual and ordinary preventive steps taken in order to reduce the risk of infection. Such practices are to be taken in all cases, not only when an inmate or staff member is known to have a communicable disease (e.g., the use of gloves when handling blood).

Urinalysis: A drug screening test that analyzes an individual's urine. This is one of several security procedures used to ensure institutions are as free of drugs as possible.

Value engineering: A construction term for a cooperative effort among the owner, architects, and construction manager to use the most cost-effective methods in building a new facility.

Visiting room: The area of a correctional facility designed for friends and family members to visit an inmate.

Voluntary commitment: Noncompulsory institutionalization for treatment or care that does not require a court order and is the patient's choice.

Walnut Street Jail: A jail facility in Philadelphia, Pennsylvania, that was modified to become one of the first prison facilities in the United States.

Warden: The chief executive officer of a correctional institution who is responsible for all aspects of inmate and staff management: budget, security, support operations, programs, and personnel.

Warrant: A court order that directs law enforcement or correctional personnel to perform a specific task.

Wergild: The concept of paying compensation to the relatives of a victim of murder; this old practice was developed to prevent blood feuds between families.

Workhouses: England's penal institutions (gaols) that were used for lesser crimes.

Work release: A correctional program that permits an inmate to leave a correctional facility to work in the community, but the prisoner must return to the prison or jail facility during nonwork hours.

Writ: A court order that specifies an activity that must be performed.

Yard: The recreation area of a prison or jail, usually a large space of several acres with ball fields and leisure-time equipment.

Zimmer Amendment: A federal law passed in 1996 that severely curtails the use of taxpayers' money to support certain types of prison recreation in the Federal Bureau of Prisons. This no-frills legislation prohibits the purchase and replacement of strength-enhancing weight equipment and electronic musical instruments and had a very negative effect on institutional recreation programs.

Index

A

ABA. *See* American Bar Association

Abstinence syndrome, negative aspects of, 404

Abt Associates, 516

Abu Ghraib prison scandal, 192, 240–241, 523

Abuse, 275
  institutional leadership and, 192
  investigation of, 270
  of medications, 104–105
  preventing, 65, 208
  in sex offenders' backgrounds, 380

Abuse of power, sexual misconduct and public awareness about, 282–283

ACA. *See* American Correctional Association

Academic education, 78

Access
  to information, prison work and, 396
  to mental health care, 115–116
  special needs offenders and, 364–365
  technology and, 56
  visiting areas and, 388

Accountability, 194, 205
  accreditation, 208
  benchmarking, 208
  compliance audits, 207
  creating, 207–209
  establishing, 270
  financial, 166, 168
  identification of corruption, 208
  inmate, 62–63
  policy, 207
  primary avenues to, 206
  strategic planning, 208–209
  training, 207

Accounting
  practices in public and private sectors, 514, 515
  standards, 270

Accreditation, 208, 530
  accountability and, 208
  correctional education programs and, 88–89
  healthcare delivery and, 21
  protective custody standards and, 327
  special mental health care procedures and, 121

ACFSA. *See* American Correctional Food Service Association

ACLS. *See* Advanced cardiac life support

Acquired immune deficiency syndrome. *See* AIDS

ACRIPA. *See* Civil Rights of Institutionalized Persons Act

Activities, performance management and, 202

Activities of daily living, special supports for, 367

Actual injury requirement, imposing, 419–421

ADA. *See* Americans with Disabilities Act

Addiction, 527–528. *See also* Substance abuse
  chemical detoxification and, 403–404

Administrative assistant, *198*

Administrative review, suicidal inmates and, 340

Administrative rule, institutional financial matters and, 163

Administrative segregation, 322, 322*t,* 333

Administrators, 197
  community, media coverage and, 176
  legal oversight and, 300
  media relations and, 173–174
  tours of prisons/jails by, 238
  work policies and role of, 392

Admissions, sex offenders and, 378–379

Advanced cardiac life support, 103

Advanced trauma life support, 103

Advocacy groups
  staff sexual misconduct and, 280–282
  victim notification and, 26

Affirmative action, 226

African Americans
  correction employees, 244
  death penalty and, 348
  sentencing guideline disparities and, 505
  suicides and, 338

After-action reviews, sexual misconduct and, 286*t*

Age, suicide and, 337, 338

Agendas, for community relations boards, 183

Aggressive inmates, managing, 67. *See also* Predatory inmates

Aging, in prisons, 107, 265

AIDS, 20, 265
  detoxification and, 404
  medical segregation and, 362
  screening for, 100
  tuberculosis and, 105–106

Airborne pathogens, jail operations and, 25

Albany County Correctional Facility (New York), 34

Alcatraz Island, United States Penitentiary on, 47, 62

Alcohol, 65
  corrections in 21st century and, 527–528
  withdrawal from, 141
  working under influence of, 232

Alcohol abuse, 20, 104
  health care in jails and, 23–24
  reentry and, 453
  screening for, 100

Alcoholics Anonymous, 104, 293, 296, 406

Alimony, inmate work wages applied to, 395

AMA. *See* American Medical Association

America. *See also* United States
  first prisons in, 8
  regional differences in prisons in, 10

American Association for Adult and Continuing Education, 85

American Bar Association, 349
  constitutionality of private imprisonment and, 516, 517

American colonies, punishment in, 7–10

American Correctional Association, 10, 14, 21, 89, 130, 369–370, 387
  accreditation and, 208
  education programs through, 226
  ethical training and, 271
  food service standards and, 151
  media access policies and, 174

Wk 1

Ch 1 & 2

Wk 2

Ch 6, 7, 8, 9, 10, 30, & 32

Rec
  Ch 29

Wk 3

Ch 14, 18, 19, 20, 21, 22, 24
  Doc ID 150 3001311

Wk 4

27, 28, 31, 35, 40, 41

Wk 5

43, 46, 47, 48

Doc ID 146 4310481
        154 5625724